λ = flux linkage

ω = angular frequency

ω_m = rotor angular speed (mech. rad/s)

ω_r = angular speed of rotor field in space

ω_s = synchronous angular speed (mech. rad/s)

ϕ = magnetic flux, total flux per pole

ϕ_{Ad} = direct axis armature-reaction flux

ϕ_{Aq} = quadrature axis armature-reaction flux

ϕ_b = backward-rotating field

ϕ_f = forward-rotating field

ϕ_m = mutually linked flux

ϕ_r = resultant flux of rotor current, rotor flux

ϕ_s = resultant field of stator currents, stator flux

τ_p = pole pitch

θ = angle between rotor and stator field axes

θ = angular displacement in electrical radians

θ_p = angular displacement in mechanical radians

$\dot{\theta}$ = rotor speed

Φ_A = fundamental armature-reaction flux

Φ_l = equivalent leakage flux

Φ_R = resultant airgap flux per pole

$\triangleq$ = defined equal to

A = area of cross section

AT_d = demagnetizing ampere-turns

B = magnetic flux density

B_{av} = average flux density

B_m = equivalent susceptance

C = transformation or connection matrix

D = rotor diameter

D = friction factor

DOCR = Direct-on-line current ratio

DOTR = Direct-on-line torque ratio

E = electric field intensity

E = induced emf

E_a = dc machine emf in the armature winding

E_{aNL} = no-load generated voltage

E_{ar} = emf due to armature reaction flux

E_r = emf due to resultant air gap flux

F = mechanical force

F = mmf

F_2 = rotating mmf of two-phase winding

F_3 = rotating mmf of three-phase winding

F_A = resultant armature reaction mmf

$F_{a,b,c}$ = armature reaction mmf of phase a, b, or c

F_d = direct axis component of mmf

F_e = force of electromagnetic origin

F_F = field mmf

F_q = quadrature axis component of mmf

F_R = force of repulsion

F_R = resultant rotor mmf

G = total number of rotor slots

G_c = G matrix for commutator primitive

G_m = equivalent conductance

G_s = motional inductance matrix

(continued on back)

ENERGY CONVERSION
Electric Motors and Generators

SAUNDERS COLLEGE PUBLISHING
A Division of Holt, Rinehart and Winston, Inc.

SERIES IN ELECTRICAL ENGINEERING

M.E. Van Valkenburg, *Senior Consulting Editor*
Adel S. Sedra, *Series Editor/Electrical Engineering*
Michael R. Lightner, *Series Editor/Computer Engineering*

ALLEN AND HOLBERG   *CMOS Analog Circuit Design*

BELANGER, ADLER, AND RUMIN   *Introduction to Circuits with Electronics: An Integrated Approach*

BOBROW   *Elementary Linear Circuit Analysis*

BOBROW   *Fundamentals of Electrical Engineering*

CHEN   *Linear System Theory and Design*

CHEN   *System and Signal Analysis*

CILETTI   *Introduction to Circuit Analysis and Design*

COMER   *Digital Logic and State Machine Design*

COMER   *Microprocessor Based System Design*

GHAUSI   *Electronic Devices and Circuits: Discrete and Integrated*

HOSTETTER, SAVANT, AND STEFANI   *Design of Feedback Control Systems, 2/e*

JONES   *Introduction to Optical Fiber Communication Systems*

KENNEDY   *Operational Amplifier Circuits: Theory and Applications*

KUO   *Digital Control Systems*

LASTMAN AND SINHA   *Microcomputer-Based Numerical Methods for Science and Engineering*

LATHI   *Modern Digital and Analog Communication Systems, 2/e*

McGILLEM AND COOPER   *Continuous and Discrete Signal and System Analysis, 2/e*

NAVON   *Semiconductor Microdevices and Materials*

PAPOULIS   *Circuits and Systems: A Modern Approach*

SADIKU   *Elements of Electromagnetics*

SCHWARZ AND OLDHAM   *Electrical Engineering: An Introduction*

SEDRA AND SMITH   *Microelectronic Circuits, 2/e*

SINHA   *Control Systems*

VAN VALKENBURG   *Analog Filter Design*

VRANESIC AND ZAKY   *Microcomputer Structures*

WOLOVICH   *Robotics: Basic Analysis and Design*

ENERGY CONVERSION
Electric Motors and Generators

Raymond Ramshaw

R. G. van Heeswijk

Electrical Engineering Department
University of Waterloo
Ontario, Canada

Saunders College Publishing
A Division of Holt, Rinehart and Winston, Inc.

Philadelphia   Fort Worth   Chicago   San Francisco
Montreal   Toronto   London   Sydney   Tokyo

Text Typeface: Times Roman
Compositor: Progressive Typographers
Acquisitions Editor: Robert Argentieri
Managing Editor: Carol Field
Project Editor: Marc Sherman
Manager of Art and Design: Carol Bleistine
Text Designer: Dorothy Chattin
Art and Design Coordinator: Doris Bruey
Cover Designer: Lawrence R. Didona
Text Artwork: GRAFACON
Director of EDP: Tim Frelick
Production Manager: Charlene Squibb

Printed in the United States of America

Energy Conversion: Electric Motors and Generators

ISBN 0-03-003399-3

Library of Congress Catalog Card Number: 89-043043

0 1 2 3   039   9 8 7 6 5 4 3 2 1

Preface

Traditionally, the study of the theory, performance, and design of electric machines was a major part of the electrical engineering curriculum. Today, energy conversion, which is studied in courses such as electric machinery, electromechanics, and electromechanical energy conversion, makes up only 10 percent of the electrical engineering core program. Given the accelerated growth rate of electrical technology and society's increasing demands for electrical energy, however, the study of motors and generators is no less important than before. Consequently, we have written *Energy Conversion: Electric Motors and Generators* to detail the operation and performance of transformers and electric machines.

Our belief in the importance of studying energy conversion is founded on the understanding that an electric machine is, after all, the link between electrical and mechanical energy. As such, it is an integral part of a power system of active devices, loads, and controls. Determining the behavior of the power or control system, therefore, depends mainly upon the student's ability to identify the character of these individual components and to understand how they affect the performance of the whole.

Unlike the design, construction, special applications, and maintenance of machines, the principles of *machine operation* are permanent. Therefore, this book focuses on basic interactions, laws, and modeling. Material characteristics and practical engineering techniques are described only occasionally, primarily to illustrate principles and indicate the limitations of technology. Wherever possible, the range of applications—from conventional to computer to robotic—are described for the various machines.

Starting from these basic principles, this text deals with the theory of operation and the performance of transformers and electric machines. It has been designed to give students a comprehensive understanding of these important concepts. Unlike alternate treatments—in which the physical picture is sometimes lost in mathematical manipulation—it also provides the basic knowledge needed to interpret a generalized theory. Steady-state operation is described, thereby equipping students for the analytical treatment that follows. This steady-state description of transformers and traditional types of rotating machines also provides insight into the physical interactions that take place in the energy transformation and conversion processes.

A discussion of the broad energy conversion concepts that can be applied to an electromechanical device is presented. These concepts are applicable to both transient and steady-state operation. Circuit analysis is applied to determine machine performance. Since the circuits discussed here are not static but involve relative motion between the material parts, the term *dynamic circuit analysis* has been applied to the unified theory. The circuit model is emphasized because circuit analysis is well developed and provides a powerful tool for the solution of machine problems. This analysis is applied to rotating machines and is described as an introduction to the practical use of the generalized theory of machines.

In the Appendix, a brief review of three-phase circuits is offered. This specialized topic of networks is seldom included in circuit-theory courses, but it is important for understanding the operation of polyphase machines. Also found in the Appendix is a discussion of specific magnetic field calculations that will be useful throughout the text, and there are applicable problems. For the reader who wants to do more advanced study, there is a bibliography at the end of each chapter. Where appropriate, books on the practical control of machines are named. A general bibliography follows Chapter 10.

The text is arranged so that it is a simple matter to choose material for a one-, a two- or a three-term course of study in electrical machines. The following is a suggested list of text sections for a two-term course. The remaining sections would, of course, then be used for reference.

1. **Energy Conversion**
   1.3: Electromechanical energy conversion; 1.4.1 and 1.4.2: Faraday's law and the Lorentz force.

2. **Energy Conversion and Dynamic Circuits**
   2.3.1 and 2.3.2: Electric circuit relations for rotating systems; 2.4.3.C and 2.4.4: Coenergy and conversion to mechanical energy; 2.5.1 and 2.5.2: Torque; 2.5.4: Synopsis of multiply-excited rotating systems.

3. **Transformers**
   3.3.3: Steady-state behavior of the transformer; 3.4.3: Approximate equivalent circuit; 3.6: Transformer performance.

4. **Introduction to Rotating Machines**
   4.2.1.A and B: Induction torque and rotating magnetic field; 4.2.2: Reluctance torque; 4.3: Excitation torque.

5. **DC Machines**
   5.3.1 and 5.3.2: Armature winding emf and the torque equation; 5.4.1 and 5.4.2: Armature reaction and compensation; 5.5.1.B and C: Reactance voltage and

interpoles; 5.7.1: Speed regulation; 5.7.3: Motor model; 5.7.4.A: Motor speed; 5.7.5: Motor speed control; 5.8: Efficiency.

6. **Induction Motors**
6.3: Slip; 6.4.1.B and C: Emfs; 6.5.1.A and B: Exact equivalent circuit and approximate equivalent circuit; 6.5.2: Parameter tests; 6.6: Motor performance; 6.7.1: Starting; 6.7.2: Speed control.

7. **Single-Phase Induction Motors**
7.3.1: Equivalent circuit; 7.3.3: Performance.

8. **Synchronous Machines**
8.5.1: Armature winding emf; 8.5.3: Synchronous impedance; 8.5.4: Equivalent circuit; 8.5.6: Voltage regulation; 8.5.7: Load angle; 8.6.2: Two-reactance theory; 8.7.1 and 8.7.2: Motor analysis and load angle.

Appendix A.   Magnetic Circuits
Section A5: Magnetic circuit computations.

Appendix D.   Three-Phase Circuits
Section D9: Summary.

Appendix E.   Per Unit Values
Section E2: Definitions.

Appendix F.   Rotating Magnetic Field
Section F5: Summary.

Appendix G.   Windings
Section G2: Winding terms.

The authors would like to thank the following persons for their helpful comments: Harit Majmudar, Worcester Polytech; Linda Laub, University of Missouri, Rolla; F.C. Brockhurst, Rose Hulman Institute of Technology; Martin Kaplan, Drexel; Darrell Schroder, University of El Paso; Charles Gross, West Point; Constantine Alafodima, University of Texas, El Paso; Bob Eghert, Wichita State University; Mohammed El-Hawary, Technical University of Nova Scotia; Arthur Hopkins, University of California, Berkeley; Carlos Trevino, University of Arkansas; Roger Webb, Georgia Tech; Edgar Tacker, University of Tulsa; William Schott, University of California, Los Angeles; Mac Van Valkenburg, University of Illinois, Urbana; Donald Thorn, University of Akron; Adly Girgis, North Carolina State University; and Paul McCormick, University of Pittsburgh.

Raymond Ramshaw
R.G. van Heeswijk
August 1989

Contents

Chapter One
ENERGY CONVERSION   1

  1.1.   Introduction   1
  1.2.   Electric Energy Processes   2
      1.2.1.   Sources of Energy   2
      1.2.2.   Electric Power System   3
  1.3.   Electromechanical Energy Conversion   5
  1.4.   Motors and Generators   9
      1.4.1.   Faraday's Law and Induced Emf   10
      1.4.2.   Lorentz Force   12
      1.4.3.   Basic Generator Operation   14
      1.4.4.   Basic Motor Operation   15
      1.4.5.   Continuous Electromechanical Energy Conversion   15
  1.5.   Summary   20
  1.6.   Problems   20
  1.7.   Bibliography   24

Chapter Two
ENERGY CONVERSION AND DYNAMIC CIRCUITS   26

  2.1.   Introduction   26
  2.2.   Symbols in Electric Machine Representation   28
      2.2.1.   Stator and Rotor Excitation   28
  2.3.   Electric Circuit Relations   33
      2.3.1.   Singly-excited Rotating System   33
      2.3.2.   Multiply-excited Rotating System   36
      2.3.3.   Translational System   38

2.4.   Conversion of Energy   **41**
    2.4.1.   Conservative System   **42**
    2.4.2.   Electric System Energy   **43**
    2.4.3.   Coupling Field Energy   **45**
    2.4.4.   Conversion to Mechanical Energy   **49**
2.5.   Torque   **53**
    2.5.1.   Torque and Magnetic Stored Energy   **53**
    2.5.2.   Torque and Coenergy   **54**
    2.5.3.   Reluctance Machine with AC Excitation (A Case Study)   **56**
    2.5.4.   Multiply-excited Rotating Systems   **59**
2.6.   Translational Magnetic Devices   **64**
2.7.   Electrostatic Devices   **66**
    2.7.1.   Singly-excited Translation System   **66**
    2.7.2.   Electrostatic Machine with AC Excitation (A Case Study)   **69**
2.8.   Mechanical System Relations   **71**
    2.8.1.   Mechanical Systems   **72**
2.9.   Dynamic Circuit Analysis   **72**
    2.9.1.   Simulation   **73**
2.10.   Summary   **76**
2.11.   Problems   **77**
2.12.   Bibliography   **89**

Chapter Three
TRANSFORMERS   **91**

3.1.   Introduction   **91**
3.2.   Ideal Single-phase Transformer   **92**
    3.2.1.   Steady-state Characteristics   **92**
3.3.   Real Single-phase Transformer   **96**
    3.3.1.   Effect of Winding Resistance   **97**
    3.3.2.   Components of Magnetic Flux   **98**
    3.3.3.   Steady-state Behavior of the Transformer   **102**
3.4.   Transformer Equivalent Circuits   **105**
    3.4.1.   Exact Equivalent Circuit   **105**
    3.4.2.   Transformer Frequency Response   **108**
    3.4.3.   Approximate Equivalent Circuit of a Power Transformer   **111**
    3.4.4.   Phasor Diagrams   **113**
3.5.   Parameter Determination   **116**
    3.5.1.   Transformer Open-circuit Test   **117**
    3.5.2.   Winding Resistances   **119**
    3.5.3.   Transformer Short-circuit Test   **120**
3.6.   Transformer Performance   **123**
    3.6.1.   Efficiency   **123**
    3.6.2.   Regulation   **126**

3.7.   Three-phase Transformers   **131**
    3.7.1.   Three-phase Connections   **131**
    3.7.2.   Three-phase Transformer Units   **133**
3.8.   Autotransformers   **138**
3.9.   Instrument Transformers   **140**
    3.9.1.   Potential Transformers   **141**
    3.9.2.   Current Transformers   **141**
3.10.   Summary   **143**
3.11.   Problems   **143**
3.12.   Bibliography   **150**

Chapter Four
INTRODUCTION TO ROTATING MACHINES   **151**

4.1.   Introduction   **151**
4.2.   Stator or Rotor Excitation   **152**
    4.2.1.   Constant Air Gap   **153**
    4.2.2.   Saliency   **157**
4.3.   Stator and Rotor Winding Excitation   **161**
    4.3.1.   Stator and Rotor Windings with Direct Current   **162**
    4.3.2.   Stator and Rotor Windings with DC Excitation at the Terminals   **164**
    4.3.3.   Stator and Rotor Windings with Single-phase AC Excitation   **166**
    4.3.4.   Stator Winding DC Excited and Rotor Winding AC Excited   **166**
    4.3.5.   Stator Winding DC Excited and Rotor Winding Polyphase Excited   **169**
4.4.   Summary   **170**

Chapter Five
DIRECT CURRENT MACHINES   **171**

5.1.   Introduction   **171**
5.2.   The Basic Machine   **172**
    5.2.1.   Magnetic Circuit   **173**
    5.2.2.   Methods of Main Field Excitation   **173**
    5.2.3.   Principle of Commutation   **175**
    5.2.4.   Armature Windings   **177**
5.3.   Generated Voltage and Developed Torque   **181**
    5.3.1.   Generated Voltage   **181**
    5.3.2.   Developed Torque   **184**

5.4.   Armature Mmf   **187**
    5.4.1.   Armature Reaction   **187**
    5.4.2.   Compensating Windings   **191**
    5.4.3.   Effect of Brush Shifting   **191**
5.5.   Commutation   **194**
    5.5.1.   Commutation Action   **194**
5.6.   DC Generators   **200**
    5.6.1.   Saturation Curve   **202**
    5.6.2.   Generator Voltage Buildup   **203**
    5.6.3.   General Circuit Diagram and Steady-state Equations   **207**
    5.6.4.   Voltage Regulation   **208**
    5.6.5.   Generator Characteristics   **210**
5.7.   DC Motors   **211**
    5.7.1.   Speed Regulation   **212**
    5.7.2.   Armature Reaction, Commutation and Interpoles   **213**
    5.7.3.   General Circuit Diagram and Steady-state Equations   **214**
    5.7.4.   Motor Characteristics   **218**
    5.7.5.   Motor Speed Control   **225**
    5.7.6.   Motor Starting   **227**
    5.7.7.   Motor Braking   **230**
    5.7.8.   Special Motors   **232**
    5.7.9.   Motor Comparison and Applications   **239**
5.8.   Losses and Efficiency   **240**
    5.8.1.   Losses   **240**
    5.8.2.   Determination of Losses   **241**
    5.8.3.   Efficiency   **242**
5.9.   Machine Rating and Dimensions   **243**
5.10.   Summary   **245**
5.11.   Problems   **246**
5.12.   Bibliography   **254**

Chapter Six
POLYPHASE INDUCTION MOTORS   **255**

6.1.   Introduction   **255**
6.2.   Principles of Action   **256**
6.3.   Synchronous Speed and Slip   **258**
6.4.   Induced Emf and the Rotating Field   **259**
    6.4.1.   Induced Emf   **262**
6.5.   Induction Motor Analysis   **267**
    6.5.1.   Circuit Model Development   **268**
    6.5.2.   Equivalent Circuit Parameters   **277**
6.6.   Performance of Induction Motors   **280**
    6.6.1.   Rotor Power   **280**
    6.6.2.   Torque   **282**

6.6.3. Losses and Efficiency   **290**
6.6.4. Output Equation   **295**
6.7. Induction Motor Control   **296**
6.7.1. Induction Motor Starting   **296**
6.7.2. Induction Motor Speed Control   **301**
6.7.3. Induction Motor Braking   **312**
6.7.4. Special Motors   **313**
6.8. Summary   **319**
6.9. Problems   **320**
6.10. Bibliography   **326**

Chapter Seven
SINGLE-PHASE INDUCTION MOTORS   **327**

7.1. Introduction   **327**
7.2. Cross-Field Theory   **327**
7.3. Double Revolving-Field Theory   **330**
7.3.1. Equivalent Circuit   **331**
7.3.2. Parameter Measurement   **332**
7.3.3. Motor Performance   **333**
7.4. Motor Starting   **336**
7.4.1. Resistance-start Split-phase Motor   **336**
7.4.2. Capacitor Motor   **337**
7.4.3. Shaded-pole Motor   **339**
7.5. Summary   **340**
7.6. Problems   **340**
7.7. Bibliography   **341**

Chapter Eight
SYNCHRONOUS MACHINES   **342**

8.1. Introduction   **342**
8.2. Basic Structure and Operation   **342**
8.3. Elementary Synchronous Machines   **344**
8.3.1. Single-phase Generator   **344**
8.3.2. Three-phase Generator   **346**
8.4. Practical Synchronous Machines   **351**
8.5. Synchronous Generator (Round Rotor)   **353**
8.5.1. Induced Emf on No Load   **353**
8.5.2. Armature Reaction   **354**
8.5.3. Synchronous Impedance   **359**
8.5.4. Equivalent Circuit and Phasor Diagrams   **360**
8.5.5. Parameter and Loss Separation   **361**
8.5.6. Voltage Regulation   **370**
8.5.7. Power and Load Angle   **378**

8.6.   Synchronous Generator (Salient Poles)   **381**
    8.6.1.   Armature Reaction   **382**
    8.6.2.   Two-reactance Theory   **382**
    8.6.3.   Power and Load Angle   **389**
    8.6.4.   Determination of Synchronous Reactances   **392**
    8.6.5.   Voltage Regulation   **394**
8.7.   Synchronous Motor   **396**
    8.7.1.   Motor Analysis   **397**
    8.7.2.   Power Torque and Load Angle   **399**
    8.7.3.   Motor Starting   **401**
    8.7.4.   Special Motors   **402**
8.8.   Efficiency   **408**
8.9.   Summary   **410**
8.10.   Problems   **411**
8.11.   Bibliography   **415**

Chapter Nine
SINGLE-PHASE AC COMMUTATOR MOTORS   **417**

9.1.   Introduction   **417**
9.2.   Series AC Commutator Motor   **418**
    9.2.1.   Analysis of Simple Series Motor   **419**
9.3.   Repulsion Motor   **423**
9.4.   Summary   **426**
9.5.   Problems   **426**
9.6.   Bibliography   **427**

Chapter Ten
DYNAMIC CIRCUIT ANALYSIS OF ROTATING MACHINES   **428**

10.1.   Introduction   **428**
10.2.   Two Coupled Coils   **429**
10.3.   Slip-Ring Primitive   **431**
    10.3.1.   Voltage Equation of the Slip-ring Primitive   **431**
    10.3.2.   Torque   **434**
    10.3.3.   Power   **435**
    10.3.4.   Conditions for Average Power Conversion   **438**
10.4.   Distributed Windings   **440**
10.5.   Commutator Primitive   **443**
    10.5.1.   Commutator Primitive Voltage Equations   **445**
    10.5.2.   Commutator Primitive Torque and Power   **447**
10.6.   Link Between Circuits and Machines   **451**
    10.6.1.   The Real Machine   **452**
10.7.   DC Machine and Dynamic Circuit Analysis   **453**
    10.7.1.   Power Flow   **457**

10.7.2.   Steady-state Back Emf   **458**
10.7.3.   DC Shunt Motor Operating at Steady State   **459**
10.8.   AC Machines and Linear Transformations   **461**
10.8.1.   Three-phase to Two-phase Transformation   **461**
10.8.2.   Transformation from Rotating to Fixed Axes   **465**
10.8.3.   Real Machine to Commutator Primitive Transformation   **467**
10.8.4.   Transformation and Power Invariance   **468**
10.8.5.   A Quadruply-excited Machine with Saliency   **469**
10.9.   Polyphase Induction Motor   **480**
10.9.1.   Steady-state Analysis   **486**
10.10.   Polyphase Synchronous Motor   **490**
10.10.1.   Steady-state Analysis   **490**
10.11.   Summary   **495**
10.12.   Problems   **497**
10.13.   Bibliography   **506**

GENERAL BIBLIOGRAPHY   **507**

APPENDIX A   MAGNETIC CIRCUITS   **513**

A1.   Introduction   **513**
A2.   Description of Terms   **513**
A3.   Ampere's Magnetic Circuital Law   **517**
A4.   Reluctance   **518**
A4.1.   Reluctances in Series and Parallel   **519**
A5.   Magnetic Circuit Computations   **521**
A6.   Magnetic Flux Leakage and Fringing   **527**
A7.   Summary   **531**
A8.   Problems   **532**

APPENDIX B   MECHANICAL SYSTEM RELATIONS   **539**
B1.   Introduction   **539**
B2.   Inertial Effects   **539**
B3.   Elastic Effects   **541**
B4.   Damping Effects   **541**
B5.   Mechanical Systems   **542**
B5.1.   Translational Mechanical System   **543**
B5.2.   Rotational Mechanical System   **544**

APPENDIX C   AC CHARACTERISTICS OF FERROMAGNETIC CIRCUITS   **547**

C1.   Introduction   **547**
C2.   Magnetic Material Characteristics   **547**

C3. Voltage and Time-varying Magnetic Flux   **549**
C4. Iron Losses   **552**
    C4.1. Hysteresis Loss   **552**
    C4.2. Eddy Current Loss   **555**
    C4.3. Separation of Iron Losses   **557**
    C4.4. Dynamic Hysteresis Loop   **558**
C5. AC Excitation Characteristics of Ferromagnetic Circuits   **559**
C6. Electric Circuit Representation   **563**
C7. Inrush Current   **567**
C8. Summary   **569**
C9. Problems   **569**

APPENDIX D   THREE-PHASE CIRCUITS   573

D1. Introduction   **573**
D2. Generation of Three-phase Voltages   **574**
    D2.1. Three-phase Generator and Load Connections   **576**
D3. Balanced Voltage Systems   **578**
    D3.1. Y Connection   **578**
    D3.2. Δ Connection   **580**
D4. Three-phase Loads   **580**
    D4.1. Delta-connected Load   **580**
    D4.2. Four-wire Y-connected Load   **582**
    D4.3. Three-wire Y-connected Load   **584**
D5. Single-phase Representation   **585**
D6. Equivalent Y and Δ Loads   **586**
D7. Phase Sequence   **588**
D8. Measurement of Three-phase Power   **589**
    D8.1. The Wattmeter   **589**
    D8.2. Three-wattmeter Method   **589**
    D8.3. Two-wattmeter Method   **591**
D9. Summary   **594**
D10. Problems   **594**

APPENDIX E   PER UNIT VALUES   597

E1. Introduction   **597**
E2. Definitions   **597**
E3. Application to Transformers   **598**
E4. Transient Calculations   **600**
E5. Summary   **600**

APPENDIX F ROTATING MAGNETIC FIELD   601

F1.   Introduction   **601**
F2.   Mmf Pattern of a Concentrated Coil   **602**
F3.   Mmf Pattern for a Distributed Winding   **602**
F4.   Mmf Pattern of a Three-phase Winding   **606**
F5.   Summary   **609**

APPENDIX G WINDINGS   611

G1.   Introduction   **611**
G2.   Winding Terms   **612**
G3.   Basic Types of Modern DC Armature Windings   **616**
G4.   AC Windings   **621**
        G4.1.   Reduction of Harmonic Effects   **621**

APPENDIX H CONSTANTS AND CONVERSION FACTORS   625

H1.   Constants   **625**
H2.   Prefixes   **625**
H3.   Conversion Factors   **626**

INDEX   627

CHAPTER ONE

Energy Conversion

1.1. INTRODUCTION

Electrostatics is the study of electric charges at rest. *Electric circuit analysis* is the study of the constrained flow of electric charge in a system represented by lumped parameter networks. *Electromagnetics* is the study of the effects of the motion of electric charge. Each of these science topics is a subject in its own right. Combining electric circuit analysis and electromagnetics results in a whole new engineering field of energy transformation and conversion. In this form it has become the electrical engineering study of designed responses and harnessed forces in the processing of electrical energy for the comfort and convenience of humanity.

By their experiments and observations, pioneers like H. C. Oersted, A. M. Ampere, M. Faraday, and J. Henry laid the foundations upon which the principles of electrotechnology are based. In 1820, Oersted demonstrated to his students the association between electric current and magnetism when he showed the deflection of a compass needle in the vicinity of a conductor carrying electric current. Conversion of electrical energy to mechanical energy was implicitly illustrated by the motion of the permanent magnet of the compass. This was developed further by the thinker and analyst Ampere in 1892. His experiments showed that two conductors carrying currents in the same direction attracted each other. But the origins of the transformer, the generator and the motor of today came from the creative researches of the independent workers Faraday and Henry in their efforts to produce electricity from magnetism. This was spurred by their knowledge that magnetism could be produced from electricity. It was in 1831 that the means to produce electricity from magnetism was discovered. The induction of current was found to be due to a changing magnetic field.

Many ways of inducing currents in conductors were investigated by Faraday but he did not exploit their application. It was left for others to develop the practical means for the production of electricity in generators and its distribution with transformers, both of which rely upon a changing magnetic field for their operation. High-efficiency generators were designed at the end of the nineteenth century and this led to the development of public supply systems. Since then innovations to improve performance and reliability have been continual.

This book is mainly about electric machines, both motors and generators.

In this chapter we want to show how machines fit into the general context of energy processing. Then we want to show how motoring and generating comes about by using two basic principles of physics described by the Lorentz force equation and Faraday's law of induction.

A source of electrical energy allows us to do work conveniently in almost any location. If the work is of a mechanical nature then the electric motor provides the means for electromechanical energy conversion. There are many types of motors to suit many types of applications. We need constant-speed, low-speed, high-speed, and adjustable-speed motors at powers from milliwatts to megawatts. We need motors to suit the environment of the home, industry, and office. We need motors to satisfy the varying requirements of space vehicles, robotic systems, computer disk drives, pumps, fans, and mills, just to name a few. Hopefully, readers will know how to choose a machine and predict its behavior after reading this book.

1.2. ELECTRIC ENERGY PROCESSES

Electric energy processes involve conversion, transfer, transmission, control, and storage. The electric power system exemplifies all these processes, and puts into perspective our main concern: the conversion of energy.

1.2.1. Sources of Energy

Of the many forms of energy used to supplement or replace human energy, electric energy is the most important and practical form, because of the following reasons.

1. Electric energy can be generated in any quantity by reliable processes.
2. Electric energy can be converted into virtually any other desired form of energy without great difficulty or risk and with reasonable efficiency.
3. Electric energy can be transmitted and distributed reliably and with high efficiency to any location.
4. Electric energy is, for certain applications, the only form of energy that can be used. Examples are long-distance communications (audio, visual, or otherwise) and data processing by computer.

Electric energy does have one major disadvantage. It cannot be stored on a large scale for later use.

Since electric energy is not directly available in nature, it has to be generated or produced from other forms of energy. These other forms of energy, which are directly available in nature, or which can be made available, are called primary energy sources. The reliable and efficient transmission and distribution of electric energy means that the generation of electricity is not restricted to any one particular location. The optimum location is determined by the availability of the primary energy source, or the cost of transportation of the medium, and the cost of transmission and distribution of the generated electric energy.

Natural primary energy sources can be divided into two basic categories. The first consists of natural sources immediately usable upon creation. They are:

1. The potential energy of water resulting from accumulated rain water in lakes or rivers. Streaming water could cover 75% of world energy consumption but only 2% is utilized.
2. The potential and kinetic energy of sea water resulting from tides. This could supply 50% of the total world energy demand.
3. The kinetic energy of wind.
4. Solar energy.

The second group of primary sources produced by nature are those that have been stored in the earth for millions of years. They are:

1. Energy liberated in the form of heat when a chemical reaction between fossil fuels, such as coal, oil or gas and air takes place.
2. Energy liberated in the process of nuclear fission or fusion.
3. Geothermal energy.

The energy stored in the primary source is not available for direct use. Primary energy undergoes conversion processes to become a secondary source of energy. It is the secondary source of energy in the form of electric energy that is of concern in this book. There are two main processes for large-scale conversion from primary source to electric energy. Heat is produced to create stored potential energy in steam and this potential energy is converted to mechanical kinetic energy in a steam turbine. Potential energy of an elevated water basin is converted to kinetic energy of streaming water for conversion to mechanical kinetic energy in a water turbine. In these processes the mechanical kinetic energy is converted to electric energy by means of electromechanical energy converters called *generators*.

1.2.2. Electric Power System

Figure 1.1 illustrates diagrammatically a simplified form of a power system involving the production, transmission and utilization of electric energy. The methods of producing the mechanical kinetic energy have been mentioned in the previous section.

A generating station may have an electric output of up to 1500 megavoltamperes at voltages not exceeding 36 kV. To transmit this power economically over

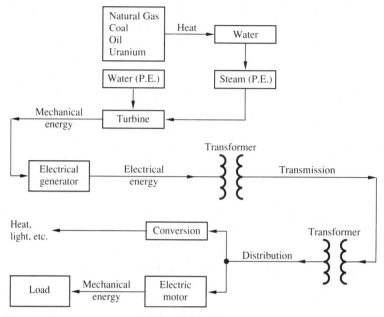

Figure 1.1 Simplified power system outline.

large distances to load centers requires transformers to raise the transmission voltage level. At the load centers the voltage is stepped down by transformers and the energy is distributed at a reduced voltage level for industrial, commercial, and domestic use. For its ultimate use the energy is converted to heat, light, mechanical energy, etc.

It may seem a complex procedure to obtain electric energy as an intermediate form of energy, but it is the efficiency and reliability of transmission of large amounts of this clean and easily controlled energy and the convenience of its utilization which has made electric energy the fastest growing secondary energy source today. Note from the figure that electromechanical energy conversion takes place at least twice in all cases.

The aerospace industry and a search for new electric energy-producing processes have stimulated the development of electricity production directly from energy sources such as solar, nuclear, and chemical sources. Each method has been found useful in a small way, and is attracting much attention as fossil fuels become depleted. In the foreseeable future, the contribution of the amount of direct energy conversion toward the total amount of electric energy generated and required is minor compared with electromechanical energy conversion, so its technological aspects are not reviewed in this text.

In this text we restrict the presentation to the basic steady-state and dynamic theory of those devices that convert mechanical energy to electric energy or vice versa.

1.3. ELECTROMECHANICAL ENERGY CONVERSION

Voltage and current are associated with electric power. Force and velocity are associated with mechanical power. But what is the link that enables one form of power to be converted to another? The key to the link lies in the Lorentz force equation, which describes the force experienced by electric charge in externally applied fields, electric and magnetic.

The vector equation is

$$\mathbf{F} = q(\mathbf{E} + \mathbf{v} \times \mathbf{B}) \qquad (1.3.1)$$

where $\mathbf{F}$ = electromagnetic force (newtons)

q = charge (coulombs)

$\mathbf{E}$ = external electric field (volts/meter)

$\mathbf{v}$ = velocity of the charge (meters/second)

and $\mathbf{B}$ = external magnetic field (tesla).

Therefore there are two basic mechanisms for generating a force and hence producing electromechanical energy conversion. One is to have a charge in an electric field. The second is to have motion of charge in a magnetic field. Motion of charge constitutes a current so that the Lorentz force equation may be expressed as

$$\mathbf{F} = q\mathbf{E} + I(\mathbf{l} \times \mathbf{B}) \qquad (1.3.2)$$

where I is the current in a conductor (amperes) and $\mathbf{l}$ is the effective length (meters) of the conductor subjected to the magnetic field.

Wherever there is a field acting on charge it can be expected that energy conversion is possible. It does not matter whether the charge is associated with one kind of charge carrier like the electrons in the conduction band of metals and certain liquids, or with both kinds of carriers, electrons and ions, in ionized gases. In each and every case there must be an electric or magnetic field present. Therefore it can be said that the fields $\mathbf{E}$ and $\mathbf{B}$ are the catalysts that enable electromechanical energy conversion to take place. Without the presence of an electric or a magnetic field no electromechanical energy conversion is possible.

Examples of electromechanical energy conversion devices are electric motors and generators, pickups, microphones, voltmeters, and ammeters, in which forces are exerted on free-moving charges in the conduction band of metal conductors. An application using a liquid is a liquid-metal pump. Figure 1.2 indicates that the electromechanical energy conversion process is reversible. The most common exam-

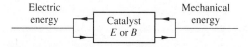

Figure 1.2 Energy conversion with electric or magnetic field.

ple of this is the ability of an electric machine to operate in either the motor or generator mode. In magnetohydrodynamics (MHD) forces are generated on the charge carriers of ionized gas. In cathode ray oscilloscopes and particle accelerators, for example, forces are exerted on a single type of charged particle. Most of the latter examples listed above are special applications, but the basic principle of operation is the same, and can be described by the Lorentz force equation.

EXAMPLE 1.1

Deflection and acceleration of charge in electric and magnetic fields is the action of particle accelerators, in which there is energy conversion of the motoring kind. A cyclotron is used for this example and is illustrated in Fig. 1.3. **(a)** An electron, whose kinetic energy is 1 eV, is introduced at right angles to a uniform magnetic field B of 0.1 T. See Fig. 1.3a. What is the electron orbit radius and the period of revolution? **(b)** If the cyclotron (Fig. 1.3b) has an oscillator frequency of 10^9 Hz, what is the value of the magnetic field that is needed for satisfactory operation?

Solution

(a) A charged particle that is introduced with a velocity **v** at right angles to a uniform magnetic field **B**, will experience a force **F** of magnitude $F = qvB$. The force **F** is perpendicular to **B** and the path of travel. Consequently the force does not change the kinetic energy of the charged particle. As long as **B** remains constant, the magnitude of the velocity and therefore the magnitude of the force remain constant. However, the constant sideways-deflecting force causes the trajectory of the particle to have a

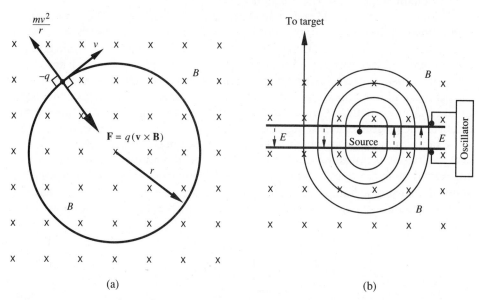

(a)   (b)

Figure 1.3 Deflection of charge in a B field. (a) Circular motion, (b) cyclotron action.

constant centripetal acceleration v^2/r. The magnetic and centripetal forces balance, so that

$$qvB = \frac{mv^2}{r}.$$

The particle velocity is found from $\frac{1}{2}mv^2 = 1.6 \times 10^{-19}$ J, since 1 eV $= 1.6 \times 10^{-19}$ J. With $m = 9.11 \times 10^{-31}$ kg, the velocity is $v = 5.93 \times 10^5$ m/s. The orbit radius is

$$r = \frac{mv^2}{qvB} = \frac{2 \times 1.6 \times 10^{-19}}{1.6 \times 10^{-19} \times 5.93 \times 10^5 \times 0.1} = 0.34 \times 10^{-4} \text{ m}.$$

(b) If an electric field **E**, collinear with the path of the electron and of proper polarity, is introduced in a narrow region, as shown in Fig. 1.3b, the charged particle will be accelerated by a force $\mathbf{F} = q\mathbf{E}$ while in this region. The increased velocity increases the radius of the circular path of the particle in the magnetic field.

The time t_p for the particle to traverse a half-circular path is

$$t_p = \frac{\pi r}{v} = \frac{\pi r m}{r q B} = \frac{\pi m}{q B}.$$

This time is independent of v and r.

The reversal of the electric field must be synchronized by an oscillator so that one half of a period of the alternating electric field equals the travel time t_p. (The time interval that the particle spends in the narrow electric field region is ignored.)

The particle rotational frequency f is

$$f = \frac{1}{2t_p} = \frac{qB}{2\pi m}$$

so the magnetic field $B = \dfrac{2\pi f m}{q} = \dfrac{2\pi \times 10^9 \times 9.11 \times 10^{-31}}{1.6 \times 10^{-19}} = 0.036$ T.

The constant magnetic field is used to bend the path of travel so that the charged particle may be subjected to the accelerating electric field again and again. In this manner, the charged particle can obtain high kinetic energy. At some stage the particle is deflected out of the accelerator and is directed toward a target.

EXAMPLE 1.2

Figure 1.4 illustrates electrostatic beam deflection in a cathode-ray tube (CRT).

The electron gun produces a focused beam of electrons, that are accelerated to a high velocity v through a voltage V_a. There are two pairs of electrostatic deflection plates in a cathode-ray tube. One pair is shown here. A voltage V_d applied to the plates causes an electric field of intensity E_d to deflect the beam, because any charge in an electric field experiences a force. The position where the beam of electrons strikes the fluorescent screen depends on the value of the plate voltage V_d. For the given data find the value of the voltage V_d that must be applied to the parallel deflection plates

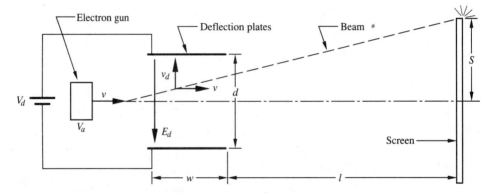

Figure 1.4 Electrostatic deflection of electrons in a cathode-ray tube.

in order that an electron beam will be deflected to one side of the CRT screen. The beam accelerating voltage V_a is 2000 V.

$$d = 0.2 \text{ cm}, \quad w = 2 \text{ cm}, \quad l = 30 \text{ cm} \quad \text{and} \quad S = 10 \text{ cm}.$$

Solution

Let v be the velocity of the injected beam and let v_d be the beam velocity toward the deflection plate as a result of the applied electric field E_d.

The time t_w spent by an electron between the plates is $t_w = w/v$. Combining the Lorentz force and the balancing inertial force[1]

$$eE_d = m \frac{dv_d}{dt}.$$

Upon integration over the time t_w we obtain the expression for the electron velocity in the vertical direction. That is

$$v_d = \frac{eE_d}{m} \frac{w}{v}.$$

The electron flight time t_l to reach the screen after exiting the deflection plates is given by $t_l = l/v$. Since the plate separation is small compared with the screen height,

$$S \approx v_d t_l = \frac{eE_d}{m} \frac{w}{v} \frac{l}{v} = \frac{eE_d wl}{mv^2}.$$

We need to know the values of the electric field intensity E_d and the beam velocity v toward the screen. The electric field intensity between parallel plates is V_d/d. From the definition of potential[2] we can equate the acquired kinetic energy obtained from the accelerating voltage V_a to the loss of potential energy in the form $\frac{1}{2}mv^2 = V_a e$.

---

[1] Frictional and gravitational forces are neglected.
[2] Potential is defined as the work done per unit charge $V \triangleq W/q$.

Therefore $S = \frac{1}{2} \frac{V_d}{V_a} \frac{wl}{d}$.

That is $V_d = 2V_a \frac{Sd}{wl} = 2 \times 2000 \times \frac{0.1 \times 2 \times 10^{-3}}{0.02 \times 0.3} = 133$ V.

Machine design is governed by the application of the physical laws of the energy conversion process, but any design has to be optimized according to other criteria. For example, consideration must be given to economics for the low cost of production and to efficiency for the low cost of operation. What kind of electric machine satisfies these critieria? Is it one that employs the electric field or is it one that uses the magnetic field for the conversion of energy? The answer to this question follows from a comparison of the stored energy densities that can be obtained economically using commercially available materials. Stored energy density leads to force density of electromagnetic origin.

The force density of magnetic field machines is about 25,000 times greater than the force density in electric field machines. This is the reason why virtually all electromechanical energy converters utilize the magnetic field as the medium for conversion and this is sufficient reason to devote the greater part of this text to magnetic field devices.

In the rest of this chapter we will devote our attention to the principles governing the behavior of magnetic field machines. These machines are called electric motors and generators. The term "electric" is the descriptive term indicating that electric energy is the converted energy even though magnetic fields permit the conversion.

1.4. MOTORS AND GENERATORS

An electric machine utilizing magnetic fields may convert mechanical energy into electric energy or vice versa. In the first case the machine is called a *generator,* in the latter case a *motor.* Either type of electromechanical energy conversion involves two basic empirical laws of physics. The laws are

- Faraday's law, which states that the electromotive force induced in a coil of wire is proportional to the rate of change of magnetic flux linking the coil and

- Lorentz's law, which states that a current carrying conductor will experience a mechanical force when placed in a magnetic field. This force is proportional to the magnitudes of the current, the magnetic field and the effective length of conductor subjected to the magnetic field. (We have considered the electric field term in eq. (1.3.2) to be negligibly small.)

If a source of mechanical energy causes a conductor to move in a magnetic field, the induced emf could give rise to the flow of electric current. This would be *generator action.* If a source of electric energy produces a current in a conductor that is exposed to an external magnetic field, the Lorentz force could give rise to motion. This would be described as *motor action.* In both these cases a magnetic field and conductors in relative motion are required for energy conversion.

For relative motion electric machines consist of two parts, one that is stationary and one that moves. The latter (rotor) normally rotates within the first (stator) for continuous motion. Conductors wound in the form of coils can be placed on either the stator or the rotor or on both members. Magnetic fields are produced by these coils when they are connected to a dc or an ac source of current. A large number of combinations of stator and rotor shape and methods of current excitation is possible.

1.4.1. Faraday's Law and Induced Emf

Consider the arrangement depicted in Fig. 1.5. The single-turn coil has straight sides. These coil sides are perpendicular to the field lines of a stationary magnetic field of constant density B, which is assumed to be uniform.

An external force F is applied to the coil and causes the coil to move with a constant velocity v in a direction that is in the plane of the coil and perpendicular to the straight coil sides. The assumed relative directions of the conductors, the magnetic field, and the velocity may appear to restrict the discussion of Faraday's law to a special case. This is indeed so, but this arrangement of conductors, velocity, and magnetic field is the one that is normally encountered in electric machines.

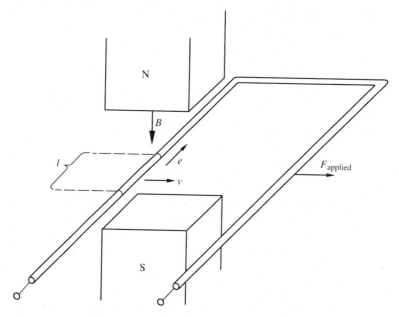

Figure 1.5   Motion of a conductor in a magnetic field.

Faraday's law states that the coil-induced electromotive force (emf) is given by

$$e = -\frac{d\lambda}{dt} \tag{1.4.1}$$

where λ is the total flux linkage with the coil. Since we consider a one-turn coil, the flux linkage λ is equal to the flux ϕ that penetrates the plane of the coil. If the length of the single straight conductor situated in the magnetic field is l meters and if this conductor moves during a time interval of dt seconds with a velocity of v meters per second to the right, then the change (reduction) in flux linking the coil is

$$-d\phi = B \times d(area) = Blv\, dt. \tag{1.4.2}$$

From eq. (1.4.1) it follows that the emf is

$$e = Blv \text{ volts} \tag{1.4.3}$$

where B is expressed in tesla. The negative sign, in eq. (1.4.1),[3] indicates that the orientation of the induced emf is such that a current resulting from this emf will oppose the change in the flux linkage that induced this emf. This is in agreement with Lenz's law. The algebraic sign, which is needed for mathematical bookkeeping in analytical work, can be dropped when the proper direction of the emf is indicated by other means, such as by the arrow in Fig. 1.5.

For the general case we should have assumed that the conductor is at some arbitrary angle α to the magnetic field lines and that the conductor is moved in a direction that makes an arbitrary angle β to the direction of the magnetic field. In that case the magnitude of the induced emf would be

$$e = Blv \sin\alpha \sin\beta. \tag{1.4.4}$$

In vector form

$$e = \mathbf{l} \cdot (\mathbf{v} \times \mathbf{B}). \tag{1.4.5}$$

From this expression it should be clear that in order to induce the maximum emf in a conductor of a certain length moving at a certain velocity in a given magnetic field, mutual perpendicularity of the three vectors $\mathbf{B}$, $\mathbf{l}$, and $\mathbf{v}$ is required so that $\sin\alpha$ and $\sin\beta$ attain their values of 1. Initially we assumed mutually perpendicular directions for $\mathbf{B}$, $\mathbf{l}$, and $\mathbf{v}$, since this arrangement is used in electric machines for economic reasons.

---

[3] We have chosen this form of Faraday's law, because college physics texts use it, and we feel the best description of machines is by means of physics. Some books on machines omit the use of the minus sign. This is all right as long as the change of flux linkage is considered to be a voltage drop in the circuit and not a voltage rise.

A. POLARITY OF INDUCED EMF

The direction of the emf induced in a conductor that moves in a magnetic field depends on the relative direction of motion.

Fleming's right-hand rule for generators assumes that the field is stationary and that the conductor moves with respect to the field. The actual situation in practice has to be reduced to the condition for which Fleming's rule applies. When the thumb of the right hand points to the relative direction of motion of the conductor and the index finger points in the direction of the field, the middle finger will point in the direction of the positive polarity of the induced emf. This is shown in Fig. 1.6. It should be pointed out that the conventional current direction is used and this current is opposite to the direction of electron movement. The direction of the induced emf can also be found from the direction in which a right-hand corkscrew would progress when turned in the same direction as would be required to line up the direction of motion with the positive field direction. Similarly the cross product in eq. (1.4.5) indicates the emf polarity.

1.4.2. Lorentz Force

It is beneficial to consider the Lorentz force exerted on a current-carrying conductor situated in a magnetic field in order to appreciate the basic operation of all machines. In the most general case the current-carrying conductor has an arbitrary direction, which encloses an angle α with the direction of the magnetic field. If a straight conductor of length l meters carries a current of i amperes and is situated in a uniform magnetic field of density B tesla, the Lorentz force exerted on the conductor is

$$F = Bil \sin\alpha \text{ newtons.} \tag{1.4.6}$$

In vector form this is

$$\mathbf{F} = i(\mathbf{l} \times \mathbf{B}). \tag{1.4.7}$$

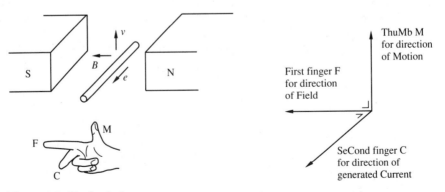

Figure 1.6 Emf polarity.

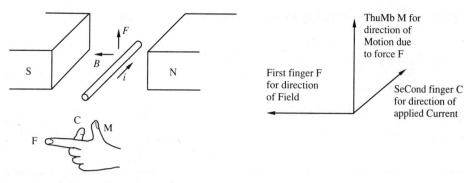

Figure 1.7 Left-hand rule for the Lorentz force.

It is usual to arrange the position of the current-carrying conductor such that it is perpendicular to the direction of the magnetic field. In this case $\sin\alpha$ is unity and the maximum force is developed for a given length of conductor, current, and magnetic flux density. Accordingly eq. (1.4.6) becomes

$$F = Bil. \tag{1.4.8}$$

A. DIRECTION OF LORENTZ FORCE

The direction of the Lorentz force is that in which a right-hand corkscrew would progress when turned in the manner to line up the positive current direction with the positive field direction. The direction of the force can also be found using the left-hand rule for motoring as shown in Fig. 1.7. Use of the cross product in eq. (1.4.7) gives the direction of the force also.

EXAMPLE 1.3

Figure 1.8 shows a diagram of a plasma motor suitable for space propulsion. A pulse of plasma of length 3 cm carries a current of 1000 amperes. Due to the external magnetic field of 2 tesla, there is a force to eject the plasma. By Newton's third law

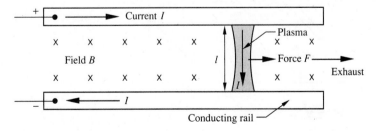

Figure 1.8 Plasma motor.

there would then be a small thrust on the spacecraft. What is the value of the force of magnetic origin?

Solution
The action of this motor is another example of the application of Lorentz's law. In elementary terms the magnitude of the force F, causing plasma motion, is

$$F = BIl = 2 \times 0.03 \times 1000 = 60 \text{ N}.$$

This force drives the plasma along the fixed, conducting rails, that enable current to flow through the plasma, until the latter is exhausted into space.

1.4.3. Basic Generator Operation

Figure 1.9 shows a one-turn coil that moves with a constant velocity v as a result of an externally applied force F_a in a uniform magnetic field of flux density B. The directions of the conductor, the motion, and the magnetic field are all mutually perpendicular so that an emf of magnitude

$$e = Blv \tag{1.4.9}$$

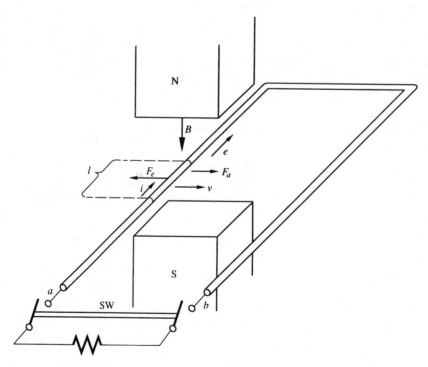

Figure 1.9 Generator action.

is induced in the conductor. The polarity is indicated, but check this by any of the rules described in subsection A of section 1.4.1. This voltage is now available at the terminals a and b of the coil (b is at a positive potential with respect to a). Upon closure of the double-pole switch SW, a resistor is connected to the coil terminals, a closed electric circuit is formed, and a current will flow in the direction of the induced emf. (This current of i amperes will flow in the resistor from b to a.)

We now have a current-carrying conductor situated in a magnetic field. Thus a Lorentz force F_e, which is of electromagnetic origin, will be developed in the direction indicated in the figure. It is to be noted that the electromagnetic force F_e is in a direction opposite to that of the externally applied force F_a. This is in agreement with Lenz's law. If the external magnetic field and the speed of motion have constant values, the generated or induced voltage has a constant value. The greater the electric energy demand, the larger the current delivered to the load, the larger the retarding force and the larger the applied force that is required to maintain the motion. Thus, the more electric energy is delivered by the generator the more mechanical energy has to be supplied.

1.4.4. Basic Motor Operation

Consider an arrangement similar to the one shown in Fig. 1.9 as depicted in Fig. 1.10 in which the resistor is replaced by a battery. A current in the direction of the externally applied voltage will flow through the coil upon closure of the switch. A current-carrying conductor is then situated in an external magnetic field. An electromagnetic Lorentz force F_e will be developed and this causes the coil to move with a velocity v in the direction of the force. As a result of the motion the coil will experience a change of flux linkage, and an emf will be induced in the direction indicated. We note that the direction of the induced emf is opposite to that of the applied voltage. This is Lenz's law at work again. In this case the induced emf e is usually called a *counter* or *back emf*. It is also known as the *motional* or *speed emf*.

Any arrangement of conductors and magnetic field in which a voltage is induced as a result of externally imparted relative motion between the two, such as in the case shown in Fig. 1.9, can be said to operate in a *generator mode*. Any arrangement of conductors and magnetic field in which a force is developed as a result of voltage application to the conductors, such as in the case shown in Fig. 1.10, can be said to operate in a *motor mode*. From the foregoing discussion and study of Figs. 1.9 and 1.10, we note that both motor and generator action take place simultaneously. This means that there is an energy balance. We cannot get more electric energy out than the mechanical energy input and vice versa.

1.4.5. Continuous Electromechanical Energy Conversion

The linear motion arrangement, such as shown in Figs. 1.9 and 1.10, is not practical for continuous electromechanical energy conversion processes. Continuous relative

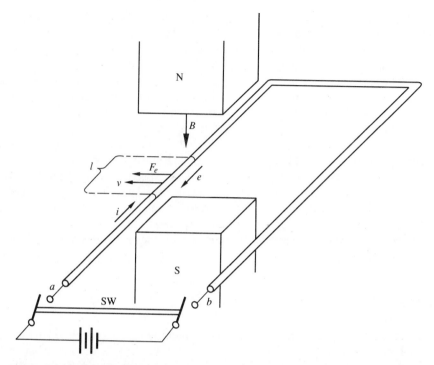

Figure 1.10 Motor action.

motion between conductors and magnetic field is required. This can be realized if either the conductors rotate with respect to the magnetic field or the magnetic field rotates with respect to the conductors.

A basic arrangement with rotation is shown in Fig. 1.11. Two parallel conductors are shaped to form a rectangular loop or coil. The coil rotates about its axis, which is perpendicular to the direction of a uniform magnetic field. Each conductor moves with the same peripheral velocity v and has the same effective length l situated in the uniform magnetic field B between poles. The directions of the induced emfs in the conductors are such that the individual conductor emfs in the coil sides are additive. In their movement the conductors remain at right angles to the magnetic field but the velocity direction encloses a varying angle θ with the field direction. When the plane of the coil is perpendicular to the field direction, the angle between the velocity and field directions is zero and the value of the induced voltage is zero. When the plane of the coil is parallel with the field lines, the angle between the conductor velocity and field is $\pi/2$ and the induced emf has its maximum value. We can write the instantaneous value of the induced coil emf as

$$e = 2Blv \sin\theta. \qquad (1.4.10)$$

The coil of Fig. 1.11 has to be mechanically supported. A good shape for a mechanical supporting structure for coils that have to rotate is cylindrical. In order to

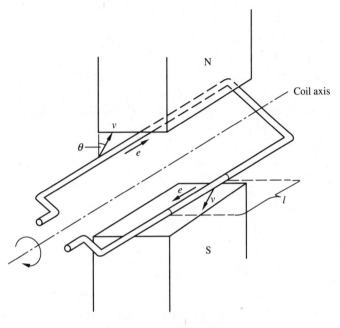

Note: v is perpendicular to
the plane of the coil

Figure 1.11 Rotating conductors.

reduce the reluctance of the magnetic circuit and attain high values of flux density, the cylindrical coil support or rotor is made of ferromagnetic material. To further reduce the reluctance the pole faces are given a hollow cylindrical shape and the coil sides are embedded in slots in the rotor surface. The poles are mounted in a cylindrical frame, called the yoke, which also provides a path for the magnetic flux. Figure 1.12a shows a cross section of such an arrangement. A narrow constant air gap now exists between the rotor surface and the pole-face surface. Because these surfaces are magnetic equipotential surfaces the field lines will cross the narrow air gap in a radial direction except at the edge of the poles, where unavoidable fringing of the magnetic flux takes place. The conductors of the rotating coil will now have a direction of motion that is always perpendicular to the field lines. Consequently, the expression for the instantaneous value of the emf induced in the single-turn coil is

$$e = 2Blv \qquad\qquad (1.4.11)$$

where B is the value of the magnetic field at the location of the coil sides.

A. WAVESHAPE AND FREQUENCY OF INDUCED VOLTAGE

For the machine depicted in Fig. 1.12a the instantaneous voltage induced in a conductor varies with time because the value of the flux density varies along the

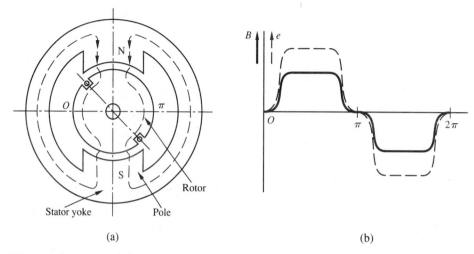

Figure 1.12 A practical arrangement for conductors. (a) Machine cross section, (b) magnetic field and emf as a function of coil position.

periphery of the rotor. (See eq. (1.4.11).) This is illustrated in the Fig. 1.12b. As the rotor revolves one complete turn at constant speed, the voltage waveform goes through one complete cycle. Thus, in the case of the two-pole arrangement the electrical frequency of alternation of the induced voltage is equal to the speed of the rotor in revolutions per second.

In Fig. 1.12b we see that the flux density distribution (and therefore the induced voltage) has an almost rectangular waveshape. To further reduce the reluctance of the magnetic circuit, but mainly to extend the influence of the magnetic field and to control the form of the flux density distribution around the rotor surface, the pole faces are enlarged to form so-called pole shoes. By judicious variation of the air-gap length under the pole shoe the gap reluctance can be varied in such a manner to give a flux density distribution that is very nearly sinusoidal. Consequently the induced voltage wave would also approach a sinusoidal form.

In many electric machines more than two poles are used. Since a north-polarity pole is always associated with a south-polarity pole, the total number of poles has to be even. A four-pole arrangement is shown in Fig. 1.13. The two coil sides, which are parallel to the shaft, must move through magnetic fields of opposite polarity for the emfs induced in the conductors to be additive. For the conductor emfs to be additive all the time, the distance between coil sides has to be equal to the distance between pole centers (or the pole pitch). Coils that satisfy this requirement are called *full-pitch coils*. (In many cases coils are used with a spread less than a pole pitch.) Conductors 1 and 1′ indicate the position of coil 1 and conductors 2 and 2′ indicate a second full-pitch coil. If the distance between conductors 1 and 2 is equal to a double pitch the instantaneous values of the emfs induced in coil 1 and coil 2 are the same at all times. Connecting the two coils in series will double the induced voltage, or connecting the coils in parallel will double the current-carrying capacity of this elementary generator. Upon one complete revolution of the rotor, coil 1 will pass the magnetic field of two pairs of poles. The voltage wave will then consist of two full cycles. If the machine has p pole pairs, the voltage wave would have p complete cycles for each revolution of the rotor. If the rotor rotates with a speed of n revolutions per second, the coil induced emf would have pn complete cycles per second. Thus, the frequency f of the induced emf is related to the number of pole pairs and the speed of rotation by the expression

$$f = pn \text{ hertz.} \tag{1.4.12}$$

Often, the rotation of a coil through a displacement is expressed in electrical radians. From Fig. 1.12, a coil that rotates through a complete electromagnetic cycle (a north pole, a south pole, and back to the north pole) of the 2-pole machine has gone through 2π electrical radians and 2π mechanical radians. From Fig. 1.13, a coil that rotates through a complete electromagnetic cycle (a north pole, a south pole, and the next north pole) of the 4-pole machine has gone through 2π electrical radians and π mechanical radians. In general, a coil that goes through 2π electrical radians (from a north pole to a south pole and on to the next north pole) of a machine with p pole pairs has gone through $2\pi/p$ mechanical radians.

We have used Faraday's law to describe an elementary generator and we have used the Lorentz force equation in order to show how an elementary motor works. The next step is to determine how well the motors and generators work. Faraday's law is very useful in this regard. However, the Lorentz force equation becomes too cumbersome to apply to all except the simplest cases. One well-proven way to help us determine the behavior of electric machines is to consider the electric circuit relations and then apply energy principles. We will do this in Chapter 2. A basic requirement is to be able to manipulate the equations linking current and magnetic flux linkage. A brief review of simple magnetic circuit calculations for engineering applications is given in Appendix A. This review covers the background so that the reader can follow

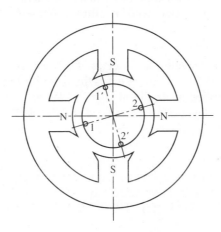

Figure 1.13 Four-pole machine.

the discussion of dynamic circuits and electromechanical energy conversion in Chapter 2 and transformers in Chapter 3.

This book uses graphic symbols to represent device windings, armatures, yokes, slip rings, and commutators, etc. In order to become familiar with the actual components of devices, there is no substitute for a visit to the machines' laboratory. Look at as many machines as possible, hopefully some in section, and identify both the parts and the type. If in doubt ask the laboratory instructor. It will be rewarding and it will certainly help to clarify the figures that follow.

1.5. SUMMARY

Electromechanical energy conversion is associated with magnetic fields, currents, and motion. The interaction of these variables for motoring action is described by the Lorentz force equation $\mathbf{F} = I(\mathbf{l} \times \mathbf{B})$. The interaction for generator action follows from Faraday's law in the form $e = \mathbf{l} \cdot (\mathbf{v} \times \mathbf{B})$. These equations can be applied only to simple configurations of elementary devices.

1.6. PROBLEMS

Section 1.3

1.1. A proton experiences no electromagnetic force while moving in the z-axis direction, even though there is a uniform magnetic field of 2T. However, if the proton moves with the same speed in the $+y$-axis direction, the force experienced is in the $+x$-axis direction. Determine the direction of the magnetic field.
(*Answer:* $+z$-axis direction.)

1.2. An electron travels in the $+x$-axis direction in an electric field E, that is in the $+y$-axis direction. In what direction must a magnetic field **B** be applied, so that the magnetic force on the electron minimizes the resultant electromagnetic force?
(*Answer:* $+z$-axis direction.)

1.3. A charged particle, traveling at a speed v of 10^6 m/s, is directed at right angles to an electric field E of 10^6 V/m. What is the magnitude of the magnetic field **B**, that must be applied mutually at right angles to v and E, in order that the particle does not deviate from its initial path?
(*Answer:* 1.0 T.)

Section 1.4

1.4. A conductor of length l moves with speed v through a constant and uniform magnetic field B. **B**, **l**, and **v** are mutually at right angles. What is the value of the emf induced in the conductor, if $B = 1.0$ T, $l = 20$ cm, and $v = 8.2$ m/s?
(*Answer:* 1.64 V.)

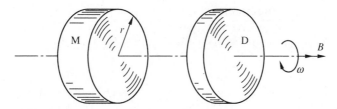

Figure P1.6

1.5. A conducting, cylindrical shell of radius r and length l rotates at n r/s about its axis and experiences the same radial magnetic field B over the whole of the surface. This type of machine is called a *homopolar generator*.[4] Calculate the emf induced in this machine for the parameters given below. In order to connect this generator to an external circuit, where would the sliding contacts be placed?

$$r = 20 \text{ cm}, \quad l = 20 \text{ cm}, \quad n = 40 \text{ r/s}, \quad \text{and} \quad B = 1.5 \text{ T}.$$

(*Answer:* 15.1 V.)

1.6. Figure P1.6 represents a primitive model of a homopolar generator. M is a permanent magnet, whose flux density B is uniform, constant and axial. D is a copper disc, whose axis is coincident with M's axis. The two discs have the same diameter and are very close to each other. M is made of conducting material. Is there an induced emf in M or D for the following cases? **(a)** M stationary and D stationary, **(b)** M stationary and D rotating at ω, **(c)** M rotating at ω and D stationary, **(d)** M rotating at ω and D rotating at ω, **(e)** M rotating at $-\omega$ and D rotating at ω, and **(f)** M rotating at ω and D rotating at $-\omega$. **(g)** If any of the answers are affirmative, calculate the induced emfs in terms of r (radius), ω (angular speed), and B (magnetic flux density).
(*Answer:* $\frac{1}{2}B\omega r^2$ V.)

1.7. Figure P1.7 overleaf illustrates the principle of a moving-coil ammeter. The current to be measured is allowed to flow through a coil wound on an armature, that can rotate between the poles of a permanent magnet. For the current direction shown in the figure, in which direction will the indicator tend to rotate?
(*Answer:* counterclockwise.)

1.8. A straight, current-carrying conductor is to be levitated in a horizontal plane by balancing the magnetic force against the gravitation force. If the external magnetic field **B**, of value 1.5 T, is directed horizontally at right angles to the conductor, what current must be passed through the conductor? The cross-

---

[4] The homopolar generator gets its name from the fact that the conductors of such a generator experience the same polarity of magnetic field at all times during their motion.

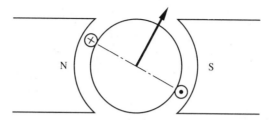

Figure P1.7

sectional area A of the conductor is 8×10^{-6} m^2 and its density ρ is 8960 kg/m^3. The gravitational constant g is 9.81 m/s^2.
(*Answer:* 0.47 A.)

1.9. Electric energy can be converted to mechanical energy to cause a conducting liquid to flow. This is illustrated in Fig. P1.9. If a current of 100 A flows through the liquid between two electrodes 1 cm apart and if a magnetic field of 2 T is applied at right angles to the current, what is the force of magnetic origin acting on the fluid?
(*Answer:* 2 N.)

1.10. Figure P1.10 shows a cylinder on an inclined plane. Current I flows through the N turn coil that is wound on the cylinder of radius r, length l, and mass M. The coil experiences an externally applied magnetic field B in the vertical direction. What is the value of the current so that the torque of magnetic origin balances the gravitational torque to prevent the cylinder rolling? The system parameters are

$$\alpha = 30°, \quad \beta = 45°, \quad l = 20 \text{ cm}, \quad M = 0.5 \text{ kg},$$
$$g = 9.81 \text{ m/s}^2, \quad N = 20, \quad \text{and} \quad B = 1.0 \text{ T}.$$

(*Answer:* 0.43 A.)

1.11. Figure P1.11 shows a conductor of mass M, resistance R, and length l sliding down inclined rails that are perfect conductors and frictionless. The rails are

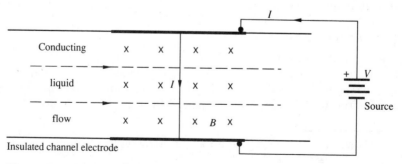

Figure P1.9 Liquid metal pump.

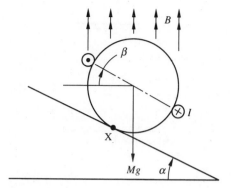

Figure P1.10

connected by a perfect conductor at the bottom of the incline. A uniform magnetic field B is applied in the vertical direction. For the system parameters given below, what is the steady-state speed v of the sliding conductor?

$\alpha = 30°$, $\quad l = 20$ cm, $\quad M = 0.5$ kg,
$g = 9.81$ m/s², $\quad B = 1.0$ T, $\quad$ and $\quad R = 0.1 \ \Omega$.

(*Answer*: 8.2 m/s.)

1.12. Consider the system in Problem 1.11 and Fig. P1.11. The inclination α is made zero, but the other conditions remain the same. The constant speed of the sliding conductor is mechanically maintained at 8.2 m/s. Calculate **(a)** the emf induced in the sliding conductor, **(b)** the value of the current circulating in the loop created by the conductor and the rails, **(c)** the thermal power dissipated by the conductor, **(d)** the electromagnetic force on the conductor due to its mo-

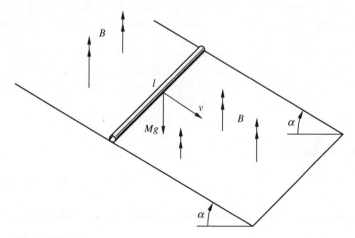

Figure P1.11

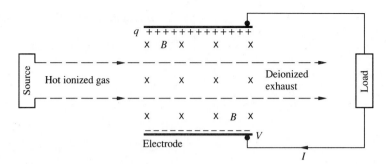

Figure P1.14

tion in the magnetic field, **(e)** the mechanical force needed to sustain motion and **(f)** the mechanical power to drive the conductor.
(*Answer:* (a) 1.6 V, (b) 16.4 A, (c) 26.9 W, (d) 3.28 N·m, (e) 3.28 N·m, (f) 26.9 W.)

1.13. A generator is placed in series with the rail at the bottom of the incline of the system shown in Fig. P1.11. In which direction must the current flow in the rails and the conductor so that the "motor" force tends to drive the conductor up the incline? If the generator current is constant at 20 A, find the speed of the conductor up the incline after 5 seconds. Assume the conductor is at rest at time $t = 0$ and the system parameter values are the same as those given in Problem 1.11.
(*Answer:* 10.1 m/s.)

1.14. The production of electricity by the motion of a gas in a magnetic field is illustrated in Fig. P1.14. The charge carriers (of elementary charge 1.6×10^{-19} C) of a thermally ionized gas are injected at 10^4 m/s into a region, in which a magnetic field $B = 2$ T exists, and experience a deflecting magnetic force. Electrodes, at a distance 1 cm apart and with their planes parallel to the B field, collect the charges and an electrostatic potential difference is established. If connections are made from an external load to the electrodes, the generated voltage will give rise to current flow. Thus the kinetic energy of the gas has been partially converted to electric energy. Calculate **(a)** the force of magnetic origin on the charged particles and **(b)** the induced emf between the electrodes.
(*Answer:* (a) 3.2×10^{-15} N, (b) 200 V.)

1.7. BIBLIOGRAPHY

Del Toro, Vincent. *Electric Machines and Power Systems.* Englewood Cliffs, N.J.: Prentice-Hall, Inc., 1985.

Elgerd, Olle. *Basic Electric Power Engineering.* Reading, Mass.: Addison-Wesley Publishing Co., 1977.

Gehmlich, D. K., and S. B. Hammond. *Electromechanical Systems.* New York: McGraw-Hill Book Co., 1967.

Halliday, David, and Robert Resnick. *Fundamentals of Physics,* 2nd ed., extended version. New York: John Wiley & Sons, Inc., 1986.

Matsch, Leander W. *Capacitors, Magnetic Circuits and Transformers.* Englewood Cliffs, N.J.: Prentice-Hall, Inc., 1964.

Nasar, S. A. *Electric Machines and Electromechanics.* Schaum's Outline Series in Engineering. New York: McGraw-Hill Book Co., 1981.

Nasar, S. A. (Ed.). *Handbook of Electric Machines.* New York: McGraw-Hill Book Co., 1987.

Say, M. G. *Introduction to the Unified Theory of Electromagnetic Machines.* London: Pitman Publishing, 1971.

Skilling, H. H. *Electromechanics.* New York: John Wiley & Sons, Inc., 1962.

CHAPTER TWO

Energy Conversion and Dynamic Circuits

2.1. INTRODUCTION

General principles will be introduced in this chapter so that we will know how to go about predicting the force, motion, and current responses of electromechanical devices.

Why must we know the behavior of a device? Consider for instance a microprocessor-controlled electric motor that drives a floppy disk. The motor performance has to be specified within fine limits so that precise information can be transferred from the disk to the memory of the computer. At the design stage an engineer has to confirm that the specifications can be met. General principles are invoked to formulate particular equations. The calculations involving these equations produce results from which it can be determined how or if the motor will meet the needs of the application. Consequently it is usual to want to know not just that a device works but how well it works. This chapter describes how we can set about learning how well it works.

The reader will have had courses in electric circuit analysis, in electricity and magnetism and most probably in mechanics. We will combine principles from all three courses to arrive at equations that describe the behavior of electromechanical devices. The conservation of energy principle underlies any energy conversion process.

Early in the chapter, electric circuit analysis is applied to devices so that we have relations for current responses. In order to determine forces of magnetic origin we need to know something about energy stored in the magnetic field and the relation between currents and magnetic fields. The necessary magnetic field calculations are reviewed in Appendix A. Finally we put together the dynamic equations of motion.

Appendix B contains a review of that part of mechanics that is useful for our purposes.

The title of this chapter contains the words "dynamic circuits." The word *dynamic* implies that there is a change in the values of electric circuit parameters that are associated with mechanical movement. This can be illustrated by a simple example. Consider a solenoid shown diagrammatically in Fig. 2.1. If the switch S in the electric circuit is closed a current flows. This causes a magnetic field to be produced in the air gap. A force will now be exerted on the iron armature A and, if it is allowed to move, the mass M will be lifted off the platform by that force. Mechanical work is done. In terms of energy conversion the electrical energy input is converted in part to energy stored in the magnetic field and in part into mechanical energy. The magnetic field is at the heart of the energy conversion process and couples the electrical and mechanical systems. The movement implies that the system geometry changes and that causes the value of the inductance to change. This change in the electric circuit parameter is described as *dynamic,* and the term is applied to the analysis used in this chapter.

The conservation of energy principle provides the bridge between the electrical and mechanical sides of a device. The electromechanical energy conversion device is considered to be a set of linked electrical, magnetic, and mechanical systems as shown in Fig. 2.2. The performance equations either for steady-state or transient conditions are derived from the dynamic circuit analysis involving three sets of relations:

1. Electric circuit relations
2. Energy conversion relations
3. Dynamic relations of the mechanical system.

The solution of (1) to obtain the currents for a particular system configuration leads to the computation of the force from (2), so that the mechanical response can be found from (3).

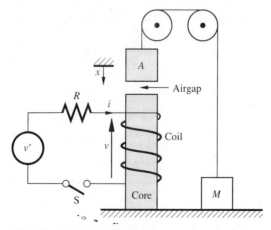

Figure 2.1 Solenoid actuator.

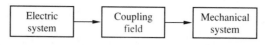

Figure 2.2 Linked systems.

It is common to categorize electromechanical energy converters into three general groups.

1. Small signal transducers (e.g., microphones, pickups, loudspeakers, and electromechanical measuring instruments).
2. Force-producing devices (e.g., magnetic relays and solenoid actuators).
3. Continuous conversion devices (e.g., rotating machines).

Dynamic circuit analysis has the advantages that it unifies the study of all these devices and provides the base for the generalized theory of rotating electrical machines. From dynamic circuit analysis the performance of an electromechanical system, either steady state or transient, can be obtained.

2.2. SYMBOLS IN ELECTRIC MACHINE REPRESENTATION

An electric machine comprises a complex arrangement of conductors, magnetic materials, insulation, and air gaps. The entire configuration can be represented by simple symbols so that the discussion of principles of operation can be made easier. This section includes a description of the symbols that will appear throughout the book. The symbols are introduced by using elementary magnetic devices.[1]

A magnetic field can be created by coils carrying currents or by permanent magnets. A general term for the production of a magnetic field is *excitation*. For example, we say that a coil has constant excitation if a dc voltage is applied to the terminals of the coil. Another case is a *doubly-excited* system. This form of excitation involves independent sources of voltage causing currents to flow in two coils. In general we have dc excitation, or ac excitation, or a combination of both to produce magnetic fields. The interaction of the excitation components gives rise to energy conversion.

2.2.1. Stator and Rotor Excitation

Figure 2.3 illustrates a singly-excited system. A fixed coil (stator) is excited by a current *i*. The coil is symbolized as having its turns concentrated close together.

---

[1] An elementary magnetic device can consist of a combination of coils carrying currents and a magnetic circuit of simple shape.

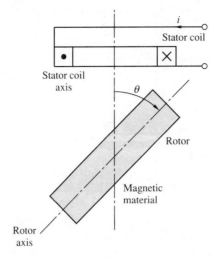

Figure 2.3 Primitive singly-excited rotational system.

Current direction is shown by a *dot* and a *cross*.[2] A member (rotor) of magnetic material is free to rotate about an axis that is perpendicular to the coil axis. There is a torque tending to bring the axis of the rotor into alignment with the coil axis.

If a coil is isolated as shown in Fig. 2.4a, the current in the coil produces a magnetic field which is symmetrical in space about the axis of the coil. The field is shown by broken lines with arrows indicating the north (N) and south (S) poles. As shown in Fig. 2.4b, if a magnetic material is placed in the vicinity of the coil, the magnetic field pattern is altered. The magnetic material is a preferential flux path to air because of its lower reluctance. The main field of the coil magnetizes the ferromagnetic rotor. Figure 2.4b shows the natural polarity of the magnetization of the rotor. This provides a physical picture of magnets acting on magnets. The south pole S of the rotor is attracted to the north pole N of the coil's field. If the rotor were allowed to move in the counterclockwise direction, it would line up with the stator axis. Alignment of the rotor and stator axes would provide a stable equilibrium position at which no torque exists.

If the rotor moves from its original position of angular displacement θ to $\theta = 0$, the reluctance of the magnetic circuit reduces, until at $\theta = 0$ it is at a minimum. Thus, the torque that exists in this singly-excited system tends to bring about the lowest possible reluctance. Accordingly, this torque has been named the reluctance torque T_R. At $\theta = 0$ and $\theta = \pi$ the torque is zero. There is also zero torque at $\theta = \pi/2$ and $\theta = 3\pi/2$, but these equilibrium positions are unstable; any small displacement would result in a torque tending to bring θ to zero or θ to π.

---

[2] The dot represents the tip of an arrow and the cross represents the feathers, so the dot indicates current out of the plane of the paper and the cross indicates the opposite.

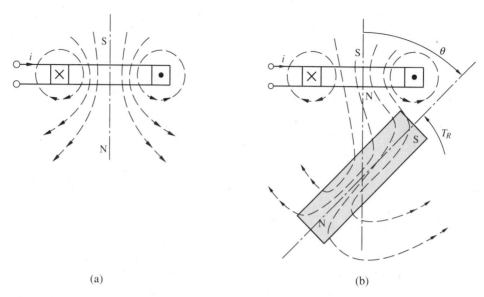

(a)   (b)

Figure 2.4 Singly-excited system. (a) Field pattern of isolated coil, (b) torque from rotor magnetization.

Newton's third law tells us that where there is action there is a reaction. Consequently, any torque on the rotor means that there is an equal and opposite torque on the coil. That is, the coil must be constrained to be stationary and is called the stator. From this simple argument we conclude that if we changed the reference frame and kept the magnetic member fixed, and allowed the coil to be free, the torque would make the coil rotate. In either case the relative movement is the same in angular displacement, velocity, and acceleration. It does not matter in principle which is the rotor and which is the stator. It is a matter of application and convenience of construction which member is taken for the rotor. If the coil were allowed to rotate, it would be necessary to provide sliding contacts (slip-rings or a commutator) in order that a stationary voltage source could supply current to a moving coil.

A practical arrangement for the exciting coil would be to have a stator of magnetic material to reduce the overall reluctance of the magnetic circuit. This would define the magnetic path more clearly and also reduce the magnetizing current for a required magnetic flux ϕ. There are two forms the stator can take. One form is a stator with a cylindrical bore. The other is a stator with salient poles. Figure 2.5 shows 2-pole configurations but any practical number of pole pairs can be formed. Concentrated coils are shown in the figure. Only the salient-pole stator would have a concentrated exciting coil; that is, the turns of the coil would be tightly formed together to make the coil as concentrated as possible. The round or cylindrical bore stator would not have all the coil sides bundled together. The turns would be distributed in slots around the bore as in Fig. 2.6. This is not exactly a cylindrical bore but for all practical purposes it can be considered to be so. The distributed winding utilizes the available space better without distorting the intended structural shape too much. It also helps

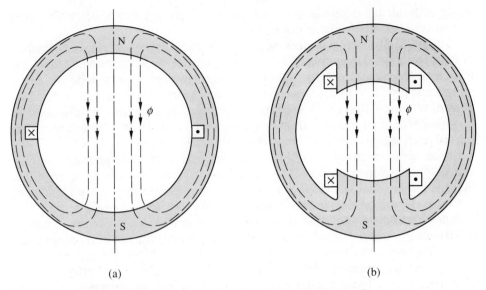

(a)   (b)

Figure 2.5 Stator configurations. (a) Cylindrical bore, (b) salient pole.

to provide a desired magnetic field pattern in the air gap between the stator and rotor. Also, there are two types of rotor shapes, the cylindrical rotor and the salient-pole rotor.

Air gaps and their geometry in rotating machines are important. The very low reluctance of stator and rotor in the magnetic circuit means that the magnetic energy can be considered to be stored in the air gap. To keep the magnetizing current low, the

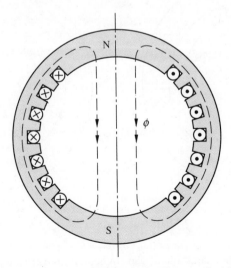

Figure 2.6 Distributed winding to provide a cylindrical bore.

length of the air gap is kept as short as possible. (It can vary from 0.05 cm for small machines to 0.25 cm for large machines.)

Figure 2.7 illustrates two machines with cylindrical stators and rotors. No matter in what position the cylindrical rotor is, the reluctance does not change. Therefore, there is no reluctance torque T_R. However, there can be electromagnetic torques of other origin produced by a singly-excited system if the rotor is cylindrical. Combinations of rotating systems with exciting coils on only one member are discussed in Chapter 4. The outer member always represents the stator, and as the name implies it is fixed relative to a reference frame which is usually the ground.

From the symbolized coil of Fig. 2.3 we have started to develop machine rotor and stator configurations in order to illustrate the forms they can take. With this information in mind we will return to primitive or elementary representations of machines. Figure 2.8 depicts the symbols used for a doubly-excited machine. The coils of the stator and rotor can be concentrated or distributed. The current excitation can be steady (dc or ac) or transient. The parts of the machine can have salient poles or not. We keep everything general until we specify a problem. We adhere to a *dot* convention that applies to magnetic field directions and the sign of mutual inductance coefficients. The *dot* convention is an aid to indicate the direction in which a coil is wound. If current enters the end of a coil marked by a dot, the magnetic field polarity is as shown in Fig. 2.8. In this case the stator and rotor fields aid each other so the mutual inductance coefficient is positive. Since we have drawn in the magnetic

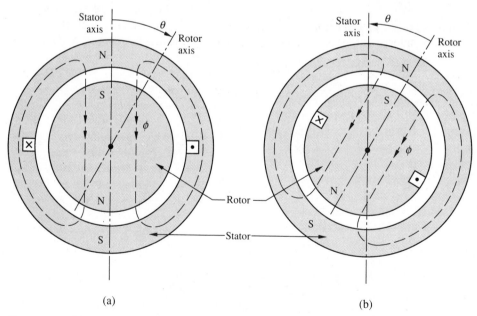

(a)   (b)

Figure 2.7 Singly-excited, constant air-gap configurations. (a) Stator-coil excitation, (b) rotor-coil excitation.

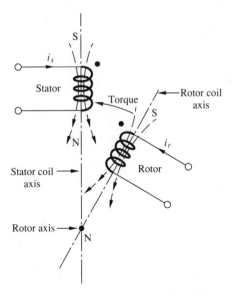

Figure 2.8 Doubly-excited primitive machine.

polarities that follow from the convention, it is possible to see that the rotor shown in Fig. 2.8 would tend to move in a counterclockwise direction.

The symbols in Fig. 2.8 are used throughout Chapter 2.

2.3. ELECTRIC CIRCUIT RELATIONS

The first step in the procedure of dynamic circuit analysis is to find a suitable form for the electric circuit relations in order to facilitate the calculation of torque in a rotating system or force in a translational system.

2.3.1. Singly-excited Rotating System

The electric circuit relationship in terms of the field quantity flux linkage λ is given by Faraday's law, which states that the emf e induced in a coil is equal to the rate of change of flux linkage of that coil. Figure 2.9 shows the symbolic representation of a singly-excited rotating energy converter called a *reluctance machine.*

If ideal conditions exist such that the magnetic potential drop in the iron is negligible compared with the drop in the air path, a simple expression can be derived to describe the flux linkage in terms of current. It follows that

$$\lambda = N\phi = NBA = N\mu_0 HA = N^2\mu_0 \frac{A}{l} i \qquad (2.3.1)$$

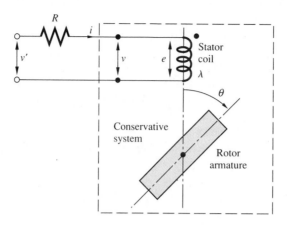

Figure 2.9 Singly-excited rotating system.

where N = number of coil turns

ϕ = magnetic flux ideally linking all turns (Wb)

B = magnetic flux density (T)

A = area of cross section of the flux path (m²)

μ_0 = permeability of free space ($4\pi \times 10^{-7}$ H/m)

H = magnetic field intensity only present in the air path because $\mu_{iron} = \infty$ (AT/m)

and l = mean length of the magnetic circuit in air (m).

Equation (2.3.1) shows the relationship of the flux linkage λ in terms of the current i and the geometry. In rotating systems the geometric variable is the angular displacement θ of the rotor armature with respect to the stator coil axis. Real systems are far from ideal, but the flux linkage is always a function of current i and angle θ, generally expressed as

$$\lambda = \lambda(i,\theta). \tag{2.3.2}$$

By definition, in a conservative or ideal system there are no losses, so the voltage v at the terminals of the coil and the induced emf are related by

$$v + e = 0 \tag{2.3.3}$$

so that

$$v = \frac{d\lambda(i,\theta)}{dt}. \tag{2.3.4}$$

Unless the conductors of the coil are superconducting there are always some electrical losses due to the circuit resistance. In order to confine analysis to the conservative part of the system, all the resistance is lumped and placed external to the conservative system. There are losses in the magnetic core due to eddy currents and hysteresis if the flux ϕ changes rapidly. The effect of this loss can be modeled by a lumped equivalent resistance external to the system to be analyzed. In this way all losses can be ignored until such performance factors as efficiency and voltage regulation are investigated. The resistance R in Fig. 2.9 models all electrical losses so that the components within the box represent the conservative system.

Within the conservative part of the electromagnetic system the voltage equation can be expanded to give

$$v = \frac{d\lambda(i,\theta)}{dt} = \frac{\partial\lambda}{\partial i}\frac{di}{dt} + \frac{\partial\lambda}{\partial\theta}\frac{d\theta}{dt}. \tag{2.3.5}$$

This voltage v has two components. One component depends on a time-varying current, and is called a *transformer voltage* v_t. The second component depends on a time-varying displacement (the angular speed of the rotor), and is called the *motional* or *speed voltage* v_m. So

$$v = v_t + v_m. \tag{2.3.6}$$

The rate of change of flux linkage per ampere is defined as the incremental inductance $L(i,\theta)$, that is

$$L(i,\theta) = \frac{\partial\lambda(i,\theta)}{\partial i}. \tag{2.3.7}$$

In a linear magnetic system, in which there is no magnetic saturation or loss, the flux linkage is directly proportional to current and the incremental inductance is the same as the inductance, which is independent of current. Thus,

$$L(\theta) = \frac{\lambda(i,\theta)}{i} \tag{2.3.8}$$

or

$$\lambda(i,\theta) = L(\theta)i. \tag{2.3.9}$$

Combining eqs. (2.3.5) through (2.3.9)

$$v = \frac{d\lambda(i,\theta)}{dt} = L(\theta)\frac{di}{dt} + i\frac{dL(\theta)}{dt} = L(\theta)\frac{di}{dt} + \dot{\theta}\frac{dL(\theta)}{d\theta}i = v_t + v_m \tag{2.3.10}$$

where $\dot{\theta} = d\theta/dt =$ rotor angular speed. Equation (2.3.10) represents the electric circuit relation for a singly-excited rotating system.

In a static circuit, eq. (2.3.10) can be solved for the unknown current since the circuit inductance would be constant and the displacement θ is unvarying (i.e., $\dot{\theta} = 0$). However, in a dynamic circuit there are two variables, current and displacement. At this stage we do not have enough information for a solution.

2.3.2. Multiply-excited Rotating System

Figure 2.10 shows a doubly-excited magnetic system in which the two coils are excited independently. The stator coil is restrained and the rotor coil is constrained to rotational movement. The same approach that was applied to the singly-excited system, can be applied to this system. A conservative or lossless system is modeled within the dashed rectangle by placing the loss parameters externally. The flux linkage provided by the mutual coupling of the two excited coils must be included in the electric circuit relations. Use is made of the notation in the figure to write the voltage equations

$$v_1 = \frac{d\lambda_1}{dt} \quad \text{and} \quad v_2 = \frac{d\lambda_2}{dt} \tag{2.3.11}$$

where $\lambda_1 = \lambda_1(i_1,i_2,\theta) = L_{11}i_1 + L_{12}i_2$ and $\lambda_2 = \lambda_2(i_1,i_2,\theta) = L_{21}i_1 + L_{22}i_2$.

L_{12} is the mutual inductance coefficient and is defined as the flux linkage in one coil (due to the current in the second coil) per ampere of current in the second coil. The mutual inductance may be a function of the coils' angular displacement. It is usual to find that the function is of the form

$$L_{12} = L_{21} = M_{\text{max}} \cos\theta. \tag{2.3.12}$$

This expression is only a first approximation.

The voltage equations and the flux linkage equations above define the electrical circuit relations for a doubly-excited system. The voltage equations are in terms of field quantities, but if it is required that they be in terms of currents, then substitution can be made from the flux linkage equations.

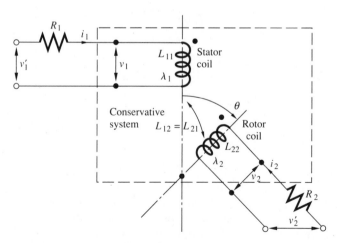

Figure 2.10 Doubly-excited rotating system.

It is convenient to express these equations in the matrix form

$$\begin{bmatrix} v_1 \\ v_2 \end{bmatrix} = \frac{d}{dt} \begin{bmatrix} \lambda_1 \\ \lambda_2 \end{bmatrix} \tag{2.3.13}$$

and

$$\begin{bmatrix} \lambda_1 \\ \lambda_2 \end{bmatrix} = \begin{bmatrix} L_{11} & L_{12} \\ L_{21} & L_{22} \end{bmatrix} \cdot \begin{bmatrix} i_1 \\ i_2 \end{bmatrix}. \tag{2.3.14}$$

From this form we can readily generalize the equations for any number of excited coils on the stator and rotor. The equations for a multiply-excited conservative system are

$$\underline{v} = \frac{d\underline{\lambda}}{dt} \quad \text{and} \quad \underline{\lambda} = [L]\underline{i} \tag{2.3.15}$$

where the voltage vector $\underline{v}^t = [v_1 v_2 v_3 v_4 \ldots]$, the flux linkage vector $\underline{\lambda}^t = [\lambda_1 \lambda_2 \lambda_3 \lambda_4 \ldots]$, the current vector $\underline{i}^t = [i_1 i_2 i_3 i_4 \ldots]$, and $[L]$ = inductance matrix comprising self and mutual inductances. Upon substitution of the flux linkage equation into the voltage equation

$$\underline{v} = \frac{d}{dt}([L]\underline{i}) = [L]\frac{d\underline{i}}{dt} + \frac{d[L]}{dt}\underline{i}. \tag{2.3.16}$$

Since the inductance matrix is a function of the angular displacement θ then

$$\underline{v} = [L]\frac{d\underline{i}}{dt} + \dot{\theta}\frac{d[L]}{d\theta}\underline{i} = \underline{v}_t + \underline{v}_m. \tag{2.3.17}$$

Thus the voltage vector has been resolved into the vector of transformer induced voltages $\underline{v}_t$ and the vector of motionally induced voltages $\underline{v}_m$ similar to the components derived for the singly-excited system.

There are two special cases to consider. First, if there is no movement ($\dot{\theta} = 0$)

$$\underline{v}_m = 0 \quad \text{and} \quad \underline{v} = \underline{v}_t = [L]\frac{d\underline{i}}{dt}. \tag{2.3.18}$$

There is no mechanical energy output and the equation describes the behavior of a multiwinding transformer. Second, if the currents are constant, then

$$\underline{v}_t = 0 \quad \text{and} \quad \underline{v} = \underline{v}_m = \dot{\theta}\frac{d[L]}{d\theta}\underline{i}. \tag{2.3.19}$$

There are only motionally induced voltages. Implicit in these cases is the fact that motionally induced voltages are directly involved in the conversion of energy, whereas the transformer-induced voltages are not.

2.3.3. Translational System

Figure 2.1 shows a singly-excited solenoid actuator. The only basic difference between this and the rotating system depicted in Fig. 2.9 is that the solenoid has translational movement in one dimension. The mechanical displacement has the variable x so that the flux linkage is a function of current i and displacement x, while the inductance L in a linear magnetic circuit is a function of displacement x only. If the same argument as for the rotating system is followed, the electrical circuit relations can be written down by inspection. The voltage equation is

$$v = \frac{d\lambda(i,x)}{dt} \tag{2.3.20}$$

and the flux linkage equation is

$$\lambda(i,x) = L(x)i. \tag{2.3.21}$$

Accordingly the voltage equation can be written in terms of current given by

$$v = \frac{d}{dt}(L(x)i) = L(x)\frac{di}{dt} + \dot{x}\frac{dL(x)}{dx}i = v_t + v_m \tag{2.3.22}$$

where v_t and v_m are the transformer and motionally induced voltages, respectively.

If the same core and armature are multiply-excited and have any number of coils, the electric circuit relations can be written down in the matrix form

$$\underline{v} = \frac{d\underline{\lambda}}{dt} \quad \text{and} \quad \underline{\lambda} = [L]\underline{i} \tag{2.3.23}$$

$$\underline{v} = \frac{d}{dt}([L]\underline{i}) = [L]\frac{d\underline{i}}{dt} + \dot{x}\frac{d[L]}{dx}\underline{i} = \underline{v}_t + \underline{v}_m. \tag{2.3.24}$$

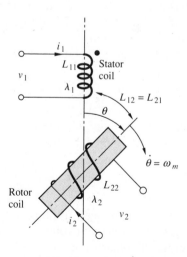

Figure 2.11 Doubly-excited rotating system.

EXAMPLE 2.1

Consider the doubly-excited rotating system represented in Fig. 2.11. The stator coil is open-circuited and the rotor coil, which is excited from a constant current source, produces a sinusoidal spatial flux density distribution. At the position $\theta = 0$, coil 1 experiences the maximum flux linkage of 10 Wb. If the speed of the rotor is 314 rad/s, find the induced voltage at the terminals of coil 1.

Solution

Ignoring the resistances of the coils does not affect the terminal voltage of coil 1. We need only consider the conservative part of the system.

The matrix voltage equation[3] is

$$\underline{v} = [Lp]\underline{i} + \dot{\theta}[dL/d\theta]\underline{i}$$

$$\text{or} \quad \begin{bmatrix} v_1 \\ v_2 \end{bmatrix} = \begin{bmatrix} L_{11}p & L_{12}p \\ L_{21}p & L_{22}p \end{bmatrix} \begin{bmatrix} i_1 \\ i_2 \end{bmatrix} + \dot{\theta}\frac{d}{d\theta} \begin{bmatrix} L_{11} & L_{12} \\ L_{21} & L_{22} \end{bmatrix} \begin{bmatrix} i_1 \\ i_2 \end{bmatrix}.$$

Hence the terminal voltage of coil 1 is

$$v_1 = L_{11}pi_1 + L_{12}pi_2 + \dot{\theta}\left(\frac{dL_{11}}{d\theta}i_1 + \frac{dL_{12}}{d\theta}i_2\right).$$

The given data are $i_1 = 0$, $i_2 = I_2 = $ constant, and $L_{12} = L_{21} = M_{max}\cos\theta$.

Also $M_{max} = \dfrac{10}{I_2}$ H and $\dot{\theta} = 314$ rad/s.

Therefore $v_1 = 0 + 0 + 314(0 - M_{max}\sin\theta\, I_2)$,

or $v_1 = -314 \times 10 \sin\theta$.

The position of the rotor as a function of time is given by

$$\theta = \dot{\theta}t + \delta = \omega_m t + \delta, \text{ where } \theta = \delta \text{ at } t = 0.$$

Accordingly $v_1 = -3140 \sin(314t + \delta)$ and the rms value is $V_1 = 3140/\sqrt{2} = 2220$ volts.

EXAMPLE 2.2

Figure 2.12 shows a cross-sectional view of a cylindrical solenoid. Coil excitation creates magnetic poles at the faces of air gap x. Calculate (a) the magnetic flux in the plunger and (b) the circuit inductance. Use the following information.

$a = 3$ cm, $b = 6$ cm, $c = 9$ cm, $d = 6$ cm, $e = 9$ cm,

$g = x = 1$ mm, $i = 10$ A, $N = 100$ turns.

Assume that the permeability of the iron is infinitely large and neglect leakage and

---

[3] The operator p is defined by $p \triangleq d/dt$.

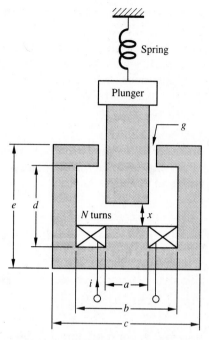

Figure 2.12   Singly-excited translational system.

fringing. **(c)** Would the force on the plunger reverse if the current direction were reversed?

Solution
(a) For the iron $\mu \to \infty$. Therefore the magnetic potential drops are only across the air gaps g and x. Ampere's circuital law for the magnetic path taken by the flux is $\Sigma\, Hl = Ni$, where H = magnetic field intensity, or $H_x x + H_g g = Ni$ where H_x = magnetic field intensity in the air gap of length x and H_g = magnetic field intensity in the air gap of length g. In terms of flux densities

$$B_x x + B_g g = \mu_0 Ni$$

and in terms of flux

$$\phi\,\frac{x}{A_x} + \phi\,\frac{g}{A_g} = \mu_0 Ni$$

where A_x and A_g are the areas at gaps x and g, respectively. Therefore

$$\phi = \mu_0\,\frac{Ni}{\left(\dfrac{x}{A_x} + \dfrac{g}{A_g}\right)}.$$

$$A_x = \frac{\pi}{4} \times (3 \times 10^{-2})^2 \qquad \text{and} \qquad A_g = 2\pi(3 + 0.1) \times 10^{-2} \times \frac{9 - 6}{2} \times 10^{-2}.$$

Upon substitution of the parameter values $\phi = 6 \times 10^{-4}$ Wb,

(b) inductance $L = N\phi/i = 100 \times 6 \times 10^{-4}/10 = 6 \times 10^{-3}$ H,

(c) the electromagnetic force between the plunger and the central core is one of attraction no matter in what direction the current flows.

2.4. CONVERSION OF ENERGY

The first step of the dynamic circuit analysis was to find the electric circuit relations. The second step is to find expressions for the mechanical force or torque developed by an electromechanical energy conversion device. To do this we will apply the principle of the conservation of energy to a singly-excited rotating magnetic field device in the motoring mode. Generalizations to cover multiply-excited systems, with a magnetic or an electric-coupling field, with rotational or translational motion, motoring or generating, will follow later. (See section 2.5.4.)

Consider that the whole system, comprising the electric and mechanical circuits and the coupling field of a magnetic device, is initially in equilibrium. Over a period of time t let there be a certain amount of electric energy supplied to the system. The total electric energy input up to the time t will have been converted to other forms of energy, but the conservation of energy principle implies that the sum of all the forms of converted energy must be equal to the energy supplied by the electric system. Consequently an energy balance can be written. Because the frequencies and velocities involved in electromechanical devices are relatively low, so-called "quasistatic" field conditions prevail. Consequently energy in the form of electromagnetic radiation is negligible in comparison with the other forms of converted energy. Thus the energy balance equation for motoring action is

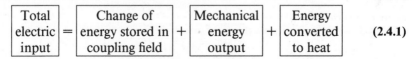

$$\boxed{\begin{array}{c}\text{Total}\\\text{electric}\\\text{input}\end{array}} = \boxed{\begin{array}{c}\text{Change of}\\\text{energy stored in}\\\text{coupling field}\end{array}} + \boxed{\begin{array}{c}\text{Mechanical}\\\text{energy}\\\text{output}\end{array}} + \boxed{\begin{array}{c}\text{Energy}\\\text{converted}\\\text{to heat}\end{array}} \qquad \textbf{(2.4.1)}$$

In the case of generator action the $\boxed{\text{Electric energy input}}$ and $\boxed{\text{Mechanical energy output}}$ terms in the above equation would have negative values making them $\boxed{\text{Electric energy output}}$ and $\boxed{\text{Mechanical energy input}}$ with positive values.

The term $\boxed{\text{Energy converted to heat}}$ is a lumped quantity representing the irreversible conversion of some of the energy into heat, which has to be dissipated in order not to exceed the safe operating temperature of the device. This is a loss of energy, which lowers the efficiency of the conversion process. There are three causes for the conversion of energy to heat.

1. Part of the electric energy is converted to heat because of the resistance of the conductors.

2. Part of the energy absorbed by the coupling field is converted to heat due to hysteresis and eddy current losses in the iron core.
3. Finally, part of the mechanical energy is absorbed by friction losses in the bearings and by windage of the moving parts (air-friction loss).

2.4.1. Conservative System

The heart of the energy-conversion process is the coupling field. The losses play no basic part in the electromechanical energy conversion processes, although they are always present. So if the losses are placed external to the coupling field system then the system is conservative. This greatly facilitates the derivation of the torque or force expressions. The ohmic losses and the windage and friction losses can be made external to the coupling field system, because they can be related directly to the electric and mechanical systems. The hysteresis and eddy current losses in the iron are associated with the coupling field, but can be assumed to be an apparent electric system loss and can be lumped with the ohmic loss. Figure 2.13 illustrates an energy chart, which separates the conservative system of the energy conversion process from the input and output of a device and its losses.

In terms of the separated losses the energy balance is

$$\begin{vmatrix} \text{Total electric} \\ \text{energy input} \\ -\text{ohmic loss} \end{vmatrix} = \begin{vmatrix} \text{Change of stored} \\ \text{field energy} \\ +\text{field loss} \end{vmatrix} + \begin{vmatrix} \text{Mechanical energy} \\ \text{output} + \text{windage} \\ \text{friction loss} \end{vmatrix} \qquad (2.4.2)$$

For the conservative system alone the energy balance is obtained from Fig. 2.13 and is

$$\begin{vmatrix} \text{Net electric energy} \\ \text{to coupling field} \end{vmatrix} = \begin{vmatrix} \text{Change of stored} \\ \text{field energy} \end{vmatrix} + \begin{vmatrix} \text{Total mechanical} \\ \text{energy output} \end{vmatrix} \qquad (2.4.3)$$

or in mathematical notation

$$W_e = \Delta W_s + W_m. \qquad (2.4.4)$$

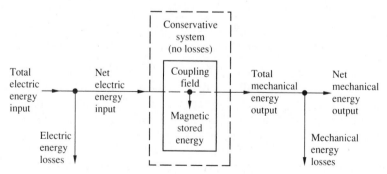

Figure 2.13   Energy conversion for motoring.

This is for motoring action and applies for any time span. For generating action the conservative energy balance is

$$
\boxed{\begin{array}{c}\text{Net mechanical}\\\text{energy input}\end{array}} = \boxed{\begin{array}{c}\text{Change of field}\\\text{stored energy}\end{array}} + \boxed{\begin{array}{c}\text{Total electric}\\\text{energy output}\end{array}} \qquad \textbf{(2.4.5)}
$$

In order to eliminate the need to know the initial conditions such as the stored energy in the system when the energy balance is set up, we can rely upon the conservation of power at any instant. For motoring action this is expressed by

$$
\boxed{\begin{array}{c}\text{Rate of change}\\\text{of net electric}\\\text{energy input}\end{array}} = \boxed{\begin{array}{c}\text{Rate of change}\\\text{of stored}\\\text{magnetic energy}\end{array}} + \boxed{\begin{array}{c}\text{Rate of change}\\\text{of gross mechanical}\\\text{energy output}\end{array}} \qquad \textbf{(2.4.6)}
$$

or in mathematical notation

$$
\frac{dW_e}{dt} = \frac{dW_s}{dt} + \frac{dW_m}{dt}. \qquad \textbf{(2.4.7)}
$$

Thus,

$$
dW_e = dW_s + dW_m. \qquad \textbf{(2.4.8)}
$$

This is an equation implying that an incremental change of electric energy input is accompanied by an incremental change of stored field energy and an incremental change of mechanical energy output. Equation (2.4.8) gives an important relationship and will be referred to as the *incremental energy balance equation* in the following sections.

2.4.2. Electric System Energy

An energy balance for the electric system is

$$
W'_e = W_{cu} + W_c + W_e \qquad \textbf{(2.4.9)}
$$

where W'_e = electric energy from the supply

W_{cu} = ohmic loss due to the conductor resistance

W_c = iron losses in the magnetic core

and W_e = net electric energy into the coupling field.

The iron loss W_c is determined empirically. Iron losses will be neglected in this analysis, since they need only be taken into account when performance factors such as efficiency are calculated. Accordingly the electric system energy balance becomes $W'_e = W_{cu} + W_e$.

We will start with a singly-excited system and develop relations for a multiply-excited system later. By using the notation in Fig. 2.9 the energy from the electric system source is calculated from

$$W'_e = \int v' i \, dt. \tag{2.4.10}$$

The ohmic loss, which is dissipated in the form of heat, is

$$W_{cu} = \int i^2 R \, dt \tag{2.4.11}$$

so that the net electric energy into the coupling field system is

$$W_e = W'_e - W_{cu} = \int (v' - iR)i \, dt = \int vi \, dt. \tag{2.4.12}$$

Since $v + e = 0$ and the induced emf is $e = -d\lambda/dt$, then

$$W_e = \int i \frac{d\lambda}{dt} \, dt = \int i \, d\lambda. \tag{2.4.13}$$

Thus an incremental change of electric input energy is

$$dW_e = i \, d\lambda \tag{2.4.14}$$

where λ is the total flux linked by the electric circuit of N turns.

In general, the flux linked with each turn varies throughout the coil. Due to the spreading of the flux and due to the radial build of a multilayer coil, some flux only partially links the complete coil. This effect can be taken into account by introducing the concept of equivalent flux ϕ, which is defined as $\phi \triangleq \lambda/N$ and is the average flux per turn. In devices with ferromagnetic cores the effect of unequal flux linkages is very small due to the high relative permeability of the core.

For the singly-excited magnetic device shown in Fig. 2.9, the incremental change of electric energy supplied to the coil can be expressed in the form

$$dW_e = i \, d\lambda = Ni \, d\phi = Fd\phi \tag{2.4.15}$$

where F is the coil mmf. This shows that a change of flux linking the winding is associated with a flow of electric energy.

The reaction of a changing field on the input circuit is an essential part of the energy conversion process. A change of field energy is caused by a change of field. A change of field induces the voltage drop v, which, through $i = (v' - v)/R$, determines the current value and thus the value of the electric energy W'_e from the supply.

We can express the incremental change of electric energy input in terms of electric circuit parameters. For the case of a linear magnetic circuit

$$dW_e = i \, d\lambda = i \, d\{L(\theta)i\} = i^2 \frac{dL(\theta)}{d\theta} \, d\theta + i \, L(\theta)di. \tag{2.4.16}$$

This indicates that motion and current changes can affect the electric energy input.

2.4.3. Coupling Field Energy

If motion of the system is restrained no mechanical work can be done and the incremental energy equation (2.4.8) becomes $dW_e = dW_s + 0$. Under this special condition the electric energy input results in a change of field energy, which is wholly absorbed by the field and $dW_s = dW_e = i\, d\lambda = F d\phi$. Consequently the total energy stored in the coupling field under the constraints of zero initial conditions and no motion is

$$W_s = \int_0^\lambda i(\lambda)d\lambda = \int_0^\phi F(\phi)d\phi. \qquad (2.4.17)$$

The notation $i(\lambda)$ and $F(\phi)$ is used to stress that the current i and mmf F are functions of flux linkage λ and flux ϕ, respectively. The exact relations between i and λ, and F and ϕ depend on the geometry of the magnetic structure and the characteristics of the materials employed.

A. STORED ENERGY AND SATURATION

For the device shown in Fig. 2.9, the relationship between current and flux linkage depends on the total reluctance of the magnetic flux path. The iron losses can be ignored since they do not play a basic role in the energy conversion process. Therefore the relation between the current and flux linkage is represented by the single-valued solid curve depicted in Fig. 2.14. This curve is the normal saturation curve for the magnetic circuit with a fixed geometry. If there is no motion, then the input energy is all stored in the magnetic field.

From eq. (2.4.14) an incremental change of field energy dW_s is $i\, d\lambda$, which is the incremental area shown to the left of the curve. If the current i is increased from zero to a maximum value i_{max}, the flux linkage increases from zero to a maximum value λ_m and the energy absorbed by the field is

$$W_s = \int_0^{\lambda_{max}} i(\lambda)d\lambda \equiv \text{Area } OabO. \qquad (2.4.18)$$

If the current is increased from some value i_1 to a value i_2, the flux linkage increases from λ_1 to λ_2 and the change of stored energy ΔW_s is

$$\Delta W_s = \int_{\lambda_1}^{\lambda_2} i(\lambda)d\lambda = \int_0^{\lambda_2} i(\lambda)d\lambda - \int_0^{\lambda_1} i(\lambda)d\lambda$$
$$= \text{Area } OdeO - \text{Area } OfgO = \text{Area } fdegf. \qquad (2.4.19)$$

The change of energy represented by the area *fdegf* is supplied by the electric system, since the mechanical system is stationary in this case.

If the current had been reduced from i_2 to i_1 the flux linkage would have decreased from λ_2 to λ_1 and the same amount of energy (area *fdegf*) would have been returned from the coupling field to the electric system.

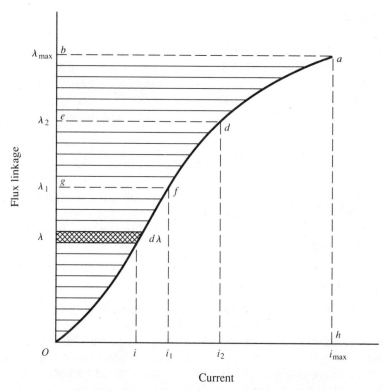

Figure 2.14 Normal saturation curve.

B. STORED ENERGY AND CONSTANT RELUCTANCE

The total stored energy W_s in a magnetic device for a particular position of the moveable part and any current i and flux linkage λ is given by the general eq. (2.4.17). Calculation of the stored energy is simplified, if it can be assumed that the reluctance of the magnetic circuit is constant. A constant reluctance signifies a linear saturation curve as shown in Fig. 2.15. Accordingly, for linear conditions the magnetic stored energy is

$$W_s = \int_0^\lambda i(\lambda)d\lambda \equiv \text{Area } OabO = \tfrac{1}{2}i\lambda. \tag{2.4.20}$$

A linear coupling field system produces a simple relation, but its use must be justified.

Unsaturated conditions in a ferromagnetic circuit give rise to an approximate linear saturation curve. This is not practical because the value of the flux would have to be low and the full potential of energy conversion would not be reached. Mechanical energy is related to motion. In order to permit relative motion of mechanical members, the magnetic circuits of energy converters have some form of air gap. If it can be proved that the magnetic energy stored in the air gap is very much greater than that stored in the iron, system linearization is accomplished by neglecting the magnetic energy stored in the iron. The reluctance of the iron path would be considered negligible and the only magnetic potential drop would be across the air gap.

Consider the simple and basic circuit shown in Fig. 2.16. The mmf F to produce the flux ϕ is given by Ampere's circuital law. That is,

$$F = Ni = \oint Hdl = \int H_i dl_i + \int H_a dl_a = H_i l_i + H_a l_a \qquad (2.4.21)$$

where the subscripts i and a refer to the iron part and air gap of the magnetic circuit, respectively. If the flux leakage is neglected, the flux in the iron core is equal to the flux in the air gap. The total stored energy, given by eq. (2.4.17), becomes

$$W_s = \int_0^\phi (H_i l_i + H_a l_a) d\phi = \int_0^{B_i} H_i l_i A_i \, dB_i + \int_0^{B_a} H_a l_a A_a \, dB_a$$

$$(2.4.22)$$

$$= Volume_i \int_0^{B_i} \frac{B_i}{\mu_i} \, dB_i + Volume_a \int_0^{B_a} \frac{B_a}{\mu_0} \, dB_a.$$

In this argument fringing and saturation are not excluded, so the equation for W_s holds in general. The flux densities in the iron and air gap are essentially the same

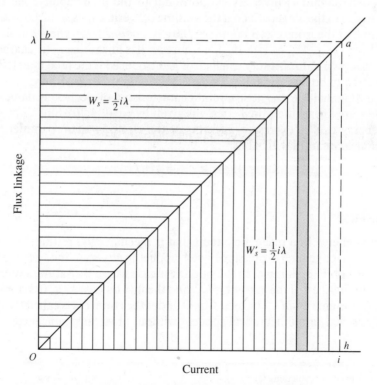

Figure 2.15 Linear saturation curve (constant permeability).

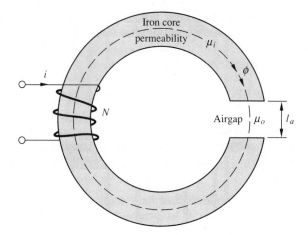

Figure 2.16 Simple magnetic circuit with air gap.

unless the air gap is long. At the usual flux densities the permeability of the iron is several thousand times (e.g., 10,000) that of air. Since the stored energy in each part of the magnetic circuit is inversely proportional to the permeability, the energy is mainly stored in the air gap. Even if the volume of the air gap is small compared with the volume of the iron (e.g., $Volume_a = 0.01 \times Volume_i$), the energy in the air gap predominates (e.g., $W_{si} = 0.01\ W_{sa}$). Because air has linear magnetic characteristics and is a lossless medium, a very small error is introduced if the total magnetic circuit is considered to be linear and lossless.

Assuming that the magnetic circuit characteristics of electromechanical energy converters are linear, the stored energy is given by eq. (2.4.20). Since $F = \phi \mathcal{R}$,[4] the stored energy can be expressed in terms of the electric system parameters or the coupling field quantities. That is

$$W_s = \tfrac{1}{2}\phi^2 \mathcal{R} = \tfrac{1}{2}\lambda^2/L. \tag{2.4.23}$$

C. COENERGY

The stored energy in a magnetic circuit that is mechanically restrained is given by (2.4.17). This is a general expression, which applies to any coupling system whether it is linear or not. For example, in the nonlinear case where the saturation curve is as depicted in Fig. 2.14, if the current is i_{max} and the flux linkage λ_{max}, the stored energy is given by the area $OabO$. The area $OahO$ under the curve Oa is called the coenergy W_s' and is strictly a mathematical concept without physical significance.

---

[4] $\mathcal{R} \triangleq$ magnetic circuit reluctance.

For the conditions i_{max} and λ_{max} the coenergy is given by

$$W'_s \equiv \text{Area } OahO = \text{Area } OhabO - \text{Area } OabO$$

$$= \int_0^{i_{max}} \lambda(i)di = i_{max}\lambda_{max} - \int_0^{\lambda_{max}} i(\lambda)d\lambda = i_{max}\lambda_{max} - W_s. \tag{2.4.24}$$

In general, for any condition i and λ

$$W_s + W'_s = i\lambda \tag{2.4.25}$$

where $W_s = \int_0^\lambda i(\lambda)d\lambda$ and $W'_s = \int_0^i \lambda(i)di$. That is, the stored field energy is explicitly a function of λ (or ϕ), whereas the coenergy is an explicit function of i (or F).

In particular, if the magnetic circuit is assumed linear as illustrated in Fig. 2.15, Area $OhaO = $ Area $OabO$. This indicates that the magnitudes of stored energy and coenergy are equal for linear conditions, and so the coenergy W'_s is given by

$$W'_s = \tfrac{1}{2}i\lambda(i) = \tfrac{1}{2}Li^2 = \tfrac{1}{2}F^2/\mathscr{R}. \tag{2.4.26}$$

The use of the concept of coenergy facilitates the calculation of torque or force in electromagnetic systems, as will be seen below.

2.4.4. Conversion to Mechanical Energy

We have already found expressions for the electric input energy and the magnetic stored energy of a conservative coupling system. Consequently the mechanical energy output can be derived from the energy balance eq. (2.4.8).

Let us consider again the device that is illustrated in Fig. 2.9. For simplicity let the electric system be excited by a dc source of constant voltage V'. If the iron rotor armature is allowed to move from a position θ_1 to a position θ_2 under the influence of the electromagnetic torque, mechanical work is done, and the alteration of the mechanical system geometry signifies that there is a change of magnetic stored energy.

The normal saturation curve of the device with the armature held at position θ_1 is given by the curve Oa in Fig. 2.17. For the particular value of voltage V' the values of the flux linkage λ_1 and current i are given by the operating point a on the magnetization curve. The abnormal saturation curve with the armature held at position θ_2 is given by the curve Ob in Fig. 2.17 and the values of flux linkage λ_2 and current i are given by the operating point b on the curve. The initial and final values of the current, going from θ_1 to θ_2, will be the same. This is because the applied voltage V' is constant. While the armature is restrained, there is no change of flux linkage and no induced emf. So if the armature is held in any position the current is determined solely by the applied voltage and the effective resistance R of the coil.

During the motion of the armature the flux linkage increases. The rate at which the flux linkage increases determines the value of the coil induced emf e, which tends to oppose the change of flux linkage. As a result the net voltage $(V' - v)$ is decreased

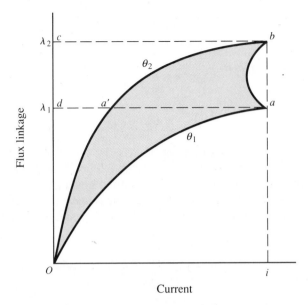

Figure 2.17 Initial and final saturation curves.

and causes a reduction of the current relative to the initial and final values. The path taken by the operating point going from a to b in Fig. 2.17 is determined by the speed at which the armature moves. If the movement is accomplished extremely slowly, the path will be essentially vertical, because the induced emf e in the coil resulting from the rate of change of flux linkage is sensibly zero. On the other hand, if the movement takes place so rapidly that the armature is at position θ_2 before the flux linkage has changed appreciably, the operating point will move along a virtually horizontal path aa' and then follow the saturation curve Ob to the point b. In general the movement takes a finite time. Accordingly the path of the operating point is some curved line between a and b. The exact determination of the curved path ab is difficult, because it is affected by the time constants of the electric and mechanical systems.

From the curves in Fig. 2.17 the initial energy W_{sa} stored in the magnetic field, if the armature is at position θ_1, is given by

$$W_{sa} = \int_0^{\lambda_1} i \, d\lambda \equiv \text{Area } OadO \qquad (2.4.27)$$

where the functional relation between i and λ is given by the curve Oa. The final stored energy W_{sb} with the armature at position θ_2 is given by

$$W_{sb} = \int_0^{\lambda_2} i \, d\lambda \equiv \text{Area } ObcO \qquad (2.4.28)$$

where the functional relation between i and λ for this case is given by the curve Ob.

Thus the change of stored energy ΔW_s brought about by the motion is

$$\Delta W_s = W_{sb} - W_{sa} \equiv \text{Area } ObcO - \text{Area } OadO. \qquad (2.4.29)$$

The net electric energy W_e supplied to the coupling system during the course of the motion is

$$W_e = \int_{t_1}^{t_2} vi \, dt = \int_{\lambda_1}^{\lambda_2} i \, d\lambda \equiv \text{Area } abcda. \qquad (2.4.30)$$

The integral follows the actual path of the operating point between a and b, and the functional relation between i and λ is given by the curve ab.

The energy balance equation for the conservative part of the whole system is given by eq. (2.4.4). It follows that the energy converted to mechanical energy W_m during the motion is

$$
\begin{aligned}
W_m = W_e - \Delta W_s &= \int_{\lambda_1}^{\lambda_2} i \, d\lambda - \int_0^{\lambda_2} i \, d\lambda + \int_0^{\lambda_1} i \, d\lambda \\
&\equiv \text{Area } abcda - (\text{Area } ObcO - \text{Area } OadO) \\
&= \text{Area } abcda + \text{Area } OadO - \text{Area } ObcO \\
&= \text{Area } OabO.
\end{aligned}
\qquad (2.4.31)
$$

Accordingly the mechanical energy is represented in Fig. 2.17 by the shaded area that is enclosed by the initial saturation curve, the final saturation curve and the path of the operating point.

This discussion has shown that energy conversion takes place when the stored energy is influenced by a change of geometry. The coupling field can be considered to be a reservoir of energy, supplying energy to the mechanical system and being supplied with energy from the electric system.

EXAMPLE 2.3

Consider the cylindrical solenoid described in Example 2.2, illustrated in Fig. 2.12.
(a) Find the magnetic stored energy in terms of the geometry and the applied current.
(b) What proportion of this energy is stored in the air gap of length x?

Solution
All the magnetic energy is stored in the air gaps x and g if the permeability of the iron is considered to be infinitely large. The magnetic stored energy W_s is $W_s = \frac{1}{2}\phi^2\mathcal{R}$.

(a) The flux ϕ is $\phi = \dfrac{Ni}{\mathcal{R}}$. The reluctance $\mathcal{R}$ of the series circuit is

$$\mathcal{R} = \mathcal{R}_x + \mathcal{R}_g = \frac{x}{\mu_0 A_x} + \frac{g}{\mu_0 A_g}$$

where $\mathcal{R}_x$, $\mathcal{R}_g$ are the reluctances of the air gaps of lengths x and g, respectively.

Hence $W_s = \frac{1}{2}\mu_0 N^2 i^2 \dfrac{A_x}{\left(x + g\dfrac{A_x}{A_g}\right)}$.

(b) $W_s = W_{sx} + W_{sg} = \frac{1}{2}\phi^2(\mathcal{R}_x + \mathcal{R}_g)$.

Therefore the proportion of the total energy stored in the air gap x is

$$\frac{W_{sx}}{W_s} = \frac{\mathcal{R}_x}{\mathcal{R}} = \frac{1}{\left(1 + \dfrac{gA_x}{xA_g}\right)}.$$

For the dimensions given in Example 2.2, $W_{sx} : W_s = 2 : 3$.

EXAMPLE 2.4

A contactor, illustrated in Fig. 2.22, is energized from a constant 25-V dc source. The coil of 20 turns has a resistance of 5 ohms. Initially the air gaps are set such that $x = 10$ mm and in this position the flux per ampere turn is 0.2 mWb/AT. Leakage, fringing, and the reluctance of the iron are neglected. The armature is allowed to move very slowly through a distance of 5 mm in the direction of the electromagnetic force. Determine the mechanical work done and the electric energy, in excess of the heat dissipated in the coil, that is provided by the source due to the motion.

Solution

Since the armature moves very slowly, the rate of change of flux during motion can be assumed to be negligibly small. Consequently the induced emf in the coil can be considered negligible and hence the current is constant. That is

$$i = I = V/R = 25/5 = 5 \text{ A}.$$

For constant current the electric energy input ΔW_e during the change in x is

$$\Delta W_e = \int_{\lambda_1}^{\lambda_2} i\, d\lambda = I \int_{\lambda_1}^{\lambda_2} d\lambda = I(\lambda_2 - \lambda_1)$$

where the subscripts 1 and 2 refer to conditions before and after the motion. Since the saturation curve is linear the change of stored field energy is

$$\Delta W_s = \frac{1}{2}I\lambda_2 - \frac{1}{2}I\lambda_1 = \frac{1}{2}I(\lambda_2 - \lambda_1).$$

The mechanical work done follows from

$$\Delta W_m = \Delta W_e - \Delta W_s = \frac{1}{2}I(\lambda_2 - \lambda_1) = \frac{1}{2}NI\phi_2 - \frac{1}{2}NI\phi_1.$$

From $F = \phi_1 \mathcal{R}_1 = \phi_2 \mathcal{R}_2 = \phi_2 \mathcal{R}_1/2$ (from the motion 10- to 5-mm gap) it follows that

$$\phi_2 = 2\phi_1 = 2 \times 0.2 \times 10^{-3} \times 5 \times 20 = 4 \times 10^{-2} \text{ Wb}.$$

This allows the calculation of the stored energies W_{s1} and W_{s2}.

$W_{s1} = \phi_1 NI/2 = 2 \times 10^{-2} \times 20 \times 5/2 = 1$ J.

$W_{s2} = \phi_2 NI/2 = 4 \times 10^{-2} \times 20 \times 5/2 = 2$ J.

Hence the change of stored energy $\Delta W_s = W_{s2} - W_{s1} = 2 - 1 = 1$ J. The mechanical work done $\Delta W_m = 1$ J. The change of electric energy is

$\Delta W_e = \Delta W_s + \Delta W_m = 1 + 1 = 2$ J.

In general, for the case of constant current half the electric energy input is stored in the magnetic field and half is converted to a mechanical output.

2.5. TORQUE

2.5.1. Torque and Magnetic Stored Energy

The discussion will be restricted to a simple, singly-excited, magnetic device with rotational freedom, as depicted in Fig. 2.9, for instance. We wish to find an expression for the electromagnetic torque from the foundations laid down in the last section.

The method of using the energy balance equation directly requires a mechanical energy term. Mechanical energy results from motion, but the torque T_e cannot be determined by considering motion over a finite angular displacement θ, because the torque varies with displacement. Since we want to know the torque at any particular position of the armature, it is necessary to consider the motion of the armature through an elemental displacement $d\theta$. We can study the effect of this infinitesimally small movement on the energy balance. This method is called the *principle of virtual work*.

The stored energy in the magnetic field can be written as a function of flux linkage λ and displacement θ. For incremental changes $d\lambda$ and $d\theta$, the total change of stored magnetic energy dW_s can be expressed as

$$dW_s = dW_s(\lambda,\theta) = \frac{\partial W_s}{\partial \lambda}\, d\lambda + \frac{\partial W_s}{\partial \theta}\, d\theta. \tag{2.5.1}$$

The increment of electric energy is $dW_e = i\, d\lambda$. The increment of mechanical work done is $dW_m = T_e\, d\theta$. From the incremental energy balance eq. (2.4.8)

$$dW_s = i\, d\lambda - T_e\, d\theta. \tag{2.5.2}$$

The stored energy has been expressed in terms of the four variables i, λ, T_e, and θ. The above two equations imply that the flux linkage λ and position θ have been chosen as the independent variables. It follows that current i and torque T_e can be expressed in terms of λ and θ. Since λ and θ are independent, the coefficients of $d\lambda$ and $d\theta$ in two equations can be equated. The results are

$$i = \frac{\partial W_s}{\partial \lambda} = \frac{dW_s(\lambda,\theta)}{d\lambda}\bigg|_{\theta=\text{constant}} \quad (2.5.3)$$

and

$$T_e = -\frac{\partial W_s}{\partial \theta} = -\frac{dW_s(\lambda,\theta)}{d\theta}\bigg|_{\lambda=\text{constant}} \quad (2.5.4)$$

It is important to note that the stored field energy W_s must be expressed in terms of the system variables λ and θ, when the current and torque are calculated from these equations.

2.5.2. Torque and Coenergy

The relation between torque and stored field energy was obtained, because the independent variables of the system were chosen to be flux linkage λ and the mechanical position θ. Current i and position θ could equally well have been chosen to be the independent variables.

Consider that λ and T_e take the functional form

$$\lambda = \lambda(i,\theta) \quad \text{and} \quad T_e = T_e(i,\theta). \quad (2.5.5)$$

The incremental energy balance equation was obtained in the form $dW_s = i\,d\lambda - T_e\,d\theta$. In order to obtain the energy balance equation in terms of i and θ we make use of the fact that

$$d(i\lambda) = i\,d\lambda + \lambda\,di \quad \text{or} \quad i\,d\lambda = d(i\lambda) - \lambda\,di \quad (2.5.6)$$

so that

$$dW_s = d(i\lambda) - \lambda\,di - T_e\,d\theta \quad \text{or} \quad d(i\lambda - W_s) = \lambda\,di + T_e\,d\theta. \quad (2.5.7)$$

However, the coenergy W_s' is defined by eq. (2.4.25) to be $W_s' = i\lambda - W_s$. Consequently the incremental energy balance equation is transformed to

$$dW_s' = \lambda\,di + T_e\,d\theta. \quad (2.5.8)$$

For incremental changes di and $d\theta$ in a conservative system the total change of coenergy dW_s' can be expressed as the sum of two components.

$$dW_s' = dW_s'(i,\theta) = \frac{\partial W_s'}{\partial i}\,di + \frac{\partial W_s'}{\partial \theta}\,d\theta. \quad (2.5.9)$$

It follows from the two eqs. (2.5.8) and (2.5.9) for coenergy that

$$\lambda = \frac{\partial W_s'}{\partial i} = \frac{\partial W_s'(i,\theta)}{\partial i}\bigg|_{\theta=\text{constant}} \quad (2.5.10)$$

and

$$T_e = \frac{\partial W_s'}{\partial \theta} = \frac{\partial W_s'(i,\theta)}{\partial \theta}\Bigg|_{i=\text{constant}} \tag{2.5.11}$$

If the flux linkage λ or the electromagnetic torque T_e is to be calculated from these equations, the coenergy must be expressed in terms of the variables i and θ.

The torque that is derived using the concept of coenergy must necessarily be the same as the torque found as a function of the stored energy. Thus the torque at any angular position is given by

$$T_e = -\frac{\partial W_s(\lambda,\theta)}{\partial \theta} = \frac{\partial W_s'(i,\theta)}{\partial \theta}. \tag{2.5.12}$$

In general, the torque T_e at any position is

$$\begin{aligned} T_e &= -\frac{\partial W_s(\lambda,\theta)}{\partial \theta} = -\frac{\partial}{\partial \theta}\int_0^\lambda i(\lambda,\theta)d\lambda \\ &= \frac{\partial W_s'(i,\theta)}{\partial \theta} = \frac{\partial}{\partial \theta}\int_0^i \lambda(i,\theta)di. \end{aligned} \tag{2.5.13}$$

In most cases it is adequate to base the coil inductance $L(\theta)$ and the magnetic path reluctance $\mathcal{R}(\theta)$ on the air-gap conditions alone and neglect the effect of the iron on these parameters. In that case the saturation curve is assumed to be linear. The stored energy and coenergy have equal values, given by

$$W_s(\lambda,\theta) = \tfrac{1}{2}i\lambda = \tfrac{1}{2}\frac{\lambda^2}{L(\theta)} = \tfrac{1}{2}\phi^2\mathcal{R}(\theta) = \tfrac{1}{2}\frac{\lambda^2\mathcal{R}(\theta)}{N^2} \tag{2.5.14}$$

and

$$W_s'(i,\theta) = \tfrac{1}{2}i\lambda = \tfrac{1}{2}L(\theta)i^2 = \tfrac{1}{2}\frac{F^2}{\mathcal{R}(\theta)} = \tfrac{1}{2}\frac{N^2i^2}{\mathcal{R}(\theta)} \tag{2.5.15}$$

where $\lambda = N\phi$, $F = Ni$, and $\lambda = L(\theta)i$. Thus the torque is expressed by

$$\begin{aligned} T_e &= -\frac{\lambda^2}{2}\frac{d}{d\theta}\left(\frac{1}{L(\theta)}\right) = -\frac{\phi^2}{2}\frac{d\mathcal{R}(\theta)}{d\theta} = \frac{i^2}{2}\frac{dL(\theta)}{d\theta} \\ &= -\frac{\lambda^2}{2N^2}\frac{d\mathcal{R}(\theta)}{d\theta} = \frac{N^2i^2}{2}\frac{d}{d\theta}\left(\frac{1}{\mathcal{R}(\theta)}\right) = \frac{F^2}{2}\frac{d}{d\theta}\left(\frac{1}{\mathcal{R}(\theta)}\right). \end{aligned} \tag{2.5.16}$$

Consequently in linear systems the torque acts in such a direction to tend to

- decrease the stored energy at a constant flux linkage
- increase the coenergy at a constant current
- decrease the reluctance
- increase the inductance.

All the torque equations derived in the foregoing sections apply to motoring action. For generating action the incremental energy equation must be written as

$dW_m = dW_s + dW_e$, where the mechanical energy is now an input and the electric energy is an output. The torque equations resulting from this energy balance would have the opposite sign to that of the motoring action torque, indicating that it is a counter torque that has to be overcome by the applied mechanical torque.

Returning to motoring action, if there is motion of the armature we have from eq. (2.3.10) that the coil's motionally induced voltage v_m is

$$v_m = i\dot{\theta}\,\frac{dL(\theta)}{d\theta} \tag{2.5.17}$$

so that the torque can be expressed in the form

$$T_e = \tfrac{1}{2}i^2\,\frac{dL(\theta)}{d\theta} = \tfrac{1}{2}\,\frac{iv_m}{\dot{\theta}}. \tag{2.5.18}$$

Accordingly the instantaneous mechanical output power p_m at any position θ is

$$p_m = T_e\dot{\theta} = \tfrac{1}{2}iv_m. \tag{2.5.19}$$

This indicates the importance of the motional component of the induced voltage in the energy conversion process.

2.5.3. Reluctance Machine with AC Excitation (A Case Study)

An application of the above theory is the singly-excited reluctance motor,[5] which is schematically shown in Fig. 2.18. A stationary yoke of laminated ferromagnetic material (stator), which carries a coil of N turns, is terminated in two specially shaped poles, between which a noncylindrical laminated ferromagnetic body (rotor) can rotate. The reluctance of the magnetic circuit varies with the position θ_m of the rotor, and this is why the machine is called a reluctance motor. The angle θ_m between the rotor-pole axis and the stator pole axis is measured in a clockwise direction. The stator-pole axis is called the direct axis or the d-axis. A second stationary axis, perpendicular to the direct axis, is called the quadrature axis or the q-axis.

This case study is described in detail to show the complexity that can be dealt with by means of energy principles. However, simplifications are made at the end in keeping with the general example format.

The reluctance of the magnetic circuit is a periodic function of the angle θ_m. It has a minimum value $\mathscr{R}_d$, when $\theta_m = 0$, π, 2π, etc., and a maximum value $\mathscr{R}_q$ for $\theta_m = \pi/2$, $3\pi/2$, etc. A plot of the reluctance as a function of θ_m may be like that shown in Fig. 2.19. The reluctance can be expressed by a Fourier series as

$$\mathscr{R}(\theta_m) = \mathscr{R}_0 + \mathscr{R}_1\cos(2\theta_m) + \mathscr{R}_3\cos3(2\theta_m) + \cdots \tag{2.5.20}$$

---

[5] The motor of an analog-type electric clock has this form because the rotational speed is in synchronism with the supply frequency.

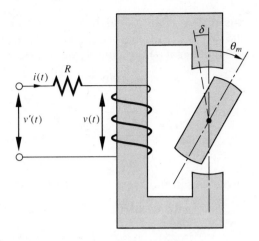

Figure 2.18 Reluctance machine.

The developed torque for any position θ_m of the rotor can be determined from one of the expressions in eq. (2.5.16). We select

$$T_e = -\tfrac{1}{2}\phi^2 \frac{\partial \mathcal{R}(\theta_m)}{\partial \theta_m}. \tag{2.5.21}$$

Thus the instantaneous value of torque is

$$T_e = -\tfrac{1}{2}\phi^2[-2\mathcal{R}_1 \sin(2\theta_m) - 6\mathcal{R}_3 \sin3(2\theta_m) - \cdots]. \tag{2.5.22}$$

In practice a sinusoidal voltage $v'(t)$ is applied to the terminals of the coil. The voltage equation for the coil is

$$v'(t) = i(t)R + v(t). \tag{2.5.23}$$

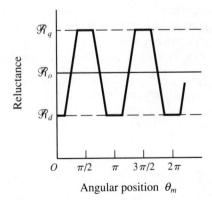

Figure 2.19 Reluctance variation with rotor position.

The magnetic flux versus current relation may be assumed to be linear due to the predominant effect of the two air gaps. This means that all three terms of the voltage equation are sinusoidal. Knowing $v'(t)$, $i(t)$, and R, we can write

$$v(t) = V_{max} \sin\omega t \tag{2.5.24}$$

from which it follows[6] that

$$\phi(t) = -\Phi_{max} \cos\omega t \tag{2.5.25}$$

where $\Phi_{max} = \dfrac{V_{max}}{\sqrt{2} \times 4.44 \, Nf}$. Normally $V'_{rms} \approx V_{rms}$ because the coil resistance is very small. Since $\phi(t)$ is known, we can substitute this in the torque equation to obtain

$$T_e(t) = \tfrac{1}{2}\Phi_{max}^2 \cos^2\omega t[2\mathcal{R}_1 \sin2\theta_m + 6\mathcal{R}_3 \sin6\theta_m + \cdots]. \tag{2.5.26}$$

This can be rewritten as follows[7]

$$\begin{aligned}
T_e(t) = \tfrac{1}{4}\Phi_{max}^2[&2\mathcal{R}_1 \sin2\theta_m + 6\mathcal{R}_3 \sin6\theta_m + \cdots \\
&+ 2\mathcal{R}_1 \cos2\omega t \sin2\theta_m + 6\mathcal{R}_3 \cos2\omega t \sin6\theta_m + \cdots].
\end{aligned} \tag{2.5.27}$$

Assume now that the rotor rotates with an angular velocity of ω_m rad/s. At time $t = 0$ we assume that the rotor position is given by the angle δ. For a particular time t the angular position of the rotor is given by the angle θ_m, where $\theta_m = \omega_m t - \delta$. We substitute this expression for θ_m in eq. (2.5.27) to get

$$\begin{aligned}
T_e(t) = \tfrac{1}{4}\Phi_{max}^2[&2\mathcal{R}_1 \sin(2\omega_m t - 2\delta) + 6\mathcal{R}_3 \sin(6\omega_m t - 6\delta) + \cdots \\
&+ 2\mathcal{R}_1 \cos2\omega t \sin(2\omega_m t - 2\delta) \\
&+ 6\mathcal{R}_3 \cos2\omega t \sin(6\omega_m t - 6\delta) + \cdots].
\end{aligned} \tag{2.5.28}$$

This can be rewritten[8] as

$$\begin{aligned}
T_e(t) = \tfrac{1}{4}\Phi_{max}^2 [&2\mathcal{R}_1 \sin(\omega_m t - 2\delta) + 6\mathcal{R}_3 \sin(6\omega_m t - 6\delta) + \cdots \\
&+ \mathcal{R}_1 \sin\{2(\omega_m + \omega)t - 2\delta\} + \mathcal{R}_1 \sin\{2(\omega_m - \omega)t - 2\delta\} \\
&+ 3\mathcal{R}_3 \sin\{2(3\omega_m + \omega)t - 6\delta\} \\
&+ 3\mathcal{R}_3 \sin\{2(3\omega_m - \omega)t - 6\delta\} + \cdots].
\end{aligned} \tag{2.5.29}$$

We see that T_e as a function of time consists of many component torques.

For the device to operate as a motor there must be an average torque T_{eav} developed. T_{eav} is nonzero when $\omega_m = \omega$ or $3\omega_m = \omega$ or $5\omega_m = \omega$, etc. Consider the important case that $\omega_m = \omega$. The instantaneous torque as a function of time becomes

$$\begin{aligned}
T_e(t) = \tfrac{1}{4}\Phi_{max}^2 [&2\mathcal{R}_1 \sin(2\omega t - 2\delta) + 6\mathcal{R}_3 \sin(6\omega t - 6\delta) + \cdots \\
&+ \mathcal{R}_1 \sin(4\omega t - 2\delta) + \mathcal{R}_1 \sin(-2\delta) + \cdots \\
&+ 3\mathcal{R}_3 \sin(8\omega t - 6\delta) + 3\mathcal{R}_3 \sin(4\omega t - 6\delta) + \cdots].
\end{aligned} \tag{2.5.30}$$

---

[6] Equation (2.5.25) is derived from Faraday's law $\phi = \dfrac{1}{N} \int v(t)dt$ and $\omega = 2\pi f$, where f is the supply frequency.

[7] Note that the $\cos^2\omega t = \tfrac{1}{2}(1 + \cos2\omega t)$.

[8] Note that $\sin A \cos B = \tfrac{1}{2}[\sin(A + B) + \sin(A - B)]$. In this case $A = 2\omega_m t - 2\delta$ and $B = 2\omega t$.

$T_e(t)$ contains a constant term and many pulsating components of 2, 4, 6, 8, etc. times the frequency of the supply voltage. Since the sinusoidally pulsating torque components do not contribute to the average torque, we find that the average torque[9] is

$$T_{eav} = -\tfrac{1}{4}\Phi^2_{\max}\mathscr{R}_1 \sin 2\delta. \tag{2.5.31}$$

$\mathscr{R}_1$ is a negative quantity[10] so that T_{eav} is a positive torque, that is, T_{eav} is in the direction of positive ω_m. Now $\omega_m = 2\pi n$, where n is the number of revolutions per second, and $\omega = 2\pi f$. Hence average torque exists for a rotor speed of $n = f$. This speed is called the *synchronous speed*.

Average values of developed torque will also exist for $3\omega_m = \omega$, $5\omega_m = \omega$, etc. Consider the case that $3\omega_m = \omega$. This condition is satisfied if $n = f/3$ r/s. At this *subsynchronous* speed the average *subsynchronous* torque is

$$T_{eav} = -\tfrac{3}{4}\Phi^2_{\max}\,\mathscr{R}_3 \sin 6\delta. \tag{2.5.32}$$

$\mathscr{R}_3$ is much smaller than $\mathscr{R}_1$ but the value of this subsynchronous torque may be just large enough to overcome the loss torque of the machine (friction and windage).

A final point can be made concerning energy conversion that occurs if the rotor speed is at the synchronous speed. The incremental energy eq. (2.4.8) was instrumental for determining the torque at any position. However, if the energy balance eq. (2.4.4) is considered over one cycle of the rotor motion, $W_e = 0 + W_m$, since the stored energy at the beginning and end of the cycle is the same and the change is zero. Therefore the conclusion is that the stored energy permits electromechanical energy conversion to take place, but for the conservative part of the system the total output is equal to the total input when conversion is continuous.

This case study of a primitive reluctance motor has demonstrated how we can use electric circuit relations and energy principles to find the conditions for energy conversion and continuous motion. It has allowed the theory to bring to light the existence of subharmonic torques in a practical device, and this itself indicates desirability for sinusoidal field distribution in the air gap.

2.5.4. Multiply-excited Rotating System

Most electromechanical energy converters have more than one coil. Normally one group of coils is mounted on the stator and another group is mounted on the rotor. A device with more than one exciting winding is called a *multiply-excited system*. The analysis of multiply-excited systems and the development of the torque equations are similar to those found in the discussion on singly-excited systems. A nonlinear

[9] The average torque is found from

$$T_{eav} = \frac{1}{T}\int_0^T T_e(t)dt.$$

[10] This follows from a Fourier analysis of the waveform shown in Fig. 2.19.

system is difficult to analyze. Consequently we will confine the discussion to a linear system with the justification that the air-gap effects predominate.

Independent of the number of windings the incremental energy eq. (2.4.8) holds true. Hence it follows from the arguments for a conservative system in sections 2.4 and 2.5 that the torque eq. (2.5.12) holds true also. The electric energy input and the magnetic stored energy are now the sum of the contributions made by all the windings. In matrix form the incremental electric energy input dW_e is

$$dW_e = \underline{i}^t \, d\underline{\lambda} \tag{2.5.33}$$

where $\underline{i}^t = [i_1 i_2 i_3 \, \cdots \, i_k \, \cdots \, i_n]$

$\underline{\lambda}^t = [\lambda_1 \lambda_2 \lambda_3 \, \cdots \, \lambda_k \, \cdots \, \lambda_n]$

and $\lambda_k = \lambda_{k1} + \lambda_{k2} + \cdots + \lambda_{kk} + \cdots + \lambda_{kn}.$

Since the system is considered to be linear the general expressions for the stored energy W_s and coenergy W_s' are

$$W_s = W_s' = \tfrac{1}{2} \underline{i}^t \underline{\lambda}. \tag{2.5.34}$$

In order to find the torque the stored energy must be expressed in terms of flux linkages or the coenergy must be expressed in terms of currents. It is more convenient to work with electric circuit quantities. Because $\underline{\lambda} = [L]\underline{i}$, where the inductance matrix is a function of angular position θ, the coenergy is

$$W_s'(i,\theta) = \tfrac{1}{2} \underline{i}^t [L(\theta)] \underline{i} \tag{2.5.35}$$

and the electromagnetic torque at any θ is

$$T_e = \frac{\partial W_s'(i,\theta)}{\partial \theta} = \tfrac{1}{2} \underline{i}^t \left[\frac{dL(\theta)}{d\theta} \right] \underline{i}. \tag{2.5.36}$$

For given values of currents it is necessary to express the elements of the inductance matrix in terms of the angular position θ for the torque to be determined.

A special case is when the currents are maintained constant. The incremental input energy is

$$dW_e = \underline{i}^t \, d\underline{\lambda} = \underline{i}^t \, d([L]\underline{i}) = \underline{i}^t[dL]\underline{i} = \underline{i}^t \left[\frac{dL}{d\theta} \right] \underline{i} \, d\theta. \tag{2.5.37}$$

The incremental stored energy is

$$dW_s = d(\tfrac{1}{2} \underline{i}^t \underline{\lambda}) = \tfrac{1}{2} \underline{i}^t[dL]\underline{i} = \tfrac{1}{2} \underline{i}^t \left[\frac{dL}{d\theta} \right] \underline{i} \, d\theta. \tag{2.5.38}$$

and the incremental mechanical energy output is

$$dW_m = T_e \, d\theta = \tfrac{1}{2} \underline{i}^t \left[\frac{dL}{d\theta} \right] \underline{i} \, d\theta = \tfrac{1}{2} \underline{i}^t[dL]\underline{i}. \tag{2.5.39}$$

This satisfies the incremental energy equation $dW_e = dW_s + dW_m$ and shows that, if the currents are kept constant, the electric input energy is shared equally between the change of stored energy and the mechanical energy output.

For all cases the instantaneous power output p_m of the device is given by

$$p_m = T_e\dot{\theta} = \tfrac{1}{2}\underline{i}^t \left[\frac{dL}{d\theta}\right] \underline{i}\dot{\theta} = \tfrac{1}{2}\underline{i}^t\underline{v}_m \tag{2.5.40}$$

where $\underline{v}_m$ is the vector of motionally induced voltages derived in section 2.3.2.

EXAMPLE 2.5

Apply the foregoing matrix relations to a doubly-excited system as shown in Fig. 2.20. Determine the coenergy, electromagnetic torque, and gross mechanical power output in terms of the circuit parameters. For the special case of constant currents show that exactly half of the total electric power input is converted to mechanical power output.

Solution
From an inspection of the figure the self-inductances L_{11} and L_{22} of the two coils as well as their mutual inductances L_{12} and L_{21} are functions of the position θ of the rotating part of the magnetic circuit. This dependence is indicated by writing

$$L_{11} = L_{11}(\theta), \; L_{12} = L_{21} = L_{12}(\theta) \quad \text{and} \quad L_{22} = L_{22}(\theta).$$

The expression for the coenergy at position θ for the linear system is

$$W'_s = \tfrac{1}{2}\underline{i}^t\underline{\lambda} = \tfrac{1}{2}\underline{i}^t[L]\underline{i} = \tfrac{1}{2}[i_1 \quad i_2]\begin{bmatrix} L_{11} & L_{12} \\ L_{21} & L_{22} \end{bmatrix}\begin{bmatrix} i_1 \\ i_2 \end{bmatrix}$$

$$= \tfrac{1}{2}L_{11}i_1^2 + L_{12}i_1i_2 + \tfrac{1}{2}L_{22}i_2^2.$$

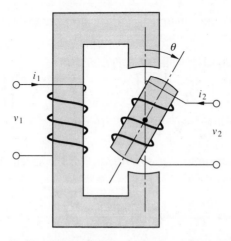

Figure 2.20 Doubly-excited rotating system.

Since the above expression for coenergy is in terms of currents and circuit parameters that are functions of θ, the torque is obtained from

$$T_e = \frac{\partial W_s'(i,\theta)}{d\theta} = \tfrac{1}{2}i^t \left[\frac{dL}{d\theta} \right] i = \tfrac{1}{2}\frac{dL_{11}}{d\theta} i_1^2 + \frac{dL_{12}}{d\theta} i_1 i_2 + \tfrac{1}{2}\frac{dL_{22}}{d\theta} i_2^2.$$

The gross mechanical power output is

$$p_m = T_e\dot\theta = \tfrac{1}{2}\frac{dL_{11}}{dt} i_1^2 + \frac{dL_{12}}{dt} i_1 i_2 + \tfrac{1}{2}\frac{dL_{22}}{dt} i_2^2.$$

The flux linkages of the coils are given by

$$\lambda = [L]i$$

or

$$\begin{bmatrix} \lambda_1 \\ \lambda_2 \end{bmatrix} = \begin{bmatrix} L_{11} & L_{12} \\ L_{21} & L_{22} \end{bmatrix} \begin{bmatrix} i_1 \\ i_2 \end{bmatrix}$$

so that $\lambda_1 = L_{11}i_1 + L_{12}i_2$ and $\lambda_2 = L_{21}i_1 + L_{22}i_2$. In terms of the electric circuit relations the induced coil voltages are

$$v = \frac{d}{dt}(\lambda) = \frac{d}{dt}([L]i) = [L]\frac{d}{dt}(i) + \left[\frac{dL}{d\theta} \right] i \frac{d\theta}{dt} = v_t + v_m$$

where v_t and v_m are the transformer and motional voltage vectors, respectively. Upon expansion the induced voltages due to a change of excitation and motion are

$$v_1 = L_{11}\frac{di_1}{dt} + L_{12}\frac{di_2}{dt} + i_1\frac{dL_{11}}{d\theta}\dot\theta + i_2\frac{dL_{12}}{d\theta}\dot\theta$$

and

$$v_2 = L_{21}\frac{di_1}{dt} + L_{22}\frac{di_2}{dt} + i_1\frac{dL_{21}}{d\theta}\dot\theta + i_2\frac{dL_{22}}{d\theta}\dot\theta.$$

The incremental electric energy dW_e that is supplied to the conservative coupling system during an incremental displacement $d\theta$ is

$$\begin{aligned} dW_e &= i^t\, d\lambda = i^t\, d([L]i) = i^t[dL]i + i^t[L]di \\ &= [i_1 \;\; i_2] \begin{bmatrix} dL_{11} & dL_{12} \\ dL_{21} & dL_{22} \end{bmatrix} \begin{bmatrix} i_1 \\ i_2 \end{bmatrix} + [i_1 \;\; i_2] \begin{bmatrix} L_{11} & L_{12} \\ L_{21} & L_{22} \end{bmatrix} \begin{bmatrix} di_1 \\ di_2 \end{bmatrix} \\ &= i_1^2 dL_{11} + 2i_1 i_2 dL_{12} + i_2^2 dL_{22} + i_1 L_{11} di_1 \\ &\quad + i_1 L_{12} di_2 + i_2 L_{21} di_1 + i_2 L_{22} di_2. \end{aligned}$$

The total electric power input is $p_e = dW_e/dt$. For the special case of constant currents, the electric power input is

$$p_e = i_1^2\frac{dL_{11}}{dt} + 2i_1 i_2\frac{dL_{12}}{dt} + i_2^2\frac{dL_{22}}{dt}.$$

Hence $p_e = 2p_m$.

EXAMPLE 2.6

Figure 2.21 depicts a doubly-excited rotating system. One coil is on the stator and is excited by a direct current of 10 A. The second coil is on the rotor and is excited by a direct current of 2 A. The coil self-inductances L_{11} and L_{22} are constants and their mutual inductance L_{12} varies with rotor position according to the expression $L_{12} = 0.1 \cos\theta$ henry. If the rotor is stationary determine the electromagnetic torque on the rotor for coil-axis displacements of 0, 30, 60, and 90 degrees.

Solution

It can be assumed that the system is linear. The coenergy W'_s of the system is given by

$$W'_s = \tfrac{1}{2}L_{11}i_1^2 + L_{12}i_1i_2 + \tfrac{1}{2}L_{22}i_2^2.$$

Torque $T_e = \dfrac{\partial W'_s(i,\theta)}{\partial \theta} = \tfrac{1}{2}i_1^2\dfrac{dL_{11}}{d\theta} + i_1i_2\dfrac{dL_{12}}{d\theta} + \tfrac{1}{2}i_2^2\dfrac{dL_{22}}{d\theta}.$

But L_{11} and L_{22} are not functions of θ, so

$$T_e = i_1i_2\frac{d}{d\theta}(0.1\cos\theta) = -10 \times 2 \times 0.1 \sin\theta = -2 \sin\theta.$$

At $\theta = 0°$, $T_e = 0$.
At $\theta = 30°$, $T_e = -1$ N·m (the torque is against the direction of increasing θ).
At $\theta = 60°$, $T_e = -1.732$ N·m.
At $\theta = 90°$, $T_e = -2$ N·m. Note the positions for maximum and minimum torques.

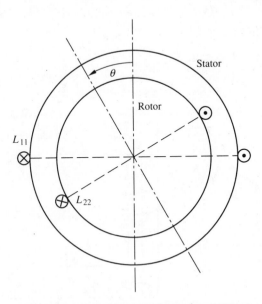

Figure 2.21 Doubly-excited rotating system.

2.6. TRANSLATIONAL MAGNETIC DEVICES

Most translational devices have a single degree of freedom of motion, portrayed by the x-direction in Fig. 2.1. Some applications are solenoid actuators, relays, and electric razors of the vibrating type. The action of the magnetic field produced by a current is to tend to cause motion in the direction x, because of the force of attraction f_e between the induced magnetic poles. The process of electromechanical energy conversion and the determination of the electromagnetic force at any position is the same as for the rotational magnetic devices with torque T_e replaced by force f_e and angular position θ replaced by translational position x. Examples of singly-excited and multiply-excited translational systems will serve as a summary of the principles described in detail in the foregoing sections.

EXAMPLE 2.7

A 60-Hz sinusoidal voltage of 100 V (rms) is connected to the coil of the contactor shown in Fig. 2.22. The area A of each pole is 6.5×10^{-4} m^2. The resistance of the coil and the reluctance of the iron may be ignored. What is the number of turns required for the contactor to develop an average force of 4.5 newtons?

Solution

Force $f_e = \dfrac{\partial W'_s(i,x)}{\partial x} = \tfrac{1}{2}i^2 \dfrac{dL}{dx}$.

Inductance $L = N^2 \mu_0 A / 2x$.

Let the voltage be $v = \hat{V} \cos\omega t$.

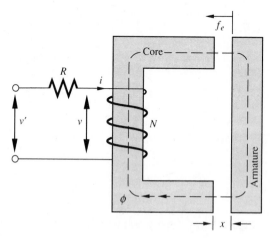

Figure 2.22   Singly-excited electromagnet.

If x is constrained (i.e., $\dot{x} = 0$ so there is no motional emf), $v = L\, di/dt$, and current $i = \dfrac{\hat{V}}{\omega L}\sin\omega t$.

Hence the instantaneous value of the electromagnetic force is

$$f_e(t) = -\frac{\hat{V}^2}{\omega^2 N^2 \mu_0 A}\sin^2\omega t.$$

The only variable is t and the average value of $\sin^2\omega t$ is 0.5. Therefore the average value of the developed force is

$$f_{e\,av} = \frac{V^2}{\omega^2 N^2 \mu_0 A}.$$

Substitution of the given data into this equation yields the required number of turns N to be 4380. Note that the force is independent of the gap setting. This is because the flux in the air gaps is determined by the ac voltage.

EXAMPLE 2.8

The electromagnet shown in Fig. 2.23 is excited by two identical current sources. Find the force of attraction between the poles in terms of current I and the geometry. If the current were reversed in one coil, what is the force between the poles?

Solution
Consider the system to be linear by assuming that the reluctance of the iron is negligibly small.

Electromagnetic force $f_e = -\dfrac{\partial W_s}{\partial x} = -\tfrac{1}{2}\phi^2\,\dfrac{d\mathcal{R}}{dx} = -\tfrac{1}{2}\phi^2\,\dfrac{d}{dx}\left(\dfrac{2x}{\mu_0 wd}\right).$

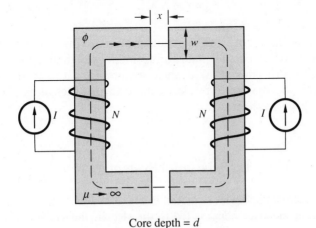

Core depth $= d$

Figure 2.23   Doubly-excited translational system.

Therefore $f_e = -\dfrac{\phi^2}{\mu_0 wd}$.

From Ampere's circuital law and the directions of the mmfs $H \times 2x = 2NI$.

But $H = \dfrac{B}{\mu_0} = \dfrac{\phi}{\mu_0 wd}$. Hence $\phi = \dfrac{\mu_0 wd}{x} NI$.

Consequently $f_e = -\dfrac{\mu_0 wd N^2 I^2}{x^2}$. The negative sign indicates a force of attraction.

Consider the current in one coil to be reversed. For the ideal case there is no leakage or fringing. If the current in one coil were reversed, the net mmf would be zero. So the flux ϕ would be zero and no force would exist. This would not be found in practice. In reality there would be parallel leakage paths for the fluxes produced by the mmf of each coil. Adjacent poles would have like polarity and a force of repulsion would exist. The problem is difficult to solve. In the limit, if the leakage paths had negligibly small reluctances the force has the same value as in the first part.

2.7. ELECTROSTATIC DEVICES

Just as energy can be stored in the magnetic field so energy can be stored in the electric field. Therefore electric energy input to an electric coupling field, whose system has some mechanical degree of freedom, can give rise to electromechanical energy conversion. Whereas in magnetic devices the important circuit parameter is the inductance, the capacitance is the important circuit parameter of electrostatic devices.

Analysis of electrostatic devices is straightforward since electrostatic devices can be considered the duals of magnetic devices. It is a reasonable approximation to consider the relation between the electric circuit variables, charge q and voltage v, to be linear. By definition this relation is

$$q \triangleq Cv \tag{2.7.1}$$

where C is the capacitance of a device's configuration. In the following discussion the device is assumed to be a parallel plate arrangement even though there may be some deviation from this in practice. With these assumptions it is possible to set about determining the force developed by electrostatic devices such as microphones, pressure transducers, and voltmeters.

2.7.1. Singly-excited Translational System

Consider the electrostatic device depicted in Fig. 2.24a with its characteristic shown in Fig. 2.24b. The electric energy is fed to the electric field of intensity E, which is assumed to be uniform and confined between the capacitor plates. One of the plates is free to move in the x direction.

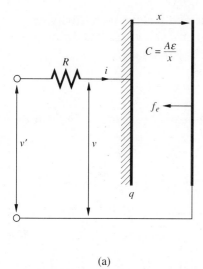
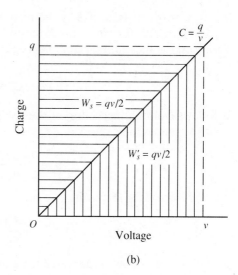

(a)   (b)

Figure 2.24 Electrostatic device. (a) Singly-excited translational system, (b) electric circuit characteristic.

The treatment is similar to that for magnetic field devices, so that for the conservative part of the system the electric energy input W_e over some period of time is

$$W_e = \int vi \, dt = \int v \, dq \tag{2.7.2}$$

since $i = dq/dt$. Consequently the incremental change of electric energy input is

$$dW_e = v \, dq. \tag{2.7.3}$$

The incremental mechanical energy output dW_m, when the free plate is allowed to move through an incremental displacement dx, is given by

$$dW_m = f_e \, dx \tag{2.7.4}$$

where the electrostatic force f_e arises from the attraction of the two plates which are charged with opposite polarities. From the incremental conservative energy balance equation $dW_e = dW_s + dW_m$ we can express the incremental stored energy in the form

$$dW_s = v \, dq - f_e \, dx. \tag{2.7.5}$$

If the free plate is held at x there is no mechanical work done so that

$$W_s = W_e = \int v \, dq = \int \frac{q}{C} \, dq = \tfrac{1}{2} \frac{q^2}{C} = \tfrac{1}{2} qv. \tag{2.7.6}$$

In terms of q and x the stored energy is

$$W_s(q,x) = \tfrac{1}{2} \frac{q^2}{C} = \tfrac{1}{2} \frac{q^2 x}{\epsilon A} \tag{2.7.7}$$

where $C = q/v = A\epsilon/x$, if A is the area of a plate and ϵ is the permittivity of the medium between the plates. Since the coupling system is conservative and q and x are variables, the incremental stored energy can also be written as

$$dW_s = \frac{\partial W_s}{\partial q} dq + \frac{\partial W_s}{\partial x} dx. \tag{2.7.8}$$

It follows from an inspection of the coefficients of eqs. (2.7.5) and (2.7.8) that the voltage v across the device is

$$v = \frac{\partial W_s(q,x)}{\partial q} \bigg|_{x=\text{constant}} \tag{2.7.9}$$

and the force f_e for the position x is

$$f_e = - \frac{\partial W_s(q,x)}{\partial x} \bigg|_{q=\text{constant}} \tag{2.7.10}$$

where W_s is expressed in terms of the variables q and x only. Therefore the force is

$$f_e = - \frac{\partial W_s(q,x)}{\partial x} = - \frac{\partial}{\partial x} \left(\tfrac{1}{2} \frac{q^2}{A\epsilon} x \right) = - \frac{q^2}{2A\epsilon}. \tag{2.7.11}$$

In practice it is more convenient to handle v and x as the independent variables. Consequently we must consider the coenergy W_s' of the system. Coenergy is defined from Fig. 2.24b by

$$W_s' \triangleq qv - W_s. \tag{2.7.12}$$

Accordingly the incremental energy balance can be written as

$$dW_e = d(qv - W_s') + dW_m \quad \text{or} \quad dW_s' = q\,dv + f_e\,dx. \tag{2.7.13}$$

The latter was derived using eq. (2.7.3). Coenergy is a function of the independent variables v and x so that

$$dW_s' = \frac{\partial W_s'}{\partial v} dv + \frac{\partial W_s'}{\partial x} dx. \tag{2.7.14}$$

It follows from these last two equations that the charge q on the plates is given by

$$q = \frac{\partial W_s'(v,x)}{\partial v} \bigg|_{x=\text{constant}} \tag{2.7.15}$$

and the electrostatic force f_e can be expressed by

$$f_e = \frac{\partial W_s'(v,x)}{\partial x} \bigg|_{v=\text{constant}.} \tag{2.7.16}$$

From the linear characteristic in Fig. 2.24b for the electrostatic device at a position x, $W_s' = W_s = \tfrac{1}{2}qv$, so that

$$W_s'(v,x) = \tfrac{1}{2}Cv^2 = \frac{v^2 A\epsilon}{2x}. \tag{2.7.17}$$

Hence the force is

$$f_e = \tfrac{1}{2}v^2 \frac{dC}{dx} = -\frac{v^2 A \epsilon}{2x^2}. \tag{2.7.18}$$

For a particular voltage the force is shown to be inversely proportional to the square of the distance between the plates of the capacitor. This only holds for x above a certain minimum value, below which a discharge would occur.

The two expressions for the force between the plates of the capacitor give the same results. This is seen from

$$f_e = -\frac{q^2}{2A\epsilon} = -\frac{v^2 C^2}{2A\epsilon} = -\frac{v^2 A \epsilon}{2x^2}. \tag{2.7.19}$$

According to eq. (2.7.18) the force acts in a direction tending to increase the capacitance and hence to decrease x. For example, if a solid dielectric (of relative permittivity $\epsilon_r > 1$) were partially introduced between the plates there would be an electrostatic force tending to pull the dielectric material into the interplate space until the maximum value of capacitance were attained. Similarly, if two parallel plates were maintained at the same potential and restrained, and if a third parallel plate at a different potential were introduced between the first two plates, then there would be a force tending to pull the third plate into alignment, in which position maximum capacitance occurs. This is the principle of an electrostatic synchronous motor and is analogous to the reluctance motor described in section 2.5.4.

If an electrostatic device were constrained to rotational movement, then the mechanical system variables are torque T_e and angular position θ. For such a case the torque is given by

$$T_e = -\frac{\partial W_s(q,\theta)}{\partial \theta} = \frac{\partial W_s'(v,\theta)}{\partial \theta}. \tag{2.7.20}$$

2.7.2. Electrostatic Machine with AC Excitation (A Case Study)

Consider the singly-excited rotational system depicted in Fig. 2.25. This is the dual of the reluctance motor described in section 2.5.4. Let the applied voltage vary sinusoidally with time and let the capacitor plates be shaped so that the capacitance C varies with position θ as shown in Fig. 2.26. We wish to find whether motion can be sustained and if energy conversion is continuous.

The capacitance can be expressed in the form

$$C(\theta) = C_0 + C_2 \cos 2\theta. \tag{2.7.21}$$

The maximum value of C is C_d and this occurs at $\theta = 0$ and π. The minimum value of C is C_q which occurs at $\theta = \pi/2$ and $3\pi/2$. So $C_0 = (C_d + C_q)/2$ and $C_2 = (C_d - C_q)/2$.

For any position θ, the coenergy W_s' is

$$W_s'(v,\theta) = \tfrac{1}{2}v^2 C(\theta) \tag{2.7.22}$$

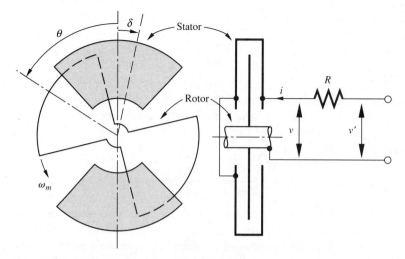

Figure 2.25   Rotational electrostatic machine.

so the torque can be found from

$$T_e = \frac{\partial W'_s(v,\theta)}{\partial \theta} = \frac{\partial}{\partial \theta}[\tfrac{1}{2}C(\theta)v^2] = -v^2 C_2 \sin 2\theta. \qquad (2.7.23)$$

The electric circuit relation is

$$v'(t) = i(t)R + v(t). \qquad (2.7.24)$$

All three terms are sinusoidal since the relation between v and q is linear. Since $v'(t) = V'_{max} \sin \omega t$, then $v = V_{max} \sin \omega t$. Upon substitution in the torque equation the instantaneous value of the torque becomes

$$T_e = -V^2_{max} C_2 \sin^2 \omega t \sin 2\theta. \qquad (2.7.25)$$

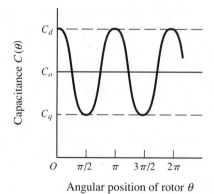

Figure 2.26   Capacitance variation with position.

We must express θ in terms of t. Let the angular speed of the rotor be given by $d\theta/dt = \omega_m$, so that $\theta = \omega_m t - \delta$, if the position of the rotor is given by $\theta = -\delta$ at $t = 0$.

After some manipulation the instantaneous value of the torque can be expressed by

$$T_e = \tfrac{1}{4}V^2_{max}C_2[-2\sin2(\omega_m t - \delta) + \sin2(\omega t + \omega_m t - \delta) \\ - \sin2(\omega t - \omega_m t + \delta)]. \tag{2.7.26}$$

The average value of a sinusoidal time varying function is zero, so the condition for an average torque to exist only occurs when the coefficient of t is zero. This suggests that for sustained motion $\omega_m \pm \omega = 0$, that is, rotation must only be at synchronous speed. For this case the average torque is

$$T_{eav} = -\tfrac{1}{4}V^2_{max}C_2 \sin2\delta. \tag{2.7.27}$$

The angle δ is known as the *torque angle*. As the mechanical load increases, the angle δ increases so that the electrostatic torque balances the load torque up to a maximum value of $\delta = 45°$, beyond which the machine pulls out of synchronism.

The average mechanical power output P_m is

$$P_m = T_{eav}\dot{\theta} = \tfrac{1}{4}V^2_{max}\omega C_2 \sin2\delta. \tag{2.7.28}$$

Accordingly continuous energy conversion is possible as long as a difference exists between the maximum and minimum values of the capacitance. Over each revolution of the rotor the net change of stored energy is zero, so that all the electric energy input is converted to mechanical energy. This demonstrates that the coupling field system acts as the catalyst in the continuous energy conversion process.

Since the force or torque densities of magnetic field devices are about 25,000 times greater than practically obtainable densities of electric-field devices the electrostatic machines do not find general application in practice except as incremental motion transducers.

2.8. MECHANICAL SYSTEM RELATIONS

We have already investigated the electrical and the coupling field systems. The third step in the procedure of dynamic circuit analysis is to study the relations between the variables of the mechanical system. Motion for all mechanical systems in this book is limited to one degree of freedom, angular displacement θ for rotational movement and displacement x for translational movement. Although displacement is a variable, speed is chosen as a variable where it is convenient. The lumped elements in a mechanical system are due to the effects of inertia, elasticity, and viscous damping.[11]

---

[11] See Appendix B for details.

2.8.1. Mechanical Systems

A rotary electromagnetic device, that is motoring, will have the electromagnetic torque T_e balanced by the inertial, friction, torsional spring, and load torques. That is

$$T_e = J\ddot{\theta} + D\dot{\theta} + K\theta + T' \tag{2.8.1}$$

where $\qquad J =$ moment of inertia

$\qquad\qquad \theta =$ displacement between rotor and stator axes

$\qquad\qquad D =$ friction coefficient

$\qquad\qquad K =$ spring constant

and $\qquad T' =$ load torque.

For a translational device the force balance is

$$f_e = M\ddot{x} + D\dot{x} + Kx + f' \tag{2.8.2}$$

where $\qquad M =$ mass

$\qquad\qquad x =$ displacement

and $\qquad f' =$ load force.

These dynamic equations of motion must be solved for position θ or x, and speed. The electromagnetic torque, or force, is a function of currents and position, so we cannot solve the equation directly. For an analysis, we combine the results of the whole chapter in the next section.

2.9. DYNAMIC CIRCUIT ANALYSIS

Investigation of the electric circuit, the coupling field, and the mechanical system of an electromechanical device has provided equations which describe the behavior of the energy conversion process. The equations are linked, because, for motoring action in a rotational system, the torque T_e of electromagnetic origin is the torque applied to the mechanical system. Thus the problem is formulated by the general equations

$$\underline{v}' = [R]\underline{i} + [L(\theta)]\frac{di}{dt} + \left[\frac{dL(\theta)}{d\theta}\right]\underline{i}\dot{\theta} \tag{2.9.1}$$

$$T_e = \tfrac{1}{2}\underline{i}^t\left[\frac{dL(\theta)}{d\theta}\right]\underline{i} \tag{2.9.2}$$

and

$$T_e = J\ddot{\theta} + D\dot{\theta} + K\theta + T' \tag{2.9.3}$$

so that the dynamic circuit can now be analyzed. For generator action the electromagnetic torque T_e and the load torque T' have negative signs. For translational systems torque is replaced by force, position becomes x and moment of inertia is replaced by mass in the equations.

Since there are products of variables, the equations are nonlinear. In order to solve for the currents i and the rotor position θ it is common practice to solve by digital simulation employing numerical analysis techniques on a rearranged set of first order differential equations.

A number of simplifications can be made to the equations. Since the electrical time constants are usually short compared with the mechanical time constant, the device can be considered to have constant speed if the electrical transients are to be investigated. Also the electrical system can be considered to be in steady state, if the mechanical oscillations are to be studied. The differential equations can be linearized by using such techniques as small perturbations about an operating point, from which information about stability, transient response, or frequency response can be derived.

Once satisfactory approximations have been made and the parameters of the device have been determined there are many computer *packages* available to solve the problem. We use a *package* in the following simulation, which describes the method of solution of a typical problem associated with a magnetic, translational position controller.

2.9.1. Simulation

Figure 2.27 illustrates schematically an electromechanical position controller. In the steady state the force acting on the spring determines the position x according to Hooke's law. The applied force has a magnetic origin and the magnitude depends

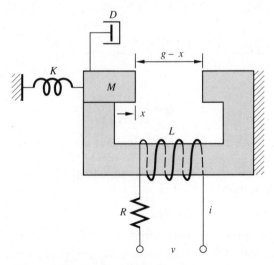

Figure 2.27 Position controller.

upon the current in the coil of the solenoid. Consequently there is a simple functional relationship between current and position.

In the transient or dynamic case (i.e., when there are changes taking place) the position x of the mass depends upon the additional mechanical parameters, inertia and friction (damping) and the spring constant, as well as the electrical parameters. Modified eqs. (2.9.1) to (2.9.3) have to be solved in order to predict the response (e.g., position as a function of time) of the controller to some forcing function (e.g., a voltage suddenly applied to the terminals of the solenoid). The type of system depicted to Fig. 2.27 offers a means to exemplify dynamic circuit analysis. We can write down the electric circuit relations, the force equation and the dynamic equation of motion that were developed in sections 2.3.3, 2.6, and 2.8, respectively. The parameters have to be measured from the practical apparatus or be calculated from design data.

From eq. (2.3.24) the electric circuit relation is described by

$$v = Ri + L\frac{di}{dt} + \dot{x}\frac{dL}{dx}\,i. \tag{2.9.4}$$

Equation (2.9.4) cannot be solved directly because it comprises the independent variables v and t and the two dependent variables i and x. The inductance L is a nonconstant coefficient of this nonlinear differential equation since it is a function of position x.

The force f_e created by the current i is given by

$$f_e = \tfrac{1}{2}i^2\frac{dL}{dx}. \tag{2.9.5}$$

The expression for the inductance is

$$L = \frac{N^2}{\mathscr{R}_a + \mathscr{R}_i} \tag{2.9.6}$$

where $\mathscr{R}_a = \dfrac{g-x}{\mu_0 A}$ is the reluctance of the air gap, and $\mathscr{R}_i$ is the reluctance of the iron path. Thus,

$$f_e = \frac{\mu_0 A i^2 N^2}{2(b-x)^2} \tag{2.9.7}$$

where $b = g + \mu_0 A\mathscr{R}_i = $ constant. Note that $b \approx g$, if $\mathscr{R}_i \ll \mathscr{R}_a$ (a common situation).

Finally we must account for the inertial, damping and spring effects in the dynamic equation of motion. That is

$$f_e = M\ddot{x} + D\dot{x} + Kx. \tag{2.9.8}$$

There is no load force other than that included in the friction force. Equations (2.9.4), (2.9.5), and (2.9.8) lead us to two nonlinear equations in two unknowns i and x with nonconstant coefficients and with v as the forcing function. These are

$$M\ddot{x} + D\dot{x} + Kx = \tfrac{1}{2}\frac{dL}{dx}i^2 \qquad (2.9.9)$$

and

$$v = Ri + L\frac{di}{dt} + \dot{x}\frac{dL}{dx}i. \qquad (2.9.10)$$

The solution for i and x is usually obtained by numerical techniques. This is often facilitated by formulating the problem into the following set of first-order differential equations.

$$\dot{x} = y \qquad (2.9.11)$$

$$\dot{y} = -\frac{1}{M}[Dy + Kx - \tfrac{1}{2}L'i^2] \qquad (2.9.12)$$

and

$$\frac{di}{dt} = \frac{1}{L}[v - Ri - L'yi] \qquad (2.9.13)$$

where $L' = dL/dx$. Now we are in a position to determine the behavior of the controller by finding a method to solve the last three equations.

EXAMPLE 2.9

Consider the position controller illustrated in Fig. 2.27. Referring to the previous section the controller parameters are as follows.

$$R = 10\ \Omega, \quad L = \frac{1}{(2 \times 10^{-2} - x)}\ \text{mH}$$
$$M = 0.04\ \text{kg}, \quad K = 250\ \text{N/m}, \quad \text{and} \quad g = 2 \times 10^{-2}\ \text{m}.$$

For a step input of 5 volts to the solenoid coil determine the current i and displacement x responses given the initial condition that $x = 0$. Consider the two cases where $D = 7$ Ns/m and $D = 1$ Ns/m.

Solution

Any suitable numerical technique may be used to solve the nonlinear eqs. (2.9.11) to (2.9.13) by computer. Figure 2.28 shows the current and displacement curves obtained by WATAND.[12] The two cases show overdamping ($D = 7$ Ns/m) and underdamping ($D = 1$ Ns/m) with its attendant oscillation. The gain of the system X/I is 2.9 mm/A which is relatively low for an electromechanical system. However, it is

---

[12] WATAND (WATerloo Analysis 'N Design) is an interactive computer program for simulating linear and nonlinear electric circuits.

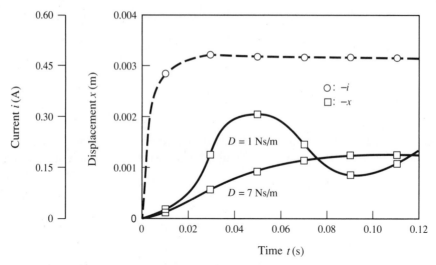

Figure 2.28   Response to a step input of 5 volts.

conventional to have the apparent mechanical time constant (0.025 s) greater than the apparent electrical time constant (0.004 s). This shows that the simplifications stated in section 2.9 can be made. That is, the mechanical displacement can be considered constant while electrical transients are investigated. Further, the electrical variables can be considered to be in steady state while the mechanical response is being determined.

2.10. SUMMARY

A *dynamic circuit* is defined as an electromagnetic system that has mechanical motion that causes the values of the electrical parameters to change.

The electric circuit relations are $\underline{v} = \dfrac{d\lambda}{dt}$ and $\underline{\lambda} = [L]\underline{i}$, where the flux linkage $\underline{\lambda}$ is a function of the currents and the position.

Conversion of energy in a conservative system that is considered to be motoring takes place according to the incremental energy balance equation, $dW_e = dW_s + dW_m$.

Energy W_s stored in the magnetic field and coenergy W_s' are related by $W_s + W_s' = i\lambda'$. For a singly-excited system at rest $W_s' = \int_0^i \lambda \, di$ and $W_s = \int_0^\lambda i \, d\lambda$. In terms of stored energy and coenergy the general expressions for the electromagnetic torque are

$$T_e = -\left.\frac{\partial W_s(\lambda,\theta)}{\partial \theta}\right|_{\lambda=\text{constant}} = \left.\frac{\partial W_s'(i,\theta)}{\partial \theta}\right|_{i=\text{constant}}.$$

For translational systems replace T_e by force f_e and θ by position x.

The electromechanical relations for an electrostatic system can be obtained from the duals of the magnetic system.

An electromechanical system has its forces in balance. For motoring action the dynamic equation of motion is $T_e = J\ddot{\theta} + D\dot{\theta} + K\theta + T'$.

The general problem of determining the behavior of any electromechanical energy converter has been formulated by considering the electrical and mechanical circuits separately, then linking them by a conservative coupling system, which is the heart of the conversion process. The aim of the formulation has been to prepare the way to find the dynamic circuit response. A solution of the mechanical equation of motion is possible only if the torque or force developed magnetically can be found from the stored energy or coenergy. In turn, the stored energy must be found in terms of flux linkage and geometry, while the coenergy must be found as a function of current and geometry. Finally to find the current and flux linkage the principles of circuit theory and electromagnetism must be applied. The method of solution of the electric circuit and mechanical equations becomes the choice of the investigator.

2.11. PROBLEMS

Section 2.3

2.1. Consider the doubly-excited system described in Example 2.1 and let the same considerations prevail except that $i_2 = \hat{I}_2 \sin\omega_m t$ and $\theta = 0$ at $t = 0$. Determine **(a)** the instantaneous value of the motional and transformer induced emfs at the terminals of coil 1, **(b)** the frequency of alternation of coil 1 terminal voltage and, **(c)** the resultant average value of the coil 1 terminal voltage. (*Answer:* (a) $-3140 \sin^2 314t$, $3140 \cos^2 314t$, (b) 100 Hz, (c) 0.)

2.2. Figure P2.2 overleaf shows a light piece of soft iron pivoted between the poles of the permanent magnet. When in the vertical plane, the net flux linkage of the coil is zero. When the stylus vibrates, the magnetic flux linking the coil changes in magnitude and direction. The changing flux induces in the coil an emf which is amplified to operate a speaker. Find the emf as a function of the speed of motion of the stylus.

Let g = average length of each air gap = 2 mm

x = displacement of soft iron tip, $(x \ll g)$

F = magnetic potential drop across the total air gap = 300 AT

A = cross-sectional area of each gap = 1 cm^2

N = coil turns = 100

and $\mu_0 = \infty$ for soft iron and the permanent magnet.

What is the speed of the stylus when the coil emf is 3.14 mV? (*Answer:* 0.33 × 10^{-2} m/s.)

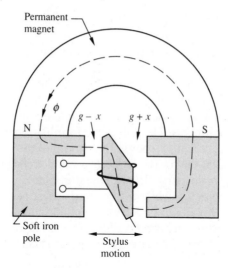

Figure P2.2

Section 2.4

2.3. An electromagnet with two air gaps is depicted in Fig. P2.3 with the dimensions as shown. The value of the permeability of the iron can be assumed to approach infinity and the lengths of the air gaps are small enough that fringing can be neglected. What is the magnetic flux in each air gap and what is the total flux linkage associated with the central core whose coil has 1000 turns and carries 1.0 ampere? What is the magnetic stored energy in this system?
(*Answer:* 10^{-4} Wb, 10^{-4} Wb, 0.2 Wb, 0.1 J.)

2.4. The solenoid depicted in Fig. P2.4 has a rectangular core section of depth d. Find **(a)** the flux through the plunger, **(b)** the circuit inductance and **(c)** the stored energy in terms of the geometry and the current i.
Assume that (i) the core permeability is high enough so that only the air-gap

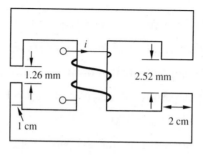

Core depth = 1 cm

Figure P2.3

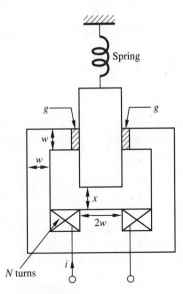

Figure P2.4

mmfs need be considered, (ii) $g \ll w$ and $x \ll w$ so that fringing can be ignored, and (iii) all the flux in the core passes through the air gaps x or g so that the leakage flux can be neglected.

2.5. The magnetization curve of the ferromagnetic material of the reactor shown in Fig. P2.5 is given by the following table.

B (tesla)	0.48	0.96	1.42	1.62	1.7	1.75
H (AT/m)	80	160	240	320	400	480

If the flux density is 1.2 tesla, what is the energy stored in the magnetic field and what is the inductance of the reactor?
(*Answer:* 60.12×10^{-3} J, 4.78 mH.)

2.6. Figure P2.6a represents a reactor, whose idealized *BH* curve is shown in Fig. P2.6b. **(a)** What is the energy stored in the magnetic field if the flux density is 1.5 tesla? **(b)** What is the ratio of the contributions of stored energy provided by the iron and the air gap? **(c)** What is the inductance of the reactor if the flux density is 1.5 tesla? **(d)** If an approximation were made that the iron had infinite permeability, what would be the error in the value of the inductance? (There are three definitions of inductance.)
(*Answer:* (a) 0.19 J, (b) 1:27, (c) 568 μH, 438 μH, 605 μH, (d) 10.2%, 17%, 3.67%.)

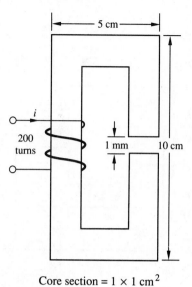

Core section = 1×1 cm^2

Figure P2.5

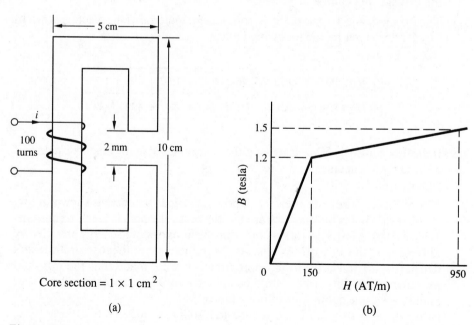

Core section = 1×1 cm^2

(a)

(b)

Figure P2.6

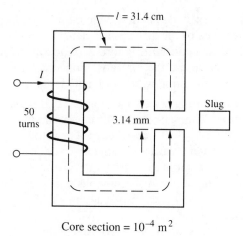

Core section = 10^{-4} m^2

Figure P2.7

2.7. The relative permeability of the iron core shown schematically in Fig. P2.7 is 10,000. The current in the coil is maintained constant at 1 ampere under all conditions. How much work is done in pulling the slug of iron into the air gap? It can be assumed that the slug has a constant relative permeability of 10,000 and is large enough to fill the air-gap volume.
(*Answer:* 4.9×10^{-3} J.)

2.8. A conservative system comprising two air-cored coils a and b are constrained to translational movement in the x direction. Their self- and mutual inductances are given by $L_a = (3 + x)$H, $L_b = (6 + 2x)$H, and $L_{ab} = (4 - x)$H. If the currents, whose values are $I_a = 2$A and $I_b = 1$ A, are kept constant while x increases from 0 to 1 meter, find **(a)** the mechanical work done and **(b)** the electrical energy input or output of the two coils.
(*Answer:* (a) 1 J, (b) 2 J, inputs 2 J, 0 J.)

Section 2.5

2.9. Consider the singly-excited system shown in Fig. 2.18. If it can be assumed that the coil inductance L varies with the rotor position θ_m according to the expression $L = L_1 + L_2 \cos 2\theta_m$ where L_1 and L_2 are constants, find **(a)** the torque on the rotor in terms of the current i and rotor position θ_m and **(b)** the value of the average torque, if the rotor has a constant angular velocity ($\theta_m = \omega t$). Let the current $i = \hat{I} \cos(\omega t + \delta)$.

2.10. Consider the singly-excited system shown in Fig. 2.18 and assume the nonlinear magnetization curve can be expressed by $i = (A_0 - A_2 \cos 2\theta_m)\lambda^2$, where A_0 and A_2 are constants and λ is the flux linkage of the coil. Find in terms of λ and θ_m **(a)** the magnetic stored energy, **(b)** the coenergy, and **(c)** the instantaneous value of the electromagnetic torque.

2.11. Consider the singly-excited system shown in Fig. 2.18. Express the electromagnetic torque in terms of the air-gap flux density and the geometry if the coil current is maintained constant. Ignore leakage and fringing and assume the flux passes radially across the air gaps. The stator and rotor poles may be considered to have the same width and axial length.

2.12. A doubly-excited rotating system (see Fig. 2.21) has a coil on the stator and a coil on the rotor. The self- and mutual inductances are given by $L_{11} = L_{22} = 0.5$ henry and $L_{12} = 0.2 \cos\theta$ henry. The rotor coil is short-circuited and the stator coil is excited from a constant current source such that $i_1 = 10 \sin\omega t$ amperes. Neglect the rotor-coil resistance. If the rotor is constrained to be stationary, determine **(a)** the instantaneous and average values of the electromagnetic torque at a coil-axes displacement of $15°$ and **(b)** the equilibrium positions of the rotor coil.

(*Answer:* (a) $2 \sin^2\omega t$ N·m, 1 N·m, (b) Stable: $\pi/2$, $3\pi/2$. Unstable: 0, π.)

2.13. Two coils in space have self- and mutual inductances L_{11}, L_{22}, and L_{21} and are not constrained. Assume the system to be conservative. **(a)** What is the total electrical power into the system in terms of inductances and currents? **(b)** What is the stored energy and rate of change of stored energy in terms of inductances and currents? **(c)** If the currents in both coils are kept constant, how much of the electrical power input is converted to mechanical power? (*Answer:* (c) half.)

Section 2.6

2.14. The total air-gap length of the electromagnet shown in Fig. P2.14 is g. What is the net horizontal force acting on the iron pellet in the air gap in terms of i, if the permeability of the iron approaches infinity and the fringing is neglected?

2.15. The total air-gap length of the electromagnet shown in Fig. P2.15 is g. What is the net horizontal force acting on the iron pellet in the air gap in terms of i, if the permeability of the iron approaches infinity and the fringing is neglected?

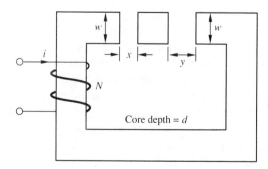

Figure P2.14

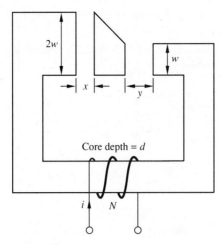

Figure P2.15

2.16. The electromagnet depicted in Fig. P2.16 is to be capable of supporting a mass
M. Show that the current required is given by the expression

$$I = \frac{2x}{N} \left[\frac{Mg}{\mu_0 wd} \right]^{1/2} .$$

where g is the gravitational acceleration. Ignore the reluctance of the iron.

2.17. The electromagnet, depicted in Fig. P2.17, controls the operation of a relay
switch. Assume that the reluctance of the iron can be neglected. **(a)** If the
air-gap length is 1 mm, what is the current required to develop a force of 31.4
newtons? **(b)** If the air-gap length is 2 mm, what is the current required to

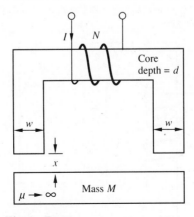

Figure P2.16

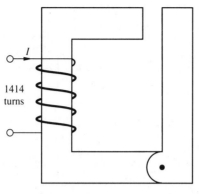

Core section = 2×2 cm^2

Figure P2.17

develop the same force? **(c)** What is the energy stored in the air-gap volume for the conditions in part **(b)**?

(*Answer:* (a) 0.25 A, (b) 0.5 A, (c) 0.063 J.)

2.18. Figure P2.18 depicts a model of an electromagnetic relay, in which the additional air gap of fixed length *l* accounts for the leakage flux, if the reluctance of the iron is neglected. If the armature air gap has a particular value *x* and the coil of 1000 turns has a current of 5 A, the leakage flux is 1.0 mWb and the armature flux is 2.0 mWb. If the armature air-gap length is increased to 1.1*x* but the armature flux kept at the same value, find **(a)** the coil current, **(b)** the leakage flux, **(c)** the energy stored in the armature air-gap volume, **(d)** the stored energy associated with the leakage flux, **(e)** the circuit inductance, **(f)** the energy stored in the inductance of the coil, **(g)** the mechanical work done to increase the air-gap length, and **(h)** the change of electrical energy input. **(i)** If *x* = 6.4 mm what is the electromagnetic force on the armature and

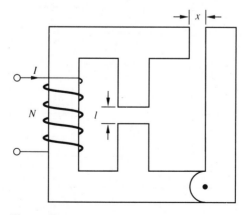

Figure P2.18

(j) what is the value of the flux density in the armature air gap? **(k)** If it took 0.01 second to increase the air-gap length, estimate the average voltage induced in the coil.

(*Answer:* (a) 5.5 A, (b) 1.1 mWb, (c) 5.5 J, (d) 3.02 J, (e) 0.56 H, (f) 8.52J, (g) 0.5 J input, (h) 0.52 J output, (i) 790 N, (j) 0.5 T, (k) 10 V.)

2.19. The solenoid shown in Fig. P2.4 has a coil of 200 turns, a plunger area of 100 cm^2, and an air-gap length x which is 0.5 and 1.0 cm at the limits of travel of the plunger. A constant current of 2 A is supplied from a dc source to the coil. Neglect leakage and fringing and ignore the reluctance of the iron and the gaps g. Determine **(a)** the coil inductance and stored energy as functions of x, **(b)** the external force developed to keep the plunger in the maximum air-gap position, **(c)** the electromagnetic force on the plunger if it is in the minimum air-gap position, and **(d)** the mechanical work done if the plunger is allowed to move slowly through the entire limit of travel.

(*Answer:* (b) 10 N, (c) 40 N, (d) 0.1 J.)

2.20. Consider the system of coils described in Problem 2.8. The coils are restrained at a position $x = 1.0$ and both coils are excited by a 50 Hz voltage of rms value 314 volts. What is the average value of the electromagnetic force acting on the coils?

(*Answer:* 0.016 N.)

2.21. Figure P2.21 represents two transformer coils, one of which is fixed and the other is free to move up the core leg with the limitation imposed by the spring. The self-inductances are constants and the mutual inductance can be assumed to be given by $M = M_0(1 - x)$, where M_0 is a constant. **(a)** Find the electromagnetic force between the two coils in terms of the coil currents and inductances. **(b)** If coil 1 is excited by an alternating current and coil 2 is connected across a resistive load, would the electromagnetic force be one of attraction or repulsion?

2.22. Figure P2.22 shows the cross-section of a doubly-excited cylindrical solenoid. The coils have the same number of turns so that when the currents are the same the plunger assumes a position $x = 0$. **(a)** Find the coil inductances for any

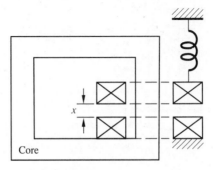

Figure P2.21

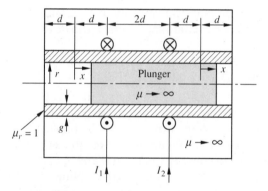

Figure P2.22

position $(-d < x < d)$ in terms of the geometry. Assume $g \ll r$. For a particular case of current imbalance, if $I_2 = 2I_1 = 2$ A, $N = 100$ turns per coil and $r = 2d = 20\ g = 0.02$ m, **(b)** find the electromagnetic force when the plunger is in the symmetrical position and find the equilibrium position x and the coil flux linkages for this condition.
(*Answer:* (b) 1.2 N, (c) 3.3 mm, 16 mWb, 32 mWb.)

2.23. An electromagnetic vibrator used in some types of electric shaver is shown in Fig. P2.23. If the reluctance of the iron is considered to be negligibly small and flux fringing is ignored, determine the horizontal and vertical forces acting on the yoke in terms of current and geometry.

2.24. The singly-excited electromechanical system shown in Fig. P2.24 is constrained to move only in the horizontal direction. Neglect winding resistance, iron reluctance, leakage and fringing. **(a)** Calculate the instantaneous value of the electromagnetic force on the movable iron member. **(b)** If the voltage

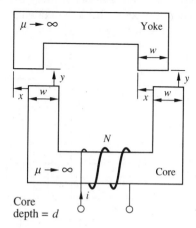

Figure P2.23

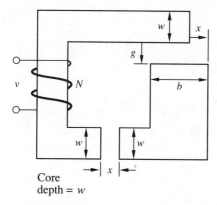

Figure P2.24

excitation is given by $v = \hat{V} \cos t$, what is the average force in terms of the applied voltage and the geometry?

Section 2.7

2.25. Determine the average value of the electrostatic force of attraction between the plates of a parallel-plate capacitor, if the capacitor is connected to a source of voltage whose value is 100 $\sin\omega t$ volts. The plate area is 100 cm^2, the dielectric has a relative permittivity of 2 and a thickness of 1 mm.
(*Answer:* 4.4×10^{-4} N.)

2.26. There is a force tending to draw the dielectric into the parallel plate capacitor as shown in Fig. P2.26. Determine the force in terms of the voltage and the parameters shown for the following two cases. **(a)** The capacitor remains connected to a constant source of potential V volts. **(b)** The capacitor is first charged to a potential V and then disconnected from the supply. (This is a case of constant charge.)

2.27. Figure P2.27 shows schematically an electrostatic voltmeter in which the electrostatic torque is balanced by the spring torque in the equilibrium position. If the spring constant is K and the capacitance varies with position θ according to the relation $C = C_0 + C_1\theta$, where C_0 and C_1 are constants, find the deflection θ as a function of the applied dc voltage V.

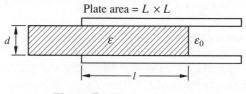

Plate area = $L \times L$

Figure P2.26

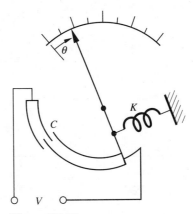

Figure P2.27

Section 2.8

2.28. Figure 2.21 shows schematically a dynamometer type ac ammeter or voltmeter, in which the electromagnetic torque is balanced by a spring torque in the equilibrium position. The coils are connected in series and carry a current $i = \hat{I} \sin \omega t$.

The self- and mutual inductances have the following values: $L_{11} = 0.4$ H, $L_{22} = 0.2$ H, and $L_{12} = 0.1 \cos\theta$ H. The restraining spring has a torque function $T = 0.64(\pi/2 - \theta) \text{N} \cdot \text{m}$ for $0 \leq \theta \leq \pi/2$. Find the rms value of the current for rotor deflections $\pi/2, 5\pi/12, \pi/3, \pi/4$, if the damping is sufficient to prevent rotor oscillations.

(*Answer:* 0, 1.32 A, 2 A, 2.74 A.)

Section 2.9

2.29. Figure P2.29 shows the cross section of a singly-excited, cylindrical solenoid, in which the plunger of mass M moves vertically in brass guide rings. The plunger is supported by a spring whose constant is K newtons per meter and whose unstretched length is l_0. A mechanical force F_m is applied to the plunger. Assume that the frictional force is linearly proportional to the plunger velocity and that the coefficient of friction is D newton seconds per meter. The coil has N turns and a resistance R ohms. A voltage v is applied to the coil terminals. The reluctance of the iron can be ignored and leakage and fringing can be considered to be negligible. Derive the electric circuit equation and the dynamic equation of motion in terms of the variables, coil current i and plunger displacement x.

2.30. Consider the electromagnetic position controller depicted in Fig. 2.27 and described in Example 2.9. Determine the frequency response about the operating conditions I and X for the case where the voltage is given a small perturbation about a steady value of 5 V. Use $D = 1$ Ns/m and then use $D = 7$ Ns/m. For a small perturbation eq. (2.9.9) becomes a linear second-order differential

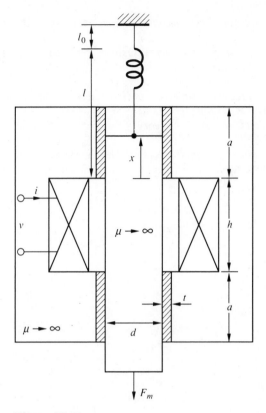

Figure P2.29

equation. The transfer function in the frequency domain is given by

$$\frac{X_1(j\omega)}{I_1(j\omega)} = \frac{K_1}{(1 - \omega^2/\omega_n^2) + j2\xi\omega/\omega_n}$$

where ω = angular frequency of the perturbation

$\omega_n = \sqrt{K/M}$ = the natural frequency

and $\xi = D/2\sqrt{KM}$ = the damping ratio.

Plot current and displacement perturbations over a frequency range of 1 to 60 Hz.

2.12. BIBLIOGRAPHY

Ellison, A. J. *Electromechanical Energy Conversion.* London: George G. Harrap & Co., Ltd., 1970.

Matsch, Leander, W. *Capacitors, Magnetic Circuits and Transformers.* Englewood Cliffs, N.J.: Prentice-Hall, Inc., 1964.

Meisel, Jerome. *Principles of Electromechanical Energy Conversion*. New York: McGraw-Hill Book Co., 1966.

Messerle, Hugo K. *Dynamic Circuit Theory*. Oxford: Pergamon Press, 1965.

Nasar, S. A. *Electric Machines and Electromechanics*. Schaum's Outline Series in Engineering. New York: McGraw-Hill Book Co., 1981.

O'Kelly, D., and S. Simmons. *Introduction to Generalized Electrical Machine Theory*. London: McGraw-Hill Publishing Co., Ltd., 1968

Say, M. G. *Introduction to the Unified Theory of Electromagnetic Machines*. London: Pitman Publishing, 1971.

Seely, S. *Electromechanical Energy Conversion*. New York: McGraw-Hill Book Co., 1962.

Smith, Steve. *Magnetic Components: Design and Applications*. New York: Van Nostrand Reinhold Co., Inc., 1985.

CHAPTER THREE

Transformers

3.1. INTRODUCTION

Transformers allow ac power to be transmitted at different voltage levels. Therefore they play an important role in most practical ac circuits, and are particularly important in power transmission and distribution systems.

In this chapter the theory and performance of transformers are described. Models are developed to aid analysis. These models, or equivalent circuits, enable the study of problems associated with the transformer as an individual device or with the transformer as one of a number of elements in a system.

A transformer is a static device — it has no moving parts. There are one, two, or more electric windings wrapped around a common core that can be air but is most usually of ferromagnetic material. Its purpose is to transfer electric energy from one or more windings by means of electromagnetic induction[1] to other windings. All these windings have a common magnetic circuit[2] and they are thus magnetically coupled.

Transformers can be grouped into categories that are based on the field of application. These groups are as follows.

Power and distribution transformers. These are used in power transmission and distribution systems and allow

1. the electric power to be generated at the most economic generator voltage level
2. power transmission at the most economic transmission voltage
3. power utilization at the most suitable or safe voltage level.

---

[1] Ac excitation of magnetic materials is reviewed in Appendix C.
[2] Magnetic circuit calculations are reviewed in Appendix A.

Instrument transformers. These are used

1. to measure high voltages and large currents with standard small-range volt-meters and ammeters
2. to transform voltages and currents to activate relays for control and protection.

Isolating transformers. These are used to electrically isolate electric circuits.

Communication transformers. These are used in electronic devices (impedance matching for maximum power transfer, isolation of dc current, etc.).

These types of transformers differ in their design because of the different specifications that have to be met, but their operating principles are the same. We will restrict our discussion mainly to power frequency transformers, such as power and distribution transformers.

3.2. IDEAL SINGLE-PHASE TRANSFORMER

An ideal transformer is a hypothetical electromagnetic device. A discussion of the ideal transformer is a useful aid to understand the physical phenomena and to derive a first approximation of the behavior of a real transformer. Also, an ideal transformer is often used as a model for a real transformer in electric network analysis.

3.2.1. Steady-state Characteristics

Consider the magnetic core and coil arrangement shown in Fig. 3.1. This represents a basic single-phase transformer. A single-phase source of power is connected to coil 1 of N_1 turns. Coil 1 is called the *primary winding.* The magnetizing current that flows

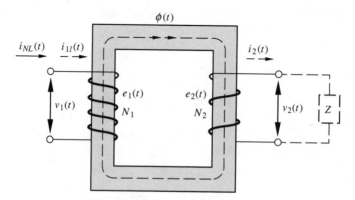

Figure 3.1 Basic transformer arrangement.

in this coil creates a magnetic flux ϕ in the core. Since the flux is time varying, emfs will be induced in any coil linking the flux. There will be an induced emf e_1 in the primary winding. Also there will be an induced emf e_2 in the coil 2 of N_2 turns. Coil 2 is called the *secondary winding*. If a load of impedance Z is connected to the secondary winding, the emf e_2 will cause a current to flow in the load. Thus power will be absorbed by the load. This power must come from the source. We cannot get out from the device any more than we put in.

A property of an ideal transformer is that it is *loss free*. This means that the windings have zero resistance. The voltage equations are

$$v_1(t) = \frac{d\lambda_1(t)}{dt} = -e_1(t) \tag{3.2.1}$$

and

$$e_2(t) = -\frac{d\lambda_2(t)}{dt} = v_2(t) \tag{3.2.2}$$

where v_1, λ_1, and e_1 are the instantaneous values of the source voltage, flux linkage, and induced emf, respectively, of the primary winding.

A second property of an ideal transformer is that all the flux ϕ links all turns of all windings. It follows that

$$e_1 = -\frac{d\lambda_1(t)}{dt} = -N_1 \frac{d\phi(t)}{dt} \tag{3.2.3}$$

and

$$e_2 = -\frac{d\lambda_2(t)}{dt} = -N_2 \frac{d\phi(t)}{dt}. \tag{3.2.4}$$

Consequently

$$\frac{e_1(t)}{e_2(t)} = \frac{N_1}{N_2} \triangleq a \tag{3.2.5}$$

where the constant a is called the *ratio of transformation*.

This last equation applies to the ratio of the instantaneous values of the induced voltages. The significance of this is that the wave shapes of the two induced voltages are the same since a is a constant. Therefore we can also write the ratio of the rms values of the voltages as

$$\frac{V_1}{V_2} = \frac{E_1}{E_2} = \frac{N_1}{N_2} = a. \tag{3.2.6}$$

A single-phase source voltage is normally sinusoidal. Therefore the induced emfs are also sinusoidal. The rms value of induced sinusoidal voltages[3] are given by

$$V_1 = E_1 = 4.44\hat{\Phi}fN_1 \qquad (3.2.7)$$

and

$$V_2 = E_2 = 4.44\hat{\Phi}fN_2. \qquad (3.2.8)$$

Thus, if N_1 and the frequency f are constant, the flux $\phi(t)$ is determined by the applied voltage $v_1(t)$.

If a load impedance Z is connected to the terminals of the secondary winding, the induced voltage $e_2(t)$ causes a current $i_2(t)$ to flow through the turns of this winding. The direction of $i_2(t)$ is, according to Lenz's law, such that the flux resulting from the mmf $N_2 i_2(t)$ will oppose the flux $\phi(t)$. Thus $i_2(t)$ tends to modify the flux in the magnetic circuit. The original flux $\phi(t)$ has to be maintained, because the applied voltage is not changed. This can only be accomplished by an additional current component $i_{1l}(t)$ in the primary winding, which creates a magnetomotive force $N_1 i_{1l}(t)$ to cancel the demagnetizing effect of the secondary current $i_2(t)$. Two conclusions can be made. First, the primary current consists of two components; thus

$$i_1(t) = i_{NL}(t) + i_{1l}(t) \qquad (3.2.9)$$

where $i_{1l}(t)$ is called the *load component* of the current in winding 1 and $i_{NL}(t)$ is called the *no-load component* that occurs when the secondary winding has no load. Second, there has to be a balance of mmfs such that

$$N_1 i_{1l}(t) = N_2 i_2(t). \qquad (3.2.10)$$

We now introduce the third property of an ideal transformer. The value of the permeability of the magnetic circuit of an ideal transformer is infinitely high. The exciting current[4] consists of two components, the magnetizing current component i_m and the power loss component i_{h+e}. An ideal transformer is loss free; thus i_{h+e} is zero and from the third property it follows that i_m is also zero. Thus, $i_{NL} = 0$ and the current in winding 1 is $i_1(t) = i_{1l}(t)$. Hence we can write for an ideal transformer that

$$N_1 i_1(t) = N_2 i_2(t) \qquad (3.2.11)$$

or

$$\frac{i_1(t)}{i_2(t)} = \frac{N_2}{N_1} = \frac{1}{a}. \qquad (3.2.12)$$

In terms of rms values

$$\frac{I_1}{I_2} = \frac{N_2}{N_1} = \frac{1}{a}. \qquad (3.2.13)$$

---

[3] See section C3 of Appendix C for the derivation.
[4] This is reviewed in Section C6 of Appendix C.

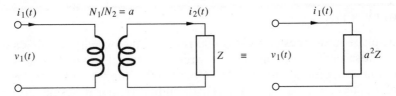

Figure 3.2 Impedance transformation.

The product of the instantaneous voltage and current ratios provides the relationship

$$\frac{v_1(t)}{v_2(t)} \times \frac{i_1(t)}{i_2(t)} = a \times \frac{1}{a} = 1 \tag{3.2.14}$$

or

$$v_1(t)i_1(t) = v_2(t)i_2(t). \tag{3.2.15}$$

Thus, in the case of a loss-free, ideal transformer, the power input equals the power output at all instants of time. This is the *power invariance principle* and means that the volt-amperes are conserved.

We can also divide the instantaneous voltage ratio by the instantaneous current ratio. We get

$$\frac{v_1(t)/v_2(t)}{i_1(t)/i_2(t)} = a^2 \tag{3.2.16}$$

or

$$\frac{v_1(t)}{i_1(t)} = a^2 \frac{v_2(t)}{i_2(t)} = a^2 Z \tag{3.2.17}$$

where Z is the load-impedance connected to the terminals of winding 2. Therefore, if an ideal transformer is connected between a source and a load impedance Z, the apparent load impedance presented to the source is a^2 times the actual load impedance. This equivalence is schematically shown in Fig. 3.2. We say that we have transferred the load impedance from the secondary winding side of the transformer to the primary winding side. This is an important development which is used in the discussion of real transformers.

The ideal transformer has been shown to be characterized by the fact that the power output is equal to the power input. The voltage ratio is equal to the turns ratio of the windings; the current ratio is the inverse of the turns ratio, and an impedance connected across the secondary winding can be transferred to the primary side using a factor equal to the square of the turns ratio.

EXAMPLE 3.1

A single-phase ideal transformer has 400 turns on the primary winding and 1000 turns on the secondary winding. The active cross-sectional area of the core is 60 cm². If the primary winding is connected to a 60-Hz, 500-V supply, find **(a)** the maximum value of the flux density in the core and **(b)** the induced emf in the secondary winding.

Solution

(a) The transformer voltage equation for the primary winding is
$V_1 = E_1 = 4.44 \hat{\Phi} f N_1$.

Therefore $\hat{\Phi} = \dfrac{500}{4.44 \times 60 \times 400} = 4.7 \times 10^{-3}$ Wb.

The maximum value of the flux density in the core is

$$B_{max} = \frac{\hat{\Phi}}{A} = \frac{4.7 \times 10^{-3}}{60 \times 10^{-4}} = 0.783 \text{ T}.$$

(b) The induced voltage in the secondary winding is

$$E_2 = V_2 = \frac{N_2}{N_1} V_1 = \frac{1000}{400} \times 500 = 1250 \text{ V}.$$

3.3. REAL SINGLE-PHASE TRANSFORMER

A real transformer is different from an ideal transformer for the following reasons.

1. A real transformer has losses. The windings have resistance that causes I^2R losses. Hysteresis and eddy current losses occur in the magnetic circuit, that is, in the core of the transformer. The power output is, therefore, not quite equal to the power input.
2. The permeability of the core material is finite. Therefore a real transformer has a measurable exciting current, which has a magnetizing component as well as a loss component. Since the exciting current is the only current that flows when no load is connected to the transformer, this current is also called the no-load current.
3. The windings have capacitance, but at normal operating conditions and low frequencies (less than say 1 kHz) the effect of winding capacitance is negligible.
4. Not all flux is linked with all turns of all windings. There are flux components that are linked with one winding but not with another. This has an important effect on the operating characteristics of transformers and requires a more detailed discussion.

We will study the effects of the above to determine a more accurate picture of the performance of a transformer than is obtained from an analysis of the ideal transformer.

3.3.1. Effect of Winding Resistance

Real transformers have winding resistance; its effect is depicted as a lumped parameter R. If the secondary winding is left as an open circuit, expressions for the input and output voltages are

$$v_1 = i_{NL}R_1 + N_1 \frac{d\phi}{dt} \quad \text{and} \quad e_2 = -N_2 \frac{d\phi}{dt}. \tag{3.3.1}$$

An initial assumption is that the magnetization curve is linear. Hence, all variables in the above equations are sinusoidal.

It frequently occurs that transformers used in communication networks do not have sinusoidal applied voltages. For such cases the input signal can be resolved into a Fourier series of sinusoidal terms. Each sinusoid can be considered separately and on transformation each term remains a sinusoid. However, the waveshape of the resultant transformed voltage will not resemble the waveshape of the input voltage, since each term will undergo a different phase shift because of the presence of the primary winding resistance.

Upon relaxation of the assumption of linearity of the magnetic circuit, even if v_1 is varying sinusoidally, the other individual terms in eq. (3.3.1) cannot be sinusoidal. The resistive voltage drop is not sinusoidal because i_{NL} cannot be sinusoidal.[5] Therefore, $N_1 d\phi/dt$ cannot be sinusoidal either. Since the output signal e_2 is directly proportional to $N_1 d\phi/dt$, this output voltage will not vary sinusoidally with time. Thus, the nonlinear characteristic of the core contributes considerably to the distortion of the output waveform.

To reduce the distortion to a minimum, transformers for communication networks must operate at a maximum flux density value well below the knee of the saturation curve. With such low values of flux density, the dynamic hysteresis loop is narrow and a virtually linear relation between ϕ and i_{NL} will exist. Further, the winding resistance must be kept low.

The use of low maximum flux-density values in power or distribution transformers leads to a prohibitive increase in cost and size of these transformers. For optimum design one has to maintain the highest maximum flux-density values. The exciting current i_{NL} will then be peaked and asymmetrical. However, the winding resistance is made as small as economically feasible. This also reduces the heat produced by the ohmic losses and increases the efficiency.[6] The non-sinusoidal resistive voltage drop due to the exciting current in the primary winding has, therefore, a negligible effect on the waveforms of the induced voltages in a practical design.

---

[5] See Fig. C9 in Appendix C.
[6] The resistive voltage drop caused by the exciting current in the primary winding of a power transformer is of the order of 0.01% of the rated voltage.

3.3.2. Components of Magnetic Flux

To introduce this topic we consider three concentrated coils that are at some distance from each other in air. This is shown in Fig. 3.3. Coil 1 carries a current i_1, which produces a flux ϕ_{11}. This flux links all turns of coil 1 and the total flux linkage of coil 1 is $\lambda_{11} = N_1\phi_{11}$.

Figure 3.3 shows three types of flux lines which are labeled a, b, and c. We see that only part of the flux ϕ_{11} links the turns of coil 2. Typical flux lines that link coil 2 are those labeled a and b. That part of the flux ϕ_{11} that links coil 2 we call ϕ_{m12}; this is a mutual flux linking both coil 1 and coil 2. That part of the flux ϕ_{11} that does not link the turns of coil 2 and that is represented by a typical flux line such as c is called the leakage or stray flux ϕ_{l12} of coil 1 with respect to coil 2. We have the relation

$$\phi_{11} = \phi_{m12} + \phi_{l12}. \tag{3.3.2}$$

Similarly, only a part of ϕ_{11} links with coil 3. This part ϕ_{m13} is represented by the typical flux line a. Typical flux lines b and c are not linked with coil 3 but are part of the leakage flux ϕ_{l13} of coil 1 with respect to coil 3. We can now write

$$\phi_{11} = \phi_{m13} + \phi_{l13}. \tag{3.3.3}$$

When we speak of leakage flux we must state which coils we consider, since ϕ_{l12} is not equal to ϕ_{l13}.

If coil 2 carries current, we can define mutual flux components ϕ_{m21} and ϕ_{m23} of coil 2 with coil 1 and coil 3, respectively. We can also define the leakage fluxes ϕ_{l21} and ϕ_{l23} as those parts of the flux ϕ_{22} that are linked with coil 2 but not with coil 1 and coil 3, respectively. In this case the relationships are

$$\phi_{22} = \phi_{m21} + \phi_{l21} \quad \text{and} \quad \phi_{22} = \phi_{m23} + \phi_{l23}. \tag{3.3.4}$$

When the coils are situated in air and both coil 1 and coil 2 carry currents, we can

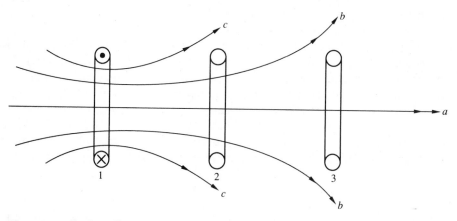

Figure 3.3   Leakage flux.

write, according to the principle of superposition,

$$\phi_1 = \phi_{11} + \phi_{m21} = \phi_{m12} + \phi_{l12} + \phi_{m21} \tag{3.3.5}$$

$$\phi_2 = \phi_{22} + \phi_{m12} = \phi_{m21} + \phi_{l21} + \phi_{m12} \tag{3.3.6}$$

and

$$\phi_3 = \phi_{m13} + \phi_{m23} \tag{3.3.7}$$

where ϕ_1, ϕ_2, and ϕ_3 are the total fluxes linking coils 1, 2, and 3, respectively. The induced voltages in the coils are

$$e_1 = -N_1 \frac{d\phi_1}{dt}, \quad e_2 = -N_2 \frac{d\phi_2}{dt} \quad \text{and} \quad e_3 = -N_3 \frac{d\phi_3}{dt} \tag{3.3.8}$$

and obviously depend on the leakage flux components. Therefore the concept of leakage flux is an important one in all cases where magnetically coupled circuits are involved.

The foregoing simple description of flux components applies to all magnetic field devices. Since no material or medium is a perfect magnetic insulator, flux will never be restricted entirely to the main ferromagnetic circuit.

Let us now consider the flux components in a real transformer. To simplify the discussion we will restrict ourselves to a two-winding transformer with cylindrical core-type windings. One leg of the transformer carrying two concentric windings is shown in Fig. 3.4. First consider the case that only the inner winding, winding 1, carries current. In a real transformer a time-varying and complex flux distribution will be set up. The distribution is determined by the height and width of the coil, the

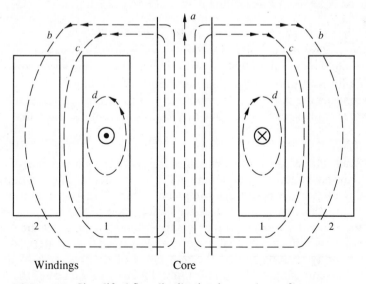

Figure 3.4 Simplified flux distribution in a real transformer.

neighboring ferromagnetic core and the presence of other iron parts required for constructional purposes.

We simplify matters by only considering typical flux lines that are representative of the actual flux distribution. These typical lines, labeled a, b, c, and d, are shown in Fig. 3.4. Their direction follows from the assumed current direction in winding 1 and is indicated. Flux lines such as a, b, and c link all the turns of winding 1. However, lines such as d link only part of the turns of winding 1. We conclude that not all turns of winding 1 link the same amount of flux. From an operational point of view, however, we are only interested in the total flux linkage λ_{11} of coil 1 with its own flux, since it is the time variation of this total flux linkage which determines the induced voltage. We may, for analytical purposes, introduce a hypothetical flux ϕ_{11}, which is linked by each and every turn of winding 1. This equivalent or average flux ϕ_{11} has to satisfy $\lambda_{11} = N_1\phi_{11}$, because in that case

$$e_1 = -\frac{d\lambda_{11}}{dt} = -N_1\frac{d\phi_{11}}{dt}. \tag{3.3.9}$$

Flux lines such as a, which constitute the main part of the total flux, are linked by all turns of winding 2. Only part of the turns of winding 2 link flux lines such as b. Flux lines such as c and d are not linked at all by any turn of winding 2. We conclude that differing amounts of flux link the different turns of winding 2. Thus, the mutual flux linkage is not sharply defined. However, for analytical purposes we can again introduce an equivalent mutual flux component ϕ_{m12} that links all the turns of winding 2. The voltage induced in winding 2 is determined by the time rate of change of the total flux linkage λ_{m12}. It follows that ϕ_{m12} has to satisfy $\lambda_{m12} = N_2\phi_{m12}$.

Flux lines such as b, c, and d contribute to leakage flux which is different for different turns. But having defined the equivalent flux components ϕ_{11} and ϕ_{m12} we can define an equivalent leakage flux component ϕ_{l12} such that

$$\phi_{11} = \phi_{m12} + \phi_{l12}. \tag{3.3.10}$$

We can then replace the actual flux pattern by a simplified equivalent one, shown in Fig. 3.5, in which the mutual flux ϕ_{m12} is restricted to the magnetic circuit and the leakage flux links all turns of winding 1 and none of winding 2.

Now consider the case that only winding 2 carries current. In a similar fashion we can introduce equivalent flux components ϕ_{22}, ϕ_{m21}, and ϕ_{l21} such that

$$\lambda_{22} = N_2\phi_{22}, \quad \lambda_{m21} = N_1\phi_{m21}, \quad \text{and} \quad \phi_{22} = \phi_{m21} + \phi_{l21}. \tag{3.3.11}$$

Almost all of the actual mutual flux in an iron-core transformer is confined to the core due to the high relative permeability of the core material. The flux in the core is, in general, not linearly proportional to the magnetomotive force that produces this flux because of the nonlinear magnetic characteristic of the iron. However, the leakage flux path is for a considerable length through air or oil or the copper and insulation of the windings. The relative permeability of all these materials is constant and close to unity. It follows that the reluctance of the iron part of the path is negligibly small compared with the constant reluctance of the rest of the path. As a result of this, the reluctance of the leakage flux path can be considered to be constant.

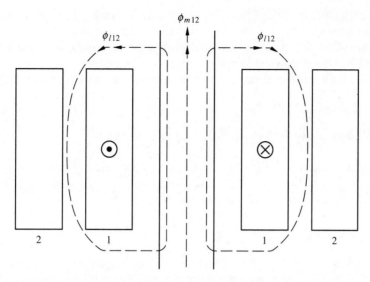

Figure 3.5 Equivalent flux components.

Therefore, in spite of the magnetic nonlinearity of the ferromagnetic core, the leakage flux is very nearly directly proportional to the current or magnetomotive force by which it is produced. This is an important property of leakage flux paths. It simplifies the analytical treatment of those devices that have ferromagnetic circuits such as transformers.

In general, there are currents in both windings. The resultant magnetic field distribution at any moment is determined by the instantaneous values of the individual currents, the level of saturation, and the physical arrangements of the core, windings and iron structural components. It is convenient to resolve this complex field distribution into the equivalent flux components produced by each current separately. These component fluxes can be combined in two different ways, each of which will lead to a different method of analysis of a two-winding transformer.

The first method assumes linearity of the magnetic circuit; that is, no saturation of the magnetic circuit occurs. In this case the principle of superposition can be applied to the component fluxes. This method leads to the classical *circuits* point of view of magnetically coupled circuits and is useful for analyzing transient phenomena. If it is applied to electromechanical energy conversion devices, this approach is called the *dynamic circuit analysis* of electromechanical devices. This method was used in Chapter 2.

The second method of combining component fluxes leads to the conventional treatment of electromagnetic devices. This approach allows physical interpretation of the devices in terms of resultant magnetic fluxes. This method is most suitable for the study of the steady-state behavior of electromechanical devices and takes saturation effects into consideration.

The terminal characteristics of a device as determined by each method are equivalent for a specific operating point on the saturation characteristic of the magnetic circuit of the device. Thus, these methods are complementary to each other and we should be able to convert from one approach to the other.

We use the second and more common method of analysis in this chapter.

3.3.3. Steady-state Behavior of the Transformer

Since a power system's voltages and currents vary sinusoidally with time, we can use phasors to represent these quantities and develop an equivalent circuit to model the steady-state behavior of the practical transformer.

A. TRANSFORMER ON NO LOAD

The no-load condition of a single-phase or two-winding transformer means that the primary winding is excited by a constant rms voltage and the secondary winding is open-circuited. There is no load across the secondary terminals so that there is no current flowing in the secondary winding.

The primary terminal voltage V_1 will cause an rms current I_{NL} to flow in the primary winding. This current is called the no-load current. A conventional circuit analysis would define I_{NL} to be limited by the reactance and resistance of the coil. Here the no-load current $\mathbf{I}_{NL}$ is defined as the phasor sum of two components $\mathbf{I}_m$ and $\mathbf{I}_{h+e}$. I_m is the current that causes the mmf $I_m N_1$ to produce a flux ϕ, which links the two windings and it is this flux that induces the emfs E_1 and E_2. The component I_{h+e} results from the losses in the iron core.[7] Figure 3.6 shows an ideal transformer circuit with an admittance Y_o to model the magnetization and the core losses of an actual transformer.

Resistance of a winding is taken into account by assuming a resistanceless winding in series with an external resistance R_1 for the primary winding and R_2 for the secondary winding. The resistance R_1 must be placed between the terminal and the node of the magnetizing admittance because the no-load current I_{NL} flows through the winding resistance and produces a potential drop $I_{NL}R_1$. Thus, we see that the applied voltage V_1 is not equal to the winding induced emf E_1 in the real transformer. Further, we have separated another power loss $I_{NL}^2 R_1$ caused by energy dissipation in the winding resistance. This loss is called the *ohmic* loss or *copper* loss.

Because the magnetic core is not perfect ($\mu \neq \infty$) and the windings are not concentrated, all the magnetic flux is not confined to the core. Some of the flux does not link both windings. Using the results of section 3.3.2 let flux ϕ_m which links both the primary and secondary windings be the mutual flux. Flux which links only the

---

[7] In Appendix C we discuss the representation of a coil wound on a ferromagnetic circuit and obtain an equivalent electric circuit to account for losses and magnetization. See Fig. C13.

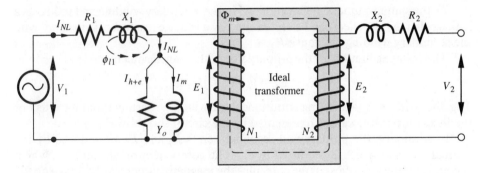

Figure 3.6 Resistance and leakage reactance.

primary turns is called the primary leakage flux ϕ_{l1}. If a current flowed in the secondary winding, there would be a secondary leakage flux ϕ_{l2} also.

We want to preserve the concept of the ideal transformer. This can be done if we can use an external element to represent the effect of the leakage flux just as we used R_1 to represent the dissipative losses in the primary winding. A stipulation of the ideal transformer is that

$$\frac{E_1}{E_2} = \frac{N_1}{N_2} \tag{3.3.12}$$

and this can hold true only if the sinusoidally varying mutual flux ($\phi_m = \hat{\Phi}_m \sin\omega\, t$) links the ideal windings so that[8]

$$E_1 = 4.44\hat{\Phi}_m f N_1 \qquad \text{and} \qquad E_2 = 4.44\hat{\Phi}_m f N_2. \tag{3.3.13}$$

This means that the voltage applied to the ideal transformer has been reduced by an amount given by

$$E_{l1} = 4.44\hat{\Phi}_{l1} f N_1. \tag{3.3.14}$$

The flux ϕ_{l1} induced the emf E_{l1}. Therefore, E_{l1} is an inductive voltage. Since we are concerned with sinusoids, the voltage is a reactive voltage and can be replaced by an equivalent voltage drop; that is, the voltage is a product of a current and a reactance.

Since the leakage flux ϕ_{l1} is proportional to the current in the primary winding, we can write the expression

$$\mathbf{E}_{l1} = j\mathbf{I}_{NL}X_1 \tag{3.3.15}$$

where $X_1 = \omega L_{l1}$ and $L_{l1} = N_1\phi_{l1}/I_{NL} = \text{constant}$. L_{l1} is independent of saturation. Thus, the element X_1, which is termed the primary leakage reactance, can be placed in series with the winding resistance R_1 to create an effective voltage reduction due to the leakage flux.

---

[8] See eq. (C3.9).

By the same argument, if current flows in the secondary winding and produces a leakage flux ϕ_{l2}, its effect can be represented by a leakage reactance X_2 in series with the secondary winding resistance R_2.

The voltage equation for the primary side of the transformer on no load becomes

$$\mathbf{V}_1 + \mathbf{E}_1 = \mathbf{I}_{NL}R_1 + j\mathbf{I}_{NL}X_1. \tag{3.3.16}$$

Together with the magnetizing admittance, the winding resistance and the effect of the flux components can be represented by the circuit shown in Fig. 3.6.

In summary, the primary and secondary emfs E_1 and E_2 are induced by the mutual flux ϕ_m, which is produced by I_m. The component of current I_{h+e}, which accounts for the power loss in the core, plus the magnetizing current I_m, make up the no-load current I_{NL}. The applied voltage V_1 is balanced by voltage drops due to the leakage reactance and winding resistance, and the ideal transformer induced emf E_1.

EXAMPLE 3.2

A single-phase transformer has 450 turns on the primary winding. If it is connected to a 120-V, 60-Hz supply, the no-load current is 1.5 A and the no-load power input P_{NL} is 82 W. Neglect the leakage reactance and winding resistance and calculate (a) the core-loss current, (b) the magnetizing current, (c) the maximum core flux, and (d) the magnetizing admittance, impedance, conductance, and susceptance of the equivalent circuit.

Solution

(a) Core loss current $I_{h+e} = \dfrac{P_{NL}}{V_1} = \dfrac{82}{120} = 0.68$ A.

(b) Magnetizing current $I_m = (I_{NL}^2 - I_{h+e}^2)^{1/2} = (1.5^2 - 0.68^2)^{1/2} = 1.34$ A.

(c) $V_1 = 4.44\hat{\Phi}fN_1$. Therefore $\hat{\Phi} = \dfrac{120}{4.44 \times 450 \times 60} = 1.0 \times 10^{-3}$ Wb.

(d) Magnetizing admittance $Y_o = I_{NL}/V_1 = 1.5/120 = 0.0125$ siemens.
Magnetizing impedance $Z_m = V_1/I_{NL} = 1/Y_0 = 80\ \Omega$.
Core loss conductance $G_m = P_{NL}/V_1^2 = 0.0057$ siemens.
Susceptance $B_m = (Y_0^2 - G_m^2)^{1/2} = 0.011$ siemens.

B. TRANSFORMER ON LOAD

If a load of impedance Z is placed across the secondary terminals of the transformer, a current I_2 will flow due to the induced emf E_2. The mmf I_2N_2 produces a flux, which, by Lenz's law, tends to oppose the mutual flux ϕ_m. From the conceptual view the only way that ϕ_m can remain approximately constant, when I_2 flows in the secondary winding, is if the secondary mmf I_2N_2 is balanced by a primary mmf $I_{1l}N_1$ similar to the case discussed in section 3.2.1; that is

$$\mathbf{I}_{1l}N_1 + \mathbf{I}_2N_2 = 0. \tag{3.3.17}$$

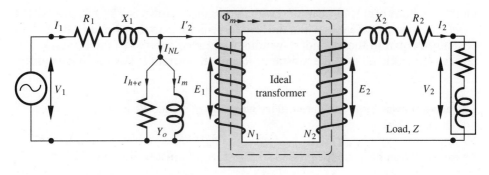

Figure 3.7   Transformer on load.

Consequently, the input current I_1 to the primary winding must contain a component I_{NL} for magnetization and power loss in the core, and a component I_{1l} to balance the secondary mmf of the ideal transformer. Thus

$$I_{1l} = \frac{N_2}{N_1} I_2 \triangleq I_2' \quad \text{and} \quad \mathbf{I}_1 = \mathbf{I}_{NL} + \mathbf{I}_2'. \tag{3.3.18}$$

It is for this reason that the magnetizing admittance Y_o is placed in parallel with the ideal primary winding. Only with two branches could I_{NL} and I_2' be separated. These currents are shown on the schematic circuit of Fig. 3.7.

3.4. TRANSFORMER EQUIVALENT CIRCUITS

To make transformer performance calculations it becomes convenient to use an equivalent circuit model which contains only the implicit concept of the ideal transformer. Figure 3.7 illustrates the ideal transformer circuit. Since there are no losses in the ideal transformer, the power input must equal the power output. This suggests that we can eliminate the ideal transformer from the circuit by referring quantities of the secondary side to the primary side. The referral is such that the power associated with the new quantities is the same as the power associated with the old quantities. In principle this is a mathematical transformation of variables while maintaining an invariance of power.

With the aid of equivalent circuits we can study problems associated with the transformer as an individual device or with the transformer as one of a number of elements in a system. In the former case detailed equivalent circuits are employed, whereas in the latter case a simplified equivalent circuit may suffice.

3.4.1. Exact Equivalent Circuit

An exact equivalent circuit must have resistances and reactances such that calculations produce values of input power, current and power factor the same as that found for the real transformer. We can start with the circuit as shown in Fig. 3.7 and proceed to *refer* the parameters R_2, X_2 and the load impedance Z to the primary side. Let R_2' be defined equal to the secondary-winding resistance referred to the primary. The power loss associated with the resistance R_2' and the current I_2' flowing in it would be

$$P'_{LOSS} = I_2'^2 R_2'. \tag{3.4.1}$$

The actual power loss in the secondary winding resistance R_2 is

$$P_{LOSS} = I_2^2 R_2. \tag{3.4.2}$$

As the power on both sides of the ideal transformer must be the same,

$$P'_{LOSS} = P_{LOSS}, \quad \text{so} \quad I_2'^2 R_2' = I_2^2 R_2, \quad \text{or} \quad R_2' = \left(\frac{I_2}{I_2'}\right)^2 R_2. \tag{3.4.3}$$

Since $I_2' N_1 = I_2 N_2$

$$R_2' = \left(\frac{N_1}{N_2}\right)^2 R_2 = a^2 R_2. \tag{3.4.4}$$

Thus, the secondary resistance referred to the primary side is the actual secondary resistance times the square of the turns ratio.

Based on reactive power and complex power the same relations hold for reactance and impedance referred to the primary; that is

$$X_2' = a^2 X_2 \quad \text{and} \quad Z' = a^2 Z. \tag{3.4.5}$$

How do we refer secondary voltages and currents to the primary side? We know that the referred secondary current I_2' must produce the same mmf as the actual secondary current I_2. That is, $I_2' N_1 = I_2 N_2$. But $I_{1l} N_1 = I_2 N_2$, so the primary load component of current and the referred secondary current are the same.

A secondary emf referred to the primary E_2' is related to the actual secondary winding emf by Faraday's law in the steady state; that is

$$E_2' = 4.44 \hat{\Phi}_m f N_1 \quad \text{and} \quad E_2 = 4.44 \hat{\Phi}_m f N_2 \tag{3.4.6}$$

so

$$E_2' = E_2 \frac{N_1}{N_2} = a E_2 = E_1. \tag{3.4.7}$$

Also

$$V_2' = a V_2 \tag{3.4.8}$$

since

$$V_2' = I_2' Z' = \frac{I_2}{a} a^2 Z = a I_2 Z = a V_2. \tag{3.4.9}$$

Having referred all the secondary quantities to the primary in this way, so that the same effects are produced in the primary as would be produced in the secondary, we

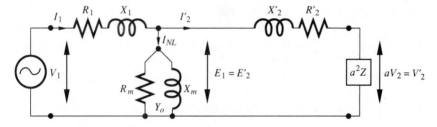

Figure 3.8 Exact equivalent circuit referred to the primary side.

can dispense with the coils of the ideal transformer because it just offers short-circuited terminals on the secondary side. The exact equivalent circuit referred to the primary is shown in Fig. 3.8.

From inspection of Fig. 3.7 we can write the voltage equations of the primary and secondary circuits

$$V_1 = I_1(R_1 + jX_1) - E_1 \tag{3.4.10}$$

and

$$V_2 = E_2 - I_2(R_2 + jX_2). \tag{3.4.11}$$

With the relations

$$I'_2 = I_2/a, \quad E_1 = E'_2 = aE_2, \quad V'_2 = aV_2$$
$$R'_2 = a^2R_2, \quad X'_2 = a^2X_2 \quad \text{and} \quad Z' = a^2Z$$

we can write the equation for the applied voltage

$$V_1 = I_1(R_1 + jX_1) + I'_2a^2(R_2 + jX_2) + V'_2. \tag{3.4.12}$$

This equation could have been written down by inspection of Fig. 3.8. It checks the validity of the equivalent circuit.

EXAMPLE 3.3

A single-phase 132/22-kV, 12,000-kVA transformer has primary and secondary winding leakage impedances of $(5 + j60)$ ohms and $(0.1 + j1.5)$ ohms, respectively, and the magnetizing admittance referred to the primary side is $(3 - j6) \times 10^{-6}$ siemens. Determine the input voltage, current, and power factor using the exact equivalent circuit, if the transformer delivers full load at rated voltage and 0.8 power factor lagging.

Solution
The ratio 132/22 kV provides the values of the rated voltages of the primary and secondary windings, respectively, at no load and therefore provides the value of the nominal turns ratio. Full load means rated volt-amperes, and in this case, is 12,000 kVA.

Let the load voltage be taken as reference.

$$\mathbf{V}_2' = a\mathbf{V}_2 = \frac{132}{22} \times 22000 \angle 0° = 132{,}000 \angle 0° \text{ V.}$$

$$R_2' + jX_2' = a^2(R_2 + jX_2) = (132/22)^2(0.1 + j1.5) = (3.6 + j54)\ \Omega.$$

$$\mathbf{I}_2' = \frac{12000}{132}(0.8 - j0.6) = 72.73 - j54.55 = 90.9 \angle -36.9° \text{ A.}$$

The voltage across Y_0 is

$$\mathbf{E}_2' = \mathbf{V}_2' + \mathbf{I}_2'(R_2' + jX_2') = 132000 + (72.73 - j54.55)(3.6 + j54) \text{ V.}$$

That is, $\mathbf{E}_2' = 135208 + j3731 = 135{,}259 \angle 1.58° \text{ V.}$

$$\mathbf{Y}_0 = (3 - j6) \times 10^{-6} = 6.71 \times 10^{-6} \angle -63.43° \text{ siemens.}$$

$$\mathbf{I}_{NL} = \mathbf{E}_2'\mathbf{Y}_0 = 135{,}259 \angle 1.58° \times 6.71 \times 10^{-6} \angle -63.43° = (0.428 - j0.8) \text{ A.}$$

Input current $\mathbf{I}_1 = \mathbf{I}_2' + \mathbf{I}_{NL} = 73.16 - j55.4 = 91.76 \angle -37.13° \text{ A.}$
The input voltage $\mathbf{V}_1$ is

$$\begin{aligned}\mathbf{V}_1 &= \mathbf{E}_2' + \mathbf{I}_1(R_1 + jX_1)\\ &= (135208 + j3731) + (73.16 - j55.34)(5 + j60) = 138{,}120 \angle 3.2° \text{ V.}\end{aligned}$$

The phase angle difference between the input voltage and current is $40.33°$. Therefore, the input power factor is $\cos 40.33° = 0.76$.

3.4.2. Transformer Frequency Response

Transformers used in communication and control systems do not normally operate at constant voltage or frequency. The analysis of their performance should be based on the analysis of the complete equivalent electric circuit model. However, it is advantageous to reduce the amount of work by simplifying the equivalent circuit, if this can be done by justifiable modifications. A modification is justified when its effect on the accuracy of the performance prediction is negligible. The design values of the elements of the exact equivalent circuit and the specific operating values of currents, voltages, and frequency determine what simplifications can be made. Whatever their specific purpose, communication transformers are required to transform an input signal from one voltage level to another without distortion of the input signal waveform. This implies that for each Fourier series component of the waveform, the ratio of primary and secondary voltage and their phase shift must be the same.

Since a wide frequency spectrum has to be considered, communication transformers are designed to have a relatively low operating core flux density in the midfrequency range. Because of this and the low currents involved, the windings consist of many turns of a small cross-section conductor and have a relatively high

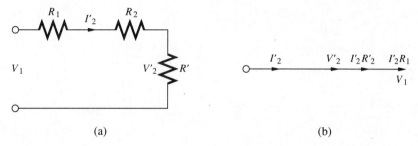

Figure 3.9 Midfrequency range. (a) Equivalent circuit, (b) phasor diagram.

resistance. Since the voltages are low, both windings are compactly wound. This results in very small leakage fluxes and therefore low-leakage inductances. In order to gain insight into the performance of a transformer as a function of frequency, three frequency ranges will be considered.

Midfrequency Range (1 kHz)
Since the flux density is low and thin laminations (approximately 0.08 mm) are used, the iron loss is extremely small. Further, the magnetizing current required to produce the low flux density is negligibly small on account of the large number of turns. Due to the small current drawn, the core admittance can be omitted from the circuit diagram. The leakage reactances are small compared with the winding resistances. The approximate equivalent circuit of the communication transformer with a resistive load reduces to that depicted in Fig. 3.9.

Low-frequency Range (30 to 60 Hz)
Assuming the same input voltage value as before, the flux density in the core is considerably higher and the magnetizing current cannot be neglected. The iron loss is still negligible; the effect of high flux density is offset by the low frequency. The leakage reactances, being proportional to frequency, are even smaller than in the previous case in which they were neglected. Thus for low frequencies the equivalent circuit is illustrated in Fig. 3.10.

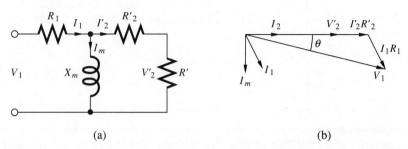

Figure 3.10 Low-frequency range. (a) Equivalent circuit, (b) phasor diagram.

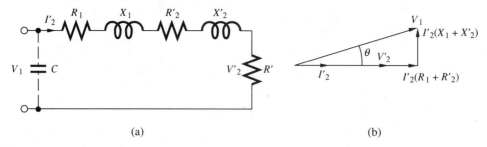

(a)   (b)

Figure 3.11 High-frequency range. (a) Equivalent circuit, (b) phasor diagram.

High-frequency Range (10 kHz)

The use of a high frequency makes the core circuit admittance so low that it can be omitted from the circuit diagram. The leakage reactances now become high enough that they cannot be neglected. Figure 3.11 shows the approximate equivalent circuit for high frequency operation of a resistively loaded communication transformer.

A. RESPONSE CHARACTERISTICS

The ratio of secondary to primary voltage as a function of frequency (frequency characteristic) and the phase angle θ between these voltages, as a function of frequency (phase characteristics), can be deduced qualitatively from the approximate circuits.

At very low frequencies X_m is small and shunts the load. The resistive voltage drop across R_1 is large; thus V_2' is considerably less than V_1. V_2' leads V_1. As the frequency increases, the value of X_m increases, the ratio of V_2' and V_1 increases and V_2' becomes more in phase with V_1. In the mid-frequency range the voltage ratio has its highest value and the phase shift is zero. The flat parts of the frequency and phase characteristics, shown in Fig. 3.12 (a is the frequency characteristic and b is the phase characteristic), indicate a distortion-free transformation of a complex input signal. At

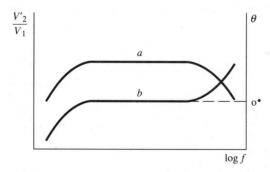

Figure 3.12 Frequency and phase characteristics.

higher frequencies the ratio of secondary to primary voltage decreases again as a result of the increased values of leakage reactance. The output voltage lags the input voltage. At very high frequencies the interwinding and turn-to-turn capacitance can be treated as a lumped capacitor C and inserted across the terminals in the circuit diagram. Resonance can occur at higher frequencies and causes a further reduction of V'_2.

It can be concluded that the presence of low- and high-frequency components of an input signal will cause distortion of the output signal.

3.4.3. Approximate Equivalent Circuit of a Power Transformer

Reasonable approximations can be used to simplify the equivalent circuit model of a power transformer. Thereby, the amount of work involved in the analysis of its performance or the performance of a system, of which the transformer is a component, is reduced. In the case of power system transformers the frequency is constant, either 50 Hz or 60 Hz, and the primary voltage is normally constant. The voltage equation of the primary winding is

$$\mathbf{V}_1 = \mathbf{I}_1(R_1 + jX_1) - \mathbf{E}_1. \tag{3.4.13}$$

Since $\mathbf{I}_1 = \mathbf{I}_{NL} + \mathbf{I}'_2$, we can write

$$\mathbf{V}_1 = \mathbf{I}_{NL}(R_1 + jX_1) + \mathbf{I}'_2(R_1 + jX_1) - \mathbf{E}_1. \tag{3.4.14}$$

The no-load current is small compared with the full-load primary input current. In practice $I_{NL} \leq 0.03\, I_1$. Further, in order to obtain a high efficiency and a small internal voltage drop, the winding resistances and leakage reactances are designed to have small values. Typically $\mathbf{I}_1(R_1 + jX_1) \leq 0.05\, V_1$. Omission of the first term in the voltage eq. (3.4.14) has a negligible effect on the value of $\mathbf{E}_1$ and therefore on $\mathbf{E}_2$ and the secondary terminal voltage $\mathbf{V}_2$. Thus, the magnetizing admittance can be connected across the input terminals. The input voltage rather than E_1 is now responsible for the magnitude of I_{NL}. It can be argued that the change in I_{NL} is so small that the resulting change in the supply current I_1 is negligible. The equivalent circuit of a power system transformer shown in Fig. 3.8 can be modified to the approximate

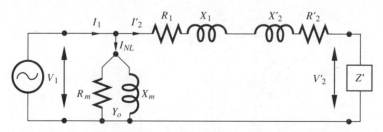

Figure 3.13 Approximate equivalent circuit referred to the primary.

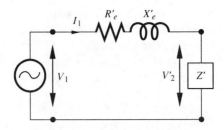

Figure 3.14 Simplified circuit referred to the primary.

equivalent circuit shown in Fig. 3.13. As a matter of definition

$$R'_e \triangleq R_1 + R'_2 = R_1 + a^2 R_2 = R_1 + \left(\frac{N_1}{N_2}\right)^2 R_2. \tag{3.4.15}$$

R'_e is called the equivalent resistance referred to the primary winding.

$$X'_e \triangleq X_1 + X'_2 = X_1 + a^2 X_2 = X_1 + \left(\frac{N_1}{N_2}\right)^2 X_2. \tag{3.4.16}$$

X'_e is called the equivalent reactance referred to the primary winding.

To simplify the study of the effect of the load on the output voltage, the magnetizing admittance may be omitted altogether and we have an equivalent circuit as shown in Fig. 3.14.

We could have referred all quantities to the secondary side. This is illustrated in Fig. 3.15 for the case of the simpler approximate equivalent circuit.

The new relations are given by

$$V''_1 = \frac{V_1}{a} \tag{3.4.17}$$

$R''_e \triangleq$ equivalent resistance referred the secondary winding

$$= R''_1 + R_2 = \frac{R_1}{a^2} + R_2 = \left(\frac{N_2}{N_1}\right)^2 R_1 + R_2 \tag{3.4.18}$$

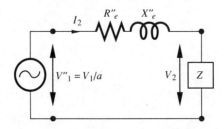

Figure 3.15 Simplified circuit referred to the secondary side.

and

X_e'' = equivalent reactance referred the secondary winding

$$= X_1'' + X_2 = \frac{X_1}{a^2} + X_2 = \left(\frac{N_2}{N_1}\right)^2 X_1 + X_2. \tag{3.4.19}$$

An approximate steady-state analysis of the transformer is readily accomplished using the simple series circuit.

EXAMPLE 3.4

Compare the results obtained from the exact equivalent circuit diagram in Example 3.3 with results obtained from using the approximate equivalent circuit diagram shown in Fig. 3.13.

Solution
Using Fig. 3.13 and the data in Example 3.3

$\mathbf{V}_2' = 132{,}000 \angle 0° \text{ V}, \quad \mathbf{I}_2' = 90.9 \angle -36.9° \text{ A}.$

Input voltage $\mathbf{V}_1 = \mathbf{V}_2' + \mathbf{I}_2'[(R_1 + jX_1) + (R_2' + jX_2')]$.
That is

$\mathbf{V}_1 = 132{,}000 + (72.73 - j54.55)[(5 + j60) + (3.6 + j54)]$
$\quad = 139{,}100 \angle 3.2° \text{ V}.$

$\mathbf{I}_{NL} = \mathbf{V}_1 \mathbf{Y}_0 = 139{,}100 \angle 3.2° \times 6.7 \times 10^{-6} \angle -63.4° = 0.0932 \angle -60.2°$
$\quad = (0.46 - j0.81) \text{ A}.$

Input current $\mathbf{I}_1 = \mathbf{I}_2' + \mathbf{I}_{NL} = 91.77 \angle -37.10° = (73.19 - j55.36) \text{ A}.$
The phase difference between the input voltage and current is 40.3°. Therefore the input power factor is cos 40.3° which is equal to 0.76. The percentage differences found for the exact and approximate equivalent circuits for input voltage, current, and power factor are 0.71%, 0.02%, and 0%, respectively. It appears that the approximate equivalent circuit is adequate.

3.4.4. Phasor Diagrams

Consider the transformer to be on load as depicted in Fig. 3.7. For this configuration we will draw the phasor diagram step by step to show the method of construction. The following phasor diagrams will be drawn from an inspection of the equivalent circuits.

1. Referring to Figs. 3.7 and 3.16 draw the output voltage $\mathbf{V}_2$ as the reference phasor.

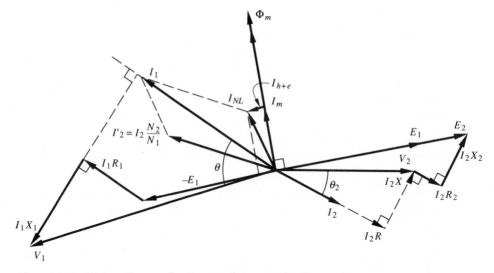

Figure 3.16 Phasor diagram for the transformer on load.

2. Since $\mathbf{V}_2 = \mathbf{I}_2\mathbf{Z} = \mathbf{I}_2(R + jX)$, $\mathbf{I}_2$ can be drawn in magnitude equal to $I_2 = V_2(R^2 + X^2)^{1/2}$ at an angle θ_2 to $\mathbf{V}_2$, where $\theta_2 = \tan^{-1}(X/R)$ and $\cos\theta_2$ is the power factor of the load. $\mathbf{I}_2$ will lag $\mathbf{V}_2$ by θ_2 if the load is inductively reactive. If the load reactance is capacitive, the current $\mathbf{I}_2$ will lead $\mathbf{V}_2$.

3. The equation for $\mathbf{V}_2$ is represented on the phasor diagram by drawing the voltage phasor $\mathbf{I}_2R$ in phase with the current and adding the voltage $j\mathbf{I}_2X$ at

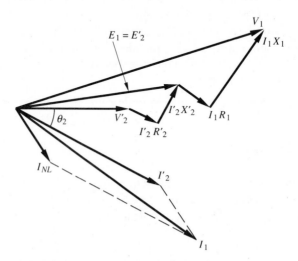

Figure 3.17 Phasor diagram of exact equivalent circuit referred to the primary side.

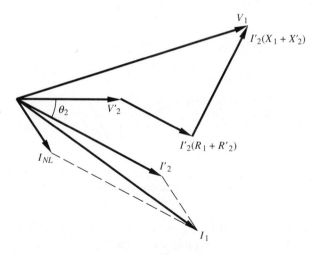

Figure 3.18 Phasor diagram of approximate equivalent circuit referred to the primary side.

right angles to it. The resultant of the two voltage drops is the terminal voltage $\mathbf{V}_2$.

4. The phasor $\mathbf{I}_2 R_2$, which is the voltage drop across the secondary winding resistance, is added to $\mathbf{V}_2$ and drawn in phase with the current $\mathbf{I}_2$. At right angles to the phasor $\mathbf{I}_2 R_2$ and leading $\mathbf{I}_2$ by 90°, the phasor $j\mathbf{I}_2 X_2$ can be drawn; this represents the voltage drop across the secondary leakage reactance X_2.

5. The phasor sum of the three voltage drops is equal to the secondary induced emf $\mathbf{E}_2$; that is

$$\mathbf{V}_2 + \mathbf{I}_2 R_2 + j\mathbf{I}_2 X_2 = \mathbf{E}_2.$$

6. The mutual flux phasor ϕ_m must lead the secondary induced emf $\mathbf{E}_2$ by 90°. Since $\mathbf{E}_1$ and $\mathbf{E}_2$ are both produced by ϕ_m, then they are cophasor. Their

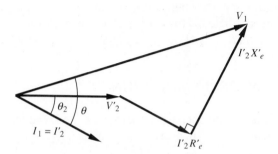

Figure 3.19 Phasor diagram of the simplified circuit referred to the primary side.

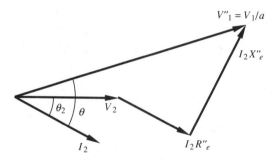

Figure 3.20 Phasor diagram of the simplified circuit referred to the secondary side.

difference in magnitude is determined by the turns ratio of the two windings so that $E_1 = E_2 N_1/N_2$. The phasor $-\mathbf{E}_1$ can be drawn $180°$ out of phase with $\mathbf{E}_1$ to represent a voltage drop in the primary winding.

7. Since the primary and secondary mmfs of the ideal transformer sum to zero, $\mathbf{I}_2' = -\mathbf{I}_2 N_2/N_1$. Therefore, the primary current component $\mathbf{I}_2'$ can be drawn $180°$ out of phase with the secondary current $\mathbf{I}_2$.

8. The phasor $\mathbf{I}_{NL}$ can be constructed so that its magnitude equals the current taken by the primary winding with an open-circuit secondary winding. $\mathbf{I}_m$ and $\mathbf{I}_{h+e}$ are the components of $\mathbf{I}_{NL}$ (see section 3.3.3.A). By adding the two phasor currents $\mathbf{I}_{NL}$ and $\mathbf{I}_2'$, the primary winding input current $\mathbf{I}_1$ can be drawn.

9. The voltage drop $\mathbf{I}_1 R_1$ across the primary winding resistance is in phase with $\mathbf{I}_1$ and is added to $-\mathbf{E}_1$. At right angles to the phasor $\mathbf{I}_1 R_1$ and leading $\mathbf{I}_1$ by $90°$, the voltage drop $j\mathbf{I}_1 X_1$ across the primary leakage reactance can be drawn. The phasor sum of the three voltage drops is equal to the terminal voltage $\mathbf{V}_1$; that is, $\mathbf{V}_1 = -\mathbf{E}_1 + \mathbf{I}_1 R_1 + j\mathbf{I}_1 X_1$.

If we can draw this phasor diagram, we have sufficient information to determine the complete performance of the transformer.

The phasor diagrams for the equivalent circuits shown in Figs. 3.8, 3.13, 3.14, and 3.15 are illustrated in Figs. 3.17 through 3.20.

We have reached a stage where the complete performance of a single-phase transformer can be predicted. However, we do need to know the values of the transformer parameters $R_1, R_2, X_1, X_2, Y_0, a$, the load impedance Z, and voltage V_2.

3.5. PARAMETER DETERMINATION

The parameters of the transformer are the resistances and reactances which combine to form the equivalent circuit. There are six parameters, $R_1, R_2, X_1, X_2, R_m,$ and X_m.[9]

---

[9] The resistance R_m and reactance X_m of the magnetizing admittance are often replaced by the conductance G_m and susceptance B_m.

The primary and secondary winding resistances are R_1 and R_2. X_1 and X_2 are the leakage reactances of the respective windings to account for the magnetic flux which does not couple all turns. The reactance X_m is associated with the mutual flux of the two windings and the resistance R_m accounts for losses due to hysteresis and eddy currents in the magnetic core.

Without expending the full power rating of the transformer it is possible to carry out simple tests to determine the approximate values of the parameters. From these determined values it is possible to predict the transformer performance at any load, as long as the frequency and the voltage of the supply are constant and do not differ significantly from their rated values.

3.5.1. Transformer Open-circuit Test

This test gives information regarding the losses in the core, the parameters R_m, X_m, and the turns ratio. Rated voltage is applied at the terminals of one winding while the other winding terminals are open-circuited. Voltage, current, and power are measured. It is usual to apply rated voltage to the low-voltage winding because a low-voltage source is more readily available. Open circuit means that the transformer is on no load.

The simple test circuit is illustrated in Fig. 3.21 while the equivalent circuit is shown in Fig. 3.22. It has been assumed that in this case the primary winding is the low-voltage winding. All quantities have been referred to the primary.

Because the secondary winding is open-circuited, the only current that flows is the no-load current I_{NL}. The no-load current is considerably smaller than the full-load primary current in power transformers. Therefore, the primary winding voltage drop $I_{NL}(R_1 + jX_1)$ can be neglected. This means that the approximate equivalent circuit, shown in Fig. 3.23, is quite adequate.

The approximate equivalent circuit indicates that the only resistance through which current flows is the equivalent core-loss resistance R_m. Hence the wattmeter reading, representing the power input, indicates the power dissipated in the core. There is some loss in the primary winding due to I_{NL} flowing through R_1 as shown in the exact equivalent circuit diagram of Fig. 3.22. Can this ohmic loss be neglected? Yes! If typically $I_{NL} \approx 0.03\,I_1$ (full load), then the I^2R loss on no load equals 0.0009

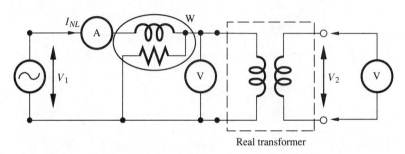

Figure 3.21 Open-circuit test circuit.

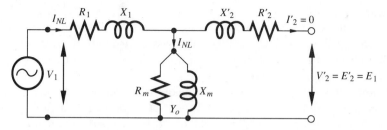

Figure 3.22 Exact equivalent circuit on no load.

times the I^2R loss on full load. The core loss is a function of the applied voltage, the frequency and the core material. These can be considered constant so the loss in the core is constant no matter what the load is. At full load the core loss is designed to be approximately equal to the total ohmic loss. Consequently at no load the ohmic loss is about 0.09% of the core loss. This is sufficient reason to assume that the power reading on the wattmeter indicates the power loss in the core.

At no load, readings are taken from the two voltmeters, the ammeter and the wattmeter (shown in Fig. 3.21). Hence V_1, V_2, I_{NL}, and the core loss P_{NL} are known.

Since the primary impedance drop $I_{NL}(R_1 + jX_1)$ is negligible, $V_1 = E_1$ and so

$$\frac{V_1}{V_2} = \frac{E_1}{E_2} = \frac{N_1}{N_2} = a. \tag{3.5.1}$$

Hence the transformation ratio is now known.

The core loss is P_{NL}, so that, by inspection of the approximate equivalent circuit in Fig. 3.23, the equivalent core resistance R_m equals V_1^2/P_{NL}.

The absolute value of the magnetizing admittance Y_0 is calculated from $Y_0 = I_{NL}/V_1$. In terms of R_m and X_m, $\mathbf{Y}_0$ is

$$\mathbf{Y}_0 = \frac{1}{R_m} + \frac{1}{jX_m} = \frac{1}{R_m} - j\frac{1}{X_m}. \tag{3.5.2}$$

Since Y_0 and R_m are now known, X_m can be calculated from the recorded measure-

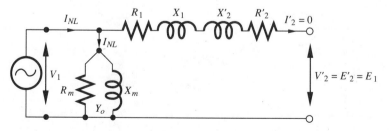

Figure 3.23 Approximate equivalent circuit on no load.

ments as follows.

$$X_m = \frac{1}{[Y_0^2 - 1/R_m^2]^{1/2}}.$$ (3.5.3)

The no-load test enables the parameters R_m, X_m, and a to be determined as well as the core loss P_{NL}. P_{NL} is a constant, very nearly, for all load conditions as long as the rms applied voltage is a constant value at the same frequency.

EXAMPLE 3.5

A single-phase 2300/230-V, 50-kVA transformer has the open-circuit test data, 2300 V, 360 W, and 0.4 power factor. If 230 V is applied to the low-voltage winding, what is the input current when there is no load on the high-voltage side?

Solution
From the no-load test data the no-load current on the high-voltage side is

$$I_{NL} = \frac{360}{2300 \times 0.4} = 0.391 \text{ A.}$$

The no-load current referred to the low-voltage side is

$$I_{NL}'' = I_{NL} \times a = 0.391 \times \frac{2300}{230} = 3.91 \text{ A.}$$

Check: Because there is power invariance, a no-load test with the high-voltage winding open-circuited would give data, 230 V, 360 W, and 0.4 power factor. Hence

$$I_{NL}'' = \frac{360}{230 \times 0.4} = 3.91 \text{ A.}$$

3.5.2. Winding Resistances

A test can be performed to determine the dc winding resistances of the transformer. Whenever direct current flows through a transformer winding, the reactances are zero and there are no losses in the core. Consequently each winding is modeled solely by its resistance when a constant direct current flows.

Any method for measuring low resistance is suitable. A low-resistance bridge method is preferred. The temperature should be noted so that corrections can be made for its influence on the resistance value. The results of this test give dc values of resistance, which are lower than the apparent ac resistances whose values are determined from the short-circuit test (see section 3.5.3). It is the ac values that have to be used for the equivalent circuit model of the real transformer. The experimental determination of the dc resistances aids the calculations used to obtain individual values of ac winding resistances. A further use of the dc test is to record the temperature rise of the windings.

Alternating currents tend to flow near the outside surface of a conductor. Therefore the effective cross-sectional area of the conductor is decreased so that the effective resistance is increased. This is called the *skin* effect and is caused by the conductor inductance, which is greater in the center region than near the surface. Current tends to flow through regions of low impedance. Thus the greater the reactance of the center region the more the current tends to flow near the surface. For power transformers the skin effect at the relatively low frequency of 50 or 60 Hz of a power system creates little change in the resistance value obtained from the dc test.

Of greater influence than the skin effect is the stray loss. When ac current flows through the windings time-varying leakage fluxes exist. (See section 3.3.2.) These are independent of the core and directly proportional to the winding currents. Eddy currents are produced in the winding copper and structural metal components of the transformer by these leakage fluxes. Additional losses, called stray losses, result from the eddy currents. Hence the total losses caused by the winding currents are the load losses, which are the sum of the ohmic loss and the stray loss.

The resistance to be used for the equivalent circuit model is defined by the general expression

$$R_{ac\,effective} \triangleq (\text{load losses})/(\text{current})^2. \tag{3.5.4}$$

The dc resistance is less than the ac effective resistance on account of the stray losses which are not present when direct current flows.

3.5.3. Transformer Short-circuit Test

This test provides an effective value for the total winding leakage impedance, which comprises

$$\mathbf{Z}_1 = R_1 + jX_1 \quad \text{and} \quad \mathbf{Z}_2 = R_2 + jX_2,$$

and the value of the load losses in the windings at full-load current. The test is to short circuit one pair of terminals, raise the ac voltage at the other pair of terminals until full-load current flows in the windings, and to record the voltage, current, and power input.

A power transformer is designed to have low winding resistance, leakage reactance, and magnetizing admittance. This is done in order that most of the applied voltage appears transformed across the load impedance and that the power to the load is very closely equal to the power input to the transformer. The voltage that is required for full-load current in the short-circuit test is determined by the low impedance of the windings and amounts to about 5% of the rated voltage of the winding to which it is applied. Usually, the low-voltage winding is short-circuited, because it is simpler to provide the full-load current to the high-voltage winding, which has the lower value of current. Further, the low voltage for full-load current in the short-circuit test is more accurately measured on the high-voltage side.

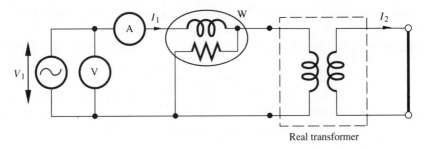

Figure 3.24 Short-circuit test circuit.

The magnetizing impedance is relatively high. So on short circuit the current I_2' is approximately equal to the current I_1 (see Fig. 3.8 for the exact equivalent circuit and note that $Z = 0$). The leakage impedances Z_1 and Z_2' are roughly the same, so that the leakage voltage drops are nearly the same. Consequently

$$|\mathbf{V}_1 - \mathbf{I}_1\mathbf{Z}_1| \approx V_1/2 = E_1 = 4.44\hat{\Phi}fN. \tag{3.5.5}$$

This means that, since the test voltage V_1 is only about 5% of the rated voltage, the core flux is only about 2.5% of its rated value. The no-load current at rated voltage is less than 3% of the transformer full-load input current. In the short-circuit test

$$I_{NLsc} < 0.05\, I_{NL} \approx 0.05(0.03\, I_1) = 0.0015\, I_1 \tag{3.5.6}$$

and can thus be neglected. The iron loss is proportional to the square of the flux in the core. If the flux has been reduced to 2.5% of its rated value, under short-circuit conditions the loss in the core has been reduced to 0.0625% of its normal value. Since full-load current is flowing there is full-load loss dissipated in the windings. If the normal core loss approximately equals the normal full-load winding loss, then in the short-circuit test the core loss will be only 0.0625% of the winding loss. The loss in the core is so small relative to the loss in the windings that the core loss element may be neglected in the circuit diagram. In general, the magnetizing impedance can be ignored for the short-circuit test calculations.

The test circuit is shown in Fig. 3.24, while the appropriate equivalent circuit is illustrated in Fig. 3.25. Readings are taken of V_1, I_1, and the wattmeter.

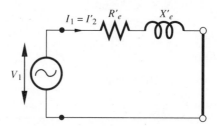

Figure 3.25 Equivalent circuit on short circuit referred to the primary side.

Since the no-load admittance Y_0 is ignored, the input power, which is the watt-meter reading W, is the load loss P_{sc} in the windings. Hence the equivalent resistance R_e' is equal to P_{sc}/I_1^2. The equivalent impedance Z_e' is equal to V_1/I_1. It follows that $X_e' = (Z_e'^2 - R_e'^2)^{1/2}$.

Now $R_e' = R_1 + a^2 R_2$. The effective ac resistances R_1 and R_2 cannot be measured separately, but they can be determined from the dc resistance measurements. If the dc test gave the results R_{1dc} and R_{2dc} then

$$R_{edc}' = R_{1dc} + a^2 R_{2dc}. \tag{3.5.7}$$

Let the calculated ratio

$$\frac{R_e'}{R_{edc}'} \triangleq b. \tag{3.5.8}$$

Then $R_1 = bR_{1dc}$ and $R_2 = bR_{2dc}$ to a first approximation. $X_e' = X_1 + a^2 X_2$ is determined from the short-circuit test also. But X_1 and X_2' cannot be separated by means of any other measurement. They are considered to be equal for analytical purposes.

Simple tests have provided the steady-state parameters R_1, R_2, X_1, X_2, R_m, and X_m. If the load and applied voltage are defined, then the equivalent circuit allows the complete circuit response to be determined for steady-state operation.

EXAMPLE 3.6

A single-phase transformer has the following test data.

TEST	V_1	V_2	I_1	I_2	INPUT POWER
Open-circuit test:	132 kV	22 kV	0.91 A	0 A	52.3 kW
Short-circuit test:	9.86 kV	0 V	91 A (full load)	546 A	71.2 kW

The dc resistance measurements at working temperature for the primary and secondary windings are 4.5 ohms and 0.09 ohms, respectively. Estimate the values of the parameters R_1, R_2, X_1, X_2, G_m, and B_m of the exact equivalent circuit.

Solution
From the open-circuit test data

$$Y_0 = \frac{I_{NL}}{V_1} = \frac{0.91}{132000} = 6.89 \times 10^{-6} \text{ siemens.}$$

$$G_m = \frac{P_{NL}}{V_1^2} = \frac{52300}{(132000)^2} = 3 \times 10^{-6} \text{ siemens.}$$

$$B_m = (Y_0^2 - G_m^2)^{1/2} = 6.2 \times 10^{-6} \text{ siemens.}$$

From the short-circuit test data

$$Z_e' = \frac{V_1}{I_1} = \frac{9860}{91} = 108.4 \ \Omega.$$

$$R'_e = \frac{P_{sc}}{I_1^2} = \frac{71200}{(91)^2} = 8.6 \ \Omega.$$

$$X'_e = (Z'^2_e - R'^2_e)^{1/2} = 108 \ \Omega.$$

$$X_1 \approx X'_2 = \tfrac{1}{2} X'_e = 54 \ \Omega.$$

$$a = \left(\frac{V_1}{V_2}\right)_{oc} = \left(\frac{I_2}{I_1}\right)_{sc} = 6.$$

$$X_2 \approx \frac{X'_2}{a^2} = \frac{X'_e}{2a^2} = 1.5 \ \Omega.$$

If it is assumed that $R_1/R_2 = R_{1\,dc}/R_{2\,dc}$, then

$$R_1 = \frac{R_{1\,dc}}{R_{1\,dc} + a^2 R_{2\,dc}} \times R'_e = \frac{4.5}{4.5 + 36 \times 0.09} \times 8.6 = 5 \ \Omega.$$

Therefore $R_2 = \dfrac{R'_e - R_1}{a^2} = 0.1 \ \Omega.$

3.6. TRANSFORMER PERFORMANCE

We have gathered sufficient information from the previous sections in this chapter to determine the specific items of performance. The important questions asked by the engineer are what is the power input for a required power output and how does the terminal voltage vary with load. These two questions are answered under the definitions of efficiency and regulation, respectively. Normalization of quantities can simplify these calculations of efficiency and regulation, if an approximate equivalent circuit for the transformer is used.

3.6.1. Efficiency

The efficiency is a measure of how well energy or power is transferred by the transformer. Conventionally, efficiency is defined by the ratio of output to input power; that is

$$\eta = \frac{P_o}{P_i} \times 100\%. \tag{3.6.1}$$

Since transformers have high efficiencies, the output power P_o is nearly equal to the input power P_i, and the use of eq. (3.6.1) can lead to large calculation errors. There is less likelihood of calculation errors if the losses are separated in order to determine the efficiency as the divergence from 100%. The losses P_{loss} are given by

$P_{loss} = P_i - P_o$. Therefore the efficiency is

$$\eta = \frac{P_i - P_{loss}}{P_i} \times 100 = \left(1 - \frac{P_{loss}}{P_i}\right) \times 100\%. \qquad (3.6.2)$$

A division of the losses is as follows: (1) heat loss in the core due to the alternating magnetic flux, and (2) heat loss due to the currents in the windings. Since the mutual flux ϕ_m varies little between no load and full load, the two tests, the open-circuit and short-circuit tests described in section 3.5, account for all the above losses. Thus at full load $P_{loss} = P_{NL} + P_{sc}$. The expression for efficiency at full load becomes

$$\eta = \left(1 - \frac{P_{NL} + P_{sc}}{P_o + P_{NL} + P_{sc}}\right) \times 100\%. \qquad (3.6.3)$$

Power output P_o is power into the load, therefore $P_o = V_2 I_2 \cos\theta_2$, where $\cos\theta_2$ is the load power factor. Accordingly, the efficiency can be put in terms of voltage and current. A transformer is rated in terms of the product volt-amperes, above which it is not recommended to exceed, other than intermittently, because of overheating. Since the terminal voltage V_2 is normally maintained constant, only the load current and load power factor vary.

Because the transformer does not operate at full-load current at all times, it is worth noting that the ohmic loss is proportional to the square of the current. If only a fraction y of full-load current is delivered to the load then the efficiency is

$$\eta_y = \left(1 - \frac{y^2 P_{sc} + P_{NL}}{y P_o + y^2 P_{sc} + P_{NL}}\right) \times 100\%. \qquad (3.6.4)$$

The rating of a transformer is given, the rated loss P_{sc} is obtained from the short-circuit test at rated current and the fixed loss P_{NL} is obtained from the open-circuit test at rated voltage. Thus the efficiency for any load power factor and any load can be calculated. Power transformer efficiencies vary from 95% to 99%, the high efficiencies being obtained from transformers with the greater ratings. For maximum efficiency

$$\frac{d\eta}{dy} = 0 \quad \text{and} \quad P_{NL} = y^2 P_{sc}. \qquad (3.6.5)$$

The maximum efficiency of a transformer occurs when the loss in the windings becomes equal to the loss in the core. It should be noted that this is independent of the power factor of the load.

EXAMPLE 3.7

A single-phase, 150-kVA transformer has a core loss of 1.4 kW and a full-load ohmic loss of 1.6 kW. Estimate the transformer efficiency at 25% and 100% full load at power factors **(a)** unity and **(b)** 0.8 lagging.

Solution

(a) The core loss is constant since the voltage is constant.

Ohmic loss $= I_1^2 R_e' \propto I_1^2 \propto$ (load current)^2.

At 0.25 full load, unity power factor:

Core loss $= 1.4$ kW. Ohmic loss $= (1/4)^2 \times 1.6 = 0.1$ kW.

Output power $= (1/4) \times 150 \times 1 = 37.5$ kW. From eq. (3.6.4)

$$\text{Efficiency} = \left(1 - \frac{0.1 + 1.4}{37.5 + 0.1 + 1.4}\right) \times 100 = 96.2\%.$$

At full load, unity power factor:

Core loss $= 1.4$ kW. Ohmic loss $= 1.6$ kW. Output power $= 150 \times 1 = 150$ kW.

$$\text{Efficiency} = \left(1 - \frac{1.6 + 1.4}{150 + 1.6 + 1.4}\right) \times 100 = 98.0\%.$$

(b) At 0.25 full load, 0.8 power factor:

Core loss $= 1.4$ kW. Ohmic loss $= 0.1$ kW. Output power $= \frac{1}{4} \times 150 \times 0.8 = 30$ kW.

$$\text{Efficiency} = \left(1 - \frac{0.1 + 1.4}{30 + 0.1 + 1.4}\right) \times 100 = 95.2\%.$$

At full load, 0.8 power factor:

Core loss $= 1.4$ kW. Ohmic loss $= 1.6$ kW. Output power $= 150 \times 0.8 = 120$ kW.

$$\text{Efficiency} = \left(1 - \frac{1.6 + 1.4}{120 + 1.6 + 1.4}\right) \times 100 = 97.6\%.$$

A. PER UNIT EFFICIENCY

The definition of efficiency η of a transformer is

$$\eta = \frac{\text{output power}}{\text{input power}} \times 100 = \left(1 - \frac{\text{losses}}{\text{output} + \text{losses}}\right) \times 100\% \qquad \text{(3.6.6)}$$

and this is a percentage value and a normalized value.

If we divide this percentage efficiency quantity by 100 we obtain a *per unit*[10] efficiency quantity. This equation can be developed further for the case where the losses are described in per unit values. Losses and output have units of power, that is, watts. If the numerator and denominator in eq. (3.6.6) are both divided by the base (or reference) VA, $(V_b \times I_b)$, then

$$\text{pu } \eta = 1 - \frac{\text{pu losses}}{\text{pu output} + \text{pu losses}}. \qquad \text{(3.6.7)}$$

The per unit losses are

$$\text{pu losses} = \frac{\text{losses (watts)}}{\text{base VA}} = \frac{P_{NL} + P_{sc}}{V_b I_b}. \qquad \text{(3.6.8)}$$

---

[10] Per unit values are defined in Appendix E. Per unit is abbreviated to *pu*.

If the transformer is delivering the fraction y of the rated volt-amperes to the load, then in general

$$\eta_y = \left(1 - \frac{P_{NL} + y^2 P_{sc}}{y V_b I_b \cos\theta_2 + P_{NL} + y^2 P_{sc}}\right) \times 100\% \qquad (3.6.9)$$

where P_{NL} and P_{sc} are the losses in the core at rated voltage and losses in the windings at rated current respectively, and $\cos\theta_2$ is the load power factor. In the per unit system

$$\text{pu } \eta = 1 - \frac{\text{rated pu core loss} + y^2\text{rated pu ohmic loss}}{y \cos\theta_2 + \text{rated pu core loss} + y^2\text{rated pu ohmic loss}}. \qquad (3.6.10)$$

The efficiency is a function of the load power factor and the fraction of the rated VA delivered to the load.

3.6.2. Regulation

The voltage regulation of a transformer is a measure of the output voltage variation as the load changes. Inspection of either of the transformer equivalent circuits in Figs. 3.22 and 3.23 shows that the voltage drops across the windings' resistances and leakage reactances increase as the load current increases. Thus the secondary terminal voltage changes with load if the primary voltage is constant.

Voltage regulation can be defined as the difference between the load and no-load voltages at the secondary terminals while the primary voltage is maintained constant. This definition leads to an expression for the percentage regulation[11] (REG), that is

$$\text{REG} = \frac{V_2(\text{no load}) - V_2(\text{load})}{V_2(\text{load})} \times 100\%. \qquad (3.6.11)$$

Since $V_2(\text{no load}) = E_2 = E_1/a$, then

$$\text{REG} = \frac{E_1/a - V_2(\text{load})}{V_2(\text{load})} \times 100\%. \qquad (3.6.12)$$

If the approximate equivalent circuit is used then $V_1 \approx E_1$ and the expression for the percentage regulation becomes

$$\text{REG} \approx \frac{\dfrac{V_1}{a} - V_2(\text{load})}{V_2(\text{load})} \times 100\%. \qquad (3.6.13)$$

---

[11] In some countries the expression for the percentage regulation uses the no-load voltage in the denominator of the eq. (3.6.11).

Regulation is based on the following:

1. Rated voltage V_2, full-load current and rated frequency.
2. If the load value is not stated, then rated load is assumed.
3. Sine-wave voltage is assumed unless specified otherwise.
4. The power factor $\cos\theta_2$ of the load must be stated, otherwise unity power factor is assumed.
5. At all loads, regulation is corrected to a reference temperature of 75°C. This is the rated full-load temperature of the windings under steady conditions, so the value of the winding resistance must be corrected to correspond to 75°C.

The given regulation equation is in a suitable form if actual voltages can be measured. In order to predict the regulation we can make use of the approximate equivalent circuit referred to the secondary and its associated phasor diagram as shown in Figs. 3.15 and 3.20. From either the circuit or the phasor diagram it can be seen that the voltage equation is

$$\frac{\mathbf{V}_1}{a} = \mathbf{V}_2 + \mathbf{I}_2(R_e'' + jX_e''). \tag{3.6.14}$$

If $\mathbf{V}_2$ is taken as the reference,

$$\frac{\mathbf{V}_1}{a} = V_2 + I_2(R_e'' \cos\theta_2 + X_e'' \sin\theta_2) + jI_2(X_e'' \cos\theta_2 - R_e'' \sin\theta_2). \tag{3.6.15}$$

In absolute terms the referred input voltage is

$$\left|\frac{V_1}{a}\right| = [V_2^2 + I_2^2(R_e''^2 + X_e''^2) + 2V_2I_2(R_e'' \cos\theta_2 + X_e'' \sin\theta_2)]^{1/2}. \tag{3.6.16}$$

Therefore the regulation is given by

$$\text{REG} = \frac{V_1/a - V_2}{V_2} \times 100 = \left(\frac{V_1/a}{V_2} - 1\right) \times 100$$

$$= 100\left[1 + \frac{I_2^2}{V_2^2}(R_e''^2 + X_e''^2) + \frac{2I_2}{V_2}(R_e'' \cos\theta_2 + X_e'' \sin\theta_2)\right]^{1/2} \tag{3.6.17}$$

$$- 100\%.$$

Let

$$x = \frac{I_2^2}{V_2^2}(R_e''^2 + X_e''^2) + 2\frac{I_2}{V_2}(R_e'' \cos\theta_2 + X_e'' \sin\theta_2). \tag{3.6.18}$$

Therefore

$$\text{REG} = [100(1 + x)^{1/2} - 100]\%. \tag{3.6.19}$$

The voltage drop across the transformer impedance is much less than the voltage drop V_2 across the load, so we can say that x is much less than unity. Hence $(1 + x)^{1/2}$ can be expanded by the Binomial series[12]

$$REG = \left[100 \left\{ 1 + nx + \frac{n(n - 1)}{2!} x^2 + \cdots \right\} - 100 \right] \%. \qquad (3.6.20)$$

Neglecting all terms of the third order and above and substituting the expression for x

$$REG \approx \left[\frac{I_2}{V_2} (R_e'' \cos\theta_2 + X_e'' \sin\theta_2) + \frac{1}{2} \frac{I_2^2}{V_2^2} (X_e'' \cos\theta_2 - R_e'' \sin\theta_2)^2 \right] \\ \times 100\%. \qquad (3.6.21)$$

This equation is sufficiently accurate for all transformers except those with very high impedances (greater than $0.2V_2/I_2$ where V_2 and I_2 are rated values).

Since the second term in eq. (3.6.21) is the square of a difference of two comparable magnitudes, it is quite small and can be neglected if an approximate but rapid calculation of the regulation is to be made. Thus

$$REG \approx \left(\frac{I_2}{V_2} (R_e'' \cos\theta_2 + X_e'' \sin\theta_2) \right) \times 100\%. \qquad (3.6.22)$$

It can be seen from the phasor diagram in Fig. 3.20 that if the angle between $\mathbf{V}_1/a$ and $\mathbf{V}_2$ is small enough to be neglected (i.e., input and output voltages are almost in phase) then, using the phasor diagram,

$$V_1/a - V_2 \approx I_2 R_e'' \cos\theta_2 + I_2 X_e'' \sin\theta_2 \qquad (3.6.23)$$

which is the regulation in volts. Changing this to a percentage value we get the approximate regulation as in eq. (3.6.22). Calculations of the regulation can be made for different load currents and power factors.

The regulation is a function of the load power factor. For all lagging power factors the regulation is positive. But as the power factor becomes leading, due to a capacitive load, the power factor will reduce until the regulation is zero, at which point $\theta_2 = -\tan^{-1}(R_e''/X_e'')$. At leading power factors below this value the regulation is negative. That is, the output terminal voltage on load is greater than the no-load voltage. In general, however, transformer loads are inductive.

The voltage at the output terminals of a transformer changes with load. There is a need to control that voltage so that the consumer may have an almost constant voltage source over a wide range of load. An adjustable turns ratio of the transformer windings accommodates this requirement. There are taps on one of the windings, and connections are brought out to terminals so that the effective turns of the winding

---

[12] The binomial series is

$$(1 + x)^n = 1 + nx + \frac{n(n - 1)}{2!} x^2 + \frac{n(n - 1)(n - 2)}{3!} x^3 + \cdots$$

if $x \ll 1$.

can be adjusted. Since $V_2/V_1 = N_2/N_1$, a change of N_2 increases or decreases V_2. For small transformers the connections to the taps may be adjusted manually, but, for high voltage and high power, tap changing is accomplished automatically. If on-load tap changers are used it will be appreciated that the switch must have a make-before-break mechanism to allow current continuity and means are provided to prevent any turns being short-circuited.

EXAMPLE 3.8

A single-phase, 2400/240-V, 50-kVA transformer has the following test data.

TEST	A	V	W
Open-circuit test	5.41	240	186
Short-circuit test	20.8	48	617

Estimate **(a)** the transformer parameters referred to the low-voltage side and **(b)** calculate the transformer regulation at full-load current, 0.8 power factor lagging.

Solution
Inspection of the nameplate information and the test data shows that the short-circuit test was performed with the meters connected to the high-voltage side. The open-circuit test data were measured from the low-voltage side.
(a) From the short-circuit test

$$R'_e = \frac{P_{sc}}{I_1^2} = \frac{617}{(20.8)^2} = 1.43 \ \Omega.$$

$$Z'_e = \frac{V_1}{I_1} = \frac{48}{20.8} = 2.31 \ \Omega.$$

$$X'_e = (Z'^2_e - R'^2_e)^{1/2} = 1.81 \ \Omega.$$

Referred to the low-voltage side

$$R''_e = \frac{R'_e}{a^2} = \frac{1.42}{(2400 \div 240)^2} = 0.0143 \ \Omega.$$

$$X''_e = \frac{X'_e}{a^2} = \frac{1.82}{(2400 \div 240)^2} = 0.0182 \ \Omega.$$

From the open-circuit test

$$Y''_0 = I_{NL}/V_{oc} = 5.41/240 = 0.0225 \text{ siemens.}$$

(b) With V_2 as reference

$$V_1/a = V_2 + I_2(\cos\theta_2 - j\sin\theta_2)(R''_e + jX''_e).$$

$$I_2 = 50000/240 = 208.3 \text{ A.}$$

Hence $\dfrac{V_1}{a} = 240 + 208.3(0.8 - j0.6)(0.0143 + j0.0181)$.

That is $\dfrac{V_1}{a} = 244.65 + j1.23 = 244.64\ \angle 0.3°$ V.

Percentage regulation is

$$\text{REG} = \frac{244.64 - 240}{240} \times 100 = 1.93\%.$$

Using eq. (3.6.22) we find that

$$\text{REG} = \frac{208.3}{240}(0.0143 \times 0.8 + 0.0181 \times 0.6) \times 100 = 1.93\%.$$

This shows that it is satisfactory to use eq. (3.6.22) to calculate the regulation.

A. PER UNIT REGULATION

The actual voltage regulation is $(E_2 - V_2)$ volts and in per unit values[13]

$$\text{pu REG} = \frac{E_2 - V_2}{V_{b2}}. \tag{3.6.24}$$

where V_{b2} is the base value for the terminal voltage.

We can use the less accurate expression for the regulation from eq. (3.6.21) to write the per unit regulation at full-load current as

$$\text{pu REG} = \frac{I_{b2}}{V_{b2}}(R_e'' \cos\theta_2 + X_e'' \sin\theta_2) + \tfrac{1}{2}\frac{I_{b2}^2}{V_{b2}^2}(X_e'' \cos\theta_2 - R_e'' \sin\theta_2)^2 \tag{3.6.25}$$

and this becomes

$$\text{pu REG} = (\text{pu } R_e \cos\theta_2 + \text{pu } X_e \sin\theta_2) + \tfrac{1}{2}(\text{pu } X_e \cos\theta_2 - \text{pu } R_e \sin\theta_2)^2. \tag{3.6.26}$$

The approximate form of the regulation in per unit quantities is

$$\text{pu REG} \approx \text{pu } R_e \cos\theta_2 + \text{pu } X_e \sin\theta_2 \tag{3.6.27}$$

where pu R_e and pu X_e are the total equivalent per unit resistance and reactance. Since they are per unit quantities it does not matter whether they are referred to the primary or the secondary windings. Both sets of values are the same.

The values of pu R_e and pu X_e are easily calculated from the short-circuit test results. If the regulation is small, so that eq. (3.6.27) holds, the regulation goes from pu R_e at unity power factor to pu X_e at zero power factor. This is a useful observation.

The load current I_2 can be any fraction y of the full-load current. In this case the

---

[13] See Appendix E for per unit definitions.

per unit regulation is

$$\text{pu REG} = \frac{yI_{b2}}{V_{b2}} (R''_e \cos\theta_2 + X''_e \sin\theta_2) + \frac{1}{2} \frac{y^2 I_{b2}^2}{V_{b2}^2} (X''_e \cos\theta_2 - R''_e \sin\theta_2)^2. \quad \textbf{(3.6.28)}$$

This becomes

$$\text{pu REG} = y(\text{pu } R_e \cos\theta_2 + \text{pu } X_e \sin\theta_2) \\ + \frac{1}{2} y^2 (\text{pu } X_e \cos\theta_2 - \text{pu } R_e \sin\theta_2)^2. \quad \textbf{(3.6.29)}$$

The approximate form of the regulation in per unit quantities for a fraction y of the rated current is

$$\text{pu REG} = y(\text{pu } R_e \cos\theta_2 + \text{pu } X_e \sin\theta_2). \quad \textbf{(3.6.30)}$$

EXAMPLE 3.9

A 10-kVA transformer supplies full load at 0.8 pf lagging. The ohmic loss of the transformer is 2% and the reactance drop is 6% of the voltage. A capacitor is connected across the load such that the regulation of the transformer at full load becomes zero. Calculate **(a)** the rating of the capacitor in terms of kVAr and **(b)** the kVA loading on the transformer when the capacitor is connected?

Solution
(a) The per unit regulation of the transformer at full load is, from eq. (3.6.27),

$$\text{pu REG} = \text{pu } R_e \cos\theta_2 + \text{pu } X_e \sin\theta_2.$$

If $\tan\theta = -\text{pu } R_e/\text{pu } X_e = -1/3$, REG $= 0$.

But $\tan\theta = \dfrac{(\text{Load kVAr} - \text{Capacitor kVAr})}{\text{Load kW}}$.

Load kW $= 10 \times 0.8 = 8$.
Load kVAr $= \sqrt{10^2 - 8^2} = 6$.

For zero regulation, $\tan\theta = -\dfrac{1}{3} = \dfrac{(6 - \text{Capacitor kVAr})}{8}$.

Therefore Capacitor kVAr $= 8.67$.
(b) Transformer loading $S = \sqrt{P^2 + Q^2}$.
$P = 8$ kW, $Q = 6 - 8.67 = -2.67$ kVAr.
Therefore $S = \sqrt{8^2 + (2.67)^2} = 8.43$ kVA.

3.7. THREE-PHASE TRANSFORMERS

3.7.1. Three-phase Connections

Virtually all large scale production of electric energy is generated by three-phase generators. The transmission and distribution of this energy to load centers is accomplished by three-phase systems.[14] These systems require transformers to raise the voltage level for economic and efficient transmission and subsequently to lower the voltage to appropriate levels for distribution and ultimate utilization.

Three-phase banks of interconnected single-phase transformers or three-phase transformer units can be used for voltage transformation in three-phase systems. In either case, each phase has a primary and secondary transformer winding associated with it. The terminals of the primary windings as well as those of the secondary windings have to be interconnected. This can be accomplished in several ways, but the two basic methods are to connect each group of three windings in either *star* (Y) or *delta* (Δ). These two types of winding connection lead to four possible combinations, Y-Y, Δ-Δ, Δ-Y, and Y-Δ.

If the primary windings are identical and if the secondary windings are identical, each set of connections forms a symmetrical three-phase arrangement. If the supply and the load are symmetrical, the whole system is balanced. This means that corresponding voltages and currents of the three phases have the same magnitude. Only phase displacements of 120° distinguish one phase from another. Thus the power, the efficiency and the regulation are also the same for each phase of a balanced system. This permits a balanced three-phase system to be analyzed on a per-phase basis, using the familiar equivalent circuit of a single-phase transformer. The line voltages, the exciting admittances and winding impedances of Δ-connected transformer windings, and the impedances of Δ-connected loads are replaced by their equivalent Y-connected values.[15] Parameters of transformers, transmission lines, and those of the connected load are normally referred to either the primary or secondary side. Referral of quantities to one side enables transformer winding impedances, transmission line impedances and load impedances to be added directly. After the analysis of a three-phase system on a per-phase basis, the per phase quantities can be converted back to the appropriate three-phase quantities. The percentage efficiency and regulation calculated on a per-phase basis is the same as that for the three-phase system.

The choice between the types of transformer winding connections to be used depends on several factors, some of which are influenced by the philosophy of the system designers. An example is whether or not to use a neutral conductor and whether or not to ground this conductor. A neutral conductor, which can only be provided when the windings are star connected, permits the simultaneous supply of phase voltage and line-to-line voltage and therefore is a preferred connection for low-voltage distribution transformers. In higher-voltage overhead transmission systems the neutral conductor can be utilized to shield the lines against lightning. If unbalanced loads can be expected and a neutral wire is undesirable, a delta-connected system is less affected by the imbalance of the load. An isolated neutral Y-Y connection inherently introduces voltage assymmetry between the phase voltages. Voltage wave distortion results due to the inability of third harmonic currents and

---

[14] See Appendix D for a review of three-phase systems.
[15] Refer to Section D6 of the Appendix.

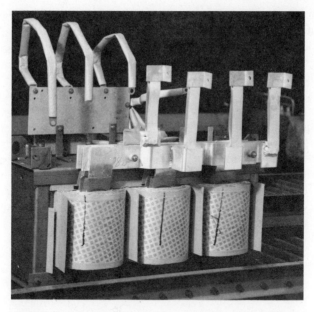

Photo 1   A General Electric Co. 30/(Three-Phase) Core & Coil assembly. (Courtesy of General Electric Co.)

multiples thereof to flow through the windings; this is not the case with a delta or grounded star connection of the phases. The presence of a delta winding in Y-Δ and Δ-Y connections overcomes the problems associated with isolated neutral star-connected windings.

Economic considerations enter any engineering decision. A Y connection gives a less expensive design for high-voltage windings, since the winding voltage is $V_l/\sqrt{3}$. A delta-connected winding, which carries a current $I_l/\sqrt{3}$, is cheaper if the line currents are high.

3.7.2. Three-phase Transformer Units

Instead of using three separate single-phase transformers in three-phase power systems, a single three-phase unit can be installed. A single three-phase transformer contains one magnetic circuit, which consists of three limbs. On each of these limbs are wound the primary and secondary windings of one phase. The ends of the three limbs are connected by yokes to complete the low reluctance flux paths.

Three-phase units with a rating in excess of several hundred kVA have certain advantages over banks of three single-phase units of the same total volt-ampere rating. Foremost is the lower cost resulting from a reduction of overall material usage.

A reduction in core material and a saving in the required number of bushings, which are high-cost items, are the main contributing factors. A three-phase unit weighs less and occupies less space than three single-phase transformers of the same total volt-ampere ratings, and these factors reduce shipping and installation cost.

The disadvantage of a single three-phase unit is that of the weight and size of the transformer is larger than that of a single-phase transformer of one third of the volt-ampere rating. Shipping limitations on weight and size may oblige one to choose three single-phase transformers if high MVA are involved. The greater cost of repair of a three-phase unit and the fact that the provision of spare capacity, or a backup unit, to maintain high reliability of a transmission system is more costly than in the case of a bank of single-phase units are other considerations to be taken into account.

The two versions of three-phase core construction normally used are the shell type and the core type. In the shell-type construction (Fig. 3.26) the windings are enclosed by the magnetic circuit. In the core-type construction (Fig. 3.27) the windings enclose the core.

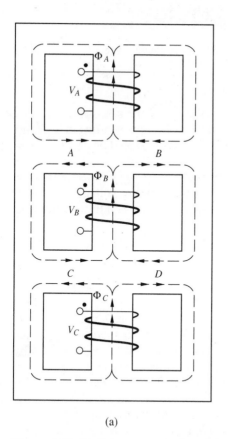

(a)

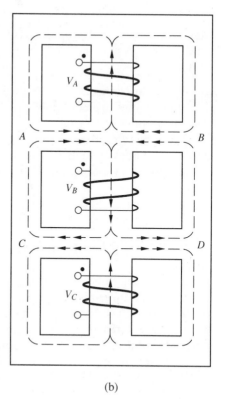

(b)

Figure 3.26 Shell-type core construction.

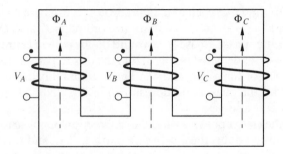

Figure 3.27 Core-type construction.

A. SHELL-TYPE CONSTRUCTION

In Fig. 3.26 the exciting windings (or primary windings) are the only ones shown, since the secondary windings play no role in the following discussions. The polarity of these windings, indicated by the dots, is determined by the way in which they are connected to the source. For the polarities shown the fluxes Φ_A, Φ_B, and Φ_C have a 120° phase displacement with respect to each other. The time displacement of the phase fluxes is shown in Fig. 3.28a.

Figure 3.26a shows that each phase flux can be considered to have its own closed magnetic circuit available. This consists of parallel paths indicated by the broken lines. A three-phase, shell-type transformer will therefore exhibit the same operational characteristics as a bank of three single-phase transformers.

In core parts A, B, C, and D (see Fig. 3.26a) the total flux is determined by the phasor difference of two equal phase fluxes Φ_{ph}. From the phasor addition carried out

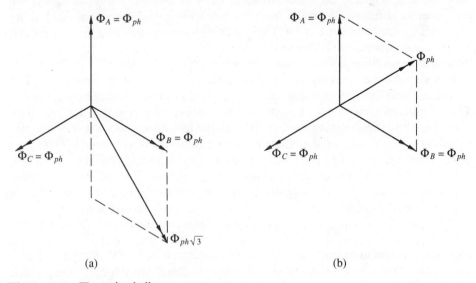

(a)   (b)

Figure 3.28 Fluxes in shell-type cores.

in Fig. 3.28a for Φ_A and Φ_B, each of magnitude ϕ_{ph}, it can be seen that the resultant flux in parts A and B is $\frac{1}{2}\sqrt{3}\phi_{ph}$. The flux in core parts C and D is also $\frac{1}{2}\sqrt{3}\phi_{ph}$. However, if the center phase, phase B, is wound on the core in a direction opposite to the other two phases, as depicted in Fig. 3.26b, the flux in parts A and B of the core is equal to half the phasor sum of Φ_A and Φ_B. The resultant flux in parts A and B is $\frac{1}{2}\phi_{ph}$ as can be seen from the phasor addition shown in Fig. 3.28b. The resultant flux in core parts C and D is also $\frac{1}{2}\phi_{ph}$.

Thus, when the center phase is wound in a direction opposite to that of the other two phases, as is always done, the flux in core parts A, B, C, and D is reduced by 42%. This allows all yoke cross sections to be equal and saves on the amount of core material required, and simplifies manufacture. Also a small reduction in the no-load loss of the transformer is obtained.

B. CORE-TYPE CONSTRUCTION

In the core-type construction, shown in Fig. 3.27, the three limbs, upon which the windings are wound, can be considered to be Y-connected by the yokes. The magnetic reluctance of the flux path of the center phase is less than that of the outer two phases. Accordingly, there exists some imbalance in the magnetic circuits of the three phases of the transformer. The consequence of this imbalance, which manifests itself in unequal magnetizing currents and their harmonic composition, will not be considered in this text.

Since the three winding limbs of the core are Y connected, the phasor sum of the core fluxes must add to zero. This is a requirement which does not have to be met in the shell-type construction. Therefore a core-type transformer will differ in character from a shell-type transformer. For instance, in the case of Y-Y connected windings with isolated neutrals, no third-harmonic exciting current components can be present in the windings. We recall that to induce a sinusoidal voltage the flux linking the windings must be cosinusoidal. The production of a cosinusoidal flux requires an exciting current with a large third-harmonic component. If the third-harmonic component of the exciting current cannot be made to flow, the missing third-harmonic mmf will cause the resulting flux to have a third-harmonic flux component. The third-harmonic flux components will induce third-harmonic voltage components. The third-harmonic flux components of a three-phase, core-type transformer are in time phase and therefore cannot flow through the low-reluctance, Y-connected ferromagnetic circuit. The missing third-harmonic mmfs, however, force the resultant third-harmonic flux to find its return path outside the core from top yoke to bottom yoke. This return path has a high reluctance, with the result that the third-harmonic flux components are considerably smaller than in a shell-type core or a bank of three single-phase transformer cores, in which they can flow through a low-reluctance ferromagnetic path. The effect of the high-reluctance, third-harmonic flux path is to reduce the magnitude of the third-harmonic flux components in three-phase, core-type transformers to at least one tenth of what it would be in a shell-type core or a single-phase transformer core. Hence the induced third-harmonic voltages, which are objectionable in the operation of an isolated neutral Y-Y connected bank of single-phase transformers and which would also be present in an isolated neutral shell-type transformer, are greatly reduced.

Similarly the imbalance of phase voltages resulting from unbalanced or single-phase loading of isolated neutral Y-Y connections is considerably reduced. This is because any unbalanced fundamental frequency flux must flow through a high reluctance nonmagnetic path outside the transformer core, greatly reducing the magnitude of this flux.

If the starpoint of the primary windings is connected to the starpoint of the source, there will be no difference between the behavior of Y-Y connected core-type, or shell-type 3-phase unit or banks of single-phase transformers.

EXAMPLE 3.10

A 3-phase, 11/3.3-kV, 1000-kVA, Y/Δ-connected transformer has primary and secondary winding leakage impedances of $(0.3 + j9)$ ohms per phase and $(0.09 + j2)$ ohms per phase, respectively. What is the applied voltage to obtain full-load current in the windings if the secondary windings are short-circuited? Calculate the total power input for this condition.

Solution
Since no data relevant to the magnetizing admittance are given, it is implied that this admittance can be neglected.
Since the three-phase system is balanced, it can be analyzed on a per-phase basis. For this, Δ-connected phase windings (of generators, transformers, and loads) must be transformed to Y-connected equivalents. In this problem the load is a short circuit, so we can concentrate on the transformation of a Y/Δ-wound transformer to a fictitious equivalent Y/Y-wound unit.
The line-to-line voltages and line currents on the primary side and the secondary side of the transformer must be the same before and after the transformation. Therefore, the Y-connected secondary phase voltages must be $3.3/\sqrt{3}$ kV. It follows that the turns ratio a' of the windings on each limb of the fictitious transformer must be

$$a' = \frac{N_1}{N_2'} = \frac{11/\sqrt{3}}{3.3/\sqrt{3}} = \left(\frac{11/\sqrt{3}}{3.3}\right)\sqrt{3} = \frac{N_1}{N_2}\sqrt{3} = 3.33$$

where N_2 is the actual number of turns of a secondary phase winding and N_2' is the number of turns of the fictitious secondary phase winding.
The equivalent Y-connected impedances of the transformer secondary windings can be found from the Δ/Y transformation.[16] Thus $Z_{2Y} = R_{2Y} + jX_{2Y} = (R_2 + jX_2)/3 = 0.03 + j0.67$ ohms. We can now determine the equivalent impedance parameters of the transformer referred to the primary on a per-phase basis.
$R_{eY}' = R_1 + a'^2R_{2Y} = 0.3 + 3.33^2 \times 0.03 = 0.663$ ohms

---

[16] See Appendix D, section D6.

and $X'_{eY} = X_1 + a'^2 X_{2Y} = 9 + 3.33^2 \times 0.67 = 16.39$ ohms.
Hence $Z'_{eY} = (R'^2_{eY} + X'^2_{eY})^{1/2} = 16.4$ ohms.

The full-load input current $I_1 = \dfrac{1000 \times 10^3}{\sqrt{3} \times 11 \times 10^3} = 52.5$ A.

Therefore, the primary voltage to obtain full-load current with the secondary side shorted is

$$V_{1sc} = \sqrt{3} I_1 Z'_{eY} = \sqrt{3} \times 52.5 \times 16.4 = 1.49 \times 10^3 \text{ V.}$$

The total input power is $P = 3 I_1^2 R'_{eY} = 3 \times 52.5^2 \times 0.663 = 5.23 \times 10^3$ W.

3.8. AUTOTRANSFORMERS

The normal single-phase transformer consists of two separate windings, electrically isolated from each other and wound on a common ferromagnetic core. Ignoring losses, the power supplied to the primary winding is transferred entirely through the medium of the mutual core flux to the secondary winding and then delivered to the load.

If the two windings are electrically connected in series, as indicated in Fig. 3.29, the transformer is known as a *single-phase autotransformer.* Although the schematic representation indicates a single coil *ac*, parts *ab* and *bc* carry currents of different magnitude and consequently have different conductor cross sections. From an inspection of this figure it follows that an autotransformer can operate as a step-up as well as a step-down transformer. In both cases winding part *bc* is common to the primary as well as the secondary side of the autotransformer and is therefore usually called the common winding. Winding part *ab* is usually called the series winding.

Autotransformers have certain advantages over the normal two-winding transformers. These advantages justify their application in power systems. Compared with

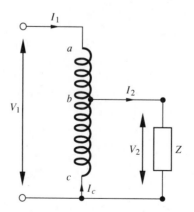

Figure 3.29 Autotransformer connections.

a two-winding transformer of the same nameplate rating an autotransformer has

1. a smaller size and weight because less copper, less insulation and therefore less core steel have been used
2. a lower cost because of the foregoing advantage
3. a greater efficiency
4. a lower short-circuit impedance and therefore a better regulation
5. a smaller exciting current.

These advantages over a two-winding transformer increase as the ratio of high and low line voltages approaches unity. If the voltage levels between two power systems, that are to be coupled, do not exceed a ratio of 2, the application of autotransformers may offer an advantage.

On the other hand, certain disadvantages are associated with the use of auto-transformers which limit their application. Because of the electrical connection between the primary and secondary circuits, each circuit is affected by any disturbance originating in the other. For example, a ground fault on the high-voltage side of an isolated system may subject the low-voltage side, and all the equipment connected to it, to a high voltage. Such danger will not be present in a Y-Y connection of auto-transformers when both starpoints are solidly grounded.

The lower short-circuit impedance of autotransformers results in higher short-circuit currents and considerably higher mechanical short-circuit stresses in the windings. To safeguard the autotransformer it can be designed to have a higher short-circuit impedance. However, before doing so, advantage can be taken from the system impedance. The combination of system impedance and autotransformer impedance may be sufficiently high to limit the short-circuit current to a value that is mechanically safe. In that case the voltage regulation does not have to be increased unnecessarily.

One very common application of low-rating autotransformers is to obtain a voltage output that is adjustable from zero to the input voltage or slightly higher (Fig. 3.30). A constant cross-section conductor is wound in a single layer on a toroidal core. The winding conductor of each turn is partly bared to allow a brush contact of

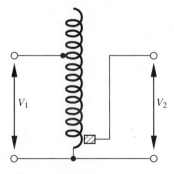

Figure 3.30 Adjustable output voltage.

the output connection to slide from one turn to another. In moving from one conductor to the next, at least one turn must be short-circuited by the sliding brush to prevent circuit interruption. Even though the voltage per turn is low, the brush is made of high-resistance carbon to limit the circulating current in the short-circuited turn. The brush is housed in copper to allow for efficient conduction of the heat generated in the brush. This type of autotransformer is extensively used as an adjustable voltage supply in laboratories or test areas. Also this type is often used in starting circuits of induction motors so that the starting currents are not excessive.

3.9. INSTRUMENT TRANSFORMERS

It is not practical or safe to apply the relatively high-power system voltages and currents directly to auxiliary equipment. Such equipment is required to measure voltage and current or to monitor the energy flow or consumption, the frequency and phase angle, or to energize protective devices and regulating or control equipment. Direct connection to power system conductors would not only create hazardous working conditions for operating personnel, but would also result in unwieldy constructions due to elaborate insulating and current carrying requirements. Therefore, for economy, accuracy, and safety, auxiliary equipment is designed for limited voltage and current ranges.

Instrument transformers provide low-magnitude voltages or currents, which are in direct proportion to and have the same time-phase relation as the actual quantities. To accomplish this, special care has to be taken to have these transformers behave as

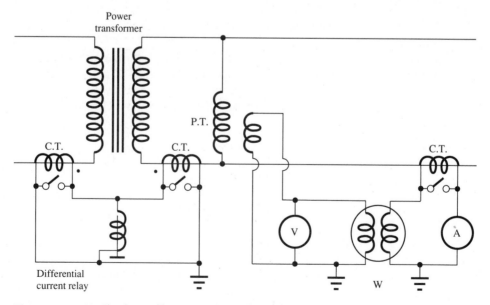

Figure 3.31 Applications of instrument transformers.

closely as possible to ideal transformers. Thus, their no-load currents should be negligible and regulation should be minimal.

Instrument transformers are classified either as potential transformers (PTs) or as current transformers (CTs) depending on their use to measure voltage or current. The design of these two types is different and is dictated by the function they perform. Figure 3.31 depicts some typical applications of instrument transformers.

3.9.1. Potential Transformers

A potential transformer performs two functions. It steps the system voltage down to a level suitable for measuring, protection and control devices. Also it electrically isolates these devices from the high potential of the primary circuit. The secondary or low-voltage side, typically designed for a rated voltage of 150 volts, is normally grounded to protect the secondary system in the event of insulation failure and to prevent the accumulation of charge due to stray capacitance coupling.

The theory of operation of potential transformers is the same as that of power transformers. From the equivalent circuit and phasor diagram it is apparent that the ratio of the terminal voltages and their phase shift are affected by the winding voltage drops. Normally a potential transformer has a constant and high impedance load which is mainly resistive. Combined with the low secondary voltage this leads to small, mainly in-phase currents. The ratio of primary and secondary voltage will approach the turns ratio very closely, if the transformer is designed to have low winding impedances. A correction for voltage ratio errors can be made simply by the addition of a few primary turns, which are then called *compensating turns*. The phase angle between primary and secondary voltages can be kept to a minimum by designing for minimum winding leakage flux by means of subdivided and interleaved windings. The exciting current should be kept small by the use of a low flux density in the core. The exciting current might otherwise adversely affect the phase displacement of the voltages; its effect on the voltage ratio is of secondary importance.

3.9.2. Current Transformers

A CT reduces the system current to a current value in its secondary winding that is suitable for measuring, control, and protection devices. Like the potential transformer, it electrically isolates the high-current system from the secondary circuit, and its secondary winding should be securely grounded.

The maximum secondary current values of current transformers are standardized in accordance with the standard current ratings of the usual measuring equipment. A secondary current rating of 5 amperes is most common. The secondary winding of a current transformer is connected to the low-impedance current coil of an ammeter or a wattmeter. The low-load impedance or *burden* is normally constant and the secondary voltage, determined by the current, is at most 2 volts. Being a current step-down transformer, the number of secondary turns is larger than the

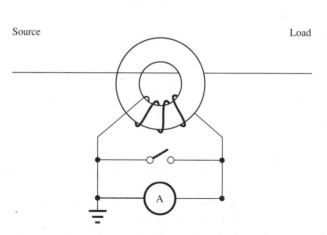

Source   Load

Figure 3.32 Series-connection current transformer.

number of primary turns, which in some instances can be just a single conductor as shown in Fig. 3.32. The secondary voltage will, when referred to the primary, be only a fraction of a volt. The current transformer, which is series connected in the primary circuit, as depicted in Fig. 3.31, will present a negligible voltage drop in this circuit.

Unlike conventional transformers the secondary voltage of a current transformer is determined by the magnitude of the primary current I_1 divided by the fixed turns ratio. The flux in the core will change as the primary current varies. Therefore a current transformer is a variable-voltage transformer. As in any transformer the primary current is the phasor sum of the exciting current and the secondary current referred to the primary side. It follows that when the primary current changes, the core flux and thus the exciting current must change. As a result of this, the ratio of and the phase angle between the primary line current and the ammeter current vary. These variations caused by the varying exciting current must be kept to a minimum by using very low flux densities in the core. Low flux densities are obtained by designing a low reluctance core. Any departure of the current ratio from the turns ratio is known as the *ratio error*.

If the secondary winding load of a current transformer is removed, the primary line current is practically unchanged. However, the opposing ampere turns (the demagnetizing effect) of the secondary winding no longer exists. The primary line current has now become the exciting current for the magnetic core. The core flux increases considerably during the first part of a half-cycle of alternation until it is limited by saturation of the iron during the major part of the half-cycle. A flat topped flux waveform results. This will give rise to an induced voltage in the open-circuited secondary winding which has a waveform with high peaks. Because the turns ratio is such that the current transformer acts as a voltage step-up transformer, the high secondary peak voltage values are damaging to insulation and dangerous to life. Therefore the secondary winding of a current transformer is supplied with a jumper connection (a shorting switch, as shown in Fig. 3.32) which should never be opened unless a meter coil is securely connected to the secondary winding.

3.10. SUMMARY

The ideal, single-phase transformer allows power to be transferred from the primary winding to the secondary winding without losses. It operates with constant flux, if the applied voltage is constant. The equations describing ideal transformer behavior are

$$V_1/V_2 = E_1/E_2 = N_1/N_2 = I_2/I_1, \quad E_1 = 4.44\hat{\Phi}fN_1$$

$$\mathbf{V}_2 = \mathbf{Z}\mathbf{I}_2, \quad \text{and} \quad P = V_2 I_2 \cos\theta_2.$$

The real, single-phase transformer can be modeled by an equivalent circuit. The circuit comprises resistances R_1 and R_2 to represent effective winding resistance, reactances X_1 and X_2 to represent the effects of leakage fluxes and an admittance Y_0 that accounts for core losses (resistance R_m) and core magnetization (reactance X_m). The open-circuit and short-circuit tests allow the transformer parameters to be determined. The importance of these parameters depends on both the operating frequency and the type of analysis to be performed. Approximations can be made to simplify the circuit calculations.

To form an equivalent circuit, secondary winding quantities are referred to the primary side or vice versa. The secondary winding leakage impedance referred to the primary winding is given by

$$Z_2' = Z_2(N_1/N_2)^2.$$

The expressions for referred values of winding resistance and leakage reactance are similar.

Efficiency and regulation are two measures of a transformer's performance. They are defined on a per-unit basis by

$$\text{pu } \eta = P_o/P_i \quad \text{and} \quad \text{pu REG} = (E_2 - V_2)/V_2.$$

Three-phase balanced transformers can be analyzed on a single-phase basis. The line-to-phase relations are $V_l = V_{ph}$ and $I_l = \sqrt{3}I_{ph}$ for Δ-connected windings, and $V_1 = \sqrt{3}V_{ph}$ and $I_l = I_{ph}$ for Y-connected windings. For both connections the total power is $P = 3V_{ph}I_{ph}\cos\theta = \sqrt{3}V_l I_i \cos\theta$.

3.11. PROBLEMS

Section 3.2

3.1. A single-phase, 2400/240-V, 60-Hz transformer is built on a core having an effective cross-sectional area of 66 cm². There are 100 turns on the low-voltage winding. Calculate **(a)** the maximum flux density and **(b)** the number of turns on the high-voltage winding.

(*Answer:* (a) 1.37 tesla, (b) 1000 turns.)

3.2. A single-phase, 100-kVA transformer, whose turns ratio is 300/24 is connected to a 2000-V, 60-Hz supply. Find **(a)** the cross-sectional areas of the winding conductors, if a current density of 1.5×10^6 A/m^2 is allowed; **(b)** the secondary winding voltage on open circuit; and **(c)** the maximum value of the core flux.
(*Answer:* (a) 33.3×10^{-6} m^2, 416×10^{-6} m^2, (b) 160 V, (c) 25 mWb.)

3.3. A load of 10 kW, which has a power factor of 0.8, is supplied by a 20-kVA transformer. What is the percentage loading of the transformer? Heating loads are to be added to the same transformer. How many kW of heating can be added before the transformer is fully loaded?
(*Answer:* 62.5%, 8.54 kW.)

3.4. A 100-kVA transformer is working at full load and the power factor is 0.6 lagging. **(a)** What reactive VArs need to be added by means of capacitors to improve the power factor to 0.8? **(b)** What percentage loading does the transformer have after the power factor improvement?
(*Answer:* 35 kVAr, 75%.)

3.5. Consider an ideal single-phase transformer. Maximum power dissipation occurs in a resistive load of 100 ohms, if the primary winding source impedance is 1 ohm (resistive). What is the transformer turns ratio? Calculate the voltage ratings of the transformer if the load current is 10 amperes. What assumptions must be made to regard the transformer as ideal?
(*Answer:* 1 : 10, 100 V, 1000 V.)

3.6. Consider the simple single-phase power system, as shown in Fig. P3.6, in which the three transformers are assumed ideal. What is the phase difference between the generated voltage and current? Determine the nominal VA output of the generator if the load voltage is 120 V.
(*Answer:* 55.8°, 3485 kVA.)

3.7. A two-winding transformer that can be considered ideal has 200 turns on its primary winding and 500 turns on its secondary winding. The primary winding is connected to a 220-V sinusoidal supply and the secondary winding supplies 10 kVA to a load. **(a)** Determine the load voltage, the secondary

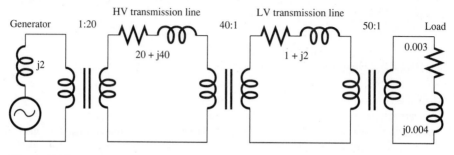

Figure P3.6

winding current and the primary winding current and **(b)** the magnitude of the impedance as seen from the supply.

(*Answer:* (a) 550 V, 18.2 A, 45.5 A, (b) 4.84 ohms.)

3.8. A heating element, whose resistance is 9 ohms, is required to radiate the maximum possible power when the ac supply voltage is 100 V rms and the source impedance is an inductive reactance of 1 ohm. Find **(a)** the turns ratio of the transformer to be connected between the supply and the load, **(b)** the voltage and current ratings of the transformer, and **(c)** the maximum power transferred. Assume an ideal transformer.

(*Answer:* (a) 1 : 3, (b) 70.8/212.4 V, 70.8/23.6 A, (c) 5kW.)

3.9. A heating element, whose resistance is 9 ohms, is to radiate 10 kW of heat. The ac supply voltage is 100 V rms so that a transformer must be connected between the source and the load. If the single-phase transformer can be considered to be ideal, calculate **(a)** the transformer turns ratio and **(b)** the currents in the transformer windings.

(*Answer:* (a) 1 : 3, (b) 100 A, 33.3 A.)

3.10. Two ideal single-phase transformers, with turns ratios 100/75 and 100/50, are connected to resistive loads, 3 and 2 ohms, respectively. The two primary windings are connected in series to a 200-V supply. **(a)** Calculate the primary winding voltages, the load currents, and the power supplied by the source. **(b)** If the resistive load of 3 ohms is replaced by a reactance load of 3 ohms, recalculate the answers for part (a).

(*Answer:* (a) 80 V, 120 V, 20 A, 30 A, 3000 W; (b) 111 V, 166.4 V, 27.7 A, 41.6 A, 3460 W.)

Section 3.3

3.11. A single-phase transformer has its primary winding supplied from a 115-V, 60-Hz supply. If the secondary winding is open-circuited the no-load current in the primary is 4 A at 0.3 power factor lagging. The number of primary turns is 100. Calculate **(a)** the magnetizing current, **(b)** the maximum value of the magnetic flux in the core, **(c)** the core loss, and draw **(d)** the phasor diagram for this condition.

(*Answer:* (a) 3.8 A, (b) 4.3 mWb, (c) 138 W.)

3.12. The primary winding of a single-phase transformer consists of two identical coils which can be connected in series or in parallel. When the coils are in series and connected to a 440-V, 50-Hz supply, the no-load current is 2 A, the no-load loss is 100 W, and the secondary voltage is 220 V. What is the total line current supplied to the transformer, what are the no-load losses and what is the secondary voltage when the coils are in parallel and connected to a 220-V, 50-Hz supply?

(*Answer:* 4 A, 100 W, 220 V.)

Section 3.4

3.13. What is the exact equivalent circuit of the two winding transformer referred to the secondary winding? Write down all the circuit relations and draw a phasor diagram.

3.14. Determine an approximate relation between the winding resistances R_1 and R_2 of the transformer.
(*Answer:* $R_1 = a^2 R_2$.)

3.15. Show that the leakage reactances X_1 and X_2' are independent of the applied voltage and that they are approximately equal.

3.16. A single-phase transformer has 90 turns on the primary winding and 180 turns on the secondary winding. The primary winding resistance is 0.067 ohm and the secondary winding resistance is 0.233 ohm. Calculate **(a)** the primary winding resistance referred to the secondary side, **(b)** the secondary winding resistance referred to the primary side, and **(c)** the transformer equivalent resistance referred to the primary side.
(*Answer:* (a) 0.268 ohms, (b) 0.058 ohms, (c) 0.125 ohms.)

3.17. A single-phase, 500/100-V, 10-kVA transformer has an equivalent leakage impedance of $(0.3 + j5.2)$ ohms referred to the primary side. An ammeter, voltmeter, and wattmeter in the primary circuit read 20 A, 500 V, and 8 kW, respectively, when the transformer is inductively loaded. Calculate the load voltage if the magnetizing current is ignored.
(*Answer:* 88 V.)

3.18. A single-phase, 2300/240-V, 100-kVA transformer has the following parameters.
$R_1 = 0.2$ ohm, $R_2 = 0.01$ ohm, $G_m = 0.05 \times 10^{-3}$ siemens, $X_1 = 0.7$ ohm, $X_2 = 0.0075$ ohm, $B_m = 6.0 \times 10^{-3}$ siemens.
Determine the values of the input voltage and power if the transformer delivers 115 kVA at 230 V and 0.8 power factor lagging.
(*Answer:* 2302 V, 95.4 kW.)

3.19. A single-phase, 6600/250-V transformer has primary and secondary winding leakage impedances of $(0.21 + j1.0)$ ohms and $(0.27 + j1.3) \times 10^{-3}$ ohms, respectively. If the secondary winding is short-circuited and 400 V is applied to the transformer, estimate the power and current input.
(*Answer:* 17 kW, 206 A.)

Section 3.5

3.20. A single-phase, 1,100/220-V, 100-kVA transformer has the following test data.

Open-circuit test:	1100 V	5.53 A	610 W
Short-circuit test:	138 V	91 A	1860 W.

The dc resistance measurements at working temperature for the primary and secondary windings are 0.18 and 0.0009 ohms, respectively. Determine the parameters of the exact equivalent circuit.

(*Answer:* $(0.2 + j0.75)$ ohms, $(0.001 + j0.03)$ ohms, 0.5×10^{-3} siemens, 5×10^{-3} siemens)

3.21. A 5-kVA, 2200/110-V transformer delivers 6 kVA to a load at 100 V. The load power factor is 0.707 leading. Calculate the power input to the transformer and the true input terminal voltage, using the following test results.

Open-circuit test:	2000 V	0.2 A	200 W
Short-circuit test:	100 V	2.0 A	141.4 W.

(*Answer:* 4.76 kW, 2006 V.)

3.22. A single-phase, 6600/330-V, 100-kVA transformer has the following test data for the secondary winding short-circuited.
Short-circuit test: 10 A, 100 V, 436 W.
If the transformer operates at full load, 330 V, 0.8 power factor lagging, what is the value of the applied voltage?
(*Answer:* 6720 V.)

Section 3.6

3.23. The following test results apply to a 10-kVA, 4000/200-V transformer.

Open-circuit test:	200 V	5.0 A	500 W
Short-circuit test:	100 V	2.5 A	300 W.

(a) Calculate the magnetizing current when rated voltage is applied to the high-voltage side. (b) Calculate the efficiency of the transformer when supplying a load of 5 kVA at rated voltage and 0.8 power factor.
(*Answer:* (a) 0.217 A, (b) 87.4%.)

3.24. A single-phase, 2300/230-V, 10-kVA transformer, whose primary and secondary winding leakage impedances are $(4 + j5)$ ohms and $(0.04 + j0.05)$ ohms, respectively, has the following test data.
Open-circuit test: 1 A, 230 V, 75 W.
(a) Draw the exact equivalent circuit referred to the primary side. (b) What is the ohmic loss at half rated output? (c) Calculate the transformer efficiency at the rated kVA, 0.85 power factor load.
(*Answer:* (b) 38 W, (c) 97.5%.)

3.25. A 33/3.3-kV, 2000-kVA transformer has the following test results.

Open-circuit test:	3.3 kV	20 A	20 kW
Short-circuit test:	800 V	50 A	20 kW.

(a) Calculate the parameters of the approximate equivalent circuit referred to the low-voltage side. (b) If the transformer supplies a load of 1000 kVA, 0.9 power factor lagging at rated voltage, estimate the input voltage to the primary winding. (c) At what loading is the efficiency of the transformer a maximum? Calculate the value of the maximum efficiency.
(*Answer:* (a) $Y_0 = 0.006$ siemens, $Z_e'' = 0.08 + j0.138$ ohms, (b) 33.4 kV, and (c) 1650 kVA, 97.6%.)

3.26. A single-phase, 2400/240-V, 30-kVA transformer has the following test data.

Short-circuit test:	18.8 A	70 V	1050 W
Open-circuit test:	0.3 A	240 V	230 W.

(a) Calculate the regulation when the load is 240 V, 12.5 A, and 0.8 power factor lagging. (b) Calculate the regulation at full load and 0.8 power factor lagging. (c) Find the transformer efficiency for the condition in part (a). (*Answer:* (a) 0.19%, (b) 1.94%, (c) 91%.)

3.27. A single-phase, 4400/220-V, 100-kVA transformer has primary and secondary winding leakage impedances of $(0.85 + j8)$ ohms and $(0.002 + j0.02)$ ohms, respectively. The magnetizing susceptance and the core loss conductance referred to the primary side are 0.00025 and 0.000045 siemens, respectively. For full load at a power factor 0.707 lagging, estimate (a) the secondary winding voltage drops, (b) the secondary winding induced emf, (c) the primary winding induced emf, (d) the applied voltage and the input power factor, and (e) the regulation. (*Answer:* (a) 0.91 V, 9.1 V, (b) 227 V, (c) 4540 V, (d) 4695 V, 0.66 power factor, (e) 6.7%.)

3.28. A single-phase, 5-kVA, 2200/110-V transformer is connected to a 4.4-kW, 110-V, 0.9 power factor (lagging) load. Test data are as follows.

Open-circuit test:	2200 V	0.1 A	0.4 power factor
Short-circuit test:	72 V	2.27 A	100 W.

Calculate (a) the input power, (b) the input power factor, (c) the efficiency, (d) the input voltage, (e) the voltage regulation at this load, (f) the per unit impedance of the transformer, and (g) the maximum short-circuit current. (*Answer:* (a) 4.58 kW, (b) 0.89, (c) 96.3%, (d) 2263 V, (e) .028 per unit, (f) 0.033 per unit, (g) 1390 A on the low-voltage side.)

3.29. A single-phase transformer has an ohmic loss of 1.0% and a reactance voltage drop of 5.0%. What is the transformer regulation at rated current when the load power factor is (a) 0.8 lagging, (b) unity, and (c) 0.8 leading? (*Answer:* (a) 3.8%, (b) 1%, (c) −2.2%.)

3.30. A single-phase, 1000/200-V, 150-kVA transformer has the following test data.

Short-circuit test:	100 V	150 A	2.5 kW
Open-circuit test:	200 V	60 A	2.0 kW.

(a) Calculate the transformer parameters R_1, R_2, X_1, X_2, and G_m and B_m of Y_0. (b) Make an exact equivalent circuit (i) referred to the high-voltage side and (ii) referred to the low-voltage side. (c) What is the voltage regulation when the power factor of the load is (i) unity power factor, (ii) 0.6 power factor lagging, and (iii) 0.6 power factor leading? (d) What is the efficiency when the current is (i) one quarter of the full-load value, (ii) one half full-load value, (iii) full-load value if the load power factor is unity, 0.8 lagging, and 0.8 leading? (e) What is the VA output of the transformer for maximum efficiency if (i) the load power factor is unity, and (ii) the load power factor is 0.8 lagging? What is the efficiency for each case in part (e)? **(f)** Calculate the per unit parameters of the transformer. **(g)** Using the per unit values calculate (i) voltage regulation at 0.8 pf lagging and rated VA to load, (ii) rated efficiency at 0.8 pf lagging, and (iii) maximum efficiency at 0.8 pf lagging.

(Answer: (a) 0.055 ohm, 0.002 ohm, 0.328 ohm, 0.013 ohm, 0.002 siemen, 0.296 siemen, (c) (i) 2.16%, (ii) 8.76%, (iii) 6.6%, (d) (i) 94.6%, 93.3%, 93.3%, (ii) 96.6%, 95.8%, 95.8%, (iii) 97.1%, 96.4%, 96.4%. (e) (i) 134 kVA, 97.1%, (ii) 134 kVA, 96.4%, (f) $R_e = 0.0167$ pu, $Z_e = 0.1$ pu, $X_e = 0.0986$ pu, (g) (i) 0.075, (ii) 0.964, (iii) 0.964.)

3.31. Power is fed from a generator along a high-voltage transmission line. A distribution transformer reduces the voltage and the power is fed to a load by means of a low-voltage feeder. Each transmission line is modeled by a series impedance. If the transformer is modeled in the same way then the network is reduced to a series circuit. The high-voltage transmission line has an impedance $(20 + j50)$ ohms. The low-voltage transmission line has an impedance $(1.6 + j2)$ ohms. The distribution transformer has the specification 10,000/ 2000 V, 100 kVA. A short-circuit test on the transformer gave measurements on the low-voltage side, 100 V, 50 A, 3 kW. Find the generator voltage V when the load takes full-load current at 0.8 power factor lagging and the load voltage is 1800 V. The solution is simplified if per unit values are used.
(Answer: 10.6 kV.)

Section 3.7

3.32. A three-phase, 66/3.3-kV, 5000-kVA transformer is connected Y/Δ. The leakage reactance of the transformer is 2% per phase. Calculate the line-line voltage in the primary circuit when the transformer is supplying a load of 4000 kVA at 0.8 pf lagging and at 3 kV line-line.
(Answer: 60.7 kV.)

3.33. A three-phase, 33/6.6-kV, 2-MVA, Δ/Y-connected transformer has primary and secondary winding resistances of 8 ohms and 0.08 ohms per phase, respectively. The equivalent leakage impedance is 0.07 per unit. Calculate the transformer efficiency and load voltage at full load, 0.75 power factor lagging. The no-load loss is 12.85 kW.
(Answer: 98%, 6.25 kV.)

3.34. A three-phase, 60/8.66-kV, 3-MVA, Δ/Y-connected transformer has primary and secondary winding resistances of 21.6 ohms and 0.15 ohms per phase, respectively. The impedance is 2% and the power input on no load is 30 kW. Calculate the secondary winding terminal voltage and the transformer efficiency at full load, 0.8 power factor lagging.
(Answer: 8.5 kV, 97.3%.)

3.35. A three-phase synchronous generator output is 1000 MVA at 20 kV line-line. A transformer bank, which consists of three single-phase units, raises the transmission voltage up to 400 kV line-line when the low-voltage windings are

connected in delta and the high-voltage windings are connected in star. **(a)** Determine the winding voltages and currents on the primary and secondary sides. **(b)** Each transformer has a per unit leakage reactance of 0.1 based on its rating. What is the synchronous generator line voltage to provide rated output at 500 kV and 0.8 power factor lagging to the transmission line?
(*Answer:* (a) 20 kV, 288 kV, 1.66 kA, 115 A, (b) 21.4 kV.)

3.36. Three single-phase transformers of 1000 kVA rating each are connected to form a 3-phase transformer bank, Δ/Δ. The transformer bank supplies a balanced load of 3000 kVA. If one of the transformers is removed for repair, **(a)** is it still possible to supply a balanced load? **(b)** What should be the reduction in load such that the ratings of the remaining transformers in the bank are not exceeded?
(*Answer:* (a) yes, (b) 1260 kVA.)

3.12. BIBLIOGRAPHY

Blume, L. F., et al. *Transformer Engineering,* 2nd ed. New York: John Wiley & Sons, Inc., 1951.

Flanagan, William M. *Handbook of Transformer Applications.* New York: McGraw-Hill Book Co., 1986.

Franklin, A. C., and D. P. Franklin. *The J & P Transformer Book,* 11th ed. London: Butterworth's, 1983.

Hochart, Bernhard (ed.). *Power Transformer Handbook.* London: Butterworth's, 1987.

Karsai, K., D. Kerenyi, and L. Kiss. *Large Power Transformers.* Amsterdam: Elsevier, 1987.

Nasar, S. A. *Electric Machines and Electromechanics.* Schaum's Outline Series in Engineering. New York: McGraw-Hill Book Co., 1981.

Still, Alfred, and Charles S. Siskind. *Elements of Electrical Machine Design.* Tokyo: McGraw-Hill Book Co., 1954.

CHAPTER FOUR

Introduction to Rotating Machines

4.1. INTRODUCTION

In Chapter 1 we read a review of electromagnetic devices in the area of electric power. Through Faraday's law and the Lorentz force equation we demonstrated electromechanical energy conversion in its simplest forms. Armed with the concepts of electromagnetic forces and motion, we developed a theory in Chapter 2 about how to determine those forces or torques in terms of energy principles. We learned a method of analysis that is applicable to any kind of general circuit configuration. Now it is time to return to energy conversion and show, in primitive form, what kind of devices will give us continuous electromechanical energy conversion, and how these devices relate to electric machines that are found in practice.

An electric machine may convert mechanical energy into electric energy or vice versa. In the first case the machine is called a *generator*. In the latter case the machine is called a *motor*. Either type of electromechanical energy conversion involves two basic empirical laws of physics. The laws are

1. Faraday's law, which states that the electromotive force induced in a coil of wire is proportional to the rate of change of magnetic flux linking the coil and
2. Lorentz's law, which states that a current-carrying conductor will experience a mechanical force when placed in a magnetic field. This force is proportional to the magnitudes of the current, the magnetic field and the effective length of conductor subjected to the magnetic field.

If a source of mechanical energy causes conductors to move in a magnetic field, the induced emfs could give rise to the flow of electric current. This would be generator

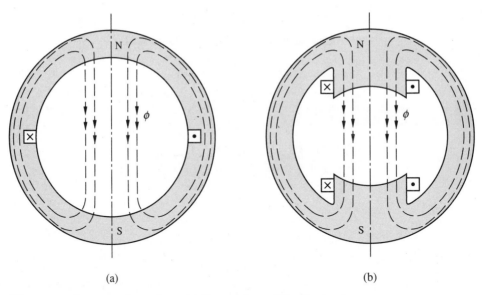

(a)   (b)

Figure 4.1 Stator configurations. (a) Smooth bore, (b) salient pole.

action. If a source of electric energy produces current in conductors that are exposed to an external magnetic field, the Lorentz force could give rise to motion. This would be described as motor action. In both these cases, conductors and a magnetic field in relative motion are required for energy conversion.

For relative motion, electric machines consist of two members, one that is stationary and one that moves. Normally the latter (rotor) rotates within the first (stator) for continuous motion. Figure 4.1 shows the two possible forms of the stator. One form has a smooth bore and the other has projecting poles called salient poles. Two-pole configurations are depicted, but any practical number of pole pairs can be used. Similarly the rotor can be cylindrical or salient. Coils can be placed on either the stator or the rotor, or on both members. Magnetic fields are produced by the mmfs of these coils when they are connected to a dc or an ac source of current. It follows that a large number of combinations of stator and rotor shape and methods of excitation is possible. In what follows we examine which combinations are conducive to electromechanical energy conversion.

4.2. STATOR OR ROTOR EXCITATION

If only one of the members (stator or rotor) is excited, a machine is said to be a *singly-excited system.* The air gap between the rotor and stator can be constant or nonuniform. Let us make a qualitative study of these configurations to see if an average electromagnetic torque can exist.

4.2.1. Constant Air Gap

Figure 4.2 illustrates two machines with a cylindrical rotor and stator, so that a constant or uniform air gap is realized. Figures 4.2a and b have the same configuration. So consideration will be given only to the configuration of Fig. 4.2a. There are two possible forms of excitation, direct current and alternating current. Since the reluctance of the magnetic flux path does not change with rotor position, there is no reluctance torque.

A. DIRECT CURRENT EXCITATION

The field pattern produced by direct current in the exciting coil of Fig. 4.2a is fixed in space as shown. Since the air gap is constant, the flux path is symmetrical about the coil, and the rotor is magnetized along the stator axis. The fact that the stator and rotor magnetic axes are aligned is a physical explanation why there is no torque.

If an external drive causes the rotor to move in the stator field ϕ_s which is fixed in space, there will be motional emfs induced in the rotor iron. Further, if the rotor is made of unlaminated conducting material, the ensuing currents, in hypothetical conducting paths as shown in Fig. 4.3, give rise to a resulting rotor flux ϕ_r, also as shown. It is assumed that rotation is anti-clockwise.

With a rotor of laminated material such that the laminations are electrically insulated from each other and in planes perpendicular to the rotor axis, the same

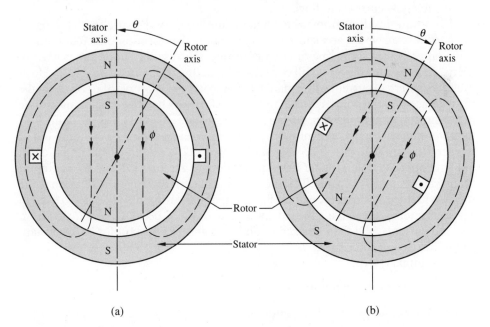

(a)   (b)

Figure 4.2 Singly-excited, constant air-gap configurations. (a) Stator coil excitation, (b) rotor coil excitation.

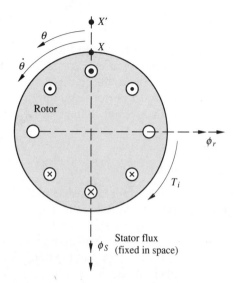

Figure 4.3 Rotor currents and flux due to rotor motion in stator field.

effect of induced currents and flux would be observed if actual conductors were placed axially on the surface of the rotor or in slots. The conductors must form closed circuits to allow currents to flow.

The torque resulting from the tendency of the flux ϕ_r to align itself with the axis of the fixed flux ϕ_s would accelerate the rotor against the direction of actual motion. This torque is then a braking or counter torque. Because the stator flux is fixed in space, the generated currents will be cyclic, alternating with a frequency $\dot{\theta}/2\pi$. However, the pattern of current will remain fixed in space. For example, the current in the rotor adjacent to the point X', fixed in space, will always have a constant magnitude in the direction out of the paper ($\dot{\theta}$ is assumed constant). However, as the point X on the rotor moves through a revolution, the current changes from a maximum out of the paper opposite X' ($\theta = 0$) to zero at a position $\pi/2$ from X' to a maximum again but with opposite polarity after half a revolution, to zero at $3\pi/2$ and to maximum again when X' and X coincide. This is one period of current and one revolution of the rotor. With the rotor current pattern fixed in space, the rotor flux axis is also fixed in space. The torque, which is called an induction torque T_i because it originated from induced currents, will then be fixed in direction and be constant for both constant stator flux ϕ_s and rotor speed $\dot{\theta}$.

In summary,

for $\dot{\theta} = 0$, the induction torque $T_i = 0$,

but for $\dot{\theta} > 0$, the induction torque $T_i < 0$ (that is, braking).

This action, producing the induction torque, describes the principle of an eddy current brake.

B. POLYPHASE ALTERNATING CURRENT EXCITATION

If the stator winding is excited by two or more phase currents, the resulting flux axis can rotate in space. The rotating flux induces motional rotor emfs whose currents interact with the stator flux to produce a torque. This is an *induction torque*. A machine that operates on these principles is called a *polyphase induction machine*.

A more detailed picture can be given of the operation of the polyphase induction machine. Consider a machine with a two-phase stator winding. Each phase is represented by a concentrated coil in Fig. 4.4a. One phase has its current 90° out of phase

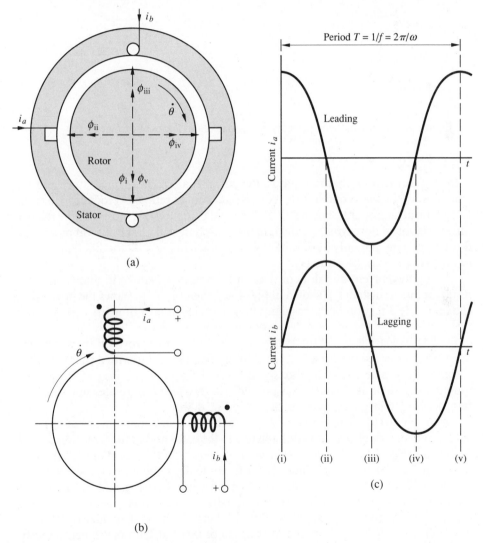

Figure 4.4 Two-phase excitation and the rotating field. (a) Resultant fluxes at different instants, (b) primitive circuit model, (c) phase currents.

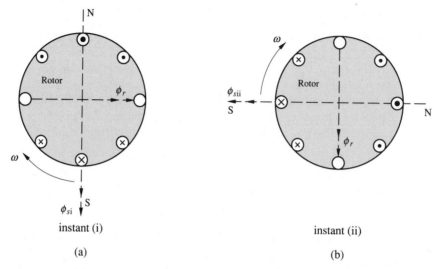

Figure 4.5 Rotor-induced currents. (a) Currents and fluxes at instant (i), (b) currents and fluxes at instant (ii).

with that in the other phase. That is, the currents are in time quadrature and the windings are in space quadrature. A two-pole configuration is produced from the two-phase arrangement, shown in Figs. 4.4a and b. Figure 4.4c illustrates the relative time variation of the currents i_a and i_b in the two-phase coils. Each current produces a flux component, but only the resultant flux is shown in Fig. 4.4a for different instants of time (i), (ii), (iii), (iv), and (v).

Using the current and dot convention such that ϕ_i (shown in Fig. 4.4a) is the resultant flux at instant (i) (shown in Fig. 4.4c), then Fig. 4.4a shows the progressive change in the direction of the resultant flux at the instants of time (ii), (iii), (iv), and (v). In one cycle of the ac, the flux axis has made one revolution. The rotational speed of the flux in space is called the *synchronous speed,* that is, f revolutions per second or ω radians per second for a 2-pole machine. For a machine with a stator winding designed for p pole pairs, the flux speed would be f/p revolutions per second. Although the individual phase fluxes pulsate, the resultant flux rotates in a direction from leading to lagging phases, as shown in Fig. 4.4a.

At instant (i) (depicted in Fig. 4.4c), with the rotor stationary, the rotating stator field moves in a clockwise direction relative to the rotor and induces motional emfs, as shown in Fig. 4.5a. Flux ϕ_r, produced by the rotor currents, is in quadrature with the stator flux. As there is a tendency for fluxes to align themselves, there will be a torque on the rotor to accelerate it. At instant (ii), as shown in Fig. 4.5b, the stator flux phasor has rotated through 90°. The motionally induced emfs generate currents as shown and they in turn produce the flux ϕ_r. There is the same tendency for alignment. The flux ϕ_r lags behind the rotating stator flux but follows it synchronously. That is, ϕ_r rotates at the same angular speed ω as ϕ_s.

With the rotor stationary there is a torque. So there is an inherent mechanism to start the motor. At starting, when the rotor is stationary, induced rotor currents alternate at a frequency equal to the rotational frequency of the stator flux. At some speed other than zero, the frequency of those currents will be less by an amount equal to the rotational frequency of the rotor. If the rotor's speed were equal to that of the speed of rotation of the stator flux in space (called synchronous speed), there would be no relative speed difference. Accordingly, there can be no motionally induced emfs in the rotor and no currents. The flux ϕ_r is zero at synchronous speed and hence there is no electromagnetic torque to drive the motor. If a mechanical load is driven, the rotor speed must be below the synchronous speed (i.e., $\dot{\theta} < \omega$).

Above synchronous speed ($\dot{\theta} > \omega$), there is again a relative speed difference between the stator field and the rotor. Motional emfs are induced in the rotor. The emf polarity and current directions will be reversed, so that ϕ_r will now lead ϕ_s and the induced torque tends to brake the machine.

The foregoing discussion did not consider the phase differences between emfs and currents. This omission was intended in order to simplify this introduction.

4.2.2. Saliency

If magnetic pole structures are formed so that the air gap between the stator and rotor is not constant, the machine is said to be *salient*. Two configurations are depicted in Fig. 4.6 to show the nature of saliency.

Consider Fig. 4.6a. If the coil were placed on the salient-pole member, the

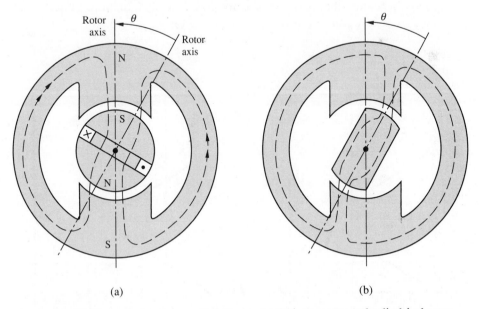

(a)   (b)

Figure 4.6 Singly-excited systems with saliency. (a) Salient motor and cylindrical rotor, (b) double saliency.

reluctance of the magnetic path would not change with rotor displacement. Consequently, with no reluctance torque, only the constant air-gap principles would apply. However, as shown with a coil on the rotor, there is a reluctance torque T_R tending to align the coil axis and the salient-pole axis. This configuration can be inverted. The stator, carrying the coil, can be round, and the rotor can be salient. The same reluctance torque would be present. The configuration in Fig. 4.6b shows both stator and rotor members to be salient. It does not matter which member has the coil (except for practical convenience); a reluctance torque can exist. The reluctance torque will be studied to see if there are useful consequences.

A. DIRECT CURRENT EXCITATION

Let direct current flow in the rotor coil of the configuration shown in Fig. 4.6a. Saliency provides a reluctance torque T_R tending to preserve an alignment such that θ tends to zero or π. This reluctance torque is only a function of θ and is independent of rotor angular speed $\dot{\theta}$. In addition to the reluctance torque T_R there is the induction torque T_i, which is a function of speed and which is in a direction opposing motion for dc excitation. Note that the induction torque exists only if there are conduction paths in the stator.

The average reluctance torque is zero for steady-state, continuous rotation and the induction torque is nearly constant (oscillating slightly because of saliency). The resultant instantaneous torque T_e is obtained by superimposing the two torques T_R and T_i. Figure 4.7 shows this torque to be an oscillating one, oscillating at twice the frequency of rotation about a dc level. The dc level is the average value of torque. This is negative, indicating that it is opposing motion. It is the induction torque which provides this dc level, since the average reluctance torque is zero.

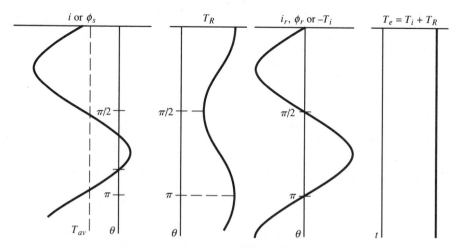

Figure 4.7 Current and torque from dc excitation in a salient pole structure.

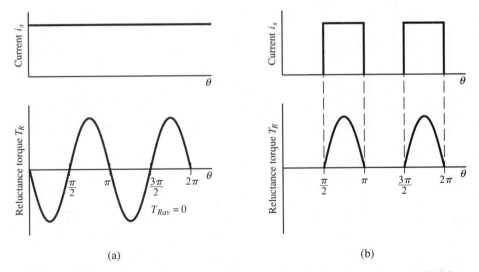

(a)   (b)

Figure 4.8 Reluctance torque from dc excitation and a salient pole structure. (a) Continuous current, (b) current chopped periodically.

B. SINGLE-PHASE ALTERNATING CURRENT

Machine saliency signifies that there is a reluctance torque if the rotor is stationary and θ is not zero.

Refer to Fig. 4.6a. If the rotor is rotating at a constant steady-state speed there is an instantaneous reluctance torque, which is a function of both the instantaneous current in the excitation coil and the displacement θ of the rotor axis. For the case of dc excitation, the average reluctance torque is zero. However, with ac excitation it is possible to find a speed of rotation at which the negative part of the reluctance torque is eliminated. This is accomplished if the excitation current i_s comes to zero when a negative torque would have existed. This is illustrated in Fig. 4.8. Figure 4.8a shows the pulsating torque T_R for dc excitation and Fig. 4.8b shows the resulting torque when the dc excitation is switched off periodically. Such a chopped waveform is an alternating one with a dc level. It is not convenient to generate such a form of excitation except for special applications. This difficulty can be overcome, since the reluctance torque is independent of the polarity of excitation. Thus, the reluctance torque remains the same if every other current pulse shown in Fig. 4.8b has a reverse polarity as depicted in Fig. 4.9a. The square wave can be approximated to a sine wave, which is the usual form of the ac supply. Figure 4.9b shows the torque if the pulsed excitation is replaced by the usual sinusoidal ac. A small negative reluctance torque exists because a stator flux exists when the rotor position is between π and $3\pi/2$. Since the current magnitude is small for $\pi \leq \theta \leq 3\pi/2$, the negative torque is small and the average reluctance torque remains positive.

For the case shown in Fig. 4.9b, the current has completed exactly one cycle when the rotor has gone through one revolution. Therefore to have a useful reluctance

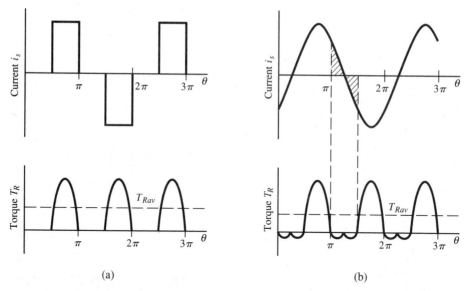

Figure 4.9   Reluctance torque from ac excitation and a salient pole structure. (a) Square-wave excitation, (b) sinusoidal excitation.

torque, the angular frequency of rotation of the rotor must be the same as the frequency of alternation of the excitation current. This type of machine is called a *reluctance machine.*

C. POLYPHASE ALTERNATING CURRENT EXCITATION

Consider Fig. 4.4 again, but let the rotor be salient. If a salient-pole machine is excited by polyphase currents in the stator winding, an induction torque T_i is present, until the rotor speed reaches the synchronous value, just as in the constant air-gap case.

Because the stator mmf produces a rotating magnetic field in the air gap, reluctance torque T_R is present. At standstill when the rotor is stationary, the displacement θ between the rotor and magnetic field axis is changing synchronously at ω radians per second (the angular speed of the stator field). Thus, the reluctance torque is alternating at 2ω radians per second, or $2f$ hertz, so the average value of the torque is zero. As the rotor increases speed, the angular frequency of the torque alternation reduces and is $2(\omega - \dot{\theta})$ radians per second. Consequently, at a rotor speed equal to synchronous speed, the alternations of the reluctance torque cease to exist. A constant reluctance torque now exists. Its value is a function of the angle θ between the rotor axis and the stator rotating field axis.

The difference between the salient-pole case and the constant air-gap case for a polyphase singly-excited system is that, in addition to induction torque at subsynchronous speeds, there exists a torque of the reluctance type for the salient-pole case at synchronous speed.

The synchronous reluctance motor is an example of a machine with polyphase excitation and a salient-pole rotor. This machine relies on induction torque for starting, and reluctance torque for running.

4.3. STATOR AND ROTOR WINDING EXCITATION

In this section a qualitative discussion is given of the electromagnetic torque that is developed if both the stator and the rotor have excited windings. Figure 4.10 depicts basic means to obtain torque. In every case the fields of each member create a torque that tends to produce alignment. The mutual reaction between the two excitations gives rise to a third kind of torque, called *excitation* torque T_E, which can be added to a possible reluctance torque T_R and a possible induction torque T_i to give a total electromagnetic torque T_e.

It can be argued that the induction torque should be part of the excitation torque. Indeed, the induction torque is a special case of the excitation torque. However, we will keep the induction and excitation torques as separate components. The induction torque is reserved for systems where one member has conducting material with a closed path and where no supply is connected.

Excitation torque can exist if the two members of the machine are excited. The different forms of excitation (direct current, single and polyphase alternating current), and the fact that the rotor can be excited through slip rings or a commutator,

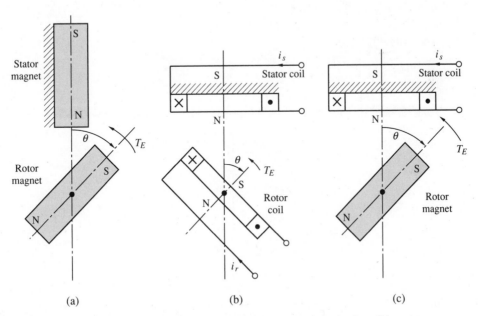

Figure 4.10 Doubly-excited systems. (a) Permanent magnet excitation, (b) current excitation, (c) combination of permanent magnet and current excitation.

produce 18 possible combinations that can lead to excitation torque for the simple configuration shown in Fig. 4.10b. The useful combinations are those that produce a net unidirectional torque over one cycle.

Some types of stator and rotor excitation are described briefly. We are giving consideration solely to the excitation torque.

4.3.1. Stator and Rotor Windings with Direct Current

Whatever real windings are used in an actual machine, the representation in Fig. 4.11 is of concentrated coils in order to show clearly the magnetic flux phasors along the

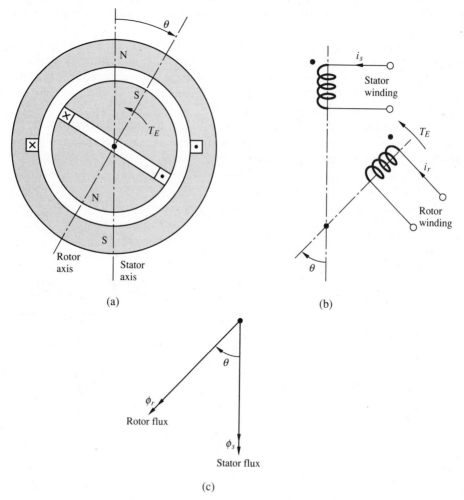

Figure 4.11 Excitation torque from direct current, doubly-excited system. (a) Schematic cross section, (b) primitive circuit, (c) flux displacement.

axes of the coils. Torque is present if there is an angular displacement between the two flux phasors. The magnetic polarities in Fig. 4.11a show the existence of torque by the attraction of unlike poles. Figure 4.11b depicts the primitive model and Figs. 4.11b and c illustrate the dot convention for flux directions.

If the rotor is stationary and if the displacement θ of the coil axes is zero, the excitation torque is zero. This is a position of stable equilibrium, because a small displacement of the rotor in either direction gives rise to a restoring torque. Another position of equilibrium is when $\theta = \pi$, but this is a position of unstable equilibrium. Any small displacement brings into play the repulsion forces between like polarities. For the general position, where $0 < \theta < \pi$, the excitation torque opposes any increase in displacement. That is, the torque is counterclockwise. Increasing the displacement still further so that $\pi < \theta < 2\pi$, inspection of Fig. 4.11 reveals that the torque has changed polarity and is clockwise in direction. Accordingly, the excitation torque is cyclic with displacement. With every revolution of the rotor, the torque goes through a complete cycle, positive and negative.

The excitation torque under steady rotation has the same form. There is an instantaneous value but the average value is zero, so there is no net electromagnetic torque. Direct current in both windings does not seem to produce a useful machine. We want to determine some of the forms of excitation that do give rise to a constant unidirectional torque, or at least a net average torque, even if the torque does vary instantaneously with displacement or time.

There is an exception to the above argument that no net average torque exists if the machine is doubly excited with direct current. It is not apparent from the machine configurations used. The geometry of all previous machines has been cyclic. Cyclic in this sense means that rotor (or stator) conductors are under the consecutive influence of stator (or rotor) S-poles and N-poles during continuous motion of the rotor. In

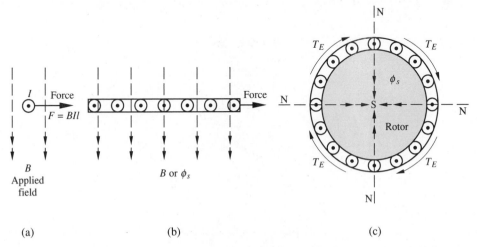

   (a)   (b)   (c)

Figure 4.12 (a) Force on a single conductor, (b) increasing the force, (c) changing the geometry to provide torque.

Fig. 4.11 the rotor conductors are under the influence of the stator south pole and the north pole in turn during motion.

We can return to first principles to determine a machine configuration with a unidirectional torque when direct current provides the double excitation. Figure 4.12 illustrates the development. It is assumed that the field excitation is constant and the conductor currents are constant. An external field, B, acting on a single conductor, produces a force as depicted in Fig. 4.12a. The force can be increased by increasing the number of conductors, as shown in Fig. 4.12b, as long as the directions of the field B and the current I remain the same. Torque and rotation are achieved by twisting the conductor configuration into the cylindrical arrangement as illustrated in Fig. 4.12c. This machine has a radial unidirectional field acting on conductors carrying unidirectional current. The result is a unidirectional torque. Such a machine has been named the *homopolar* machine, because the rotor conductors always see a magnetic field of the same polarity as they proceed through any one cycle or revolution. This is an acyclic configuration. The reader might be interested to work out some of the practical details of how to arrange the field winding and how to provide the unidirectional current in this, the only true dc machine.

4.3.2. Stator and Rotor Windings with DC Excitation at the Terminals

A stator winding excited with direct current will produce a field ϕ_s fixed in space. Figure 4.13 illustrates this for a salient-pole stator. The rotor field must also be fixed in space if the excitation torque T_E is to be unidirectional. The rotor winding is distributed around the periphery to make the best use of space and the best use of the stator field.

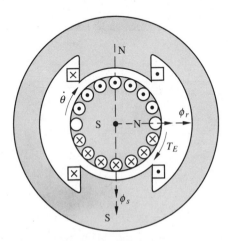

Figure 4.13   Direct current motor.

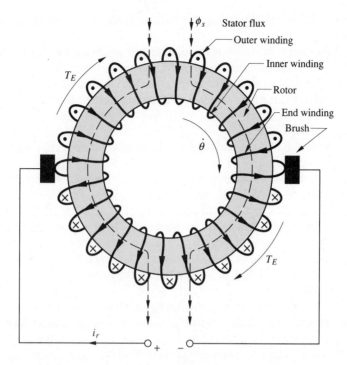

Figure 4.14 Gramme ring winding and commutation.

For motoring action, the conductors adjacent to the stator NORTH pole must always have current coming out of the paper. Therefore, conductors adjacent to the stator SOUTH pole must have current in a direction into the paper. The stator flux and the resultant rotor flux are always mutually at right angles, fixed in space and producing a unidirectional torque, if this rotor current pattern is maintained.

In order to maintain the rotor current pattern, the current in each conductor has to reverse direction every half revolution of the rotor motion. Rotor conductors with alternating current do not seem to indicate excitation from a dc voltage supply, but it can be so. This paradox is explained schematically in Fig. 4.14, which shows a Gramme ring winding (named after the inventor). The rotor consists of a continuous toroidal winding closed on itself. It is rotating. Two stationary leads from a dc source are connected to stationary brushes, which slide on the conductors of the toroid as it rotates. No matter what position the rotor takes, whether stationary or moving, the current distribution in space is as shown in the figure. That current distribution is fixed in space. The current in the outer part of the rotor conductors (marked · and ×) interacts with the stator field. For a toroid of magnetic material the stator flux is not experienced by the inner conductors, so they contribute nothing to the torque.

It is a simple step to progress from the obsolete Gramme ring winding, which has much nonutilized conductor, to the present-day drum winding which eliminates the inner part of the toroidal winding. The drum winding is discussed in Chapter 5. In practice, the switching of the conductor current is accomplished by brushes sliding over a commutator.

These principles refer to what is called the *dc machine.* One of the advantages over other types of machines is that the commutator affords the machine a torque, not only at standstill, but also at any speed, because the current pattern of the rotor conductors remains stationary relative to the main field of the stator.

4.3.3. Stator and Rotor Windings with Single-phase AC Excitation

Consider a stator winding to be excited by alternating current. What rotor excitation is needed to maintain a unidirectional torque for motoring action? As shown in Fig. 4.15a, during the first half cycle of the stator winding excitation, the current is such as to produce a field in space, downward in the sketch. Using a distributed rotor winding and supplying direct current through a commutator fixes the current pattern in space while the armature rotates. Figure 4.15b shows this pattern. It does not matter to what speed the rotor has accelerated, or what position the rotor takes, the torque remains clockwise for motoring action.

Over the next half-cycle of stator winding excitation, current reversal causes the stator flux to reverse. This is shown in Fig. 4.15c. If the rotor current were supplied from a dc source, the rotor current pattern would be the same as that depicted in Fig. 4.15b and so the resultant torque would reverse and be in a counterclockwise direction. The net torque over one cycle of current would then be zero. However, if the armature current were reversed at the same time that the stator current reversed, the torque would again be in the clockwise direction, as indicated in Fig. 4.15c. So when the armature current alternates in sympathy with the stator excitation, a useful motoring torque is produced. For the stator and armature currents to alternate in phase with each other the two windings can be connected in series.

Such a motor with single-phase alternating current in both windings is one member of a family of machines called ac commutator motors. This motor has the same structural form as the dc commutator motor. Such a machine could be operated from either a dc or an ac supply. If the windings are in series, this motor is called a *universal* motor. An example of this versatility is the motor used in some electric shavers. Other examples are domestic appliances and power tools.

4.3.4. Stator Winding DC Excited and Rotor Winding AC Excited

A rotor winding that is supplied with single-phase alternating current through a commutator has just been shown to produce a flux alternating along a fixed axis. *Alternating* means reversing direction. With a fixed stator field, no net torque can be generated.

Other means of providing alternating current to the rotor windings are by a generator on the same shaft, or, more commonly, by stationary brush contacts sliding on slip rings, which are connected to the ends of the winding. In this way the rotor

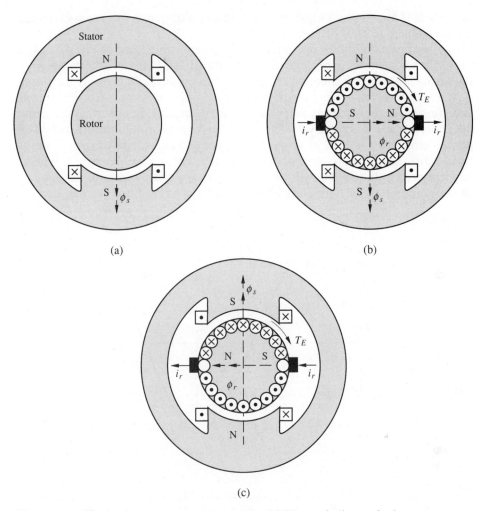

(a)   (b)

(c)

Figure 4.15 Single-phase ac commutator motor. (a) Stator winding excitation over one-half cycle, (b) rotor current pattern for clockwise torque, (c) rotor current pattern for clockwise torque and reversed stator excitation.

winding produces an alternating flux, whose axis is fixed relative to the rotor and its winding. If the rotor rotates, the flux axis rotates at the same speed.

The only way, in which an average electromagnetic torque will exist for a fixed stator field, is if the angular frequency of rotation of the rotor is the same as the frequency of alternation of the rotor winding current. Once the rotor is at such a speed, the torque will pulsate but will not reverse its polarity. As the frequency of rotation and the frequency of excitation are the same, the machine is called a *single-phase synchronous machine*. As a motor it is not self-starting and will only run at one speed, if the frequency of the supply is constant. Starting this type of motor, while not

impossible if using other torques such as induction torque, is too much of an economic problem to make its use widespread.

A description of motor operation can be given with the aid of Fig. 4.16. This
shows the rotor current to be a maximum at instant (i) and entering at terminal X. As
shown in Fig. 4.16b, this situation at instant (i) produces a resultant clockwise torque.
As the rotor revolves, the flux axis of the rotor follows at the same speed. The current
is decreasing, so the rotor flux magnitude is also decreasing. In order that the torque
does not reverse, it is necessary that the current in the rotor winding reverses when the
point X reaches a point $\pi/2$ in advance of the position at instant (i). Figure 4.16b, at
instants (ii) and (iii), shows the resultant flux just before and after X is at $\pi/2$. When X
has advanced π, the current will be a maximum again in the reverse direction, as
shown at instant (iv), having completed one-half cycle. The rotor flux has the same

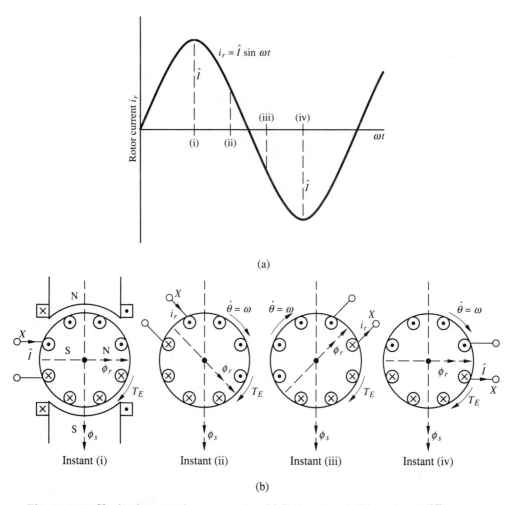

(a)

(b)

Figure 4.16 Single-phase synchronous motor. (a) Rotor current, (b) torque at different
instants of time for synchronous speed ($\dot{\theta} = \omega$) of rotor.

direction and magnitude as at instant (i). Thus, the net torque is unidirectional but it pulsates in magnitude. If the rotor loses synchronism (i.e., if $\dot{\theta} \neq \omega$), there is an alternating forward and reverse torque. That is, the average torque is zero.

4.3.5. Stator Winding DC Excited and Rotor Winding Polyphase Excited

It was shown in subsection B under section 4.2.1 that polyphase-winding excitation produces a magnetic field that travels around the air gap of the machine at ω radians per second relative to that winding. The angular frequency of rotation ω is the same as the polyphase excitation angular frequency ($\omega = 2\pi f$) in a 2-pole machine. How is the rotor field ϕ_r made to be stationary in space, lagging or leading the stator field ϕ_s, so that the excitation torque T_E remains unidirectional?

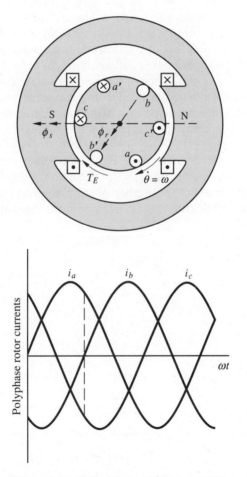

Figure 4.17 Polyphase synchronous motor.

As shown in Fig. 4.17 for motor action, a rotor field, stationary in space, occurs only if the rotor winding connections are such that the rotor field rotates synchronously in a counterclockwise direction relative to the winding when the rotor itself rotates synchronously in a clockwise direction relative to the stator.

This type of motor is called a *polyphase synchronous motor.* Because there is no inherent starting torque, the machine is used chiefly as a generator. Usually the generator is driven at constant speed to obtain a constant frequency output voltage.

4.4. SUMMARY

The electromagnetic torque, whether for motoring or generating action, has been subdivided into the three forms, reluctance, induction, and excitation, which can exist separately or in any combination.

Excitation of a machine can be on the rotor, on the stator or on both, using direct current, single-phase or polyphase alternating current, or a combination of these. Not all forms of combination produce a useful machine but many do. However, there are three basic types of machines that are most commonly used. They are the dc machine, the induction machine, and the synchronous machine, and they can involve one or more of the above variations of configuration. The induction motor excitation is different from the others in that one of the member windings (usually the rotor) is short-circuited rather than being supplied from an external source.

We have described the primitive forms of the most common machines. In reality these machines usually employ special steels for the magnetic circuit and copper for the windings carrying currents. There are many more types of machines and other kinds of materials that can be used. Some liquids and gasses can conduct electricity also. When superconductors are employed for excitation, it is not always necessary to employ magnetic materials. The number of ways to create a torque due to the interaction of magnetic fields is large. The possibility for innovation always remains.

CHAPTER FIVE

Direct Current Machines

5.1. INTRODUCTION

The advantage offered by dc machines is their versatility. The ability to develop high starting and braking torques, to make quick reversals of rotation, to maintain constant mechanical power output or to maintain constant torque and to permit continuous speed variation over a range as large as 4 : 1, make dc motors better suited for many industrial applications than their ac counterparts.

The current supplied to a dc motor by a dc voltage supply needs to be inverted to an alternating current in the conductors of the rotor winding in order to develop a unidirectional torque. This inversion is accomplished by means of a commutator mounted on the shaft and a stationary brush arrangement, which are integral parts of the conventional dc machine. The commutator and brushes make dc machines the least rugged of electric machines. Also, the added complexity of the commutator makes dc machines more expensive than ac machines of comparable rating. In spite of these disadvantages the versatility of dc machines makes them the preferred machine for many applications.

DC generators are not suited for the generation and subsequent transmission of electric energy over large distances. The ease and high efficiency with which a generated ac voltage can be transformed to any high voltage level makes dc generators suitable only for the generation of electric energy for local consumption (e.g., wind generators). As a constant voltage dc supply, the dc generator, normally driven by an ac motor, has to give way to the modern high-power semiconductor rectifier supplies on economic grounds. DC generators are used if the voltage of a dc energy source has to be varied in a specified manner over a relatively wide range.

In this chapter we will discuss some topics that have a bearing on the operation and performance of dc machines.

171

5.2. THE BASIC MACHINE

The physical processes taking place in dc motors and generators are the same. As a result, the same machine can perform either function. To generate a voltage or develop a torque a magnetic field and an arrangement of conductors are needed. A magnetic field is produced by the mmf of field coils supplied with direct current. The arrangement of conductors, which are interconnected to form a number of identical coils, is called the *armature winding*. That part of the magnetic circuit that carries the armature winding is often called the *armature*.

In dc machines, the current direction in a coil of the armature winding has to be reversed when the coil sides move from a magnetic field of one polarity into the magnetic field of opposite polarity. In conventional dc machines this commutation process is accomplished by a switching operation performed by stationary brushes sliding over the cylindrical surface of a commutator. A commutator consists of copper segments, which form a cylindrical body mounted on the machine shaft. The individual segments are electrically insulated from each other and from the shaft. Each coil end is connected to a commutator segment. Hence, in conventional dc machines, the armature winding is on the rotor to facilitate the required commutation of the coil current. The coil sides of the armature winding are contained in axial slots that are uniformly distributed around the periphery of the rotor, as schematically indicated in Fig. 5.1.

The stator of a dc machine has salient poles, on which the main field coils are mounted. With the development of strong permanent magnet materials these are also used for the poles to produce the main field of fractional-horsepower motors or very small special purpose generators.

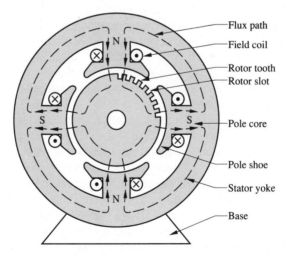

Figure 5.1 Schematic cross section of a 4-pole dc machine.

5.2.1. Magnetic Circuit

The path followed by the main-pole flux of a dc machine is indicated by the broken lines in Fig. 5.1. Ferromagnetic materials are used for the stator yoke, the poles, and the rotor. They provide a low reluctance path and they guide the flux in a desired manner across the air gap. The pole surface, which is facing the conductors on the rotor, is extended in a radial direction to form the so-called *pole shoe.* This is done to increase the number of armature coil sides that are subjected to a strong magnetic field, and this increases the output of a machine. The use of pole shoes reduces the reluctance of the air gap and this reduces the number of ampere turns required to produce a certain air-gap flux. This is an important consideration, because, even with pole shoes, 85% or more of the mmf is required at rated operating conditions to overcome the reluctance of the air gaps.

The reluctance of the air gap at the location of rotor teeth is smaller than the gap reluctance at the location of the slots. This superposes a periodic flux-density variation on the average flux density in the air gap. Rotation of the armature will cause a variation of the flux density in the pole-shoe material. To reduce the resulting eddy current loss, the pole shoes must be made of magnetic steel laminations. In general, the whole pole structure is made of laminated steel.

The stator yoke carries a unidirectional flux and is made of solid ferromagnetic material. Only fractional-horsepower machines have their complete stator structure made of laminations to facilitate economical mass production. In small, integral-horsepower machines, cast iron can be used for the stator yoke, but fabricated (rolled) steel is invariably used for the larger machines. Having about twice the permeability of cast steel only half the cross section is required to conduct the same stator flux for about the same number of ampere turns. The resulting reduction in weight is an important factor in larger machines.

5.2.2. Methods of Main Field Excitation

The main magnetic flux in dc machines is produced by the mmf of field coils that are mounted on the cores of the salient stator poles. The field coils all have the same number of turns. They are connected in series to ensure that each field coil produces the same mmf. Care is also taken that all magnetic flux paths have the same reluctance. The magnetic flux in each path is then the same. This is an important requirement to prevent the occurrence of circulating currents in the armature winding. The series-connected field coils are commonly referred to as the *field winding.*

The characteristics of a dc machine are determined by the manner in which the field winding is supplied with direct current. If the field current of a generator is supplied by some independent dc voltage source, the machine is said to be a *separately-excited* generator. In the schematic connection diagrams of Fig. 5.2, and subsequent diagrams, only the principle of the winding connections is indicated. The armature winding is represented by the circular shape of the rotor. The axes of the field winding are shown in their proper space relations to the brush axis.

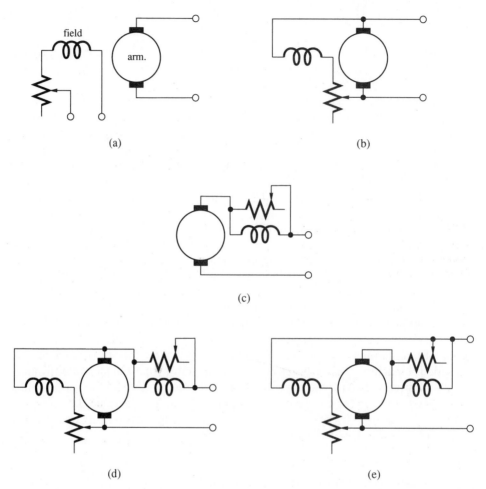

Figure 5.2 Field and armature winding connections of a dc generator. (a) Separate excitation, (b) shunt excitation, (c) series excitation, (d) short compounding, (e) long compounding.

If a generator supplies its own field current, we speak of a *self-excited* generator. Always in series with a rheostat, the field winding is connected in parallel with (or as a shunt across) the armature winding. Accordingly a self-excited generator, with its field winding connected in this manner, is called a *shunt* generator. The voltage available at the terminals of the generator supplies the excitation current, which can be varied by adjusting the field rheostat. As the name implies, the field winding of a *series* generator is connected in series with the armature winding. Therefore, the field excitation is determined by the load current to be delivered by the generator. Adjustment of the excitation current is accomplished by means of a variable resistor, often referred to as a diverter, connected in parallel with the field winding.

The connection diagrams, shown in Figs. 5.2d and e, are those of *compound* generators. In both cases, part of the excitation is provided by shunt-connected field coils and part is provided by series-connected field coils. Each stator pole carries a coil of the shunt winding, on top of which is wound a coil of the series winding. The connection diagram in Fig. 5.2d is that of a short-compound generator, whereas the diagram in Fig. 5.2e is that of a long-compound generator. The difference between the operating characteristics of the two types of connection of a given machine is small. In the long-compound connection the voltage across the shunt field is slightly lower, but the current in the series winding slightly higher than in the case of the short-compound connection.

The excitation current in the field windings of motors is always supplied by the dc voltage of the electric energy source. The different connections of the field winding relative to the armature winding depicted in Fig. 5.2 are also used for motor excitation. The same terminology is used to indicate the type of connection and to name a motor a shunt, a series, or a compound motor.

In the case of a shunt-connected motor, the field current has to be supplied by the electric energy source and constitutes that part of the supply current that does not contribute to energy conversion. The shunt-field current must therefore be kept small.[1] Since the voltage across the field winding is relatively high, a high winding resistance is required to keep the field current small. Being small, the permitted field current requires an exciting coil of many turns of small cross section to produce the ampere turns for the necessary air-gap flux.

The current that flows through a series winding is the supply current. The voltage drop across the series winding should be kept small, since it reduces the terminal voltage of a generator or the armature-winding voltage of a motor. Also the ohmic loss in the series field winding should be kept low. It follows that the field coils of a series-excited machine must be made of large cross-section conductor and need only relatively few turns to produce the necessary ampere turns.

5.2.3. Principle of Commutation

The commutator and the brushes comprise a set of mechanical switches that operate so that the interaction between the main field and the armature-winding currents provides a unidirectional electromagnetic torque. This is a requirement for motoring action. In a dc generator, the commutator acts as a rectifier, so that the output at the armature winding terminals is direct current.

The principle of commutation can be described by considering a single-turn coil rotating in a magnetic field, as shown in Fig. 5.3. In this case the commutator consists of two half-cylindrical copper segments that are mechanically secured to but electrically insulated from the rotor shaft. The two segments are also electrically insulated

---

[1] The shunt field current varies from about 0.5% to 5% of the full-load current. Normally, the larger the machine the smaller the percentage field current.

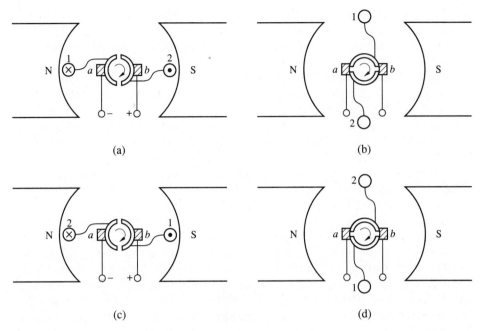

Figure 5.3 Commutation process for a single-turn coil.

from each other. Each commutator segment is connected to one coil end. Electrical connection between the ends of the rotating coil and the terminals of a dc machine is accomplished by means of two stationary carbon brushes. The commutator segments slide under these brushes.

Figure 5.3 shows four successive positions of the coil as it rotates in a clockwise direction between two stationary poles. The directions of the induced conductor voltages are indicated by crosses and dots.[2] In Fig. 5.3a, conductor 1 is in electrical contact with brush *a* and conductor 2 is in contact with brush *b*. Brush *a* is connected to the negative terminal and brush *b* is connected to the positive terminal of this elementary generator.

Figure 5.3b shows the position of the coil after it has rotated 90°. The coil sides are now in a position in which they leave the influence of one magnetic polarity to enter the influence of the opposite magnetic polarity. The magnetic flux density is zero at the location of the coil sides and the emf induced in the conductors is zero. The brushes are in electrical contact with both segments and short-circuit the coil. The terminal voltage is zero. The plane perpendicular to the magnetic field is called the *geometric neutral plane*.

---

[2] A cross indicates a direction into the plane of the paper and a dot indicates a direction out of the plane of the paper.

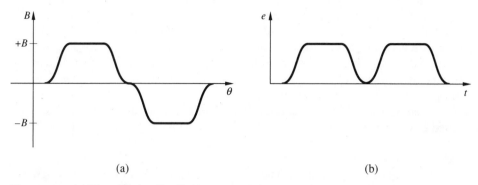

Figure 5.4 (a) Flux density distribution around the rotor, (b) generated voltage.

Figure 5.3c shows the coil after it has moved through another 90°. Now the polarity of the emfs in each of the coil sides is reversed. Because the commutator segments are fixed to the shaft of the rotor and the brushes are stationary in space, conductor 2 is now in electrical contact with brush *a* while conductor 1 is in contact with brush *b*. Thus the polarity of the voltage at the brushes and at the terminals of the machine has not changed. In Fig. 5.3d the coil has rotated another 90°, and the coil sides are again in the geometric neutral plane. The induced coil emf is zero and in this position the coil is short-circuited, since each brush contacts both commutator segments simultaneously. We see that the action of the rotating commutator and stationary brushes is to reverse the connections of the coil to the machine terminals when the polarity of the induced coil voltage reverses. Each brush is maintained at the same polarity at all times. Therefore, the voltage at the terminals of the machine is unidirectional and will be of the form indicated in Fig. 5.4b, if the magnetic flux density distribution around the rotor has the form shown in Fig. 5.4a.

If the machine is operating as a motor, the terminals are connected to a dc voltage supply and the stationary brushes have a fixed polarity. By means of the rotating commutator segments the current in the conductors is reversed when the coil sides pass through the geometric neutral plane. Thus the direction of current in the conductor moving under a specific pole face will always be the same. In this manner a unidirectional torque is developed.

A more detailed discussion of commutation is given in section 5.5.

5.2.4. Armature Windings

The excitation winding on the poles of the stator of a dc machine provides the magnetic field that allows electromechanical energy conversion to take place. The conversion is associated with the electric energy in the winding of the armature. So the question is, what form does the armature winding take, if we have a given terminal voltage and a particular power requirement?

The unidirectional terminal voltage generated by a single-turn armature coil pulsates. (See Fig. 5.4.) Further, the magnitude of this voltage is low. Several multi-turn coils, evenly distributed around the periphery of the rotor and connected in series with each other, will increase the total generated voltage at reasonable rotor speeds, flux densities, and machine dimensions. An important effect of using a number of series-connected coils, constituting the armature winding, is that the voltage pulsation is reduced. Why is this so?

The earliest developed armature winding is the Gramme ring winding. It is named after its inventor. This type of winding, which has been obsolete for over half a century, illustrates clearly some of the major requirements and basic ideas underlying modern armature windings. Figure 5.5 illustrates an end view of a 2-pole, dc genera-tor with a Gramme ring winding. The rotor body, which supports the windings, is a hollow cylinder of ferromagnetic material. The armature winding is wound spirally around the cylinder wall. The conductors on the outer surface are parallel to the axis of the cylinder. The turns of the winding are uniformly distributed and at equally spaced points, between which we may have a certain number of turns, this winding is tapped and connected to commutator segments.

If the generator is driven in a clockwise direction between the two magnetic field poles, emfs will be generated in the conductors that lie on the outer surface of the rotor. The conductors on the inner surface of the rotor cylinder are shielded from the influence of the magnetic field and will not contribute to the voltage generation. The poor utilization of conductor copper and difficulties in manufacturing and mainte-nance are the reasons that the Gramme winding has been superceded by the *drum* type of winding.

All emfs induced in conductors moving under a certain magnetic pole are in the same direction as indicated in Fig. 5.5. The direction of induced emf reverses as the conductor passes through the geometric neutral plane. Tracing the general direction

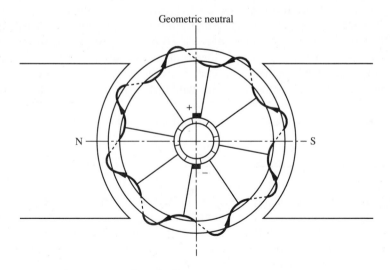

Figure 5.5 Gramme ring winding.

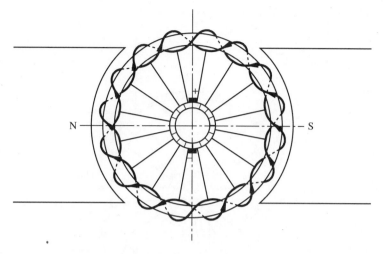

Figure 5.6 Duplex Gramme ring winding.

of conductor emfs through the winding we find one point where two emfs meet and one point where two emfs separate. Between these two points, which are 180° apart, there exists the maximum potential difference. It follows that, to obtain the maximum machine terminal voltage, the brushes should make contact with commutator segments connected to conductors that are in the geometric neutral plane.

The generator load current enters the winding through the negative brush. The current divides, part flowing through the conductors under the north pole and part flowing through the conductors under the south pole. Both current components combine at the positive brush where the load current leaves the generator winding. To maintain an equal division of the load current between the two parallel branches at all times, the winding should be closed and the resistance of each branch should be the same for any position of the rotor. These requirements, combined with the requirement that the total voltage generated in each parallel branch must be the same at all times (to prevent circulating currents in the winding), demand a truly uniform distribution of the conductors along the rotor surface.

The number of parallel paths in an armature winding depends on the winding design and the number of magnetic poles. The effect of winding design on the number of parallel paths can be understood by considering the winding shown in Fig. 5.6. In tracing the winding from the positive to the negative brush, we can proceed along four different parallel paths consisting of an identical number of conductors, which generate an equal total voltage between the brushes. This winding is called a duplex winding. With a further increase in the number of conductors and commutator segments, triplex, quadruplex, etc. windings can be designed. The number of parallel paths in these windings is three times, four times, etc. as many as the number of parallel paths in the simplex winding of Fig. 5.5.

The effect of the number of poles on the number of parallel paths can be understood by considering a four-pole machine, which has a simplex winding. (See Fig.

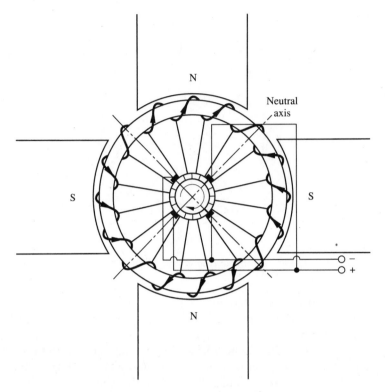

Figure 5.7 Four-pole, simplex Gramme ring winding.

5.7.) In this case, we have four geometric neutral zones or axes. Tracing the general directions of the conductor emfs of this generator, which is being driven in a clockwise direction, we see that emfs meet and separate at the geometric neutral axes. We have two positive brushes and two negative brushes. The voltage difference between each positive and negative brush is equal, because the winding is uniformly distributed and perfectly centered between magnetic poles of equal strength. Thus, we can connect electrically both positive brushes and both negative brushes. In doing so we have four identical parallel paths through the winding between the positive and the negative terminals of the machine.

The number of pairs of parallel paths between the machine terminals is generally indicated by the symbol a. One of the advantages of a larger number of parallel paths is that the machine has a larger current carrying capacity without having to resort to an excessive cross section for the conductors.

The Gramme ring winding has been used to illustrate the form of a distributed armature winding and the need for multicoil, series-parallel arrangements. However, the practical winding is a drum-type winding. This type is described in Appendix G. The drum winding has the same characteristics as the Gramme ring winding described above, but is more economical with space and materials.

5.3. GENERATED VOLTAGE AND DEVELOPED TORQUE

5.3.1. Generated Voltage

Each armature conductor of a dc machine has induced in it an emf due to the conductor motion in the main-pole field. The generated terminal voltage is equal to the total induced emf in one of the parallel paths of the armature winding. Due to the switching of coils from one path to the next by the commutator, the voltages of the coils in each parallel winding path are always additive when the brushes are on the magnetic neutral axis.

Figure 5.8 shows the flux density distribution when the armature winding carries no current and when the effect of rotor teeth is neglected. Although the following discussion applies to any type of armature winding, it may be helpful to consider the coils forming one path of a simplex lap winding of a two-pole machine. The brushes are fixed and the instantaneous voltage generated between them is the sum of the instantaneous coil voltages. As the armature rotates, the angle α, which is measured from the neutral axis, varies from zero to a maximum value of $2\pi/G$, where G is the total number of rotor slots. The angle α cannot exceed this maximum value because, when it has reached this value, coil 6 is removed from the armature path considered and is commutated to the other parallel armature path of the winding. At this moment, a new coil, positioned at $\alpha = 0$, is switched into the winding path that is being considered. The effect of the variation of α is that the generated voltage will

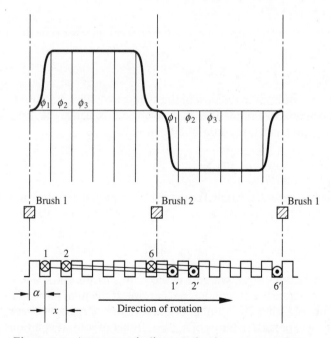

Figure 5.8 Armature winding rotating in a magnetic field.

have a pulsation or ripple superposed upon the average value. The larger is the number of slots, or distributed coils, the smaller the ripple.

The total flux per pole can be divided into a number of partial flux components, Φ_1, Φ_2, Φ_3, etc., each of which flows through a small area lx where l is the axial length of the conductors in the magnetic field and x (the slot pitch) is the distance between adjacent rotor slots. If each coil has z turns, the average voltage induced in a coil in slots $1-1'$ when traveling at a constant speed v is

$$e_1 = 2zB_{1\,av}lv \qquad (5.3.1)$$

where $B_{1\,av}$ can be found from $B_{1\,av} = \Phi_1/lx$. Thus the average voltage induced in a coil in slots $1-1'$ is

$$e_1 = \frac{2zv\Phi_1}{x}. \qquad (5.3.2)$$

In a similar manner, we find that the average value of the voltage induced per coil in slots $2-2'$ when traveling the distance x is given by

$$e_2 = \frac{2zv\Phi_2}{x}. \qquad (5.3.3)$$

If each slot contains z_s coil sides, we find that the sum of the average voltages of all coils in a double-pole pitch is

$$\sum e_{av} = \frac{z_s 2zv}{x}(\Phi_1 + \Phi_2 + \Phi_3 + \cdots). \qquad (5.3.4)$$

The sum of all the partial fluxes is equal to the total flux per pole Φ and thus

$$\sum e_{av} = \frac{2zz_s v}{x}\Phi. \qquad (5.3.5)$$

This equation not only gives the sum of the average voltages, but also shows that the value is independent of the manner in which the flux density is distributed. For a machine with p pole pairs, the sum of all coil voltages is

$$p\sum e_{av} = \frac{2zz_s vp\Phi}{x}. \qquad (5.3.6)$$

If the winding has a pairs of parallel paths, the average generated terminal voltage E_a is given by

$$E_a = \frac{p}{2a}\sum e_{av} = \frac{zz_s vp\Phi}{xa}. \qquad (5.3.7)$$

The distance x is $x = 2\pi r/G$, where r is the radius of the rotor and G is the total number of slots. There are zz_s conductors per slot. If the total number of conductors on the armature is Z, then the number of slots is Z/zz_s. Thus $x = 2\pi rzz_s/Z$. The coil velocity v can be expressed in terms of n, the number of rotor revolutions per second. That is $v = 2\pi rn$. Substitution of the relations for x and v in the expression for the average generated terminal voltage gives

$$E_a = \frac{p}{a} \Phi n Z. \tag{5.3.8}$$

If a machine operates as a motor, a counter or back emf is generated in the armature winding. Thus, regardless of the mode of operation, the average voltage generated in an armature winding of a dc machine is

$$E_a = \frac{pZ}{a} n\Phi = K_e n\Phi \tag{5.3.9}$$

where $K_e = pZ/a$ is a design constant for a particular machine.

For the dc machine operating as a generator, the current I_a in the armature winding is in the direction of the generated voltage E_a. As a motor, the generated emf E_a opposes the applied voltage and the current I_a is in a direction opposite to that of the generated voltage. If the total resistance of the armature circuit between the machine terminals is R_a ohms, the armature current I_a will cause a resistance voltage drop in the winding. Thus, in a generator, the voltage E_a will exceed the terminal voltage V, whereas the opposite will be the case in a motor. The basic armature circuit equations are then as follows.

For generator operation

$$V = E_a - I_a R_a \tag{5.3.10}$$

and for motor operation

$$V = E_a + I_a R_a. \tag{5.3.11}$$

These equations are important for calculating the armature circuit response (i.e., the value of the current).

EXAMPLE 5.1

A separately excited 4-pole, 100-kW, 250-V generator has a duplex lap winding with a resistance R_a of 0.05 ohm. There are 500 armature conductors. The flux is 0.04 weber per pole. At what speed must the armature be driven for the generator to deliver its rated power?

Solution

For a separately-excited generator the load current equals the armature current. Hence, at rated load, the armature current is

$$I_a = \frac{100000}{250} = 400 \text{ A.}$$

The generated voltage follows from

$$E_a = V + I_a R_a = 250 + (400 \times 0.05) = 270 \text{ V.}$$

For a lap winding $a = mp$, where m is the multiplex factor (for simplex $m = 1$, for duplex $m = 2$, etc.). Therefore

$$a = mp = 2 \times 2 = 4.$$

It follows that the duplex lap winding of this generator has four pairs of parallel paths. The required speed follows from

$$n = \frac{aE_a}{pZ\Phi} = \frac{4 \times 270}{2 \times 500 \times 0.04} = 27 \text{ r/s.}$$

EXAMPLE 5.2

The resistance of the load supplied by the generator of Example 5.1 is doubled. The flux per pole remains the same. What should the speed be in order to maintain the same terminal voltage?

Solution
The terminal voltage remains the same and the load resistance is doubled. Thus the armature current of the separately-excited generator is halved. That is

$$I_a = 400/2 = 200 \text{ A.}$$

The generated voltage must be $E_a = V + I_a R_a = 250 + (200 \times 0.05) = 260$ V.
The generated voltage varies directly with the speed, since the flux remains constant. At $n = 27$ r/s, E_a is 270 V.

Therefore the new speed must be $n = \dfrac{260}{270} \times 27 = 26$ r/s.

5.3.2. Developed Torque

For both generators and motors an electromagnetic torque is developed if the conductors of the armature winding carry current and are in the magnetic field of the main poles. Each conductor of the winding carries a current $I_a/2a$. The forces developed on the conductors aid each other at all times, if the brushes are placed in the magnetic neutral axis. The total instantaneous torque will have a small periodic variation of its magnitude as the rotor moves through an angular displacement equal to the distance between two slots in the rotor surface. However, the instantaneous torque of commercial machines does not deviate appreciably from the average torque value. Let us determine the average torque expression.

Consider Fig. 5.8 again. The average force $f_{1\,av}$, developed by all conductors in slot 1, is from $F = BlI$

$$f_{1\,av} = zz_s \frac{\Phi_1}{lx} l \frac{I_a}{2a} \qquad (5.3.12)$$

where z is the number of turns per coil and z_s is the number of coil sides per slot. If r is the radius of the rotor, the average partial torque developed by the conductors in slots $1-1'$ is given by

$$T_{1\,partial} = 2zz_s r \frac{I_a \Phi_1}{2ax}.$$

(5.3.13)

Similar expressions can be written for the partial torques developed by the conductors of coils in the other rotor slots. The total torque developed by all conductors per pole pair is found from summation. That is

$$T = 2zz_s r \frac{I_a}{2ax} (\Phi_1 + \Phi_2 + \Phi_3 + \cdots) = 2zz_s r \frac{I_a \Phi}{2ax}$$

(5.3.14)

where Φ is the total flux per pole. The total average electromagnetic torque T_e that is developed in a machine which has p pole pairs is

$$T_e = zz_s rp \frac{I_a \Phi}{ax}.$$

(5.3.15)

However, $x = 2\pi rzz_s/Z$. Hence

$$T_e = \frac{pZ}{2\pi a} I_a \Phi = K_T I_a \Phi$$

(5.3.16)

where K_T is a design constant for a particular machine. We notice that the developed torque is independent of the speed of the rotor. This electromagnetic torque is not equal to the shaft torque T_m. In the case of a motor, internal loss torques due to friction have to be overcome, so that in this case the useful shaft torque is

$$T_m = T_e - T_{loss}.$$

(5.3.17)

In the case of generator operation the torque of electromagnetic origin and the internal loss torques have to be overcome by the prime mover, so that the shaft torque to be delivered by the prime mover is

$$T_m = T_e + T_{loss}.$$

(5.3.18)

The basic voltage eq. (5.3.11) for the armature circuit of a motor is $V = E_a + I_a R_a$. Multiplying both sides of this equation by I_a results in a power equation

$$VI_a = E_a I_a + I_a^2 R_a.$$

(5.3.19)

The term VI_a on the left-hand side represents the total power input to the armature circuit. The second term $I_a^2 R_a$ on the right-hand side is the ohmic power loss in the armature winding. Consequently we come to the important conclusion that the term $E_a I_a$ is the net electrical power that is converted to mechanical power. The electrical power P converted to mechanical power is

$$P = T_e \times 2\pi n.$$

(5.3.20)

But

$$P = E_a I_a = K_e \Phi n I_a \qquad (5.3.21)$$

so that

$$T_e = \frac{K_e \Phi n I_a}{2\pi n} = K_T \Phi I_a \qquad (5.3.22)$$

where $K_T = K_e/2\pi$. These are important equations for determining dc motor performance.

EXAMPLE 5.3

A shunt motor draws an armature winding current of 40 A from a 120-V supply when delivering 4 kW of mechanical power at a speed of 20 r/s. The armature winding has a resistance of 0.25 ohms. (a) What is the loss torque of the machine at 20 r/s? (b) What armature current is required to supply half the mechanical power at the same speed?

Solution
(a) To determine the loss torque we have to know the electromagnetic developed torque T_e and the useful shaft torque T_m. To determine T_e we make use of the power relations

$$P = E_a I_a = 2\pi n T_e$$

since we know I_a and n, and can determine E_a from $V = E_a + I_a R_a$.
Thus $E_a = 120 - (40 \times 0.25) = 110$ V
and $T_e = \dfrac{110 \times 40}{2\pi \times 20} = 35$ N·m.

The useful shaft torque follows from $T_m = \dfrac{P_o}{2\pi n} = \dfrac{4000}{2\pi \times 20} = 31.8$ N·m.

Thus, the loss torque is $T_{loss} = T_e - T_m = 35 - 31.8 = 3.2$ N·m.
(b) Half the mechanical power is delivered at the same speed when the shaft torque is halved. Thus $T_m = 31.8/2 = 15.9$ N·m.
Since the speed is not altered, the loss torque remains 3.2 N·m. The electromagnetic torque that has to be developed is

$$T_e = T_m + T_{loss} = 15.9 + 3.2 = 19.1 \text{ N·m}.$$

The flux of a shunt machine remains constant for a given applied voltage. The developed torque is then directly proportional to the armature current. A current of 40 A develops a torque of 35 N·m. Hence, to develop a torque of 19.1 N·m, the required armature current is

$$I_a = \frac{19.1}{35} \times 40 = 21.8 \text{ A}.$$

5.4. ARMATURE MMF

5.4.1. Armature Reaction

An mmf is set up by the current-carrying armature winding and it affects the resultant flux distribution in the air gap. The generated voltage and the developed torque are directly proportional to the total air-gap flux per pole. Therefore, the armature mmf, which is called *armature reaction,* will have an effect on the operating characteristics of the machine. In order to determine this effect we analyze the mmf of the armature winding first. Then we combine the armature and main-pole mmfs to obtain the resultant flux distribution.

Consider the 2-pole machine shown in Fig. 5.9. The conductor currents are always in one direction under a north pole and in the opposite direction under a south pole. Hence, the armature winding is the magnetic equivalent of a solenoid, which has its axis stationary and coinciding with the brush axis. The axis of the armature

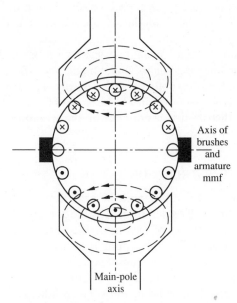

Figure 5.9 Magnetic field pattern of the armature winding mmf.

mmf is perpendicular to the axis of the main poles, if the brushes are in the geometric neutral position. The mmf gives rise to a magnetic field that is often referred to as the *armature cross field.* The same applies to multipole machines. That is, the armature mmf is perpendicular to the main pole mmf if electrical degrees are used to describe the relative position of the two mmfs.

A. ARMATURE MMF

To derive the armature mmf distribution first consider the full-pitch coil $1 - 1'$ of z turns, shown in Fig. 5.10a. The mmf of this single coil is zi ampere turns and its spatial distribution is plotted in Fig. 5.10b as a rectangular wave of height $zi/2$ ampere turns. All coils of the winding are identical to coil $1 - 1'$ and carry the same current. The mmf wave of coil $2 - 2'$, displaced by one slot pitch and also with a height of $zi/2$ ampere turns, is shown by the broken line.

If the individual mmf waves of all coils are algebraically added, the resultant armature mmf distribution is the stepped triangular wave of Fig. 5.10c. Actually, each step is multisloped and determined by the number of conductors, their arrangement in the slot, and the current per conductor. The sloped steps, combined with the fact that commercial machines have considerably more slots per pole pitch than shown in Fig. 5.10a, allow the armature mmf distribution to be represented to a close approximation by the straight-line, triangular wave shown in Fig. 5.10c. The number of conductors per pole pitch is $Z/2p$. Each conductor carries a current $i = I_a/2a$. Accordingly the total armature mmf along a brush axis is $ZI_a/4pa$ ampere turns. Half of this number of ampere turns acts at each brush location of the rotor winding.

The magnetic flux resulting from the armature mmf is determined by the reluctance of its magnetic path. The reluctance is virtually determined by the length of the air paths involved. Assuming a constant air-gap length between the rotor and pole shoe surface, the reluctance is constant under a pole shoe. Beyond the pole tips, the reluctance of the armature flux path increases sharply. The armature flux curve will not exhibit an abrupt change in magnitude, because of fringing of the flux at the pole tips. The flux density distribution resulting from the armature mmf is shown in Fig. 5.10d.

B. RESULTANT FLUX DISTRIBUTION

If both the main pole field winding and the armature winding carry current, the resultant flux distribution will be determined by the combined effect of both winding mmf distributions. If saturation of ferromagnetic parts of the magnetic circuit is ignored, the resultant flux distribution around the rotor can be found by applying the principle of superposition to the flux components.

This approach has been taken in Fig. 5.11. Curve F represents the flux distribution due to the main pole field winding only. The solid curve A represents the flux distribution resulting from the distributed armature winding mmf, which is indicated by the broken line. Curve R represents the resultant flux distribution curve. It should be noted that the resultant flux distribution curve is peaked and that the magnetic

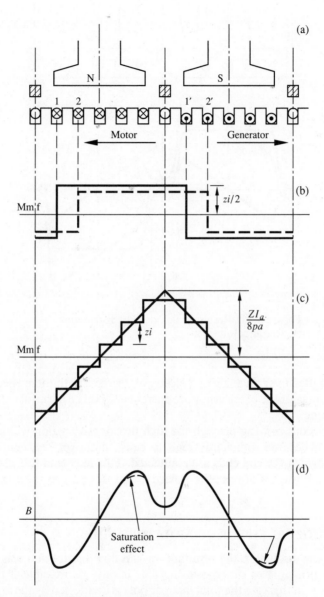

Figure 5.10 Armature mmf and flux density distributions. (a) Developed diagrams of a 2-pole machine, (b) mmf of coils $1-1'$ and $2-2'$, (c) mmf distribution of complete winding, (d) armature flux density distribution.

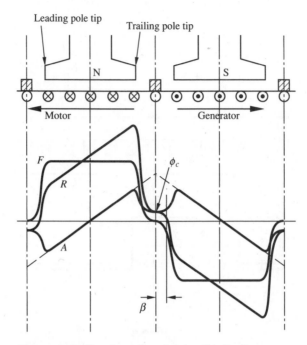

Figure 5.11 Resultant flux density distribution neglecting saturation.

neutral has shifted over an angle of β electrical degrees. Both these effects are determined by the magnitude of the armature current and can be harmful to the operation of a dc machine.

Armature coils moving through the high flux density region will have induced emfs that can become sufficiently high to cause flashover between commutator segments, to which the coil ends are connected. This may lead to a complete short circuit between brushes of opposite polarity, that is, between the terminals of the machine.

C. DEMAGNETIZING EFFECT OF ARMATURE MMF

In the foregoing discussion the resultant air-gap flux distribution was obtained by applying the principle of superposition. This is only permissible if the magnetic circuit is linear. In this case the total flux per pole is unaffected by the armature mmf since the increase in flux under one-half of a pole shoe is equal to the decrease in flux under the other half.

Economic reasons dictate operation in the semi-saturated region of the ferromagnetic circuit of a machine. As a result of the partial saturation, the increase of air-gap flux density under one-half of the pole shoe is less than the decrease of the air-gap flux density under the other half of the pole shoe. (See Fig. 5.10d.) It follows that the total flux per pole will be reduced as a result of the armature winding mmf. The reduction of the total flux per pole, referred to as the demagnetizing effect of the armature current, can be expressed as an equivalent number of demagnetizing ampere turns. These are proportional to the armature current and depend on the level of saturation of the machine. The reduction of the total flux per pole has to be taken into consideration in the design or in the analysis of machine performance, since both generated voltage and developed torque are directly proportional to the total resultant flux per pole.

5.4.2. Compensating Windings

The larger is the armture current the greater is the resulting flux density distortion and the greater is the probability of commutator flashover. It is for this reason that the larger, high-current machines are equipped with compensating windings.

A compensating winding is embedded in axial slots in the pole shoe surface. This winding has the task to cancel the effect of the armature winding mmf by providing an mmf of equal magnitude and spatial distribution, but of opposite polarity. In order to provide cancellation of the armature mmf for all load conditions, the compensating winding is connected in series with the armature winding. Then both windings will carry the same current at all times. Due to its physical location, the compensating winding can only cancel that part of the armature-mmf distribution, which is within the span of a pole shoe.

Since the conductors of the compensating winding carry the full armature current I_a and the armature winding conductors carry a current $I_a/2a$, the number of compensating winding conductors per pole shoe, Z_c, can be found from the relation

$$Z_c I_a = \frac{b}{\tau_p} \frac{Z I_a}{4pa} \tag{5.4.1}$$

where b/τ_p is the ratio of pole-shoe arc to pole pitch.

The reader can consider whether the flux produced by the compensating winding has any effect on the energy conversion process.

5.4.3. Effect of Brush Shifting

The reason for shifting the brushes from the geometric neutral axis over a small angle in the direction of rotor rotation for a generator (in the opposite direction for a motor) is to improve commutation. This will be discussed in section 5.5. At this point we will consider the effect of brush shifting on the resultant flux density distribution.

Figure 5.12 shows a 2-pole generator driven in a clockwise direction, or a 2-pole motor rotating in a counterclockwise direction. The brush axis BB is rotated with respect to the geometric neutral axis over a small angle α. Since the total armature current enters and leaves the armature winding through the brushes the axis of the armature mmf is shifted over the same angle α from the geometric neutral. Curve F of

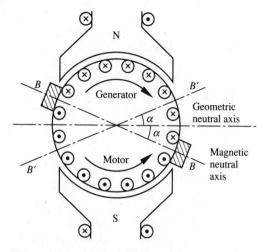

Figure 5.12 Brush shifting and demagnetizing effect of the armature winding mmf.

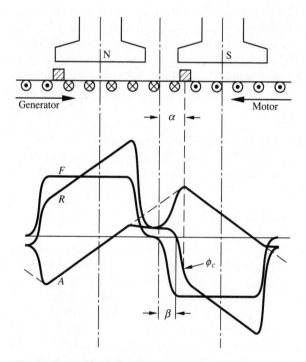

Figure 5.13 Field distribution for brush shift.

Fig. 5.13 represents the flux density distribution that would exist if the main pole field winding were the only current-carrying winding. The triangular wave represents the distributed armature winding mmf. The peak value of the armature mmf coincides with the brush location and has a value $ZI_a/8pa$ ampere turns. The solid curve A represents the flux distribution resulting from the armature winding mmf. The resultant flux distribution is represented by the curve labeled R. The larger is the armature current the larger is the angle β, over which the true magnetic neutral is shifted from the geometric neutral.

In Fig. 5.11, for generator operation, coil sides, which move through a resultant magnetic field of one polarity, are still moving through a weak field of the same polarity when they are short-circuited by a brush and undergoing commutation. With a sufficiently large shift of the brush axis in the direction of rotation, as is shown in Fig. 5.13, the generator coils will have their coil sides move through a field of opposite polarity while being commutated.

In the case of motor operation Fig. 5.13 represents a brush shift against the direction of rotation. Figure 5.11 shows that coils being commutated by brushes, which are positioned on the geometric neutral axis, rotate through a weak magnetic field which has a polarity opposite to that of the field through which the coil sides are moving prior to reaching the brushes. If the brushes are shifted over a sufficiently large angle α against the direction of rotation, the coil sides, which are being commutated, move through a field having the same polarity as the resultant magnetic field from which they just came. The resultant field at the brush axis is called the commutating field Φ_c and its magnitude and polarity are of great importance for satisfactory commutation. The magnitude of Φ_c is determined by the magnitude of the armature current as well as by the angle of brush displacement α.

The net area enclosed by a resultant flux distribution curve R, with the horizontal line representing the rotor surface between two adjacent brushes, is a measure of the total flux linked with the coils of one armature winding path. This area in Fig. 5.13 is smaller than that in Fig. 5.11. Therefore the demagnetizing effect of the armature-winding mmf is larger if the brush axis is rotated over an angle α, as seen in Fig. 5.12. The armature winding can be resolved into two parts by drawing a line $B'B'$, which, like the brush axis, encloses an angle α with the neutral axis. The conductors enclosed between the two lines BB and $B'B'$ can be considered to form a coil, which produces an mmf in direct opposition to the mmf of the main-pole field winding. The demagnetizing ampere turns of this part of the armature winding are

$$AT_d = \frac{1}{2}\frac{4\alpha}{360}\frac{Z}{2}\frac{I_a}{2a} = \frac{\alpha}{180}\frac{ZI_a}{4a}. \tag{5.4.2}$$

The remainder of the armature winding conductors forms a cross-magnetizing mmf given by

$$AT_c = \frac{1}{2}\left(\frac{360-4\alpha}{360}\right)\frac{ZI_a}{4a}. \tag{5.4.3}$$

These two equations give the ampere turns per air gap.

5.5. COMMUTATION

5.5.1. Commutation Action

Commutation is the process of switching the current direction in armature coils, as the coils move from a magnetic field of one polarity to the influence of a magnetic field of the opposite polarity. The time interval required for this switching is known as the *commutation period*.

The actual switching process is illustrated in an elementary form in Fig. 5.14, which shows three instants of the current reversal as it takes place in coil B of this figure. To simplify the discussions in this chapter a brush width is chosen equal to the width of one commutator segment. In general, the brush width is greater than the width of a commutator segment.

Figure 5.14 shows that the current per armature path, and therefore the current in each coil of such a path, is $I_a/2a$ amperes. In Fig. 5.14a the current of two winding paths is collected by a brush, which is shown to rest on a single commutator segment (segment number 3). Coils B and C belong to two different armature winding paths and each rotate through a magnetic field of opposite polarity. The current directions in the paths, to which these coils belong, are therefore shown to be opposite to each other. An instant later the situation exists as shown in Fig. 5.14b. The stationary brush is now in contact with commutator segments 2 and 3 to which the ends of coil B are connected. Coil B is short-circuited. It should be noticed that the total current collected by the brush must remain the sum of the currents in two armature winding paths. Figure 5.14c shows the moment at which the brush contact with commutator segment 3 is interrupted. Coil B has now become part of the armature path to which coil C belongs and the current direction in coil B has been reversed. If the reversed current in coil B has not attained the value $I_a/2a$ at the moment that direct contact between commutator segment 3 and the brush is broken, the difference between $I_a/2a$ in coil C and the actual coil B current at that moment has no other choice than to flow from commutator segment number 3 to the trailing tip of the brush in the form of an arc. Arcing or sparking is destructive to both the commutator and the brushes. Sparking may also be caused by unsatisfactory mechanical conditions such as a rough, unevenly worn commutator surface, sticking of the brushes in their brush holder or chattering of the brushes. In spite of the great importance of the mechanical aspects, we will restrict ourselves to the electrical phenomena that govern the commutation process.

A. IDEAL OR LINEAR COMMUTATION

A plot of the coil current as a function of time during the commutation period T_c is called a *commutation curve*. Figure 5.15 shows a smooth curve from which we will deduce the ideal commutation curve. At the start of the commutation period ($t = 0$) the coil current has a value $+I_a/2a$. At the end of this period ($t = T_c$) the coil current is assumed to have attained the value $-I_a/2a$. At an arbitrary instant ($t = \tau$) the current in the coil has a value $+i_s$.

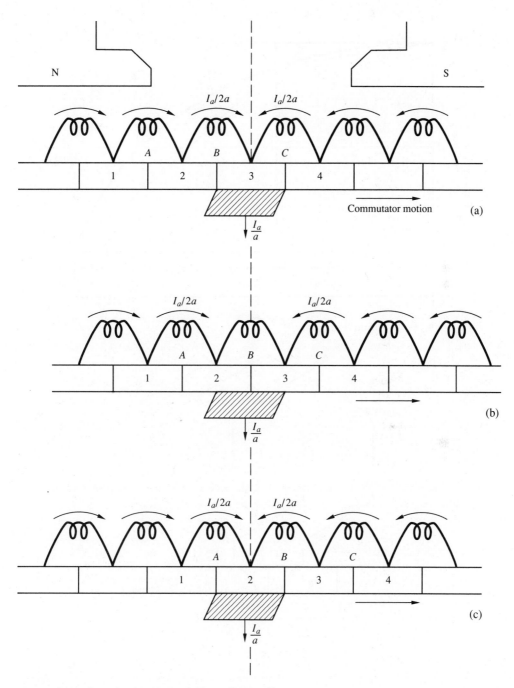

Figure 5.14 Steps in the commutation of one coil.

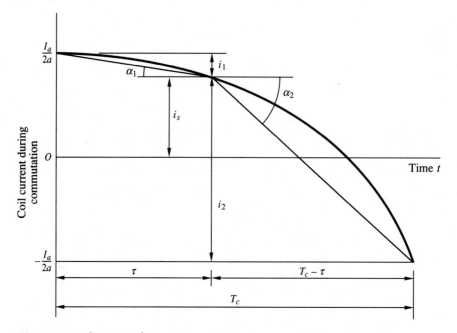

Figure 5.15   Commutation curve.

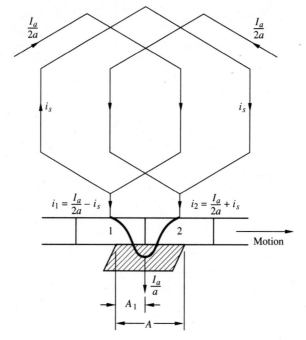

Figure 5.16   Coil of a lap winding undergoing commutation.

Assume that the current variation shown in Fig. 5.15 is that of the current in a coil shown in Fig. 5.16. The position of the brush corresponds to the instant $t = \tau$. At this moment the brush, which has the width of one commutator segment, has an area $A_1 = A\tau/T_c$.

It follows from Fig. 5.16 that the current density in the leading part of the brush is

$$J_1 = \frac{i_1}{A_1} = \frac{\dfrac{I_a}{2a} - i_s}{\tau}\frac{T_c}{A} = \frac{T_c}{A}\tan\alpha_1 .$$

The current density of the trailing part of the brush is

$$J_2 = \frac{\dfrac{I_a}{2a} + i_s}{A - A_1} = \frac{\dfrac{I_a}{2a} + i_s}{T_c - \tau}\frac{T_c}{A} = \frac{T_c}{A}\tan\alpha_2 .$$

The angles α_1 and α_2 are indicated in Fig. 5.15. The ratio T_c/A has a constant value. The current density in each part of the brush at any instant during the commutation period is given by the slopes of the lines connecting the instantaneous value of i_s with the initial and final current values in the coil. For the commutation curve shown in Fig. 5.15 (a condition called undercommutation), the current density in the trailing part of the brush is higher than that in the leading part.

Nonuniform current distribution will lead to nonuniform heating. Local over-heating of the brush and its contact area with the commutator can lead to disintegration of the brush material and cause chemical damage to the surface of the commutator segments. Also, the ohmic loss in any conductor has its lowest value when the current density distribution is uniform. Thus, for long trouble-free operation and minimum ohmic loss, the current density distribution over the brush contact area must be uniform. This ideal condition will be met if the commutation curve is a straight line. In that case $\alpha_1 = \alpha_2 = \alpha$ regardless of the position of the brush during the commutation period. The current from a commutator segment to the brush is then proportional to the contact area of the brush with the commutator segment. The contact area and therefore the current from segment to brush is zero at the moment that the electric contact is broken and no arcing can occur. If the commutation curve is ideal, that is, a straight line, we speak of *linear commutation*. Several factors affect the shape of the commutation curve and what is termed the quality of commutation. Some of these factors will be discussed in the following sections.

B. REACTANCE VOLTAGE

The commutation period T_c is a short period of time, normally less than 1 ms. The rapidly changing currents give rise to emfs of self and mutual induction in a commutating coil. The emfs of mutual induction result from the fact that in a two-layer, full-pitch winding all coil sides in a slot belong to coils that are commutated simultaneously by other brushes.

The total instantaneous voltage induced in a commutating coil is called the *reactance voltage* e_r. According to Lenz's law, the reactance voltage tends to retard the change of coil currents during commutation. The reactance voltage will lead to undercommutation, that is, it will cause the commutation curve to have the general form of curve 2 in Fig. 5.17. If the reactance voltage is large it may prevent current reversal to be completed at the end of the commutation period as shown by curve 3. This is called *undercommutation* and will lead to sparking at the trailing edge of the brushes.

C. COMMUTATING EMF

Linear commutation can be obtained if the unavoidable reactance voltage is balanced by a voltage of opposite polarity. Such a voltage can be generated in the commutating coil when it moves through a magnetic field of proper strength and polarity. The generated emf is called a *commutating emf.* The magnetic field at the brush axis is correspondingly called the *commutating field.*

Since the reactance voltage tends to maintain current in its original direction, the commutating emf must be in a direction opposite to that of the current before commutation. In the case of a generator, the commutating coil side emerging from a main field of north polarity must be subjected to a commutating field of south polarity. The reader should check for himself or herself that, in the case of a motor, the commutating field must have the same polarity as the main field from which the commutated coil is emerging.

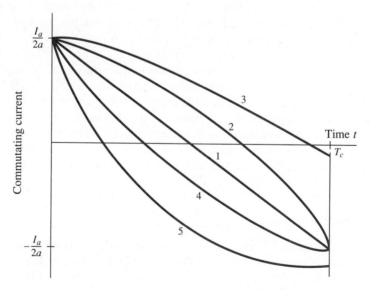

Figure 5.17 Commutation characteristics. 1: linear commutation; 2 and 3: undercommutation; 4 and 5: overcommutation.

Brush Shifting

If the brushes are positioned in the geometric neutral axis of a dc generator, the armature mmf causes the commutating field to have the same polarity as that of the magnetic field just left by the commutating coil sides (see Fig. 5.11). This commutating field will generate a commutating emf, which has the same direction as the reactance voltage and will make the conditions for commutation worse. However, if the brushes of a dc generator are shifted in the direction of rotation past the magnetic neutrals, a commutating field of opposite polarity can be obtained (see Fig. 5.13). The generated commutating emf will now counteract the reactance voltage and improve the commutation.

If the armature current increases, the reactance voltage will also increase. Ideally the commutating voltage must increase by an equal amount. However, the larger the armature current the larger the shift of the magnetic neutral is in the direction of rotation of the generator and the smaller the commutating field becomes at the location of the shifted brushes. The amount of brush shift required to obtain optimum commutation performance depends on the magnitude of the load current. It is impracticable to adjust the brush position with every change in load current for best commutation. Only in smaller machines does one rely on brush shifting to improve commutation. The setting chosen depends on the load pattern. Overcommutation will result at light loads.

The reader can check that, in the case of motor operation, the brushes have to be shifted against the direction of rotation in order to improve commutation.

Interpoles

Brush shifting has the disadvantage that a particular amount of brush shift is only optimum for one particular load condition. This method cannot be used in the case of motors that have to be capable of running in both directions (e.g., hoists and reversing steel mill drives). Another disadvantage of brush shifting is that the useful flux per pole is decreased when the brushes are shifted from the geometric neutral position and so the output of a machine is reduced.

A better but more costly method is to provide a commutating field of proper polarity and strength by means of auxiliary field poles. These auxiliary poles are called interpoles or commutating poles and are located in the geometric neutral axis of the machine. The brushes remain positioned at the geometric neutrals so that maximum output is obtained from the machine.

The interpoles are narrow poles, as indicated in Fig. 5.18, since the commutating field is only required in the small arc through which the commutating coil moves during the period of commutation. The mmf of each interpole must at all load conditions be sufficient to cancel the armature mmf at a geometric neutral and to produce a commutating field of required strength. Since the armature mmf and the reactance voltage both are proportional to the armature current, the mmf of the interpoles also has to be directly proportional to the armature current. This requirement is easily satisfied when the exciting windings of the interpoles are connected in

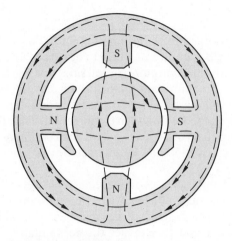

Figure 5.18 Main pole and interpole
flux paths of a generator.

series with the armature winding. In the case of a generator, the interpoles must have
the same polarity as the following main poles. In the case of a motor, the interpoles
must have a polarity opposite to that of the following main poles. The interpole flux
follows a path as illustrated. To reduce the nonlinearity of the interpole flux path, the
air gap under the interpole is made relatively large. The commutating voltage need be
induced only in one coil side. Therefore, in practice, it is possible to eliminate half the
number of interpoles. The resultant magnetic field of an interpole machine equipped
with a compensating winding is shown in Fig. 5.19. The resultant field is obtained by
superimposing the individual component fields shown in the figure (assuming no
saturation).

The type of commutation in which an active voltage is generated to balance the
reactance voltage is called voltage commutation. The great advantage of applying
interpoles is that restrictions on rotor speed, conductor current, and factors that affect
the self and mutual inductance of winding coils can be considerably relaxed. This
increases the possible output of a machine for a given frame size. Practically all
modern dc motors and generators, except those of small size, are equipped with
interpoles.

5.6. DC GENERATORS

Only for special applications are dc generators used as dc power sources. This is
because static semiconductor converters have largely superceded dc generators
today. Consequently, only a brief description of this type of generator will be given
here. We will look at the ability of a dc machine to generate electric power, and we will
give attention to the equivalent circuit for performance evaluation. This is useful
because dc motors can be used in the regeneration mode (when braking).

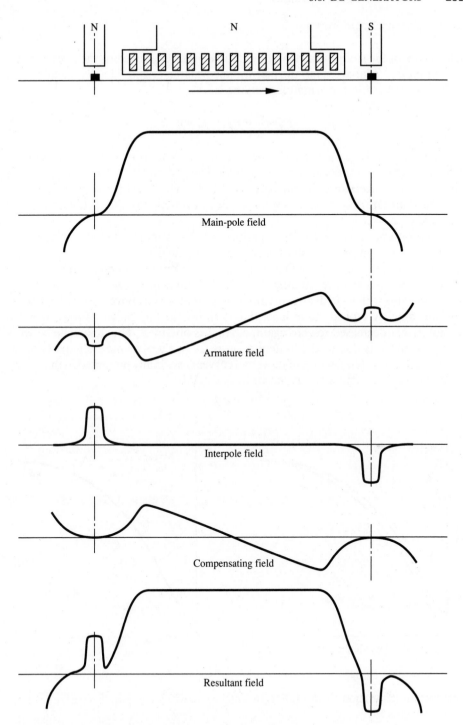

Figure 5.19 Fields of a compensated interpole machine.

5.6.1. Saturation Curve

The shape of the saturation curve (a plot of armature emf versus the field current) determines, to a large extent, the performance characteristics of a dc machine. The voltage generated in the armature winding is

$$E_a = \frac{p}{a} \Phi n Z = K_e n \Phi. \tag{5.3.9}$$

At a constant speed the generated voltage is directly proportional to the flux per pole, which in turn is a nonlinear function of the field coils' mmf. The nonlinearity is due to saturation effects in different parts of the magnetic circuit. It follows that the generated armature voltage is a similar nonlinear function of the exciting current in the field coils.

The saturation curve can be experimentally determined. For this purpose the machine is driven at a constant speed, while its field winding is connected to a separate and adjustable source of dc voltage. A voltmeter is connected across the terminals of the armature winding. The field current is progressively increased from zero to a value which yields an armature voltage that is well above the rated voltage of the machine. A typical curve labeled 1 is shown in Fig. 5.20a. If the current is progressively decreased to zero again, curve 2 is obtained. The difference between curves 1 and 2 is due to the hysteresis property of the ferromagnetic parts of the magnetic circuit. The average of these two curves is normally presented as the saturation curve of a machine and is shown in Fig. 5.20b.

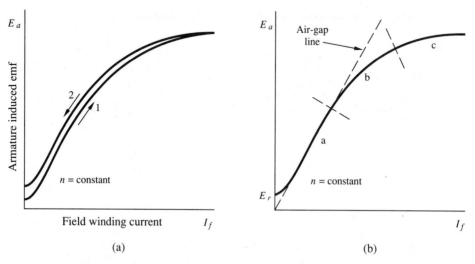

Figure 5.20 Saturation curve. (a) Experimental curve with hysteresis, (b) average single-valued curve.

Examination of the saturation curve of Fig. 5.20b shows that there is a small voltage E_r generated at zero field current due to residual flux present in the magnetic circuit once it has been magnetized. For low values of the generated voltage the required pole flux is small. The reluctance of the ferromagnetic part of the magnetic circuit is then negligible compared with that of the air gap. The lower part of the saturation curve is therefore virtually a straight line. The curve would follow the so called air-gap line if there were no saturation in the magnetic circuit. However, for increasing values of the voltage E_a the required flux densities will be such that saturation sets in, first in the rotor teeth, then in the pole cores and lastly in the armature iron.

The normal range of operation is at the knee (part b) of the curve for economy of design. Operation in the saturated part of the curve (part c) would result in a marginal increase in E_a at the cost of a significant increase in iron loss in the rotor and a considerable increase of the exciting current.

From the emf equation it follows that, if the curve has been experimentally determined for one constant speed of rotation, it can be derived for any other speed. For an equal value of exciting current I_f, but different rotational speed, we calculate the corresponding generated voltage from

$$E_{a2} = \frac{n_2}{n_1} E_{a1} \tag{5.6.1}$$

where E_{a1} and E_{a2} are generated at speed n_1 and n_2, respectively. Performing this calculation for several values of I_f will yield a number of corresponding values of voltage from which a saturation curve for speed n_2 can be plotted.

5.6.2. Generator Voltage Buildup

The no-load or open-circuit voltage of a generator is the voltage that exists between the terminals when it is not supplying any current to an electrical load. The no-load voltage of a separately-excited generator can be obtained from the saturation curve for the appropriate speed and field current.

Most generators are self-excited. Consider a shunt generator as depicted in Fig. 5.21. It is driven at its rated speed n but is not connected to a load. While the switch in the field circuit is open a small voltage E_r, the residual voltage, will be generated in the armature circuit. On closure of the field-circuit switch S, the residual voltage will be responsible for a small amount of field current. If the resulting field mmf produces a flux in the same direction as that of the residual flux, the increased total flux will generate a larger voltage in the armature winding. This results in a larger field current which augments the flux and causes a further increase of the generated voltage. This cumulative process will continue until a point of equilibrium is reached and the terminal voltage has built up to the steady no-load value.

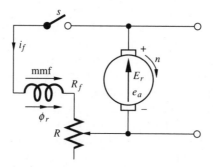

Figure 5.21 Self-excitation of a
shunt generator on open circuit.

During the voltage buildup process, Kirchhoff's voltage equation for the circuit
of Fig. 5.21 is

$$e_a = i_f(R_f + R) + L \frac{di_f}{dt} \qquad (5.6.2)$$

where e_a = instantaneous value of generated emf in the armature
winding

i_f = instantaneous value of the current in the field circuit

L = total inductance of the circuit

$R_f + R$ = resistance of the field winding and the field rheostat.

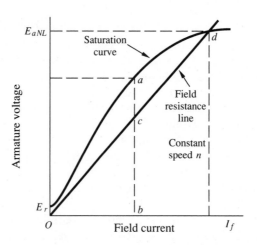

Figure 5.22 Shunt generator voltage buildup.

The armature winding resistance R_a is negligibly small compared with R_f, so it has not been considered. Also the contact voltage drop of the brushes has been neglected.

For an arbitrary instant of time during the voltage buildup the momentary value of the generated voltage is represented in Fig. 5.22 by point a on the saturation curve. The corresponding momentary field current is given by point b on the current axis. The relation between the voltage across the total field-circuit resistance and the field current is given by the straight, field-resistance line Ocd according to Ohm's law. The field current (Ob) corresponds to a resistive voltage drop (bc). The voltage difference (ac) between e_a and $i_f(R_f + R)$ is proportional to the rate of change of the current, and is the inductive voltage drop. When the steady state is reached the field current has a constant value. Its rate of change is zero and the line segment ac must have reduced to zero. That is, the point of equilibrium is given by point d. The voltage E_{aNL} associated with the intersection of the field-resistance line and the magnetization curve is the open-circuit or no-load voltage of the shunt generator. A compound-connected generator has no current in its series field winding under no-load conditions. Therefore, the foregoing applies equally well to the no-load voltage buildup of a compound generator.

An increase of the field resistance from some value R_1 to a value R_2 will increase the slope of the field resistance line as shown in Fig. 5.23. The point of intersection of this line with the saturation curve will move to a lower voltage value. Thus the no-load value of a shunt or compound generator voltage can be varied by adjusting the rheostat in the shunt-field circuit. If the field circuit resistance is increased to a large value ($R_f + R_3$) the voltage does not build up beyond a value slightly higher than the residual voltage. The value of field circuit resistance, beyond which the voltage does not build up, is given by a field resistance line with a slope approximately equal to that of the air-gap line. This value of field resistance R_c is called the critical field resistance.

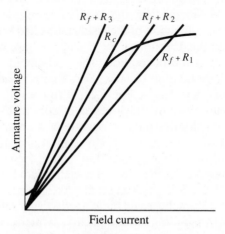

Figure 5.23 Effect of field resistance on voltage buildup.

EXAMPLE 5.4

A dc shunt generator is rated at 100 A, 250 V, 20 r/s. The armature-winding resistance is 0.1 ohm. The separately-excited saturation curve for 20 r/s is obtained from the following data.

E_a (volts)	95	158	198	230	250	265
I_f (amperes)	0.2	0.4	0.6	0.8	1.0	1.2

Determine **(a)** the field circuit resistance so that the terminal voltage is 250 V for an armature current of 100 A, if the generator is driven at 20 r/s, and **(b)** the no-load terminal voltage for the same field-circuit resistance.

Solution
(a) We need to find the field current I_f from the generator characteristic for the given conditions of $V = 250$ V and $I_a = 100$ A. Thus

$$E_a = V + I_a R_a = 250 + (100 \times 0.1) = 260 \text{ V}.$$

For $E_a = 260$ V, $I_f = 1.13$ A from the saturation curve.
Therefore, the total field-circuit resistance is

$$R_f = V/I_f = \frac{250}{1.13} = 221 \text{ ohms}.$$

(b) With $R_f = 221$ ohms and the generator on no load, there will be a new balance of I_f to support E_a. By drawing the field resistance line on the open-circuit characteristic, the point of intersection will give the point of balance. It is $E_a = 265$ V for $I_f = 1.2$ A. Hence the terminal voltage is 265 V on no load.

The ability of a shunt or compound generator to build up a voltage is entirely due to the presence of residual magnetism and an initial mmf that produces a flux that aids the residual flux. If the direction of the residual flux had been the reverse of that shown in Fig. 5.21, then the direction of the residual voltage would also have been reversed for the same direction of rotation. The field-circuit connections must be reversed to permit voltage buildup to take place. The no-load voltage will have a polarity opposite to that depicted by Fig. 5.21. It follows that the polarity of the generated voltage is determined by the direction or polarity of the residual flux.

Generators are normally designed to operate at a specific or rated speed as indicated on the nameplate of the machine. However, it is interesting to consider the effect of armature speed on the voltage buildup. Figure 5.24 shows four saturation curves for the same generator. Each curve is associated with a different speed. Let n_1 be the rated speed. With the field rheostat set at a fixed value, voltage buildup takes place. At a lower speed n_2 the voltage will build up to a lower value. At a considerably reduced speed n_3 the corresponding saturation curve is intersected by the field-resistance line at a voltage close to the residual voltage associated with the speed n_3 and no voltage buildup takes place. There is a specific speed below which no voltage buildup can take place. At this speed, the critical speed n_c, the field resistance line is approxi-

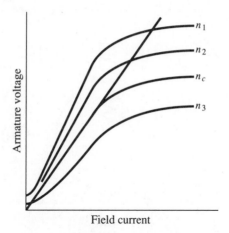

Figure 5.24 Effect of armature speed
on voltage buildup.

mately a tangent to the unsaturated part of the saturation curve for that speed. The
critical speed of a generator depends on the magnitude of the field circuit resistance.

A series generator can only build up a voltage between its terminals under load
conditions, because the load current is also the field current. In order that a terminal
voltage builds up, the flux due to the series field winding mmf must aid the residual
flux, the speed must be above the critical speed for a given load, or the load should
have a resistance less than the critical resistance for the speed of operation.

5.6.3. General Circuit Diagram and Steady-state Equations

The most general connection diagram of a dc generator is one that includes all
windings that can be present. This diagram, shown in Fig. 5.25, is that of a compound
generator equipped with interpole and compensating windings. During steady-state
operation the load current does not vary, so that the winding inductances do not play
a role. Accordingly, the equations governing the performance are

$$V = E_a - I_a(R_a + R_i + R_c + R_s) - V_b \qquad (5.6.3)$$

$$E_a = \frac{pZ}{a}\,n\Phi = K_e n\Phi \qquad (5.6.4)$$

$$V = I_f R_f \quad \text{and} \quad I_a = I_L + I_f \qquad (5.6.5)$$

where V = the terminal voltage available to the generator load

E_a = the voltage generated by the armature winding

I_a = the total armature winding current

I_L = the current supplied to the load

I_f = the current in the shunt field circuit

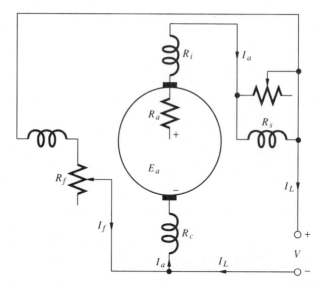

Figure 5.25   General connection diagram of a dc generator.

R_a, R_i and R_c = resistances between the terminals of the armature winding, the interpole winding, and the compensating winding, respectively

R_f = total series resistance of the shunt field winding and the rheostat

R_s = total resistance of the series field winding and the diverter in parallel

and V_b = the voltage drop due to the brush contact resistance. It has a nearly constant value over a wide range of armature current. This voltage drop is of the order of 1 to 2 volts.

The equations for any specific generator follow from the above equations by omitting the appropriate resistances and currents.

From eq. (5.6.3), it can be seen that the terminal voltage is affected by the resistive voltage drop caused by the current in the armature winding and the interpole and compensating windings. The resistance of these windings is kept small, so that the terminal voltage V is mainly determined by the voltage E_a generated in the armature winding. A low-resistance winding is also required to keep the losses and the temperature rise to satisfactory low values.

5.6.4. Voltage Regulation

The terminal voltage of a generator normally changes as the load current varies. This voltage variation is described by the voltage regulation. National institutions define the percentage voltage regulation of a generator as the change of terminal voltage (with constant field circuit resistance and driven at constant speed) when the load current is gradually reduced from rated value to zero. This change of voltage is expressed as a percentage of the rated voltage at full-load current, that is

$$\text{Regulation} \triangleq \frac{V_{NL} - V_{FL}}{V_{FL}} \times 100\% \qquad (5.6.6)$$

where V_{NL} and V_{FL} are the no-load and full-load terminal voltages, respectively.

Normally, the speed at which the generator is driven increases as the load current is reduced from full-load value to zero. The increased speed will increase the generated no-load voltage and thus increase the voltage regulation. However, the speed variation is a characteristic of the prime mover and should not be permitted to influence the measured voltage regulation of a generator. Therefore the speed must be kept constant during the change of load current.

It may be of importance to know the difference between full-load and no-load voltage of a specific motor-generator combination. The generator is then driven at rated speed while delivering full-load current and will have an increased speed at no load. In this case we speak of the overall percentage voltage regulation of the motor-generator set.

EXAMPLE 5.5

A separately-excited, dc generator is rated at 100 A, 250 V, 1200 r/min. It delivers a current of 100 A at 250 V, when it is driven at 1400 r/min and excited from a rated source of 250 V. The armature winding resistance is 0.1 ohm and the field winding resistance is 250 ohms. Determine **(a)** the resistance of the field rheostat, **(b)** the electromagnetic torque, and **(c)** the voltage regulation. Assume a linear magnetization curve.

Solution
(a) The induced emf at both speeds is the same, that is

$$E_a = V + I_a R_a = 250 + (100 \times 0.1) = 260 \text{ V}.$$

We assume that the relation between the flux Φ and the current I_f is linear, so that at any field current I_f and any speed n,

$E_a = k I_f n$, where k is a constant.

For an exciting voltage V_f and field circuit resistance R_f

$I_f = V_f / R_f.$

Therefore, if the back emf E_a is constant, for two speeds n_1, and n_2

$$\frac{R_{f1}}{R_{f2}} = \frac{n_1}{n_2}.$$

At 1200 r/min, $R_{f1} = 250$ ohms. Hence at 1400 r/min

$$R_{f2} = \frac{1400}{1200} \times 250 = 292 \text{ ohms}.$$

The added field circuit resistance is $(R_{f2} - R_{f1}) = 42$ ohms.

(b) The electromagnetic torque is

$$T_e = \frac{E_a I_a}{2\pi n} = \frac{260 \times 100}{2\pi \times 1400/60} = 177 \text{ N} \cdot \text{m}.$$

(c) The voltage regulation is $\text{REG} = \dfrac{V_{NL} - V_{FL}}{V_{FL}} \times 100\%$.

At no load $V_{NL} = E_a$. Therefore

$$\text{REG} = \frac{260 - 250}{250} \times 100 = 4\%.$$

5.6.5. Generator Characteristics

From a user's point of view, the most important characteristic of a generator is the external characteristic. This characteristic gives the variation of the terminal voltage as a function of the load current for constant values of field circuit resistance and armature speed. The general shape of the external characteristic of a dc generator can be deduced (in a qualitative manner) from the performance equations presented in section 5.6.3.

Equation (5.6.3) shows that the terminal voltage is less than the generated voltage, because of the resistive voltage drop in the armature circuit. The armature-winding current is either equal to, or almost equal to, the load current. Thus the resistive voltage reduction varies linearly with the load current.

At constant armature speed, the generated voltage E_a is determined by the total flux per pole Φ (eq. (5.6.4)). If the generator supplies a load current, the armature mmf causes a slight reduction of the flux per pole and E_a decreases slightly as the load current increases.

The field current of a separately-excited generator is independent of the terminal voltage of the generator. Hence, for a certain field-winding current, the generated voltage E_a is only affected by the armature reaction. The terminal voltage is less than E_a by the resistive voltage drop. This results in a drooping external characteristic, the terminal voltage decreasing as the load current increases.

For a constant value of the field-circuit resistance, the field-winding current of a shunt generator is determined by the generator's own terminal voltage. In the same way as for a separately-excited generator, the terminal voltage of a shunt generator decreases with increasing load current. Hence the field winding current decreases when the load current is increased. The resulting reduction of the generated voltage causes a shunt generator to have a larger reduction of terminal voltage with increasing load current, than if the machine had been separately excited.

The series generator has the field winding in series with the armature winding. So the load current determines the excitation. Increasing the load current increases the generated voltage E_a at a much larger rate than the small voltage drops in the armature circuit. This causes the terminal voltage to increase steeply with load current and to vary widely with load.

For a compound generator, the excitation is mainly provided by the mmf of the shunt field winding. The much smaller mmf of the series winding varies with the load current. If the mmfs of the series and shunt windings are aiding, the machine is said to be cumulatively compounded. If the mmfs oppose each other the machine is differentially compounded. This latter type of compounding is rarely used except for the application of dc generators used for welding purposes. In this particular case the generator has to keep constant current at the weld. As current tends to increase, the series field tends to increase. This weakens the resultant field. Hence the generated emf is reduced and this prevents the increase of current.

Consider a cumulatively-compounded generator. If at full load the mmf produced by the series winding compensates for both the armature reaction and the resistive voltage drop in the armature circuit, the generator is said to be flat-compounded. If the series mmf is larger than is required for flat compounding, the generator is overcompounded and the terminal voltage at full load is higher than at no load. If the mmf of the series winding is not sufficient to offset the effects of armature reaction and resistance, the full-load terminal voltage will be less than the no-load value. In this case the generator is undercompounded.

Cumulatively-compounded generators are normally designed to have a series winding that gives a relatively large measure of overcompounding. The actual amount of compounding required for a given situation can be obtained by adjusting the amount of current through the series winding by means of the diverter. A slightly overcompounded generator is used more than any other type of generator to compensate for the resistive drop in the supply lines to the load. Overcompounding is also used to counteract the effect that the usual decrease in speed of the prime mover with increased load has on the generated voltage.

5.7. DC MOTORS

The same machine that is used as a generator can be operated as a motor. In fact, in some practical applications dc machines operate alternately as a motor and a generator.

If a machine operates as a motor, a current is forced through the armature winding in a direction determined by the polarity of the voltage applied to the machine terminals. For the same polarity of the terminal voltage, the direction of the armature winding current is opposite to that in a generator. For the case that the main field poles have the same polarity as in the generator mode, the electromagnetically developed torque will be reversed in direction. This torque is the accelerating torque of the motor. It determines the direction of rotation of the armature. For the polarities stated, it follows that the direction of rotation is the same for motor and generator operation. In a motor the voltage E_a, generated in the armature, directly opposes the applied terminal voltage. Accordingly E_a is called the counter emf or back emf. It is

Photo 2   Dc motor.

known also as the motionally induced emf or the speed voltage. The counter emf limits the magnitude of the armature winding current to exactly the value required to drive the mechanical load.

If an auxiliary source is used for the field winding excitation, the motor is said to be separately-excited. Normally, the same source that supplies the armature-winding current also delivers the field excitation current. In this case dc motors are classified as shunt, series, or compound motors depending on the connection of the field circuit relative to the armature circuit.

5.7.1. Speed Regulation

Assuming that the flux of the main poles of a dc motor remains constant, the emf E_a is determined solely by the speed of the armature. At a fixed speed, the armature winding current determines the magnitude of the electromagnetically developed torque T_e and the power taken by the motor from the electric supply. The electric power taken by the motor is automatically adjusted to suit the mechanical load.

A mechanical load increase means that the developed torque is temporarily insufficient to balance the new load torque and the armature decelerates. As the rotor reduces its speed, the counter emf E_a also decreases and allows the armature winding current to increase. Thus, the driving torque ($T_e = K_T \Phi I_a$) increases, and the torque imbalance reduces. When the developed driving torque has increased to a value sufficient to balance the machine's loss torque and the increased load torque, a new steady-state operating condition is attained at a reduced speed.

The extent to which the rotor speed changes with varying load is an important operating characteristic of a motor and is expressed by the speed regulation. The percentage speed regulation is the difference between the rotor speeds at no load n_{NL} and full load n_{FL} expressed as a percentage of the full-load speed. That is

$$\text{Speed regulation} \triangleq \frac{n_{NL} - n_{FL}}{n_{FL}} \times 100\%. \tag{5.7.1}$$

The voltage applied at the terminals and the resistance of the field circuit are to remain constant while n_{NL} and n_{FL} are determined.

5.7.2. Armature Reaction, Commutation, and Interpoles

Current in the armature winding produces an mmf that is directed along the brush axis. This results in a distortion of the main field and a shift of the magnetic neutral against the direction of armature rotation. (See Fig. 5.11.)

If the brushes of a motor are in the geometric neutral position, a coil being commutated moves through a commutating field which has a polarity opposite to that of the field just left by the coil. Therefore, the polarity of the commutating emf generated is opposite to that of the voltage that was generated in this coil by the pole field it just left. Since the armature winding current in a motor flows in a direction opposite to the generated armature winding voltage, the commutating emf will have the same polarity as the reactance voltage induced in the commutating coil. Hence, with the brushes in the geometric neutral position, the shift of the magnetic neutral due to armature mmf will lead to undercommutation. This limits the armature-winding current that can be commutated without sparking at the trailing edges of the brushes.

Improved commutation can be obtained if the commutating coil moves through a commutating field at the brush position that has the same polarity as the main field just left by the coil. This can be accomplished by shifting the brushes against the direction of rotation, the optimum position depending on the operating condition. However, the demagnetizing effect of the armature reaction becomes more pronounced. This is undesirable for stable motor operation at high load demands.

It is preferred to employ interpoles to produce a commutating field of proper strength and polarity in the geometric neutral. The excitation coils on the interpoles are connected in series with the armature windings, so that the strength of the commutating field varies with the armature winding current. This results in the generation of a commutating voltage that counteracts the reactance voltage at all values of armature winding current. The commutation time and the reactance voltage depend on the speed of the rotor. As the reactance voltage increases with rotor speed so does the commutating emf and balance is maintained.

The polarity of the commutating emf required to force linear reversal of current in the commutating coils of a motor is opposite to that required in a generator. Hence the polarity of the interpoles of a motor must be opposite to the polarities that they must have in a generator. When a generator is to be used as a motor, the interpoles automatically reverse their polarity because of the reversal of current in the armature winding to which the interpole windings are connected in series.

The mmf of compensating windings must oppose the armature mmf. Since compensating windings are also connected in series with the armature winding, the proper relative direction of mmfs is maintained when a dc machine changes from a generator to a motor mode of operation.

In some applications, such as electric traction and rolling mill drives, a dc motor must be capable of reversing its direction of rotation. The direction of rotation is determined by the developed torque, which depends on the directions of the flux and armature winding current. Hence, to change the direction of rotation, either the polarity of the main poles or the direction of the armature winding current must be reversed. Whichever method is used, the proper reactive polarity of the interpoles with respect to the main poles and the direction of rotation must be maintained. This is automatically accomplished if the connection between the armature winding and the interpole winding is not disturbed.

5.7.3. General Circuit Diagram and Steady-state Equations

The most general connection diagram for a dc motor is one that contains all the windings that can be present in the machine. This is the connection diagram of a compound motor equipped with interpoles and employing a compensating winding. The diagram for a long-compounded motor, as shown in Fig. 5.26, depicts a shunt field rheostat and a series-field diverter resistance. The equations governing the steady-state performance are

$$V = E_a + I_a(R_a + R_i + R_c + R_s) + V_b \tag{5.7.2}$$

$$E_a = \frac{pZ}{a} n\Phi = K_e n\Phi \tag{5.7.3}$$

$$V = I_f R_f \tag{5.7.4}$$

$$I_L = I_a + I_f \tag{5.7.5}$$

$$T_e = \frac{pZ}{2\pi a} I_a \Phi = K_T I_a \Phi = \frac{K_e}{2\pi} I_a \Phi \tag{5.7.6}$$

$$T_m = T_e - T_{loss} \tag{5.7.7}$$

and

$$n = \frac{V - I_a(R'_a + R_s)}{K_e \Phi} \tag{5.7.8}$$

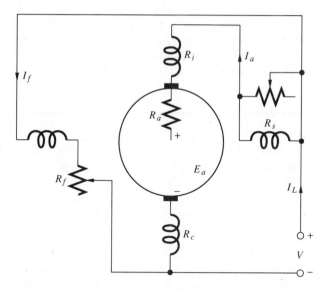

Figure 5.26 General connection diagram for a dc motor.

where V = the terminal voltage

E_a = the back emf generated by the armature winding

V_b = the brush contact voltage drop

I_a = the total armature winding current

I_f = the current in the shunt field winding

I_L = the current supplied by the dc energy source

R_a' = the resistance of the complete armature circuit, equal to the sum of R_a the resistance of the armature winding, R_i the resistance of the interpole winding, and R_c the resistance of the compensating winding

R_f = the total series resistance of the shunt field winding and the rheostat

and R_s = the total resistance of the series field winding and the diverter in parallel.

The brush-contact voltage drop V_b is not normally higher than two volts and may often be neglected when compared to the terminal voltage. Equation (5.7.8) is the speed equation for a dc motor and follows from eqs. (5.7.2) and (5.7.3). The speed n is expressed in revolutions per second in these equations.

The speed is inversely proportional to the flux per pole Φ. This observation points to the danger associated with the interruption of the field circuit when the motor is driving a load. If the load is small so that sufficient torque can be supplied by the residual field, the motor speed will increase until demolition by centrifugal forces occurs. If the load requires a torque larger than can be supplied by the residual field, the motor will slow down, the armature winding current will become excessive, and the motor may burn out if overcurrent relays are not used.

EXAMPLE 5.6

A 250-V, dc shunt motor draws a full-load line current of 100 A at the rated speed of 1200 r/min. The armature circuit resistance R_a is 0.1 ohm and the field winding resistance R_f is 250 ohms. Determine (a) the gross, full-load, mechanical power output, (b) the electromagnetic torque developed at full load, and (c) the speed regulation, if the no-load armature-winding current is 10 A.

Solution
(a) The gross mechanical power output $P = E_a I_a$.
The armature winding current $I_a = I_L - I_f = 100 - 250/250 = 99$ A.
The armature winding emf $E_a = V - I_a R_a = 250 - 99 \times 0.1 = 240.1$ V.
Hence the gross, full-load power $P = 240.1 \times 99 = 23770$ W $= 23.77$ kW.
(b) The electromagnetic torque $T_e = P/2\pi n$.

Therefore the full-load torque $T_e = \dfrac{23770}{2\pi \times 1200/60} = 189$ N·m.

(c) The speed regulation is REG $= \dfrac{n_{NL} - n_{FL}}{n_{FL}} \times 100\%$.

At no load the induced emf $E_{aNL} = V - I_{aNL} R_a = 250 - 10 \times 0.1 = 249$ V.
At full load the induced emf $E_{aFL} = 240.1$ V, from part (a).
At full load the speed is the rated value, $n_{FL} = 20$ r/s.
Since the shunt field is constant, at no load the speed n_{NL} is given by

$$n_{NL} = \frac{E_{aNL}}{E_{aFL}} n_{FL} = \frac{249}{240.1} \times 20 = 20.74 \text{ r/s.}$$

Therefore the speed regulation is REG $= \dfrac{20.74 - 20}{20} \times 100 = 3.7\%$.

This a typical value for a shunt motor.

EXAMPLE 5.7

A 350-V, dc series motor draws a full-load current of 100 A at the rated speed of 1200 r/min. The armature winding resistance R_a is 0.1 ohm, and the field winding resistance R_e is 0.1 ohm. The saturation curve (separately-excited, open-circuit characteristic) for the machine driven at 1200 r/min is obtained from the following data.

| E_a (volts) | 87 | 145 | 182 | 212 | 230 | 244 |
| I_f (amperes) | 20 | 40 | 60 | 80 | 100 | 120 |

Determine the motor speed ratio for 100% and 10% full-load currents, if the terminal voltage is 250 V.

Solution
For 100% load current $I_a = I_f = 100$ A.
The induced emf $E_a = V - I_a(R_a + R_s) = 250 - 100 \times 0.2 = 230$ V.
From the saturation curve, for $I_f = 100$ A, $E_a = 230$ V.
Accordingly the speed is 1200 r/min for 100% load current.
For 10% load current $I_a = I_f = 10$ A, the induced emf E_a is

$$E_a = V - I_a(R_a + R_s) = 250 - 10 \times 0.2 = 248 \text{ V}.$$

For $I_f = 10$ A, $E_a = 44$ V from the given saturation curve at 1200 r/min. But the emf is proportional to speed for a constant flux. Therefore, to generate $E_a = 248$ V at 10% load current ($I_f = 10$ A), the motor speed n must be

$$n = \frac{248}{44} \times 1200 = 6760 \text{ r/min}.$$

The speed ratio for 100% and 10% full-load current is $1 : 5.6$.

EXAMPLE 5.8

A 500-V, dc compound motor with interpoles has an armature winding resistance of 0.23 ohm, a series field winding resistance of 0.007 ohm, and an interpole winding resistance of 0.013 ohm. The shunt field winding, which is connected across the supply terminals, has a resistance of 250 ohms. On no load the main flux is 22 mWb, the speed is 1000 r/min, and the motor draws 6 A from the supply. On full load, the main flux is 25 mWb and the motor draws 102A from the supply. **(a)** What are the full-load speed and the speed regulation? **(b)** What is the electromagnetic torque?

Solution
(a) The shunt field current has the constant value

$$I_f = \frac{V}{R_f} = \frac{500}{250} = 2 \text{ A}.$$

The armature winding current follows from $I_a = I_L - I_f$ and the generated armature winding voltage follows from $E_a = V - I_a(R_a + R_i + R_s) = K_e n\Phi$.
For the no-load condition we find $I_{aNL} = 6 - 2 = 4$ A and $E_{aNL} = 500 - 4(0.013 + 0.007 + 0.23) = 499$ V.
For the full-load condition $I_a = 102 - 2 = 100$ A and $E_a = 500 - 100(0.013 + 0.007 + 0.23) = 475$ V.
At no load $E_{aNL} = K_e n_{NL}\Phi_{NL} = K_e \times \dfrac{1000}{60} \times 0.022 = 499$ V and at full load $E_a = K_e n\Phi = K_e n \times 0.025 = 475$ V.
Hence $n = \dfrac{475}{499} \times \dfrac{0.022}{0.025} \times \dfrac{1000}{60} = 13.97$ r/s.

The regulation is REG $= \dfrac{1000 - 60 \times 13.97}{60 \times 13.97} \times 100\% = 19.3\%$.

(b) The electromagnetic torque is $T_e = K_T I_a \Phi$,

where $K_T = \dfrac{K_e}{2\pi}$ and $K_e = \dfrac{E_{aNL}}{n_{NL} \Phi_{NL}}$.

Hence $T_e = \dfrac{499 \times 102 \times 0.025}{2\pi \times 1000 \times 0.022} = 552 \ \text{N} \cdot \text{m}$.

5.7.4. Motor Characteristics

The characteristic behavior of a motor is mainly determined by the manner in which the field excitation is accomplished. The characteristic curves of interest show how the motor speed and the developed torque vary with the armature or load current for the conditions that the applied voltage and the resistance of the field circuit remain constant. From these two characteristics the torque speed characteristic can be determined.

A. SPEED CHARACTERISTICS

The general form of the speed characteristic for the different types of motors can be deduced by argument. However, there are graphical methods available for a more accurate prediction of the variation of speed with armature-winding current. From the speed expression

$$n = \frac{V - I_a R'_a}{K_e \Phi} \tag{5.7.9}$$

it follows that the speed for a particular value of armature winding current can be controlled either by variation of the applied voltage or by changing the armature circuit resistance or by varying the magnetic flux. Load current affects the speed in an uncontrolled manner and this can be seen from the above equation.

Separately-excited Motors and Shunt Motors

If the applied voltage remains constant, there is no difference between the characteristic behavior of a separately-excited motor and a shunt motor.

The general shape of the speed characteristic can be determined from the following argument. For a fixed value of the field circuit resistance and a constant value of applied voltage, the field current is constant. The flux will have its maximum value at no load and will decrease slightly due to armature reaction as the armature winding current increases. The numerator of eq. (5.7.9) will decrease linearly with increasing armature winding current. Since R'_a is kept low to maximize the efficiency and output of the machine, the voltage drop in the armature circuit is small compared to the

applied voltage. Since the numerator of the speed equation normally decreases somewhat more than the flux in the denominator, the speed drops slightly for a load
increase from no load to full load. The separately-excited motor and the shunt motor
are virtually constant speed motors, if the applied voltage, the armature-circuit
resistance and the field rheostat setting remain unchanged.

For the graphical construction of the speed characteristic of a shunt motor refer
to Fig. 5.27. In the first quadrant, the straight line ab depicts the value of the numerator $(V - I_a R_a')$ of the speed equation as a function of armature winding current. The
full-load armature winding current is indicated by point c on the axis. In the third

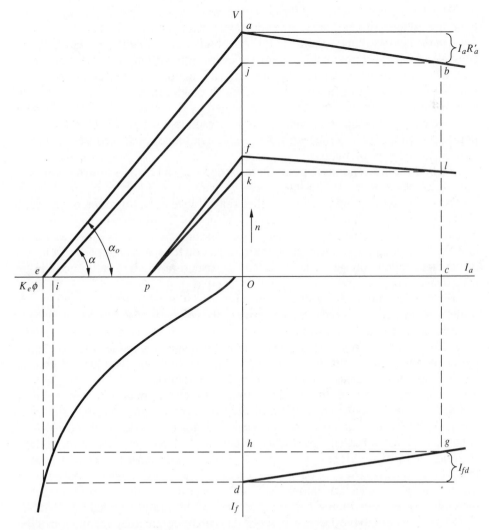

Figure 5.27 Construction of speed characteristics for a dc shunt motor.

quadrant the denominator $K_e\Phi$ of the speed equation is plotted as a function of the field current. This curve is obtained from the saturation curve of the machine. Let point d on the ordinate be the value of field current for a constant value of field circuit resistance. The no-load value of $K_e\Phi$ is indicated by point e.

The theoretical no-load speed is given by the ratio of line segments Oa/Oe or the tangent of angle α_0. Selecting an appropriate speed scale along the ordinate (also used earlier for voltage), the theoretical no-load speed is indicated by point f. The term *theoretical no-load speed* is used to indicate the speed at $I_a = 0$. In reality, when a motor does not drive an external load it still has to overcome the loss torque of the machine. To develop this torque a small armature winding current is required. The theoretical and actual no-load speeds differ by an insignificant amount.

The demagnetizing effect of the armature mmf can be thought to be caused by a hypothetical current component, the equivalent demagnetizing field current, which flows in the field winding. Assume that in the absence of compensating windings the equivalent demagnetizing field current varies linearly with armature winding current. The equivalent demagnetizing current as a function of armature winding current is given by the straight line dg in the fourth quadrant. At full-load current the equivalent demagnetizing current is represented by line segment dh. The effective field current, at full load and given by Oh, will result in a value of $K_e\Phi$ indicated by point i. At full load $(V - I_a R_a')$ is given by $0j$. The full-load speed is equal to the ratio $0j/0i$ or the tangent of angle α.

To find the corresponding point on the speed scale a line fp is drawn parallel to line ae to locate point p. The line pk drawn parallel to ij locates point k, which represents the full-load speed. Point l is then the full-load operating point of the speed characteristic. In a similar manner additional points for other values of armature winding current can be found. The resulting speed characteristic fl is almost a straight line. From the foregoing construction the influence on the speed and speed regulation of varying applied voltage, or field current, or armature winding resistance can be investigated.

If the voltage applied to a shunt motor is varied, the field current varies proportionally. At no load the speed is increased when the voltage is raised above the normal or rated value, because saturation prevents the flux increasing as much as the voltage does. If the applied voltage is reduced, the no-load speed is reduced for the same reason. As the applied voltage is reduced more and more the no-load speed decreases less because of the increased slope of the $K_e\Phi = f(I_f)$ curve.

Under load conditions the speed also follows the change in applied voltage but is affected by the demagnetizing effect of the armature winding current. At lower voltage levels the armature reaction effect causes a larger reduction in flux for a given change in applied voltage than at higher voltage levels. Hence the further the voltage is reduced at rated armature winding current the less the decrease in speed. The speed regulation reduces as the applied voltage is reduced. For a low applied voltage and in the case of pronounced armature reaction the speed may actually increase with increasing armature winding current. This causes the motor to become unstable. Thus there is a limit to the decrease in speed that can be obtained by a reduction of the applied voltage.

For a separately-excited motor a strong main field can be maintained and the applied voltage can be lowered without the danger that the demagnetizing effect of armature mmf becomes too influential. The speed of this type of motor can be varied over a much wider range than is possible with a shunt motor. The maximum speed attainable is limited by the maximum permissible voltage.

A constant voltage applied to the terminals of a shunt motor means that the field current and thus the flux is changed by adjusting the field circuit resistance. Decreasing the field current will decrease the flux and consequently raise the no-load speed of the motor. For an equal change of field current the increase in speed tends to be larger than the decrease because of the nonlinear shape of the flux curve. The numerator of the speed eq. (5.7.9) has a constant value at full-load armature winding current. The demagnetizing effect of armature reaction decreases as the magnetic circuit becomes less saturated. The lower the field current, the lower is the no-load flux and the larger is the further reduction of flux due to armature reaction. Hence the increase in speed resulting from a lower value of field current is larger at full load than at no load. In other words the speed regulation decreases when the field current is reduced. For too low a value of excitation the motor may become unstable, that is, it may have a negative speed regulation when the armature reaction is large. Thus there is a limit to the maximum speed. The saturation effect sets a lower speed limit. There is no difference between the behavior of a separately-excited motor and a shunt motor when speed is controlled by means of field current adjustment.

Also, there is no difference of behavior between the two types of motor if variation of armature-winding is used to adjust the motor speed. For constant values of applied voltage and field current the no-load speed is fixed. The total armature circuit resistance can be increased by placing a properly dimensioned resistor external to the motor in series with the machine's internal armature circuit. Only the numerator of the speed equation decreases with increasing additional resistances. It follows that the full-load speed decreases and the regulation increases with increasing total armature circuit resistance. This method of speed control is wasteful on account of the added ohmic loss, but allows for speed control at constant torque since both I_a and Φ can be kept constant while changing the speed. The speed can be reduced to zero. This is convenient for applications that require *jogging* or *inching* of a load into position.

Compound Motor

If the armature circuit does not carry current, the series winding of the compound motor does not contribute to the excitation. The theoretical no-load speed is determined by the applied terminal voltage and the flux produced by the shunt field winding as in the case of a shunt motor.

Assume a constant applied voltage and a fixed setting of the shunt field rheostat. For the full range of load current, the shunt field mmf will remain constant in the case of a long-compound motor. The series field mmf is proportional to the armature winding current in a long-compound motor and almost so in a short-compound motor.

If the motor is cumulatively compounded, the series mmf opposes the demagnetizing effect of armature reaction and normally tends to increase the total flux with increasing armature winding current. This, combined with a decreasing value of the numerator of the speed equation, means that the speed reduces as the armature winding current increases. The cumulatively-compounded motor always has a positive speed regulation (up to 25%) that is larger than that of a shunt motor (up to 8%).

For the case of differential compounding, the series mmf will aid the demagnetizing effect of armature reaction. The flux may decrease faster than $V - I_a(R'_a + R_s)$, in which case the speed rises with increasing armature winding current. The negative speed regulation does not permit stable operation.

It is possible to proportion the series-winding mmf with respect to the shunt-winding mmf in such a manner that the speed characteristic is almost flat over the normal operating range. However, on overload, the characteristic will unavoidably rise slightly. A slight increase in speed with increase in load will cause the motor to *run away*. This inherent danger, plus other problems associated with possible difficulties in starting, are the reasons that differentially-compounded motors are used infrequently.

Series Motor

The flux in a series motor depends on the armature winding current that also flows through the series-field winding. At light loads the armature winding current is low and the resulting flux is small. As the armature winding current increases the flux also increases. The numerator of the speed equation for the series motor decreases somewhat with increasing armature winding current. The speed, being inversely proportional to the flux, is mainly affected by the large variation of the flux with armature winding current. It follows that the speed is high at light loads and considerably lower at heavy loads. The series motor is therefore a variable speed motor.

Under no-load conditions the flux is the small residual flux in the magnetic circuit and the speed becomes so high that centrifugal forces may destroy the machine. For this reason series motors must always be directly coupled or geared to their load.

B. TORQUE CHARACTERISTICS

Normally a motor is supplied from a constant voltage source. Hence, as the mechanical load of a motor increases, the current drawn from the electric energy supply must also increase. The torque developed by a motor as a function of the armature winding current is called the *torque characteristic*. In the following sections we will deduce the torque characteristic for the different types of dc motor from the torque equation

$$T_e = \frac{EI_a}{2\pi n} = \frac{pZ}{2\pi a} I_a\Phi = K_T I_a\Phi \tag{5.7.10}$$

by argument and graphical construction.

Separately-excited and Shunt Motors

The field windings of separately-excited and shunt motors are connected to a constant dc voltage supply. The mmf of a field winding is therefore constant. With normal excitation the flux is slightly modified by armature reaction. If the flux is assumed to remain constant, electromagnetic torque is only dependent on the armature winding current and the torque characteristic will be a straight line.

The graphical construction of the torque characteristic is shown in Fig. 5.28. In the third quadrant the curve $K_T\Phi = f(I_f)$ is plotted. This curve can be obtained from the saturation curve $E_a = f(I_f)$, since $K_T\Phi = E_a/2\pi n$ for equal values of field current, where n is the constant speed at which the saturation curve is determined. Let Of be the actual field current value. Assume that the equivalent demagnetizing current I_{fd} of the armature reaction varies linearly with I_a and is equal to bc for the armature-

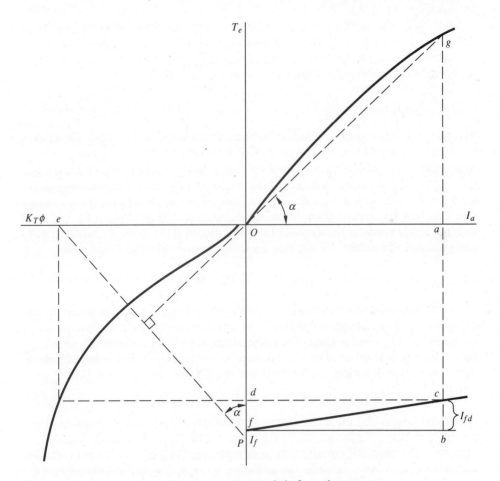

Figure 5.28 Construction of the torque characteristic for a shunt motor.

current value Oa. The effective field winding current as a function of armature winding current is then represented by the straight line fc in the fourth quadrant. For an armature current value $0a$, the effective field current value Od produces a flux which gives a value of $K_T\Phi$, represented by length Oe. Arbitrarily choose a point P on the vertically downward field current coordinate and draw the line Pe which encloses an angle α with the vertical. Through the origin draw a line perpendicular to Pe. For armature winding current Oa we find a point g on this line. We can now write

$$ag = Oa \tan\alpha = Oa \frac{Oe}{OP} = \frac{T_e}{OP} \tag{5.7.11}$$

since Oa and Oe are corresponding values of I_a and $K_T\Phi$, and the torque $T_e = K_T I_a \Phi$. Thus ordinate ag is a measure of the torque developed by the armature winding current Oa. The scale of the torque ordinate is determined by the location chosen for point P.

For other values of armature winding current, repetition of the construction using the same length OP will yield additional points needed to draw the torque characteristic. The characteristic is slightly curved because the curve $K_T\Phi = f(I_f)$ is not straight for effective field currents between Of and Od.

Compound Motor

In a cumulatively-compound motor the mmf of the series winding aids the mmf of the shunt winding. Hence, for the same mmf of the shunt field and the same armature winding current, the torque developed by a cumulatively-compound motor is larger than that of the same machine as a shunt motor. As the armature winding current increases the torque will increase. The increase of torque of the cumulatively-compounded motor is greater than that of the shunt motor because the total field mmf does not remain constant as it does in the shunt motor. If the machine is differentially compounded, the motor will develop a smaller torque than the shunt motor.

Series Motor

In a series motor the flux depends on the armature winding current, since that current is also the exciting current. At light loads the armature winding current and therefore the series winding mmf is small. The magnetic circuit is then unsaturated, so that the flux is directly proportional to the armature winding current. For light-load conditions, the torque equation can be rewritten in the form

$$T_e = K_T\Phi I_a = K_T(kI_a)I_a = KI_a^2. \tag{5.7.12}$$

Hence, as long as the magnetic circuit is unsaturated, the torque characteristic is parabolic in shape. As the mechanical load increases, the armature winding current increases. The magnetic circuit tends to saturate and the flux rises at a progressively diminishing rate. At overloads, the armature winding current is so large that the magnetic circuit is saturated and the flux becomes almost constant. For these conditions the torque increases almost linearly with the armature winding current.

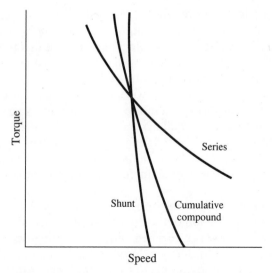

Figure 5.29 Mechanical characteristics for dc motors.

C. TORQUE-SPEED CHARACTERISTICS

The performance of a motor is best expressed by its mechanical characteristic or its torque-speed curve. To construct such a curve for a motor, corresponding values of torque and speed associated with each of a series of values of armature winding current are obtained from its torque and speed characteristics. Typical mechanical characteristics for shunt, compound, and series motors are shown in Fig. 5.29. For the sake of comparison the curves are drawn through a common point of rated torque and speed.

5.7.5. Motor Speed Control

One of the attractive features offered by dc motors is the relative ease with which the speed can be adjusted over a relatively wide range. The speed can be changed by altering the value of the flux, the applied voltage, or the total resistance of the armature circuit between the motor terminals.

Of these three methods the variation of flux is the simplest, and the least expensive one that is most applicable to shunt motors. As the value of the resistance of the rheostat in the shunt field circuit is increased the flux decreases and the speed increases. To prevent overheating of the motor the rated armature winding current should not be exceeded. Hence the torque that can be developed by a shunt motor at increased speeds is reduced below its rated value. If the maximum armature winding current is to remain the rated value then it follows from $E_a = V - I_a R'_a$ that the back emf will remain constant. The mechanical output is $2\pi n T_e = E_a I_a$. Hence, if the maximum permissible armature winding current I_a and the back emf E_a remain constant, the maximum permissible output must also remain constant over the whole range of speed variation. At low values of main-pole flux the danger exists that the machine becomes unstable (negative speed regulation) due to the demagnetizing effect of armature reaction. This limits the maximum safe speed obtainable with a regular shunt motor to approximately twice its rated speed. The addition of a series field winding, which tends to increase the flux with increasing armature winding current, stabilizes the motor and allows the speed range to be extended to 4 : 1 for medium-sized motors. Also the speed range can be increased to as much as 6 : 1 in specially designed shunt motors. The ohmic loss in the shunt field rheostat reduces the efficiency very slightly, since the shunt field current is less than 5% of the full-load current.

The connection of an adjustable resistor in parallel with the field winding of a series motor permits a reduction of the flux for a certain armature winding current. The resulting speed increase is associated with a constant gross mechanical output $E_a I_a$. Hence the developed torque is reduced. Added armature circuit resistance is more commonly used for series motors to reduce the speed. Although this method is also applicable to shunt and compound motors, the resulting large increase of speed regulation is often objectionable for shunt motor applications that require a virtually constant speed drive. Since neither the flux nor the armature winding current is affected by an increase of the total armature circuit resistance, a constant torque is maintained and the mechanical output is reduced. If an external resistor is inserted in the armature circuit of a shunt or compound motor, the shunt field should not be affected by the voltage drop across the additional resistor. A great disadvantage of this method of speed control is the considerable reduction of efficiency due to the large ohmic loss in the additional resistor which has to carry the full armature winding current. This method is therefore best used when temporary reductions in speed are required. Unlike a shunt field rheostat the additional armature-winding resistor is a rather bulky and expensive item.

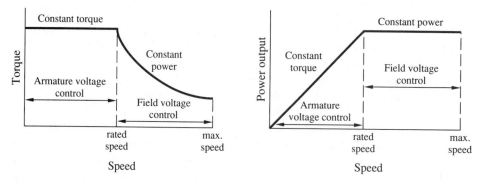

Figure 5.30 Torque and power characteristics of dc motors.

Speed control by means of varying the voltage applied to the armature terminals of a motor is the most flexible method of speed variation and is applicable to all three types of motor. A wide range of smooth speed variation as large as 10:1 can be obtained with commercially available machines. Conventionally, the motor is supplied from a separately-excited dc generator driven by a constant speed prime mover. This system, known as a Ward-Leonard system, is an expensive method of speed control on account of the cost of the variable voltage supply. If the motor field excitation can also be varied, a further extension of the controllable speed range to 40:1 can easily be obtained. The motor-generator sets, providing the adjustable voltage, are replaced by silicon controlled rectifier converters today. Figure 5.30 shows the constant torque and constant power characteristics of a separately-excited motor for speed adjustment by armature voltage and field voltage control.

5.7.6. Motor Starting

At the moment of starting a dc motor, the supply voltage V is applied to the motor terminals, but the rotor is at standstill. Under these conditions the back emf is zero and the current is only limited by the total armature circuit resistance. In all but small motors of one-half kilowatt rating or less, the armature winding resistance value is low (i.e., considerably less than 1 ohm). This means that, at the moment of starting, a current 10 to 20 times the rated full-load current will flow.

As the motor picks up speed and builds up a back emf, the current gradually reduces to the normal operating value. The time required by the motor to come up to speed is determined by the developed accelerating torque and the inertia of the rotating parts. During this time interval a current in excess of the full-load value flows. Mechanical forces associated with the high current, heating of the armature circuit, inability to commutate such high currents without destructive sparking, and limited safe current-carrying capacity of the brushes can damage the motor. Also a high starting current might produce an excessive voltage drop in the electric supply system affecting other equipment, or it might trip overload protective devices. A high starting current may also create a starting torque in excess of the torque limitation of the mechanical load.

The maximum current during starting must be limited to an acceptable value by an external starting resistor connected in series with the armature circuit. It is advantageous to have the starting current as high as permissible, because the torque is directly proportional to this parameter and the higher the starting torque the shorter is the start-up interval. Also for this reason, full terminal voltage is applied to the shunt field circuit and the rheostat is set to zero resistance to have maximum starting torque and the lowest initial speed after starting.

The starting current is generally permitted to be between 1.25 and 2 times the full-load current, so the motor will be capable of starting under load. This avoids the necessity of clutch arrangements and permits series motors to be directly coupled to their mechanical load.

On connection of the supply voltage to the motor terminals the flux in a shunt motor will build up with a time constant influenced by the relatively high inductance of the shunt field winding. This provides a sufficiently slow rise of torque to prevent shock damage to the mechanical coupling between the motor and load. In a series motor the rise of flux and armature winding current takes place quickly, since the inductance of the armature circuit is small. However, the buildup of the starting torque is not instantaneous.

As the motor accelerates from standstill, the back emf E_a builds up to oppose the applied voltage V and the armature winding current reduces. When the current drops to full-load value, part of the starting resistance must be disconnected. Letting the current reduce to less than full-load value would prolong unnecessarily the interval of starting. Therefore, as the motor picks up speed, the starting resistance is reduced in steps such that the current is limited to the maximum permissible value and does not fall below the full-load value.

Figure 5.31 shows a series and a parallel arrangement of resistors to control the starting current. In the series arrangement a sliding contact moves from contact points 1 to 5. In the parallel arrangement switches 1 to 5 are closed in turn. Thus the overall resistance in series with the armature circuit is reduced to zero in a stepped manner. The resistance values of the individual steps of a starting resistor can be

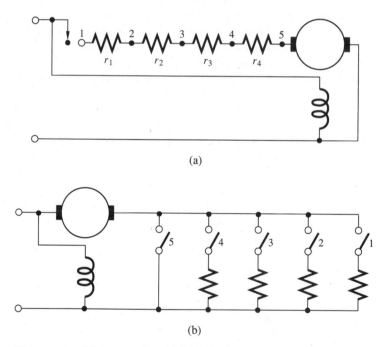

(a)

(b)

Figure 5.31 Motor starting. (a) Series-resistance starter, (b) parallel-resistance starter.

estimated by considering the general case of a series arrangement of n resistance steps between $(n + 1)$ switching contacts. The current value is not to exceed a high value I_H and is not to drop below a low value I_L. The total required series resistance at starting from zero speed is determined from $R_1 = V/I_H$, where $R_1 = r_1 + r_2 + r_3 + \cdots + R'_a$ and r_1, r_2, r_3, etc., are the resistance values of the individual steps of the starting resistor.

If inductance effects are neglected, I_H will flow as soon as the voltage supply is connected by moving the starter resistance arm to contact 1. An accelerating torque is produced, the armature starts to rotate, back emf is generated in proportion to the speed, and the current gradually reduces to the value I_L, at which point the starter arm is moved to contact 2. The current rises abruptly to the value I_H again, if the switching is assumed to take place instantaneously and inductance effects are neglected. Just before switching, the current is $I_L = (V - E_a)/R_1$ and just after switching, the back emf has had no time to increase so $I_H = (V - E_a)/R_2$, where $R_2 = r_2 + r_3 + \cdots R'_a$.

If the current ratio is kept constant at a value k, then in general

$$\frac{I_H}{I_L} = k = \frac{R_1}{R_2} = \frac{R_2}{R_3} = \cdots = \frac{R_n}{R'_a}. \tag{5.7.13}$$

The resistance values are in geometric progression and

$$k^n = \frac{R_1}{R_2} \times \frac{R_2}{R_3} \times \cdots \times \frac{R_n}{R'_a} = \frac{R_1}{R'_a}. \tag{5.7.14}$$

From this relation the number of steps n can be determined. A certain compromise may have to be made between the number of steps and the current ratio. The values of the individual resistance elements follow from

$$r_1 = R_1 - R_2 = R_1 \left(\frac{k - 1}{k} \right) \tag{5.7.15}$$

$$r_2 = R_2 - R_3 = R_2 \left(\frac{k - 1}{k} \right) = R_1 \left(\frac{k - 1}{k^2} \right), \text{ etc.} \tag{5.7.16}$$

Many different types of dc motor starters are available to suit special requirements and economic factors, but the principle of disconnecting steps of a starting resistance is common to all.

EXAMPLE 5.9

A four-step starting resistor is required for a 10-kW shunt motor which is supplied from a 100-V source. The resistance of the armature circuit is 0.05 ohm and that of the shunt-field circuit is 20 ohms. A maximum armature current of twice the rated value is permitted to flow for short periods of time. Determine the values of the four resistance steps.

Solution

Assuming that the terminal voltage remains 100 V during starting, the shunt field current is

$I_f = 100/20 = 5$ A.

The rated armature-winding current is then

$I_a = I_L - I_f = 100 - 5 = 95$ A.

Without a starting resistor, the initial value of the armature current at starting would be 2000 A or 21 times the rated value. The total value of armature circuit resistance, that is required to limit the starting current to 190 A, follows from

$R_1 = V/I_H = 100/190 = 0.53$ ohm.

From $k^n = R_1/R_a'$ it follows that $k = 1.80$, since $n = 4$.
Hence $r_1 = R_1(k-1)/k = 0.236$ ohm, $r_2 = r_1/k = 0.131$ ohm, $r_3 = r_2/k = 0.073$ ohm, and $r_4 = r_3/k = 0.040$ ohm, where the designations r_1, r_2, r_3, and r_4 correspond with those resistors shown in Fig. 5.31.

5.7.7. Motor Braking

If the electric supply to a motor were removed, the motor would coast to a stop. The only braking effect would be due to mechanical friction. The time taken to come to rest would depend on the stored kinetic energy of the motor armature and the mechanical load. Often this time is too long.

   Smooth and accelerated braking can be accomplished when the rate of retardation is increased by using the inertia of the armature and connected load to drive the machine as a generator during the stopping period. As a generator the electromagnetic torque opposes the motion of the armature and the load. The kinetic energy stored in the mechanical system is then converted into electrical energy which can either be fed to the electrical system or, more commonly, is converted to thermal energy. The braking time now depends on how quickly the kinetic energy can be transferred, that is, the magnitude of current that can be tolerated during that period. The greatest braking effort takes place at the highest speeds and the highest values of flux, because the power transferred depends on the generated emf which in turn is proportional to both these parameters.

   When a motor is changed to the generator mode of operation, the direction of the magnetic field is maintained, because the large amount of stored magnetic field energy involved makes switching impracticable. For the same direction of rotation the polarity of the generated emf is then unchanged, so that the armature current must be reversed if the direction of electromagnetic torque is to reverse and become a braking torque. A shunt motor will smoothly convert from motor to generator operation without any change in the connections of the field winding to the supply terminals. To maintain the same field direction in a series machine the connections of the series winding have to be reversed or the field winding is supplied from a separate voltage source to obtain better control of the braking process. In the case of a compound machine the series winding is often disconnected during braking and only the shunt field remains excited.

There are three methods of electric braking. These are plugging, dynamic braking, and regenerative braking. *Plugging* is used when a motor has to be brought to a halt quickly or when a fast reversal of the direction of rotation is required. In this method, the connections of the supply voltage to the armature are reversed while the direction of the magnetic field is maintained the same. During the braking period the generated voltage and the supply voltage aid each other. At the moment of plugging the current could rise to nearly twice the maximum possible starting current and would reduce to this latter value when the rotor comes to standstill. At standstill the torque is still in the same direction so that, if the voltage supply remains connected, the machine builds up speed in the reverse direction to operate as a motor once again. When this method of fast braking is used, a current limiting or plugging resistor of almost twice the value of the starting resistor must be inserted in the armature circuit to keep the current within proper limits. The kinetic energy of the armature and mechanical load to be brought to a stop will thus be dissipated as heat in the plugging resistor. All methods of electric braking have the common characteristic that, as the speed reduces, the generated emf reduces. The value of the plugging resistor must therefore be reduced as the speed drops in order to maintain a satisfactory armature winding current and retarding torque.

Dynamic braking, sometimes called rheostatic braking, is initiated by switching the armature circuit from the electric supply to a braking resistor, while the original flux direction is maintained. The machine will now act as a generator as long as it is rotating. An armature winding current will flow in a direction opposite to that of the motor mode of operation, so that the developed torque reverses and acts as a braking torque. The braking torque is a function of the armature winding current and the flux. As the armature slows down, and hence the generated voltage reduces, the armature current is maintained within certain limits by a stepped reduction of the braking resistance. This is similar to the practice used for starting. Nevertheless the electromagnetic braking action at slow speeds diminishes to zero and must be supplemented by mechanical means. All the kinetic energy of the rotating mass is dissipated as heat.

Both plugging and dynamic braking normally waste the stored kinetic energy. Regenerative braking, on the other hand, converts the kinetic energy of the mass into electric energy which is fed back to the electric system. To facilitate this action the generated voltage has to become higher than the supply system voltage so that the current reverses its direction. Hence, either the flux has to be increased, or the speed is increased above the normal speed, or a combination of both has to take place. An increase in speed results from the load increasing its kinetic energy under the influence of gravity when the load is of an overhauling character, such as hoists and elevators. Regenerative braking is used to maintain safe speeds but cannot bring a load to standstill. For this, plugging or dynamic braking plus mechanical braking are necessary.

EXAMPLE 5.10

A 230-V shunt motor takes a line current of 102.3 A while delivering its full-load torque at a speed of 25 r/s. The armature-circuit resistance is 0.1 ohm and the shunt-field circuit resistance is 100 ohms. A braking resistor of 1.0 ohm is used for dynamic braking. What is the initial electromagnetic braking torque?

Solution
The shunt-field circuit remains connected to the 230-volt supply and draws a current of

$$I_f = \frac{V}{R_f} = \frac{230}{100} = 2.3 \text{ A.}$$

While operating as a motor, the armature-winding current is

$$I_a = I_L - I_f = 102.3 - 2.3 = 100 \text{ A.}$$

A counter emf is then generated and its value is given by

$$E_a = V - I_a R'_a = 230 - (100 \times 0.1) = 220 \text{ V.}$$

The same armature voltage is generated at the initial moment of braking, since Φ is constant and n has not changed during the short interval of time required by the switching operation in the armature circuit.

The armature winding current at the initial moment of braking is

$$I_{abr} = \frac{E_a}{R_{br} + R'_a} = \frac{220}{1 + 0.1} = 200 \text{ A.}$$

This is twice as large as the full-load armature winding current. The full-load motoring torque is

$$T_e = \frac{E_a I_a}{2\pi n} = \frac{220 \times 100}{2\pi \times 25} = 140 \text{ N} \cdot \text{m.}$$

Hence the initial electromagnetic braking torque is 280 N·m.

5.7.8. Special Motors

The conventional dc motor is used for speed control over a wide speed range and for a wide range of powers. There are many cases in control systems where the power requirement is low, the controller must be simple (such as a single voltage adjustment), sliding contacts cannot be tolerated, and the mechanical response to a voltage change must be rapid. Many kinds of servomotors have been designed to attempt to satisfy one or more of these criteria.

Voltage control for a reasonable speed range is only satisfied by a dc motor. If only a single voltage is to be controlled then simplicity is accomplished by using permanent magnets for the field system of small motors. (For large motors the cost of permanent magnets is too great.) Sliding contacts (brushes and commutator) can be eliminated, if the armature winding is placed on the stator. Rapid response can be achieved if both the electrical and mechanical time constants are small. The mechanical time constant can be decreased by designing the motor to have a low moment of inertia. Inertia is kept low by eliminating rotor slots and having the winding on the surface of the core. A better way yet is to eliminate the rotor iron and have a moving-coil armature winding or else a *disc* armature. The increased air gap decreases the armature winding inductance and decreases the value of the electrical time constant.

These improvements can be made to small dc motors, some of which are described in the following sections.

A. PERMANENT MAGNET MOTOR

If a motor is supplied from a dc source and if the power demanded by the load is low (usually less than 1 kW), it is likely that the machine is a permanent-magnet dc motor. This type of motor differs from the conventional dc machine by having permanent magnets in place of the stator field coils.

There are three types of permanent magnet that are commonly used in motors. These are Alnico magnet, ferrite (or ceramic) magnet, and rare earth magnet (samarium-cobalt magnet). None of these is ideal, but, by designing the field structure with special combinations of magnets and ferromagnetic materials, high-performance (high-torque, low-volume, high-efficiency) motors can be manufactured at low cost.

The ideal permanent magnet would have a high value of remanent magnetic flux density B_r, would have a high value of flux density (equal to B_r) while demagnetization (such as that produced by armature reaction) is present, and would have a high value of coercivity H_c (the value of the demagnetization that would bring the flux density of the field to zero). Figure 5.32 shows the general major demagnetization characteristics of Alnico, barium-ferrite, and samarium-cobalt magnets. Alnico has the advantages of high remanent flux density and low cost, and the disadvantage of being easily demagnetized. Barium ferrite has high coercivity and low cost, but it has a low operating flux density. It would seem that the rare earth magnet is best in that both B_r and H_c are high. However, the cost of the material limits its use.

The characteristics of the permanent magnet motor make it suitable for use as a dc servomotor in control systems because the motor and electronic controller are inexpensive. There are many general-purpose applications in the home, automobile, and computer. Some examples in the home are blenders, carving knives, electric shavers, lawn mowers, miniature motors in toys (racing cars), power tools, tooth brushes, and vacuum cleaners. A number of electric motors in the automobile is likely to be a permanent magnet type. It is used to drive air-conditioner and heater blowers, power seats and power windows, tape decks, windshield wipers, and windshield washers. In computers this type of motor is used for capstan and tape drives.

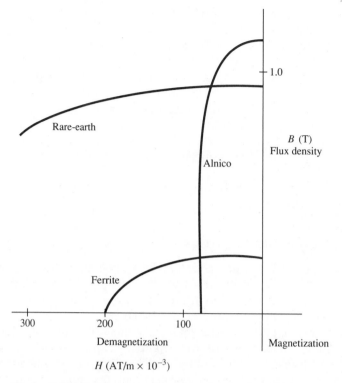

Figure 5.32 Permanent magnet demagnetization characteristics.

B. BRUSHLESS DC MOTORS

For speed control applications the dc motor is used widely. The reason is that the control circuitry is relatively simple. A wide range of speed control is accomplished by regulating the voltage at the terminals of the armature winding, or at the terminals of the field winding. The speed control of ac motors requires both frequency and voltage adjustment and this is more complex to implement.

A major disadvantage associated with dc motors is the need to have a commutator. Commutators and their brushes wear, so there is a need for more maintenance than that required for ac motors. In space, where there is little or no atmosphere, the friction between the brushes and the commutator is so high that commutators cannot be used. To maintain the simplicity of dc motor speed control and to eliminate the commutator on the rotor, the ingenuity of the designer has produced the *inverted* dc motor, that is, the brushless dc motor. This machine has its armature winding on the stator and the field system on the rotor.

In order that there are no sliding connections to the field system on the rotor, permanent magnets are used to produce the main field. This configuration means

Photo 3   Brushless dc motors.

that the main field cannot be adjusted and speed control is only through the armature. This makes the controller simpler.

The purpose of the commutator in a conventional dc motor is to maintain an
armature mmf axis at right angles to the pole axis, regardless of the rotor position. In
this way the torque is the maximum possible value. Consequently, if the armature
winding is placed on the stator, an electronic switching circuit must replace the
mechanical commutator so that the field of the armature is always ahead of the field
of the rotor poles. This enables the driving torque to be maintained.

A commutator has many segments, so that there are many switching actions per
revolution of the rotor. Cost does not allow many transistors in the electronic switching circuit of the brushless machine's armature. The design of the armature winding
must be simplified to accommodate only a few switches and yet maintain a high value
of torque. Figure 5.33 shows a diagram of the main elements of a brushless dc motor.
There are the permanent magnets on the rotor, three coils comprising the armature
winding on the stator, three switches, and a dc supply. What is not shown is how the

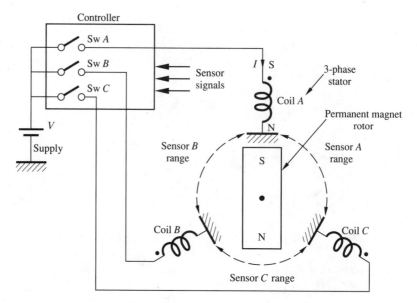

Figure 5.33 Brushless dc motor.

rotor position is sensed so that the switches are closed and opened in the right sequence.

The stator and rotor configuration of the motor depicted in Fig. 5.33 could well represent a synchronous motor, a stepper motor, or a brushless dc motor. From the point of view of a rotor magnet chasing a moving stator field, they are all the same. The differences lie in how the stator field is produced.

If the stator coils are excited from a three-phase ac supply, the machine is called a synchronous motor. Three-phase alternating currents in the stator winding produce a magnetic field that rotates in the air gap at a constant speed, that depends on the frequency of the supply. If the rotor poles can lock in by the forces of attraction, the rotor will be in *synchronism* with the stator field. For this *synchronous* motor to have an adjustable speed the frequency must be changed. This is relatively complex.

Another form of synchronous motor is the stepper motor.[3] The three-phase ac supply is replaced by a dc supply and switches as shown in the figure. Switches *A*, *B*, and *C* are switched on and off in turn in order to excite the stator coils *A*, *B*, and *C* sequentially. Exciting coil *A* causes the rotor to align itself as shown. If the next coil to be excited is coil *B*, the rotor will *step* through 120 mechanical degrees to align itself with the axis of coil *B*. A further motion of 120° ensues if coil *C* is excited alone. The speed of the motor is a function of the frequency of the switching. So the machine is not considered to be a dc motor.

The configuration in the figure can represent the form of a dc motor, if, and only

---

[3] See subsection B under section 8.7.4.

if, the switches are turned on and off by sensors that detect the rotor position similar to the action of brushes on a conventional commutator. For the sake of discussion let there be three sensors A, B, and C on the stator. These sensors (light or magnetic sensors, for example) can each detect the rotor position over a range of $120°$ as shown. If the rotor south pole lies between the axes of coils A and B, sensor B is activated and its signal amplified to turn on switch B. The torque resulting from the rotor poles and coil B's magnetic field causes the permanent magnet to rotate anticlockwise until the S pole aligns itself with the axis of coil B. This torque, that accelerates the rotor, is proportional to the strength of the coil field, which is a function of the applied voltage. As soon as the south pole aligns itself with coil B axis, sensor B is made inoperative (by a rotor shield) and sensor C becomes active over a range of $120°$. In this state switches A and B are off and switch C is turned on to maintain an anticlockwise torque to align the rotor south pole with the axis of coil C. The speed of this motion depends on the value of the torque, which, in turn, is a function of the voltage. With the alignment of the rotor south pole and coil C axis, sensor C becomes inoperative, switch C turns off, sensor A is excited and turns on switch A. Now the rotor wants to complete its first cycle by being attracted to coil A axis. The cycle is repeated to provide continuous motion at a speed that is a function of the supply voltage.

Only the principles of operation of the brushless dc motor have been described. In practice, the motor may have two phases (coils), three phases (most common), or four phases. The switches (transistors) can be arranged (i) to provide unidirectional or bidirectional current in the phases, so that the field polarity of a coil can be reversed, (ii) to provide one or more coils to be excited at any one time so that the torque characteristics can be changed, or (iii) to provide a reversal of the sequence of coil excitation, so that speed reversal is possible.

For low-power servomechanisms in space vehicles and aircraft, where adjustable speed by voltage control is a requirement, brushless dc motors find application. These motors can be made flat. This reduces the length of the drive, so they will be found in both floppy-disk and hard-disk drives.

C. DISC MOTOR

A requirement for a motor might be that it must be light in weight, or that the motor must have a short mechanical time constant so that the response is rapid, or that the axial length of the motor must be short to satisfy space considerations. A way to help meet any one of these requirements is to make the armature coreless. That is, the armature winding has no ferromagnetic core. This makes the armature light and dynamically responsive, but in a motor of conventional geometry the magnetic field paths in air would be long. A solution to this problem is to change the geometry of the armature from a drum shape to a disc.

Figure 5.34 illustrates how the developed armature winding can be formed on a printed circuit board to create a disc. The disc is double-sided to complete the winding. An axial field is produced by magnets on the stator.

Applications of the disc motor have been to provide the electric drive for an automobile radiator fan and to provide drives for robots.

Photo 4   Disc motor.

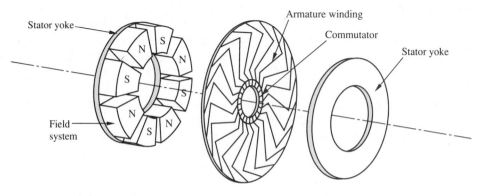

Figure 5.34   Disc motor.

5.7.9. Motor Comparison and Applications

The characteristic behavior of a dc motor is determined by the manner in which the main field excitation is accomplished. Typical mechanical characteristics of the different types of motor named after their method of excitation are shown in Fig. 5.29. Their applicability to certain types of load can be inferred from these characteristics.

It follows that shunt motors are essentially suited for loads that are to be operated at virtually constant speed. Since a shunt motor operates at a constant flux, it is capable of producing a moderate starting torque (usually limited to approximately 2.5 times its rated torque). Hence, shunt motors should be used for such loads as fans, centrifugal pumps, and conveyors, whose load characteristics are compatible with the shunt motor characteristics.

On the other hand, series-excited motors are capable of producing a high starting torque (up to five times their rated torque). They are therefore suited to high-inertia loads that require a high starting torque to accelerate the heavy load mass. Consequently series motors are used for electric traction, cranes, and hoists. The large variation of speed with torque of the series motor is also compatible with these types of load. Speed control down to zero speed at constant torque by means of a variable resistor inserted in the armature circuit allows *inching* (or *jogging*) of the load. This is ideal for the above mentioned types of applications.

To demonstrate another important difference between shunt and series motors we remember that the back emf is almost equal to the applied terminal voltage and that for a shunt motor the load current differs from the armature winding current by only a small amount due to the shunt field current. Simplifying for the sake of argument, we can write for both types of motors $VI_L \approx E_a I_a = 2\pi nT$, where VI_L is the power supplied at constant voltage by the dc source. When the mechanical load torque increases, the torque developed by the dc motor has to increase also. Since the speed of a shunt motor is virtually constant the delivered mechanical power has to increase. This is reflcted by an increased current demand from the electrical energy supply. Hence fluctuations of the load torque are met by almost equal fluctuations in the current supplied to the shunt motor. For a series motor an increase in developed torque is associated with a decrease in speed. Therefore the power delivered varies far less than the load torque fluctuates. Consequently the load current supplied by the source of electrical energy fluctuates far less when a variable-load torque is driven by a series motor. Accordingly the series motor is far more suited to cope with severe overloads.

A cumulatively-compounded motor is a combination shunt-series machine and has characteristics that are between those of a shunt and a series motor. The degree of similarity with either of these two types of motor depends on how the strength of the series winding mmf compares with the shunt winding mmf. The starting torque of a cumulatively-compounded motor is larger than that of a comparable shunt motor and it is better suited to handle overloads and loads that have a fluctuating torque demand such as shears, presses, and plunger pumps. To reduce the variation of speed of a compound motor driving a pulsating load, use is made of the load equalizing property of flywheels. The inherent instability of differentially-compounded motors makes them dangerous to use. There is no real need for this type of motor, since shunt motors normally meet the requirement of constant speed to a sufficient extent.

5.8. LOSSES AND EFFICIENCY

5.8.1. Losses

Whatever its mode of operation, the output power of a machine is always less than the input because of internal power losses. The internal losses are undesirable, since they reduce the useful output of the machine and contribute to the operating cost. However they are unavoidable. Losses are converted into heat. This has to be dissipated to maintain a maximum safe operating temperature so as not to endanger or shorten the life of the insulating materials. Special measures, taken to dissipate the heat, decrease further the efficiency and increase the cost.

The power losses in a dc machine can be classified into three groups.

1. Rotational losses, sometimes referred to as stray power losses, resulting from the rotation of the armature.
2. Copper losses, or ohmic losses, due to the current flow in the various windings of a machine.
3. Stray load losses, which are of miscellaneous origin.

A. ROTATIONAL LOSSES

When the armature rotates in the stationary field of a dc machine the rotor iron experiences a rotating flux. This is the cause of eddy current and rotational-hysteresis losses. These losses vary with the speed and the flux density. For a shunt machine operating at constant terminal voltage, the iron losses remain virtually constant for different values of the load current, since the speed remains almost constant. However, for a series machine, both the excitation and the speed change when the load varies.

Friction losses are also a part of the rotational losses, and occur in the bearings and at the sliding contact between the brushes and the commutator surface. As a first approximation, these losses can be taken to vary directly with the speed of the armature. *Windage loss* refers to the power required to overcome the air friction experienced by the slotted rotor in the air gap and the coil ends protruding beyond the rotor iron. Fan blades that are mounted on the shaft to force cooling air over the armature winding and through ventilating ducts cause additional power to be consumed. In general, windage losses vary with the third power of the speed of rotation.

B. OHMIC LOSSES

The most important contribution to the ohmic losses is made by the current in the armature winding. These ohmic losses vary with the square of the armature current.

The same can be said for the ohmic losses in the series field winding, the interpole, and the compensating winding.

Shunt field ohmic losses depend on the resistance of the shunt field circuit and the voltage, which is normally constant for shunt motor operation. The terminal voltage of a shunt generator varies with the load current. It is quite satisfactory to assume a constant rated terminal voltage for the normal range of operation when calculating efficiency.

Brush contact losses are assumed to vary directly with the armature current since the brush contact voltage drop is normally assumed to be constant at approximately 2 volts for all load currents. Hence the brush contact loss amounts to $2I_a$ watts.

C. STRAY LOAD LOSSES

Stray load losses are of miscellaneous origin and are normally a minor contribution to the total losses. Magnetic stray flux penetrates solid metal constructional parts of the rotor assembly causing a small amount of iron losses in these parts. Further, the armature mmf causes a distortion of the magnetic flux density distribution under a pole shoe if no compensating winding is employed. The flux density is increased under one-half of a pole shoe and increases the iron losses in the affected parts more than the decrease in iron losses in other parts due to a decrease of the flux density under the other half of a pole shoe. This causes a small increase of the no-load iron losses.

Leakage flux in the rotor slots reverses each 180 electrical degrees of revolution of the armature and induces a small amount of eddy current loss in the armature conductors. The relatively high frequency of armature current reversal itself gives rise to skin effect. Additional losses are also caused by the current in coils that are being commutated.

The stray-load losses are difficult to measure or calculate. Therefore their contribution to the total loss of a machine is accounted for by an extra allowance of 0.5% of the input power for a compensated machine, or 1.0% in the absence of a compensating winding.

5.8.2. Determination of Losses

Of the three categories of losses only the ohmic losses and the rotational losses have to be determined by measurement.

A. OHMIC LOSSES

Since the currents through the various windings of a machine can be calculated, it is only necessary to measure the resistances of these windings to determine the ohmic losses. Since all windings except the shunt field circuit have low resistances the measurements should be made with great care to enhance the accuracy. When measuring the resistance of the armature winding, commutator segments directly below two adjacent *plus* and *minus* brushes should be used as winding terminals to exclude the brush contact resistance, which is relatively high with the rotor at standstill. Since the winding resistances are temperature dependent, the temperature, at which the resistance measurement is made, should be recorded. The measured values can then be corrected to correspond to those at a temperature of 75°C, as required by National Standards for efficiency calculations.

B. ROTATIONAL LOSSES

A simple way to measure the rotational losses is to operate a machine as a shunt or separately-excited motor without a load. With rated terminal voltage and at rated speed the terminal voltage and the no-load armature current I_{aNL} are measured. From the power equation

$$VI_{aNL} = E_aI_{aNL} \pm I_{aNL}^2 R_a' \qquad (5.8.1)$$

and a knowledge of V, I_{aNL}, and R_a', the rotational losses can be determined as being equal to E_aI_{aNL}. The machine is run at rated speed to ensure that the friction and windage losses are those that occur at that speed. At rated terminal voltage, the iron losses are those corresponding to rated flux density and frequency.

5.8.3. Efficiency

The efficiency of a machine is defined as

$$\eta = \frac{\text{useful power output}}{\text{total power input}} = \frac{P_o}{P_i}. \qquad (5.8.2)$$

The efficiency of fractional-kilowatt machines can be determined from the measurement of input and output power. For large machines this direct method becomes expensive and sometimes impracticable. The larger the machine rating the smaller the relative difference between input and output power becomes. Small errors in the measured input or output will then cause a large percentage error in the calculated efficiency. Since the useful power output P_o is equal to the power input P_i minus the internal machine losses P_{loss}, we can rewrite the expression for machine efficiency as

$$\eta = \frac{P_i - P_{loss}}{P_i} = 1 - \frac{P_{loss}}{P_i} \text{ (per unit)}. \qquad (5.8.3)$$

This expresion shows that, if unavoidable errors are made in the determination of the losses, the efficiency can still be determined with a high degree of accuracy, since P_{loss} is such a small percentage of the input power.

The conventional method of determining the efficiency is therefore based on an accurate evaluation of the losses from measurements.

EXAMPLE 5.11

A dc shunt motor has an armature circuit resistance of 0.045 ohm, and the brush voltage drop is 2 V, when appreciable current flows. No-load tests show that the armature current is 30 A, if the supply voltage is 600 V, and is 24 A for a supply voltage of 300 V, while the field-winding current is 5 A. Find the motor efficiency, if the machine is loaded so that the armature current is 400 A at a supply voltage of 600 V for the same value of field-winding current.

Solution

The no-load armature current provides torque to overcome the friction, windage, eddy-current, and hysteresis retarding torques. At a constant value of flux the variation of the no-load torque and therefore the no-load armature current with speed can be approximated by

$$T_{eNL} \propto I_{aNL} = C_1 + k'n.$$

It is sufficiently accurate to assume that, at no load, the supply voltage V equals the back emf E_{aNL}. With the flux constant $E_{aNL} \propto n$, so that we can write

$$I_{aNL} = C + kV.$$

From the no-load tests with constant flux, we have $I_{aNL} = 30$ A when $V = 600$ V, while $I_{aNL} = 24$ A when $V = 300$ V. From this information we find the intercept $C = 18$ A and $k = 0.02$ A/V.

When the machine drives a load and the flux is the same as at no load,

$$E_a = V - I_a R'_a - V_b = 600 - (400 \times 0.045) - 2 = 580 \text{ V}.$$

With 580 V at no load we would have had

$$I_a = 18 + (0.02 \times 580) = 29.6 \text{ A}.$$

Hence, rotational loss $= 580 \times 29.6 = 17.2 \times 10^3$ W.
For the ohmic losses we find field circuit loss $= 600 \times 5 = 3 \times 10^3$ W and armature circuit loss $= 400^2 \times 0.045 = 7.2 \times 10^3$ W.
The brush contact loss $= 2 \times 400 = 800$ W. Thus the total losses are 28.2 kW.
The input power is $VI_L = V(I_a + I_f) = 600 \times 405 = 243 \times 10^3$ W.
We find for the efficiency

$$\eta = \left(1 - \frac{\text{total losses}}{\text{input}}\right) \times 100\% = \left(1 - \frac{28.2}{243}\right) \times 100\% = 88.4\%.$$

5.9. MACHINE RATING AND DIMENSIONS

Designers of electrical machines attempt to minimize the amount of material in a motor or generator for a given power output. A small volume for a machine is simply a matter of economics in such a highly competitive area. Even though there is a continual reduction of size of machines, as better materials are introduced and as more efficient forms of cooling are discovered, there remains a fundamental relation between output and dimensions of electrical machines. This relation can be derived as follows.

The armature force F_e on an armature conductor, that experiences a mean magnetic flux density B is, according to the Lorentz equation,

$$F_e = BL \frac{I_a}{2a} \qquad (5.9.1)$$

where L = the length of the armature and $\dfrac{I_a}{2a}$ = the conductor current for a winding with $2a$ parallel paths.

Thus the total work done per revolution W on all Z armature conductors is given by

$$W = ZF_e \pi D = B(\pi DL) \times \left(\frac{I_a}{2a} Z \right) \qquad (5.9.2)$$

where D = the diameter of the rotor.

The specific magnetic loading is defined as the mean flux density B and this is a measure of the utilization of the magnetic steel of a machine. The specific electric loading q, defined as the equivalent current density per unit armature circumference, is a measure of the utilization of the copper of the armature winding. From these definitions it follows that the total magnetic loading of a machine is given by

$$2p\Phi = B\pi DL \qquad (5.9.3)$$

and the total electric loading is given by

$$\frac{I_a}{2a} Z = q\pi D. \qquad (5.9.4)$$

We can now rewrite the work done per revolution as

$$W = B\pi DL \times q\pi D = \pi D^2 L \times \pi qB \qquad (5.9.5)$$

where $\pi D^2 L/4$ = the volume of the armature. Thus, the work done per revolution is proportional to the volume of the armature times the product of the specific loadings.

The energy converted in the armature per revolution is

$$W = \frac{P}{n} = \frac{E_a I_a}{n} = 2\pi T_e. \qquad (5.9.6)$$

Hence

$$E_a I_a = 2\pi n T_e = \tfrac{1}{4}(\text{armature volume}) \times \pi nqB \qquad (5.9.7)$$

$$\text{or armature volume} = \frac{\pi}{4} \frac{E_a I_a}{nqB} = \frac{8T_e}{qB} \qquad (5.9.8)$$

$$\text{and } T_e = k \times (\text{armature volume}) \times q \times B. \qquad (5.9.9)$$

We can conclude that the electromagnetic torque acting on the armature is proportional to the armature volume and the specific loading factors. For a given volume of the armature and a given speed, the higher the product of the specific loading factors the greater is the output. The practical limits set on both B and q are compromises. As B is increased, saturation of the magnetic circuit is reached and both iron losses and excitation losses become excessive, so that the efficiency falls below acceptable levels. Similarly as q is increased the ohmic losses in the armature increase at a greater rate and the efficiency reduces. In both cases, the heat to be dissipated becomes a problem. Further, commutation difficulties increase with increasing q and so does the armature reaction effect. Also, for a given power output and specific loading factors, a machine's volume decreases as its rated speed increases.

We can visualize a 10-kW machine as being large. Yet, for instance, if its rated speed were 1500 r/s, the size may only be as small as a fist. This raises other serious problems. At 90% efficiency it would be an interesting problem to have to dissipate 1 kW of heat from such a small machine. There is still room for improvements. For example, if superconductors (no resistance) become economic, then B can be increased by a factor of up to 10 and there would be no need for ferromagnetic materials.

5.10. SUMMARY

The dc machine has salient poles on the stator to provide the main magnetic field. The armature winding on the rotor is connected to a commutator to enable a uniform torque to be developed when dc voltages are present at the terminals of the windings.

The excitation winding can be series-, shunt-, or compound-connected with the armature winding. Each provides different characteristics. Excitation can be provided from an independent supply also.

The modern armature winding is the drum type. It is wound in a *lap* or *wave* configuration. A simplex lap winding has $a = p$, where a is the number of pairs of parallel paths in the armature winding and p is the number of pairs of poles on the machine. In general $a = mp$, where m is the multiplicity of the winding (for a simplex winding $m = 1$, for a duplex winding $m = 2$, etc.)

A simplex wave winding has $a = 1$ independent of the number of poles. In general $a = m$. Either modes of operation, motoring or generating, can be accomplished by the dc machine.

The average emf induced in the armature winding is $E_a = \dfrac{p}{a} \Phi n Z$.

The power associated with electromechanical energy conversion is the armature power $P = E_a I_a$. For motoring action the applied voltage V injects current I_a into the armature winding so that $V = E_a + I_a R_a$.

For generating action the induced emf E_a in the armature winding delivers a current I_a so that $V = E_a - I_a R_a$. Although the armature mmf is necessary for interaction with the main field in the energy conversion process, there is a demagnetizing effect. This effect can be reduced by using compensating windings imbedded in the pole shoes.

Commutation incurs the reversal of current in armature coils. The reactance voltage that results from this is balanced by a commutating emf that can be produced by interpoles.

Voltage regulation of a generator is defined as $(V_{NL} - V_{FL})/V_{FL}$ per unit for constant speed. Speed regulation of a motor is defined as $(n_{NL} - n_{FL})/n_{FL}$ per unit for constant terminal voltage. Shunt motors are suitable for adjustable-speed applications with the speed varying little with load. Series motors find application for driving high-inertia loads.

The cumulatively-compounded motor has characteristics somewhere between those of the shunt and series machines and is suitable for applications with fluctuating loads. The starting currents of dc motors are inherently high. These currents are reduced by effectively reducing the terminal voltage at start-up.

Speed control is accomplished by adjusting the voltage at the armature and field winding terminals. Reducing the armature winding voltage causes a speed decrease and reducing the field winding voltage tends to increase the motor speed. The versatility and wide speed range make the dc machine an important motor for adjustable-speed applications.

The torque equation is

$$T_e = \frac{E_a I_a}{2\pi n} = \frac{p}{2\pi a} \Phi Z I_a = K_T \Phi I_a.$$

The speed equation is

$$n = \frac{a E_a}{p \Phi Z} = \frac{V - I_a R_a}{K_e \Phi}.$$

For a given power rating the designed machine volume is proportional to the full-load torque. In other words, the higher is the rated speed, the lower is the machine volume.

5.11. PROBLEMS

Section 5.2

5.1. A simplex lap winding is to be installed in a 4-pole dc machine that has a rotor with 40 slots. With two conductors per coil and two coil sides per slot determine (a) the total number of conductors, (b) the number of commutator segments required, (c) the coil pitch, (d) the commutator pitch, and (e) the number of pairs of parallel paths in the winding. See Appendix G.
(*Answer:* (a) $Z = 160$, (b) $S = 40$, (c) $y_s = 10$, (d) $y_c = 1$, (e) $a = 2$.)

Section 5.3

5.2. A lap-wound armature, which has 100 slots, with four conductors per slot, rotates at 600 r/min. If the flux per pole is 60 mWb, calculate the induced emf that could be measured at the brushes.
(*Answer:* 240 V.)

5.3. A 6-pole, 500-r/min, dc generator has 100 armature slots. When the flux per pole is 60 mWb, the open-circuit voltage at the terminals is 240 V. How many conductors per slot are there?
(*Answer:* 1.6.)

5.4. A 6-pole, 500-r/min, dc generator has 100 armature slots. There are four conductors per slot, and the main excitation gives 60 mWb per pole. Compare the terminal voltages, currents, and power rating for **(a)** a lap winding and **(b)** a wave winding, if each conductor in the armature carries 100 A. It is assumed that the voltage drop due to the armature-winding resistance is 5%.
(*Answer:* **(a)** 190 V, 600 A, 114 kW, **(b)** 570 V, 200 A, 114 kW.)

5.5. A lap-wound armature with 500 conductors has an induced emf E_a of 250 V at 600 r/min. The area of the pole face is 0.05 m^2 and the air gap is 6.284 mm. What is the total number of ampere turns per pole to produce the main flux. Assume that the air gap requires 90% of the total excitation.
(*Answer:* 5555 AT.)

5.6. A 2-pole, 570-V, dc generator has 375 armature conductors and is driven at 1200 r/min. What is the useful flux per pole?
(*Answer:* 0.1 Wb.)

5.7. The number of turns on each pole of a dc machine is 1000. When the machine is excited the useful flux pole is 0.1 weber. Calculate the average value of the induced emf across each coil, if the flux decays to zero in 0.2 second when the field-winding circuit breaker is opened.
(*Answer:* 500 V.)

Section 5.5

5.8. A 4-pole, dc generator has a full-pitch wave winding on the armature, whose diameter is 0.3412 m and whose effective length is 0.2 m. The armature winding has 1000 conductors and is connected to a commutator consisting of 200 segments. The rated speed is 10 r/s. If the field winding is excited to give a maximum and average air-gap flux density of 0.75 tesla and 0.5 tesla, respectively, what is the open-circuit terminal voltage and the maximum voltage between adjacent commutator segments?
(*Answer:* 500 V, 15 V.)

5.9. The brushes of a 4-pole, 34.5-kW, 230-V series dc generator are advanced 6 mechanical degrees to improve commutation. The 240 conductors of the armature winding form a simplex lap winding. Calculate the demagnetizing and cross-magnetizing ampere turns per pole at full load.
(*Answer:* $AT_d = 75$, $AT_c = 1050$.)

5.10. A 240 r/min, dc machine has a commutator whose circumference is 7.5 m. If the brush width is 15 mm calculate the time of commutation.
(*Answer:* 0.5 ms.)

5.11. In the commutating zone a 10-turn armature coil has an inductance of 0.2 mH. What is the value of the commutation field required for linear commutation, if the armature winding current is 50 A.
(*Answer:* 1 mWb.)

5.12. A 500-kW, 500-V, 8-pole, dc generator has a lap-wound armature with 512 turns. The ratio of interpole to armature ampere turns is 1.5 in order that the interpoles provide a net flux density of 0.2 tesla at full load. Estimate the air-gap length under the interpoles.
(*Answer:* 12.6 mm.)

5.13. A 6-pole dc machine has 72 slots in the armature and each slot contains 1000 ampere conductors at full load. Calculate the necessary ampere turns per interpole, so that the net interpolar flux density is 0.3142 tesla, if the gap under the interpole is 12 mm.
(*Answer:* 9000 AT.)

5.14. A 240-kW, 600-V, 12-pole, shunt generator has a simplex lap winding with a total of 1440 conductors. A pole face covers 70% of a pole pitch. At full load the shunt field current is 5% of the armature current. Calculate the number of compensating conductors per pole and the current through them at full load.
(*Answer:* $Z_c = 7421$ A.)

Section 5.6

5.15. The magnetization characteristic of a separately-excited dc generator driven at rated speed is given by

E_a (volts)	20	50	100	300	350	400	450	500
I_f (amperes)	0	0.7	1.25	3.7	4.3	5.1	6.5	10.

(a) If the field winding is connected directly across the armature terminals, what is the armature induced emf E_a, if the field winding resistance is 50 ohms? **(b)** What additional resistance must be added in series with the field winding to reduce E_a to 450 V? **(c)** What additional resistance must be added in series to make the total field-winding resistance equal to the critical value? Neglect the effect of the armature winding resistance.
(*Answer:* (a) 500 V, (b) 19.8 ohms, (c) 30 ohms.)

5.16. A separately-excited dc generator, driven at 800 r/min, has the following open-circuit characteristic.

Armature emf	E_a (volts)	40	80	120	160	200	220	240
Field current	I_f (amperes)	0.14	0.31	0.5	0.72	1.01	1.24	1.45.

If the machine is driven at 800 r/min as a shunt generator, what is the necessary field winding resistance for the open-circuit induced emf to be 230 V? If the generator speed is reduced to 600 r/min and the shunt field resistance remains unaltered, find the value of the open-circuit induced emf E_a. Neglect the effects of the armature winding resistance.
(*Answer:* 168 ohms, 128 V.)

5.17. A dc shunt generator has an armature winding resistance of 0.025 ohm and a field winding resistance of 60 ohms. If separately-excited, the generator open-circuit characteristic is as follows.

Armature emf E_a (volts)	45	90	130	162	192	218	264	306
Field current I_f (amperes)	0.5	1.0	1.5	2.0	2.5	3.0	4.0	5.0.

Find the armature-induced emfs on full load and no load, if the full-load armature current is 400 A and if the speed is maintained constant. Ignore armature reaction. Calculate the voltage regulation at full load.
(*Answer:* 300 V, 318 V, 9.7%.)

5.18. If driven at 1200 r/min, a separately-excited, dc generator has the following open-circuit magnetization characteristic.

Armature emf E_a (volts)	100	200	300	400	500	600	700
Field current I_f (amperes)	0.6	1.2	2.0	3.2	4.6	6.5	9.3.

The machine is to be driven at 1000 r/min as a shunt-excited generator, whose field-winding resistance is 60 ohms, armature winding resistance is 0.12 ohms, and whose brush drop is 2 V on load. Estimate the terminal voltage on no load and at full load, when the armature winding current is 400 A. Armature reaction effects may be neglected.
(*Answer:* 600 V, 515 V.)

5.19. A belt-driven dc shunt generator, driven at 1000 r/min, delivers 52.5 kW to the 500-V bus bars. The machine continues to run when the belt breaks, drawing 7.5 kW from the bus bars. What is the new speed? The armature winding resistance is 0.18 ohm, the field winding resistance is 100 ohms, and the brush voltage drop is 2 V. Ignore armature reaction.
(*Answer:* 951 r/min.)

Section 5.7

5.20. A dc shunt machine is connected across a 500-V supply. The total armature winding resistance is 0.05 ohm and the field-circuit resistance is 100 ohms. Calculate the ratio of speed as a generator to speed as a motor for the line current to be 200 A in each case.
(*Answer:* 1.04.)

5.21. A dc motor develops a torque of 50 N·m. Determine the electromagnetic torque when the armature winding current is increased by 50% and the flux is reduced by 10%.
(*Answer:* 67.5 N·m.)

5.22. A shunt motor draws 26.5 A from a 120-V supply when it drives a load at 25 r/s. The resistance of the field circuit is 120 ohms and the armature winding resistance is 0.2 ohm. What is the developed electromagnetic torque?
(*Answer:* 18.3 N·m.)

5.23. A 500-V, dc shunt motor has an armature-winding resistance of 0.1 ohm. The full-load armature current is 100 A. Find the value of the resistance to be

connected in series with the armature circuit in order that the speed is 25% of the rated value if full-load current flows.
(*Answer:* 3.68 ohms.)

5.24. A 1000-V, dc shunt motor on no load runs at 600 r/min and takes 60 A. The armature circuit resistance is 0.006 ohms and the field winding resistance is 100 ohms. What is the motor speed when loaded and taking a current of 6010 A, if the armature-reaction effect weakens the field by 2%? What is the percentage speed regulation?
(*Answer:* 590 r/min, 1.7%.)

5.25. A 600-V, dc series motor has a total circuit resistance of 0.1 ohm and runs at 1120 r/min with a current of 400 A. Determine the motor speed, if 0.9 ohm is added in series with the armature winding, and the current is again 400 A.
(*Answer:* 400 r/min.)

5.26. A dc shunt motor draws a current I at a speed n. Find the current at a motor speed $3n$ for the cases **(a)** field control and **(b)** armature voltage control, if (i) the output power is constant and (ii) the torque is constant. Assume the losses are negligible and that the field current is very small compared with the armature current.
(*Answer:* (a)(i) I, (ii) $3I$, (b)(i) $I/3$, (ii) I.)

5.27. The full-load current of a 4-pole, dc shunt motor is 100 A. The lap-wound armature has 1000 conductors to give a total resistance of 0.58 ohm and the total brush drop is 2 V when armature current flows. If the motor is connected to a supply of rated voltage 660 V, the field ampere turns per pole are 5000. Find the no-load and full-load motor speed accounting for armature reaction and assuming that the pole arc is 0.7 of the pole pitch. The magnetization characteristic is given by

Field AT/pole	1000	2000	3000	4000	5000	6000	7000	8000
Air-gap flux/pole (mWb)	5.4	12.8	17.3	20.0	22.0	23.6	24.6	25.5.

(*Answer:* 1800 r/min, 1731 r/min.)

5.28. A dc shunt motor has an armature winding resistance of 0.2 ohm and drives a 9-kW load at 1200 r/min with a full-load armature current of 50 A, if the machine is connected across a 210-V supply. The loss torque due to windage and friction can be considered constant and the load torque is proportional to the square of the speed. Calculate the value of the resistance connected in series with the armature winding to reduce the motor speed to 600 r/min. Neglect the effect of armature reaction and the brush voltage drop.
(*Answer:* 6.5 ohms.)

5.29. A 600-V, fully-compensated, dc shunt motor has a no-load speed of 1200 r/min. At the full-load armature current of 400 A the speed is 1100 r/min. The load is to rotate at 1000 r/min at full-load torque. Estimate the value of the resistance added to the armature circuit in order to satisfy this condition. Assume that the load torque is independent of speed.
(*Answer:* 0.125 ohms.)

5.30. A 500-V dc series motor has a total resistance of 0.5 ohm including the field, armature, and brushes. At full load the current is 100 A at a speed of 6 r/s. At a reduced load, the current is 50 A at a speed of 10 r/s. The torque was found to be proportional to current over this range. What is the motor current when the gross mechanical torque is 786 N·m, and what value of resistance must be added for the speed to be 6 r/s at this torque?
(*Answer:* 85 A, 0.9 ohm.)

5.31. Two identical dc series motors are used to drive a locomotive. The two machines are on the same shaft, are supplied by a 1000-V supply, have a full-load current of 1000 A, and each has a resistance (including armature circuit, interpoles, and field winding) of 0.12 ohm. The locomotive travels at 50 km/h with a tractive force of 65.0 kN, if the two machines are connected electrically in series, drawing full-load current with an added resistance of 0.02 ohm in series with the armatures for speed control. Calculate the locomotive tractive power and speed, if the two machines are connected in parallel, each drawing full-load current without any added resistance. Neglect brush voltage drops and assume the windage and friction retarding force is a constant.
(*Answer:* 1882 kW, 104 km/h.)

5.32. A dc compound motor has the following magnetization characteristic if driven at 1200 r/min and the shunt winding is separately excited.

Armature emf E_a (volts)	100	200	300	400	500	600	700
Shunt field current I_f (amperes)	0.6	1.2	2.0	3.2	4.6	6.5	9.3.

The armature-circuit resistance is 0.1 ohm and includes the series field, interpole and armature windings plus the brush resistance. The long-compound connected shunt field winding resistance is 400 ohms. In order that the speed falls as the load increases the series field winding has 10 turns per pole and the shunt field winding has 1000 turns per pole. For a supply voltage of 500 V, plot the gross torque versus speed. Ignore armature reaction.
(*Answer:* Example, 449 N·m at 718 r/min at an effective excitation of 9300 AT.)

5.33. A dc compound motor has the following magnetization characteristic when driven at 1200 r/min and the shunt winding is separately excited.

Armature emf E_a (volts)	100	200	300	400	500	600	700
Shunt field current I_f (amperes)	0.6	1.2	2.0	3.2	4.6	6.5	9.3.

The interpole and armature-circuit resistance is 0.095 ohm, the series winding resistance is 0.005 ohm, and the brush voltage drop is 3 V when appreciable current flows. The series field winding has 10 turns per pole and the long-shunt field winding has 1000 turns per pole. Calculate the motor speed for a supply voltage of 500 V and with a shunt field current of 3.8 A, **(a)** when unloaded, **(b)** when the armature winding current is 400 A, and **(c)** when the armature winding current is 400 A and the series field winding is short-circuited. What is the ratio of the gross torques for cases (b) and (c)? Ignore armature reaction.
(*Answer:* (a) 1333 r/min, (b) 852 r/min, (c) 1227 r/min, 1.43.)

5.34. A 450-V dc shunt motor has an armature winding resistance of 0.1 ohm. The full-load armature current is 100 A. **(a)** What is the starting current as a percentage of the full-load value, if the motor is connected directly across the supply? **(b)** What is the value of the resistance in series with the armature circuit to limit the starting current to 150% full-load value? **(c)** What is the starting voltage across the armature circuit? **(d)** What is the starting torque as a percentage of full-load torque if the flux is reduced by 3% due to the starting current?
(*Answer:* (a) 450%, (b) 2.9 ohms, (c) 15 V, (d) 1.46.)

5.35. A 600-V dc series motor draws 130 A from the supply for half full-load torque. A locked armature test requires 12 V at the terminals for the full-load current of 200 A to flow. Calculate the values of the first two sections of the starting resistance, if, during starting, the motor operates between full-load and 1.5 full-load torque. Find the motor speeds, as a fraction of the rated speed, at which the first two resistance sections are short-circuited. Assume that the torque-current characteristic is linear between 130 A and 200 A and that the circuit inductance is negligible.
(*Answer:* 0.64 ohms, 0.53 ohm, 0.265 per unit, and 0.483 per unit.)

5.36. A 20-kW, 250-V, 30-r/s shunt motor has a total armature circuit resistance of 0.5 ohm. The resistance of the shunt-field circuit is 250 ohms. At starting, the maximum permitted armature current is twice the rated value. Determine the steps of the starting resistor and the speed attained by the armature at each step, if the effect of armature reaction can be neglected.
(*Answer:* 1.6 ohms, 0.8 ohm, 0.3 ohm; 975 r/min, 1464 r/min, 1709 r/min.)

5.37. A 45-kW output, 220-V shunt motor has an efficiency of 90%. The armature circuit resistance is 0.09 ohm and the shunt field circuit resistance is 30 ohms. What is the braking torque at the moment of plugging, in terms of the rated torque, if a plugging resistor of 1 ohm is used? Neglect armature reaction.
(*Answer:* twice full-load torque.)

5.38. Two identical dc series motors are used to drive a 6000-kg electric vehicle. The two machines, which are connected in parallel for dynamic braking, each have a combined armature, interpole, and field-winding resistance of 0.25 ohm. Tested from a supply voltage of 220 V, the following characteristics emerged.

Current per motor (amperes)	40	60	80
Vehicle speed (km/h)	32	22	18
Force per motor at wheel rim (N)	650	1650	2850.

The road resistance is approximately constant at 900 newtons and the windage and friction torque is independent of the speed over the range of interest. Estimate the steady speed of the vehicle down a 1 in 10 incline with dynamic braking at constant flux Φ and a load resistance of 1.375 ohms.
(*Answer:* 20 km/h.)

5.39. A dc hoist motor operates at a constant field excitation and the hoisting speed is adjusted by armature voltage control. When the loads are lowered, regenera-

tive braking is employed. In order to hoist a load of 1937 kg at 10 m/s, the armature voltage is adjusted to 420 V so that the motor takes 500 A. In order to lower a load of 3000 kg at 1 m/s, the armature is allowed to deliver 733 A. The armature voltage is adjusted to 220 V, when a 1500-kg load is to be hoisted. Calculate the acceleration of the load, when it reaches a speed of 5 m/s. The armature resistance is 0.04 ohm and the windage and friction force is directly proportional to the speed only.
(*Answer:* 3.19 ms^{-2}.)

Section 5.8

5.40. A 600-V, totally-compensated, dc shunt motor has a no-load armature current of 18 A and a speed of 1200 r/min. At full load, the armature winding current is 400 A and the speed is 1100 r/min. The field-winding current is 4 A. What is the full-load motor efficiency? With the limitation of the given data, what assumption is needed to arrive at the efficiency?
(*Answer:* 86.3%)

5.41. A 500-V, dc shunt motor has an armature winding resistance of 1.0 ohm and a field-winding resistance of 500 ohms. The no-load and full-load currents drawn from the supply are 2 A and 51 A, respectively. If the stray loss is 1% of the input and the voltage drop across each brush is 1 V, calculate the full-load efficiency of the motor.
(*Answer:* 84.9%)

5.42. A 600-V, dc shunt generator has a rated output of 240 kW at 1000 r/min. The total armature circuit resistance is 0.04 ohms and the field winding resistance is 360 ohms. No-load tests show that the machine requires a driving power of **(a)** 1.2 kW at 1000 r/min, if the field is unexcited, **(b)** 3.6 kW at 1000 r/min, if the field winding is separately excited at rated value, and **(c)** 1.6 kW at 500 r/min, if the field winding is separately excited at rated value. Derive the values of (i) mechanical losses, (ii) hysteresis loss, (iii) eddy-current loss, (iv) total ohmic losses, and (v) generator efficiency, all at rated output and rated speed.
(*Answer:* 1200 W, 1600 W, 800 W, 7450 W, 95.6%.)

5.43. A separately-excited dc motor was tested at no load to give the following results, while the excitation was kept constant. At 600 V the speed was 1200 r/min, the armature winding current was 30 A, the friction loss was 5.4 kW, and the windage loss was 2.2 kW. At 300 V the armature winding current was 24 A. If the excitation is adjusted so that the speed is 1500 r/min on no load at 600 V, estimate the armature current.
(*Answer:* 33.2 A.)

Section 5.9

5.44. For the stability of a dc motor the ratio of field ampere turns to full-load armature ampere turns is specified at 1.05. The 60 r/min motor has been designed with a specific electric loading of 60,000 ampere conductors per meter of armature circumference, which is 3π meters, and a flux density in the

air gap under the pole of 0.65 tesla. Find the minimum air-gap length, if the frequency of flux alternation experienced by the armature winding is not to be greater than 20 Hz for minimum loss requirements.
(*Answer:* 0.014 m.)

5.45. The gross output of a dc motor is to be 500 kW at 60 r/min. A preliminary design study has shown that, for a specific electric loading of 60,000 ampere conductors per meter and an air-gap flux density under the pole of 0.65 tesla, the ratio of pole arc to pole pitch is to be 0.65 and the armature diameter is to be 3 m. Deduce a suitable length for the armature core.
(*Answer:* 2.2 m.)

5.12. BIBLIOGRAPHY

Bose, B. K. *Microcomputer Control of Power Electronics and Drives.* New York: IEEE Press, 1987.

Carr, Laurence. *The Testing of Electrical Machines.* London: MacDonald Co., Ltd., 1960.

Dewan, S. B., G. Slemon, and A. Straughen. *Power Semiconductor Drives.* New York: Wiley-Interscience, 1984.

Fransua, A., and R. Magureanu. *Electrical Machines and Drive Systems.* Oxford: Technical Press, 1984.

Hindmarsh, J. *Electrical Machines and Drives: Worked Examples,* 2nd ed. Oxford: Pergamon Press, 1985.

Ireland, J. R. *Ceramic Permanent-Magnet Motors.* McGraw-Hill Book Co., 1968.

Kenjo, T., and S. Nagamori. *Permanent-Magnet and Brushless DC Motors.* Oxford: Clarendon Press, 1985.

Kusko, A. *Solid-State DC Motor Drives.* Cambridge, Mass.: MIT Press, 1969.

Langsdorf, A. S. *Principles of Direct-Current Machines.* New York: McGraw-Hill Book Co., 1959.

Nasar, S. A. *Electric Machines and Electromechanics.* Schaum's Outline Series in Engineering. New York: McGraw-Hill Book Co., 1981.

Say, M. G., and E. O. Taylor. *Direct Current Machines.* London: Pitman Publishing, 1980.

Sen, P. C. *Thyristor DC Drives.* New York: Wiley-Interscience, 1981.

Siskind, C. S. *Direct-Current Machinery.* New York: McGraw-Hill Book Co., 1952.

Still, Alfred, and Charles S. Siskind. *Elements of Electrical Machine Design.* Tokyo: McGraw-Hill Book Co., 1954.

_____. *Test Code for Direct-Current Machines.* IEEE Standard No. 113. New York, 1962.

CHAPTER SIX

Polyphase Induction Motors

6.1. INTRODUCTION

Of all the different types of electric machines the induction motor is the most common. The reason for this is that its characteristics closely match the requirements of the majority of general industrial drives with respect to cost, robustness, maintenance, and constancy of speed. Accordingly it finds application in spite of its imperfections of relatively low efficiency and low power factor. This asynchronous ac machine has, in general, only one source of supply and this excites the stator winding. The name induction motor arises from the nature of the developed torque. The rotor winding is not connected directly to a source of electric energy. Instead, this winding is excited by induction from the changing magnetic field set up by the stator winding currents. A primitive polyphase induction motor was first described in Chapter 4, subsection B, under section 4.2.1.

The polyphase induction machine[1] is remarkably versatile. It has found application as a generator, a frequency converter, a phase converter, and a transformer or regulator; and, in spite of it being the nearly constant speed and reliable workhorse of industry, there have been interesting developments to create an adjustable-speed motor. Power sizes range from a few kW to beyond 16,000 kW (21,500 hp) and, unlike dc machines, the induction machine can operate from supplies in excess of 10 kV.

---

[1] The word *polyphase* indicates that the machine has a stator winding that is arranged into a number of phases, usually three.

Prediction of steady-state performance is the main aim in this chapter. Motor performance involves machine currents, torque, power, efficiency, speed, and power factor for different load conditions. For this purpose we use a circuit model, which is derived by considering the induction motor to be similar to a transformer, whose secondary windings are free to rotate. The development of an equivalent circuit will follow a description of the action of the motor and a study of emfs in windings that experience traveling flux waves.

6.2. PRINCIPLES OF ACTION

Dominique François Arago communicated the results of his experiments with rotating magnets in 1825. Arago rotated a permanent magnet over a freely suspended copper disc and noticed that the disc tended to follow at a slower speed. Similarly, he noted that if the copper disc were rotated, then the magnet tended to rotate in the same direction. These are good demonstrations of the induction motor action. However, it was left to Michael Faraday to explain that the relative motion of the magnetic field and the disc caused eddy currents that produced a torque.

Galileo Ferrari understood that when two beams of plane-polarized light were phase displaced in time and space from one another, the resultant was a beam of circularly polarized light. In the 1880s, Ferrari applied this principle to magnetic fields. The mechanical motion of the permanent magnets was replaced by stationary electromagnets excited by two-phase currents. The resultant rotating field dragged the copper disc rotor after it. At the same time that Ferrari constructed his induction machine, Nikola Tesla independently built a two-phase motor with a rotating field.

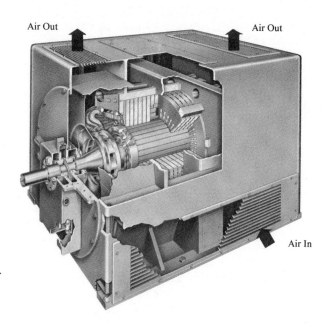

Photo 5 Squirrel cage induction motor. (Courtesy of Electric Machinery — Dresser Rand.)

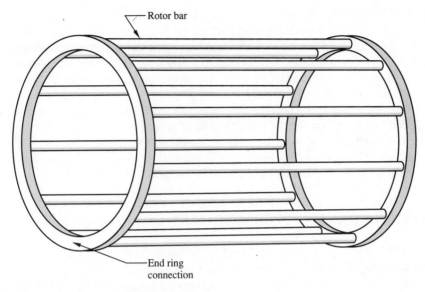

Figure 6.1 Cage winding for the rotor.

Tesla's motors became commercially practicable when he used multiple coils on an iron rotor in place of the copper disc.

Today the induction motor has a polyphase winding set in the slots of the laminated iron core of the stator. Except for special applications, there are no slip rings or commutator connected to the rotor windings. This latter characteristic alone is sufficient to make the motor popular with users, because, without brushes, there are no sliding contacts and there are no arcs and sparks that might cause damage; hence there is negligible maintenance. In most cases aluminum or copper bars embedded in the rotor slots form the rotor windings. End rings close the rotor circuit. This type of winding looks like a squirrel cage, and is given the name *cage* winding. Figure 6.1 shows the cage-like structure of the winding. The simplicity and rugged nature of the cage winding give rise to reliability and ease of manufacture, resulting in a cost advantage over other machines. There are cases where the rotor has a winding similar to the stator winding. The rotor-winding connections are brought out to terminals for control purposes. Both the stator and rotor cores are cylindrical, so there is no saliency, and the air gap between the cores is sensibly uniform, if slot effects can be neglected.

The set of polyphase currents in the stator winding produces a rotating magnetic field.[2] The rotor winding experiences this field and so there are emfs induced in the rotor conductors. Since these conductors, together with the end rings, form closed circuits, the emfs cause currents to flow. In their turn the rotor mmfs produce a

---

[2] See Appendix F for a description of rotating magnetic fields and their production.

magnetic flux pattern, which also rotates in the air gap at the same speed as the stator winding field. There is a torque to tend to align the fields and to make the rotor move in the same direction as the magnetic fields. This torque can start the induction motor from rest. The motor can be accelerated to a speed at which the electromagnetic torque is balanced by the load torque. At this point the speed is steady. Up to what speed can the motor run?

If the rotor could attain the speed of the stator field in space (called the synchronous speed because it is a function of the pairs of poles and supply frequency alone), there can be no torque, because the rotor no longer experiences a changing field, and the rotor conductors will not have current flowing in them. Consequently the motor speed must be at some value below synchronous speed.

6.3. SYNCHRONOUS SPEED AND SLIP

Synchronous speed is defined as the speed of rotation of the field in the air gap, and is related to the frequency f of the supply voltage and the number of pole pairs p created by the windings. For every cycle of supply current, a complete cycle of flux (a north and south pole, or a pole pair) moves past a fixed point in the air gap of the machine. Thus, if the machine is wound for p pole pairs, the synchronous speed of the field in space is

$$n_s = \frac{f}{p} \qquad (6.3.1)$$

Let the speed of the rotor be n revolutions per second or ω_m mechanical rad/s, where $\omega_m = 2\pi n$. The slip speed of an induction motor is defined as the relative speed of the rotor with respect to synchronous speed and tells by how much the rotor slips behind the synchronous speed. Then from this definition, the value of slip speed s is given by

$$s = (n_s - n).$$

As a normalized quantity slip can be defined by

$$s \triangleq \frac{n_s - n}{n_s} \times 100\%$$

or

$$s = \frac{n_s - n}{n_s} = \frac{\omega/p - \omega_m}{\omega/p} \quad \text{per unit} \qquad (6.3.2)$$

where ω is the angular frequency $(2\pi f)$ of the supply voltage. Inspection of eq. (6.3.2) provides the following information.

$s = 1$ signifies $n = 0$, that is, the rotor is at standstill.
$s = 0$ indicates that $n = n_s$, that is, the rotor is revolving at synchronous speed. This can only happen if direct current is injected into the rotor winding, or if the rotor is driven mechanically.
$1 > s > 0$ means that the rotor speed is somewhere between standstill and synchro-

nous speed. The machine is said to be running at asynchronous speed. The slip for a high-efficiency motor running at full load is about 0.04 per unit.

$s < 0$ represents a supersynchronous speed, that is, $n > n_s$. This can occur as a motor, if a particular voltage of the correct frequency can be injected into the rotor-winding circuit and also occurs if the machine is acting as a generator.

$s > 1$ indicates that the rotor speed is opposite to that of the rotating field. This produces a braking torque that tends to bring the machine to rest.

EXAMPLE 6.1

A 3-phase, 12-pole synchronous generator supplies electric power to an induction motor, which has a full-load speed of 1100 r/min. Calculate the induction motor slip at full load and the number of poles of the motor. Assume that the generator is being driven by a diesel engine at a constant speed of 600 r/min.

Solution

A synchronous generator produces alternating current at a frequency that is determined by the speed (called the synchronous speed) at which it is driven. The relation between the frequency and the speed is given by eq. (6.3.1). In this example the generator frequency is

$$f = n_s p = \frac{600}{60} \times 6 = 60 \text{ Hz}.$$

Induction motor synchronous speed is $n_s = f/p$.
For $p = 1$ and $f = 60$ Hz, $n_s = 60$ r/s (3600 r/min).
For $p = 2$ and $f = 60$ Hz, $n_s = 30$ r/s (1800 r/min).
For $p = 3$ and $f = 60$ Hz, $n_s = 20$ r/s (1200 r/min).
For $p = 4$ and $f = 60$ Hz, $n_s = 15$ r/s (900 r/min).
On full load the motor runs just below synchronous speed. In this case $n = 1100$ r/min. The nearest synchronous speed above 1100 r/min is $n_s = 1200$ r/min (for $p = 3$).

Therefore $s = \dfrac{n_s - n}{n_s} = \dfrac{1200 - 1100}{1200} = 0.083$ per unit, and the number of poles = $2p = 2 \times 3 = 6$.

6.4. INDUCED EMF AND THE ROTATING FIELD

Three-phase currents in a balanced 3-phase winding on the stator of an induction motor produce a traveling mmf wave[3] in the air gap. In turn, the mmf produces a magnetic field pattern that rotates. Any conductor in the machine that *sees* this

---

[3] For a mathematical derivation of the mmf wave see Appendix F.

changing field pattern will have an emf induced in it (Faraday's law). If the conductor is part of a closed circuit the induced emf gives rise to a current. This current will tend to modify the magnetic field pattern (armature reaction). In the end there is a balance between currents, the magnetic field and the induced emfs.

We want to know the values of the emfs in the induction motor windings because they have a strong bearing on the electromechanical energy conversion. A knowledge of the emfs helps us determine the motor performance.

If we idealize conditions, the field pattern will be sinusoidal. Under this condition the flux density B in the air gap is described by

$$B = \hat{B}_m \sin(\theta - \omega t) \tag{6.4.1}$$

where θ is the angular position around the air gap, measured from some reference point, and ω is the angular frequency of the supply voltage. The equation represents a traveling wave. If we could be at one point ($\theta = $ constant) in the air gap and could measure magnetic flux density, we would observe that B changed sinusoidally with time. If we could ride on the crest of the wave ($\theta - \omega t = \pi/2 = $ constant), we would be traveling at synchronous speed (position θ changes at the rate $\dot{\theta} = \omega$) around the air gap. So B has a sinusoidal pattern of amplitude $\hat{B}_m$ that travels around the air gap at synchronous speed.

The form of the flux density traveling wave in eq. (6.4.1) is general, in that the field could be produced by a stationary polyphase winding, or a set of field poles revolving at synchronous speed. This flux density form could also be the combined resultant of the flux density produced by the stator and rotor windings of the induction motor.

Consider a sinusoidal flux density traveling wave and its interaction with an arbitrary concentrated coil. Figure 6.2 illustrates this arrangement in developed form, in which the coil has a full pole pitch of π radians and the rotor has a diameter D. We will consider the machine to have p pole pairs and all angular measurement θ will be in electrical radians. Let the coil have a speed ω_m. Particular cases are when

1. $\omega_m = 0$. The coil represents a stator coil, or a rotor coil at standstill.
2. $\omega_m = \omega_m$. The coil represents a rotor coil at some arbitrary speed.

Inspection of Fig. 6.2 indicates that the element of flux $d\phi$ linking the coil is

$$d\phi = B dA = BL \frac{D}{2} d\theta.$$

Therefore, in general, the flux ϕ linked with the coil at an instant of time, at which the coil position is $\theta = \theta_0$, is

$$\phi = \int_{(\theta_0 - \pi/2)}^{(\theta_0 + \pi/2)} \frac{BLD}{2} d\theta = \int_{(\theta_0 - \pi/2)}^{(\theta_0 + \pi/2)} \frac{LD}{2} \hat{B}_m \sin(\theta - \omega t) d\theta$$

or

$$\phi = LD\hat{B}_m \sin(\theta_0 - \omega t). \tag{6.4.2}$$

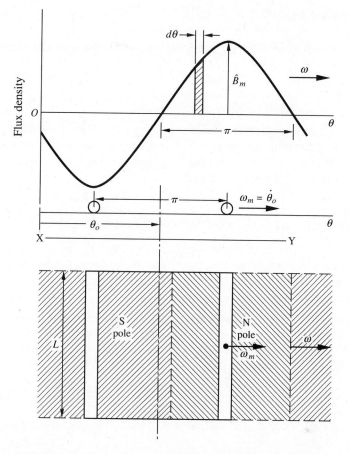

Figure 6.2 Coil flux linkage. (a) Developed diagram, (b) plan view of coil.

For a sine wave the average flux density is $B_{av} = \dfrac{2}{\pi} \hat{B}_m$, so that the total flux per pole Φ is

$$\Phi = B_{av}L\,\frac{D}{2}\,\pi = LD\hat{B}_m. \tag{6.4.3}$$

Therefore the flux linking the coil at any instant is

$$\phi = \Phi \sin(\theta_0 - \omega t). \tag{6.4.4}$$

This has the same form as the mmf, except that Φ is not a maximum value but is the total flux per pole, and both θ_0 and ωt are functions of time.

6.4.1. Induced Emf

From eq. (6.4.4), if a coil of N turns experiences a traveling field, the coil emf is given by

$$e = -N\frac{d\phi}{dt} = -N\Phi(\dot{\theta}_0 - \omega)\cos(\theta_0 - \omega t). \qquad (6.4.5)$$

In this equation two speed terms are identified: ω is the speed of the field in space and $\dot{\theta}_0 = \omega_m$ is the speed of the coil in space. By inspection, this induced emf is 90° out of phase with the flux. Let us consider the two particular cases for the coil. That is, the coil on the stator and the coil on the rotor.

A. STATOR WINDING INDUCED EMF (OPEN-CIRCUIT ROTOR WINDING)

In the general expression of the induced emf, eq. (6.4.5), we must apply the constraints that the stator winding is stationary and the rotor coil has no influence since it is open-circuited. Thus $\dot{\theta}_0 = \omega_m = 0$. Therefore, the emf induced in the stator winding by a rotating field is

$$e = N\Phi\omega \cos(\theta_0 - \omega t). \qquad (6.4.6)$$

B. ROTOR WINDING INDUCED EMF

Let us consider that the flux in the air-gap is produced by stator currents and is described by eq. (6.4.4). This implies that the rotor winding is open-circuited. The rotor winding has a speed ω_m, so that, from eq. (6.4.5), the rotor winding induced emf is given by

$$e = N\Phi(\omega - \omega_m)\cos(\theta_0 - \omega t).$$

Slip $s = (\omega - \omega_m)/\omega$ for a 2-pole machine. On substitution

$$e = N\Phi s\omega \cos(\theta_0 - \omega t). \qquad (6.4.7)$$

For the special condition that the rotor is driven at synchronous speed

$$\omega_m = \omega \quad \text{and} \quad s = 0, \quad \text{so} \quad e = 0.$$

In general, the relative angular displacement between coil and field at any time t is $(\theta_0 - \omega t)$ elec. rad. Therefore the relative angular speed is $(\dot{\theta}_0 - \omega)$ elec. rad/s, or $(\omega_m - \omega)$ elec. rad/s. Hence the frequency of alternation f_2 of the emf in the rotor winding is

$$f_2 = \frac{\omega - \omega_m}{2\pi} = \frac{s\omega}{2\pi} = sf \qquad (6.4.8)$$

where f is the frequency $(\omega/2\pi)$ of the stator winding currents.

EXAMPLE 6.2

A 3-phase, 60-Hz, 6-pole, slip-ring induction motor is used as a frequency converter. That is, the conventional supply is connected to the stator winding and the rotor is driven by external means such as an adjustable speed dc motor. 100 V at 120 Hz is obtained at the slip rings of the rotor winding. Determine the rotor speed to give 180 Hz. What will be the slip-ring voltage at 180 Hz?

Solution

This example is associated with an induction machine whose 3-phase, stator-winding currents set up a rotating field that is described by eq. (6.4.4). The rotor winding is special. It is not a cage winding. It is a winding similar to the stator winding and it is connected to slip rings. The slip rings give us a way to measure the rotor-winding emf, while the winding is open-circuited.

The rotor-winding emf is given by eq. (6.4.7) and is a function of the slip s and hence the rotor speed ω_m. The frequency f_2 of the rotor-winding emf is shown by eq. (6.4.8) to be a function of s and hence the rotor speed ω_m. Consequently, if the rotor of the induction machine is driven by a separate motor, adjustable speed ω_m gives an adjustable frequency (f_2) rms induced emf at the slip rings. In this way the induction machine is used as a frequency converter. The fixed frequency of the stator winding is converted to another frequency at the rotor winding terminals.

The synchronous speed is $n_s = \dfrac{60\,f}{p} = \dfrac{60 \times 60}{3} = 1200$ r/min.

At 180 Hz, slip $s = f_2/f = 180/60 = 3$.

From the definition of slip the rotor speed in revolutions per minute is $n_m = n_s(1 - s)$, but slip s can be positive or negative.

Hence $n_m = 1200(1 \pm 3) = 2400$ r/min. backward or 4800 r/min. forward.

From eq. (6.4.7) the rms voltage at the slip rings is proportional to slip, since all other terms making up the amplitude are constant.

At 120 Hz, slip $s = f_2/f = 120/60 = 2$, and the rms voltage is 100 V.

Therefore at 180 Hz the slip-ring voltage is $(3/2) \times 100 = 150$ V.

We have studied the rotor windings on open circuit. Now let the rotor windings be closed. In Appendix F there is a mathematical description of a rotating field being produced by a set of polyphase currents. In Chapter 4 there is a simple description of the same thing. Both treatments show that polyphase currents of frequency f set up a field whose speed of rotation is f/p revolutions per second (p is the number of pole pairs created by the coil currents) with respect to the current coils. Applying this to the closed rotor winding, the induced emfs in the polyphase rotor windings cause currents of frequency f_2 to flow. The polyphase currents in the rotor set up an mmf wave, whose fundamental component revolves at f_2/p revolutions per second with respect to the rotor winding. The rotor mmf pattern revolves at f_2/p r/s with respect to the rotor coils in the same direction as the stator field. However, the rotor revolves at ω_m

rad/s. Therefore in space the rotor field revolves at

$$\omega_r = \omega_m + \frac{2\pi f_2}{p} = \omega_m + \frac{2\pi s f}{p} = \omega_m + s\frac{\omega}{p} = \omega_m + \left(\frac{\omega}{p} - \omega_m\right) = \frac{\omega}{p}. \quad (6.4.9)$$

That is, the induced rotor field rotates synchronously in the air gap no matter what the speed of the rotor is. So the stator and rotor fields have the same speed.

We come to the conclusion that the resultant effects of the simultaneous presence of stator currents and rotor currents can be expressed in the general form

$$F = \hat{F}_m \sin(\theta - \omega t), \quad B = \hat{B}_m \sin(\theta - \omega t), \quad \text{and} \quad \phi = \Phi \sin(\theta - \omega t) \quad (6.4.10)$$

since two sinusoids of the same frequency, in phase or not, give a resultant, which is a sinusoid of the same frequency. Hence, the induced emf in the closed rotor winding has the same form as in eq. (6.4.7).

C. STEADY-STATE EMFS

Consider one phase of a three-phase stator winding. Let the turns per phase be N_1'. Instead of being a concentrated coil, the winding is considered in the practical case to have one or more configurations[4] involving spread, skewing, pitch, and connection. These special arrangements for the winding are to reduce harmonics so that the field distribution approaches, as closely as possible, a sinusoid. The disadvantage is that the induced emf is less than that in a concentrated coil by a factor called the winding factor $k_{\omega 1}$. If we apply this factor to the emf equation of the concentrated coil, we arrive at the emf equation for the distributed winding.

The stator-induced emf in eq. (6.4.6) becomes

$$e_1 = k_{\omega 1} N_1' \Phi \omega \cos(\theta - \omega t). \quad (6.4.11)$$

The terms $k_{\omega 1}$ and N_1' can be combined to give an effective number of turns per phase, N_1, so that

$$e_1 = \Phi \omega N_1 \cos(\theta - \omega t). \quad (6.4.12)$$

This voltage equation has the form

$$e_1 = \hat{E}_1 \cos(\theta - \omega t) \quad (6.4.13)$$

where $\hat{E}_1$ is the maximum value of a sinusoidally varying emf, given by $\hat{E}_1 = \Phi \omega N_1$. Thus the steady-state rms value E_1 of the stator induced emf per phase is

$$E_1 = \frac{\hat{E}_1}{\sqrt{2}} = \frac{1}{\sqrt{2}} \Phi \omega N_1 = \frac{2\pi}{\sqrt{2}} \Phi f N_1$$

that is

$$E_1 = 4.44 \Phi f N_1 \quad (6.4.14)$$

---

[4] See Appendix G for a description of windings.

where it will be remembered that Φ is the resultant total flux per pole due to all currents, both stator and rotor.

Equation (6.4.7) gives the form of the rotor winding emf per phase. If we develop the argument in the same fashion as for the stator phase emf, we arrive at the conclusion that at any slip s the rms emf E_{2s} per phase of the rotor winding is

$$E_{2s} = 4.44\Phi s f N_2 \triangleq s E_2 \qquad (6.4.15)$$

where E_2 is the standstill induced emf ($s = 1$) and is given by

$$E_2 = 4.44\Phi f N_2. \qquad (6.4.16)$$

N_2 is the effective number of turns accounting for the rotor winding factor.

The rotor and stator emf equations of the induction motor are similar to the primary and secondary emf equations of the transformer, and the expressions for the ratio of the emfs are the same, that is

$$\frac{E_1}{E_2} = \frac{N_1}{N_2} \triangleq a. \qquad (6.4.17)$$

It is with these emf equations that we can begin to develop the equivalent circuit of the induction motor. The equivalent circuit is the model from which the motor performance can be predicted.

D. ARMATURE REACTION

Armature reaction is the effect that the rotor winding currents have on the magnetic flux distribution of the machine. By intuition it would seem that the armature reaction produces a traveling wave of flux in space. Let us prove it.

The stator winding mmf produces a traveling wave pattern of flux which induces emfs in the rotor windings. These emfs produce currents and the resulting mmf in one phase of a rotor winding is given by a Fourier series (eq. (F4.6) in Appendix F). It can be assumed that the winding arrangement has been designed to make the harmonic components of mmf negligibly small. Therefore we need only consider the fundamental component of mmf, whose amplitude we have expressed as

$$F(\theta) = \frac{4}{\pi} \frac{N'i}{2} k_\omega \sin\theta = \frac{4}{\pi} \frac{Ni}{2} \sin\theta \qquad (6.4.18)$$

where $N = k_\omega N' =$ the effective number of turns (as though the phase winding were concentrated with N turns).

Let the rotor winding comprise m balanced phases with angle $2\pi/m$ elec. radians between each successive phase axis. Let the phases be numbered 0, 1, 2, . . . r, . . . $m - 1$. The resultant mmf F_r of phase r has a phase angle $2\pi r/m$ in space with respect to the phase 0, which is taken as reference. This angle is specified by the unit vector $e^{j2\pi r/m}$. Further, the instantaneous current i_r in phase r can be defined as

$$i_r = \hat{I} \cos(\omega t + \theta_t - 2\pi r/m) \qquad (6.4.19)$$

where θ_t is a time phase angle due to winding impedance and $2\pi r/m$ is the phase shift with respect to phase 0. This enables the phase mmf to be written as a function of both space and time

$$F_r(\theta,t) = \frac{4}{\pi}\frac{N\hat{I}}{2}\cos\left(\omega t + \theta_t - \frac{2\pi r}{m}\right)\sin\left(\theta - \frac{2\pi r}{m}\right) \qquad (6.4.20)$$

where θ in eq. (6.4.18) has been replaced by $(\theta - 2\pi r/m)$ to account for the fact that the mmf axis of the rth phase is displaced by $2\pi r/m$. The resultant mmf $F(\theta,t)$ of all the m phases in space and time is given by superposition.

$$F(\theta,t) = F_0 + F_1 + \cdots + F_r + \cdots + F_{m-1}$$

or

$$F(\theta,t) = \frac{4}{\pi}\frac{N\hat{I}}{2}\sum_{r=0}^{m-1}\cos\left(\omega t + \theta_t - \frac{2\pi r}{m}\right)\sin\left(\theta - \frac{2\pi r}{m}\right). \qquad (6.4.21)$$

This summation leads to

$$F(\theta,t) = \frac{4}{\pi}\frac{N\hat{I}}{2}\frac{m}{2}\sin(\theta - \omega t - \theta_t). \qquad (6.4.22)$$

This represents a traveling mmf wave with a sine wave distribution in space, of amplitude $m/2$ times the peak fundamental phase mmf, and speed ω elec. radians per second relative to the windings. The peak value of mmf is

$$\hat{F}(\theta,t) = \frac{4}{\pi}\frac{N\hat{I}}{2}\frac{m}{2} \qquad (6.4.23)$$

and as a space vector

$$\mathbf{F} = \frac{4}{\pi}\frac{N\hat{I}}{2}\frac{m}{2}e^{j(\omega t + \theta_t)} = \frac{4}{\pi}\frac{N\sqrt{2}\mathbf{I}}{2}\frac{m}{2}e^{j\omega t} \qquad (6.4.24)$$

where the rms phasor current is $\mathbf{I} = \dfrac{\hat{I}}{\sqrt{2}}e^{j\theta_t}$.

In order to separate the stator and rotor quantities, subscript 1 is to refer to stator quantities and subscript 2 is to refer to rotor quantities. The stator fundamental mmf of m_1 phase currents, which vary sinusoidally with time, is, from eq. (6.4.24),

$$\mathbf{F}_1 = \frac{N_1}{\pi}\sqrt{2}\mathbf{I}_1 m_1 e^{j\omega t} \qquad (6.4.25)$$

where ω is the angular frequency of the currents. This mmf causes a flux to rotate around the air gap in the same way as revolving field poles.

The rotor mmf wave revolves at synchronous speed with respect to the stator. See eq. (6.4.9). Using eq. (6.4.25) as a reference, the rotor mmf $\mathbf{F}_2$ can be described by the equation

$$\mathbf{F}_2 = \frac{N_2}{\pi}\sqrt{2}\mathbf{I}_2 m_2 e^{j\omega t}. \qquad (6.4.26)$$

This mmf causes a flux to rotate around the air gap in the same way as revolving field poles.

Since the stator and rotor mmfs must have the same number of poles and since they rotate at the same speed, we can find the resultant mmf $\mathbf{F}$.

$$\mathbf{F} = \mathbf{F}_1 + \mathbf{F}_2 = \frac{\sqrt{2}}{\pi} (m_1 N_1 \mathbf{I}_1 + m_2 N_2 \mathbf{I}_2) e^{j\omega t}. \tag{6.4.27}$$

That is

$$\mathbf{F} = \frac{\sqrt{2}}{\pi} m_1 N_1 \left(\mathbf{I}_1 + \frac{m_2 N_2}{m_1 N_1} \mathbf{I}_2 \right) e^{j\omega t}. \tag{6.4.28}$$

Let

$$\mathbf{I}_m = \mathbf{I}_1 + \frac{m_2 N_2}{m_1 N_1} \mathbf{I}_2. \tag{6.4.29}$$

Then

$$\mathbf{F} = \frac{\sqrt{2}}{\pi} m_1 N_1 \mathbf{I}_m e^{j\omega t}. \tag{6.4.30}$$

Thus we can say that there is a stator magnetizing current $\mathbf{I}_m$, which is the resultant of the stator phase current $\mathbf{I}_1$ and the equivalent transformed rotor current $\dfrac{m_2 N_2}{m_1 N_1} \mathbf{I}_2$, which is effectively referred to the stator just as in transformer theory.

The resultant mmf $\mathbf{F}$ gives rise to a synchronously rotating field, whose net flux is

$$\phi = \frac{\mathbf{F}}{\mathcal{R}} = \frac{\sqrt{2}}{\pi \mathcal{R}} m_1 N_1 \mathbf{I}_m e^{j\omega t} = \mathbf{\Phi} e^{j\omega t} \tag{6.4.31}$$

where $\mathcal{R}$ = the air-gap reluctance. This flux rotates at ω elec. rad/s with respect to the stator coils and $s\omega$ with respect to the rotor coils.

In this section we have determined the net flux that rotates due to both stator and rotor winding currents. This net flux gives rise to emfs induced in the stator and rotor winding phases and we have quantified their values. This is a good base on which we can develop the induction motor equivalent circuits for performance analysis.

6.5. INDUCTION MOTOR ANALYSIS

We need to predict the performance of a motor or to determine its operating characteristics without having to do practical tests on the actual machine. This is most readily achieved by modeling the motor as an equivalent electric circuit, so that the equations governing the machine's behavior can be written down by inspection. The most common quantities required to be found are phase currents, torque, speed, power, power factor, losses, and efficiency.

6.5.1. Circuit Model Development

It is assumed that there are symmetrical three-phase windings on both the stator and the rotor, as shown in Fig. 6.3, and, as a special case, all phases are identical. A further assumption is that steady-state conditions prevail, so that the rotor revolves at a constant speed ω_m mech. rad/s.

We start by drawing the two coupled coils, as in Fig. 6.4a, to represent the stator and rotor windings on a per phase basis. These coils are coupled by a flux Φ, which rotates synchronously, and can be taken as a reference direction in the phasor diagrams. Due to this rotating flux, the stator and rotor phases have induced emfs, that are given by

$$\mathbf{E}_1 = -j4.44\Phi f N_1 \quad \text{and} \quad s\mathbf{E}_2 = -j4.44\Phi s f N_2 \qquad (6.5.1)$$

respectively. In this special case of identical phases $N_1 = N_2$, and $E_1 = E_2$. These equations indicate that $\mathbf{E}_1$ and $s\mathbf{E}_2$ are in phase and lagging behind Φ by 90°. These facts are demonstrated in Fig. 6.4b. The stator induced emf E_1 has a frequency f and the rotor-induced emf sE_2 has a frequency f_2, where $f_2 = sf$ and s is the slip defined by eqs. (6.3.2) and (6.4.8).

In the ideal machine the stator winding has neither resistance nor leakage reactance, so that the stator-induced emf is exactly balanced by the applied phase voltage $\mathbf{V}_1$, no matter what currents flow. This has an important implication. Since

$$V_1 = E_1 = 4.44\Phi f N_1 \qquad (6.5.2)$$

then, if the applied voltage $\mathbf{V}_1$ has both a constant magnitude V_1 and a constant frequency f, the flux Φ must be constant and independent of all currents except the magnetizing current I_m, which is defined in eq. (6.4.29).

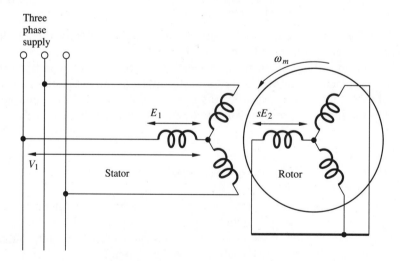

Figure 6.3 Three-phase, wound rotor induction motor.

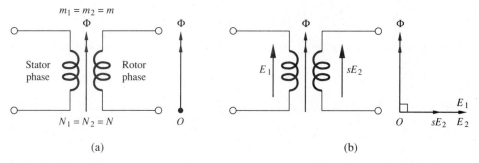

Figure 6.4 Development of simple equivalent circuit. (a) Mutual flux, (b) induced emf and phasor diagram.

Let the rotor windings be closed by short-circuiting the terminals. The induced emf sE_2 causes phase current I_2 to circulate. This current is limited by the coil impedance. Let the rotor winding resistance be R_2 ohms per phase. Let the leakage inductance be L_2 henrys per phase. The inductance L_2 is virtually constant and independent of rotor iron saturation or rotor speed, since it is due mostly to the slot leakage flux. At standstill, the rotor is stationary and the leakage reactance is $X_2 = \omega L_2 =$ constant, so that at a frequency f_2, when the rotor revolves, the leakage reactance is

$$2\pi f_2 L_2 = 2\pi s f L_2 = s\omega L_2 = sX_2. \tag{6.5.3}$$

The *leakage* impedance of the rotor winding becomes

$$\mathbf{Z}_2 = R_2 + jsX_2 \text{ ohms per phase.} \tag{6.5.4}$$

Thus the rotor current per phase is

$$\mathbf{I}_2 = \frac{s\mathbf{E}_2}{\mathbf{Z}_2} = \frac{s\mathbf{E}_2}{R_2 + jsX_2}. \tag{6.5.5}$$

The phase angle θ_2 by which $\mathbf{I}_2$ lags behind $\mathbf{E}_2$ is given by

$$\theta_2 = \tan^{-1} \frac{sX_2}{R_2}. \tag{6.5.6}$$

We can add I_2 and Z_2 to the electric circuit and $\mathbf{I}_2$ at θ_2 to the phasor diagram in Fig. 6.5. Note that in practice it is usual that X_2 is several times greater than R_2 and this results in θ_2 being a relatively large phase angle, especially at large values of slip.

The net air-gap mmf $\mathbf{F}$ to produce the flux Φ is given by

$$\mathbf{F} = \mathbf{F}_1 + \mathbf{F}_2 = \frac{\sqrt{2}}{\pi} m_1 N_1 \left(\mathbf{I}_1 + \frac{m_2 N_2}{m_1 N_1} \mathbf{I}_2 \right) e^{j\omega t}$$

$$= \frac{\sqrt{2}}{\pi} m_1 N_1 \mathbf{I}_m e^{j\omega t} \quad \text{by definition} \tag{6.4.28}$$

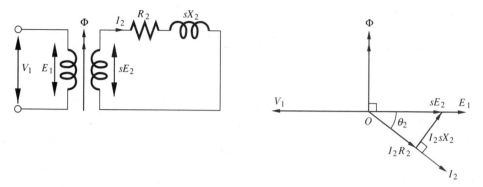

Figure 6.5   Closed rotor circuit and leakage impedance.

where I_m is the magnetizing current. In this special case, where $N_2 = N_1$ and $m_1 = m_2 = 3$, it follows that

$$I_1 = I_m - I_2 \triangleq I_m + I_2' \tag{6.5.7}$$

where $I_2' \triangleq -I_2$. Since I_m must necessarily be in phase with the flux Φ, and since I_2 has already been determined, I_1 is readily expressed on the phasor diagram in Fig. 6.6.

That part of the stator mmf $I_2' N_1$ balances the rotor mmf $I_2 N_2$ as in the transformer. In an equivalent electric circuit the relation $I_1 = I_m + I_2'$ is provided by a node. The input to the node is I_1. One output is I_2', which is associated with the rotor mmf, and hence flows in the ideal coil. The second node output is the magnetizing current I_m, which, in the ideal case, must be associated with a constant magnetizing reactance X_m, across which the supply voltage V_1 is applied. See Fig. 6.6. With a constant voltage V_1 across a constant reactance X_m, I_m, and Φ are constants. Any analysis which stems from this ideal equivalent circuit is said to derive from the *constant-flux theory.*

From Fig. 6.6, the equation for the rotor circuit is

$$sE_2 = I_2(R_2 + jsX_2). \tag{6.5.8}$$

If both sides of the equation are divided by the slip s

$$E_2 = I_2 \left(\frac{R_2}{s} + jX_2 \right). \tag{6.5.9}$$

The current I_2 is the same in both the above equations, so that the rotor mmf is preserved. However, the emf becomes E_2, which has been given by

$$E_2 = 4.44\Phi f N_2 \tag{6.5.10}$$

so that there has been a transformation not only of the rms value, but also of frequency. The emf is now associated with the supply frequency. This frequency transformation is further verified by the new value of the rotor winding leakage reactance, which is X_2, the same as the standstill value at a frequency f. The phase

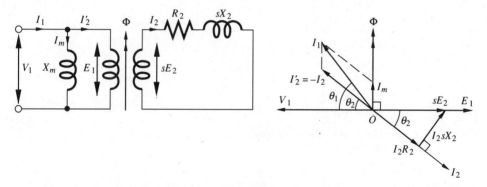

Figure 6.6 Mmf balance and magnetizing current.

angle between current and emf remains the same, that is

$$\theta_2 = \tan^{-1} \frac{sX_2}{R_2}. \tag{6.5.11}$$

Further, the transformation of frequency from f_2 to f is to be interpreted as bringing the equivalent rotor coil to a standstill. Figure 6.7 illustrates the new equivalent circuit that satisfies these relations in eqs. (6.5.9) to (6.5.11).

The power input per phase is

$$P_i = V_1 I_1 \cos\theta_1. \tag{6.5.12}$$

Since there are no stator losses in the ideal machine, this power must be the same as the power associated with the stator coil,

$$P_1 = E_1 I_2' \cos\theta_2. \tag{6.5.13}$$

P_1 is called the air-gap power, or the power that crosses the air gap from stator to rotor. Since $E_1 = E_2$ and $I_2' = I_2$,

$$P_1 = E_1 I_2' \cos\theta_2 = E_2 I_2 \cos\theta_2. \tag{6.5.14}$$

Part of this power P_1 is dissipated as heat in the winding resistance R_2 and the rest

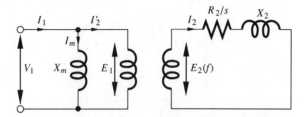

Figure 6.7 Transformation of voltage and impedance.

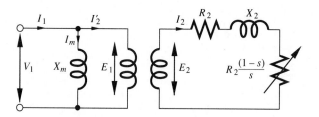

Figure 6.8 Fictitious electrical load representing the mechanical load.

must appear as mechanical power at the shaft. Thus we can write

$$P_1 = P_2 + P_m \tag{6.5.15}$$

where P_2 = the ohmic loss per phase and P_m = the power per phase converted to mechanical power. From the equivalent circuit of Fig. 6.7,

$$P_1 = E_2 I_2 \cos\theta_2 = I_2^2 \frac{R_2}{s} \tag{6.5.16}$$

and since

$$P_2 = I_2^2 R_2$$
$$P_m = P_1 - P_2 = I_2^2 R_2 (1 - s)/s. \tag{6.5.17}$$

Also, since

$$\frac{R_2}{s} = R_2 + R_2(1 - s)/s$$

we can conclude that the resistance $R_2(1 - s)/s$ represents a fictitious electrical load, which enables the mechanical load P_m to be modeled electrically. Thus the equivalent circuit can be depicted as in Fig. 6.8.

For this special case the turns ratio is unity ($N_1 = N_2$), the number of phases is the same on both sides of the air gap, the two emfs E_1 and E_2 are the same and the currents in the coils are the same. This suggests that the ideal coils are superfluous and a new single equivalent circuit can be drawn. All rotor quantities are referred to the stator. This is shown in Fig. 6.9.

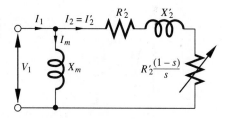

Figure 6.9 Referred values.

A. EXACT EQUIVALENT CIRCUIT

The induction motor model considered in this section is related to a real machine. The effective turns per phase are N_1, N_2 and the number of phases are m_1, m_2 for the stator and rotor circuits, respectively. We want to create an equivalent circuit of the induction motor, taking into account the development of the last section, but changing the special case of equal number of turns and phases into the general case.

Heat loss from the stator winding is accounted for by the lumped parameter R_1 ohms per phase in series with the ideal coil. The power dissipated is $I_1^2 R_1$. All stator flux not associated with mutual coupling with the rotor is associated with a leakage reactance X_1. So the leakage impedance Z_1 is given by

$$Z_1 = R_1 + jX_1 \text{ ohms per phase.} \tag{6.5.18}$$

There is no change in the rotor phase quantities. Figure 6.10 illustrates the model circuit with the leakage impedance of the stator winding. The circuit is based on per phase quantities, but the coupling coils remain because of the different turns of the stator and rotor phases and the different number of phases. The stator circuit accounts for quantities at the supply frequency and the rotor circuit quantities are associated with the frequency f_2 (where $f_2 = sf$). We would like to refer all rotor quantities to the stator in order to simplify the circuit and its analysis.

As in the special case of the last section a first transformation is to divide the rotor winding voltage and impedance by s without altering the value of the current I_2. This gives

$$E_2 = I_2(R_2/s + jX_2) \tag{6.5.19}$$

and this is equivalent to quantities at the supply frequency, so that the rotor circuit has been brought to rest, mathematically if not physically.

A second transformation is to refer the induced emf E_2 of the rotor circuit to the stator circuit. Let the referred value be E_2'. This induced emf must have a value given by

$$E_2' = 4.44\Phi f N_1 \text{ volts per stator phase} \tag{6.5.20}$$

because it is supposed to be induced in N_1 turns by a flux Φ with alternations f hertz.

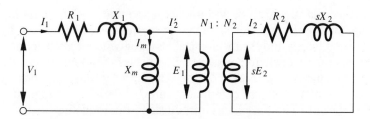

Figure 6.10 Induction motor circuit per phase.

Since the actual standstill induced emf in a rotor phase is

$$E_2 = 4.44\Phi f N_2 \text{ volts per rotor phase} \tag{6.5.21}$$

$$E_2' = E_2 \frac{N_1}{N_2} = aE_2 = E_1. \tag{6.5.22}$$

The referred value E_2' is the same as the actual stator value E_1 and is the actual rotor value E_2 times the turns ratio a.

A third transformation is needed to refer the rotor phase current I_2 to the stator circuit. From eqs. (6.4.28) and (6.4.29) we quote the value of the net air-gap mmf $\mathbf{F}$, which is the sum of the total stator and rotor mmfs of all phases, $\mathbf{F}_1$ and $\mathbf{F}_2$, respectively. That is

$$\mathbf{F} = \mathbf{F}_1 + \mathbf{F}_2 = \frac{\sqrt{2}}{\pi} m_1 N_1 \left(\mathbf{I}_1 + \frac{m_2 N_2}{m_1 N_1} \mathbf{I}_2 \right) e^{j\omega t}. \tag{6.5.23}$$

But from the equivalent circuit in Fig. 6.10 $\mathbf{I}_1 = \mathbf{I}_2' + \mathbf{I}_m$. The value of the rotor phase current referred to the stator is dictated by the need for total mmf balance. Hence

$$I_2' = \frac{m_2 N_2}{m_1 N_1} I_2. \tag{6.5.24}$$

The ohmic transformation of R_2 and X_2, obtained in the form R_2' and X_2' and referred to the stator coil, is calculated from the voltage and current transformations. If we write the frequency transformed rotor equation again

$$\mathbf{E}_2 = \mathbf{I}_2(R_2/s + jX_2) \tag{6.5.19}$$

we can substitute the transformed values of $\mathbf{E}_2$ and $\mathbf{I}_2$ to get

$$\mathbf{E}_2' = \mathbf{I}_2' \frac{m_1}{m_2} a^2 (R_2/s + jX_2). \tag{6.5.25}$$

But the transformed circuit must obey the equation

$$\mathbf{E}_2' = \mathbf{I}_2'(R_2'/s + jX_2'). \tag{6.5.26}$$

Hence a comparison of eqs. (6.5.25) and (6.5.26) provides the relations

$$R_2' = a^2 \frac{m_1}{m_2} R_2 \quad \text{and} \quad X_2' = a^2 \frac{m_1}{m_2} X_2. \tag{6.5.27}$$

One further quantity must be accounted for and that is the total no-load loss of the machine. This loss is associated with the core or iron losses, comprising hysteresis, and eddy current losses, and the mechanical losses of windage and friction. A resistance R_m across the magnetizing reactance X_m satisfies the modeling condition that these losses are roughly independent of load. The branched current is now defined as the no-load current $\mathbf{I}_{NL}$, whose components are $\mathbf{I}_m$, the magnetizing current, and I_{h+e}, the loss current.[5] There are errors introduced by this modeling because the losses

---

[5] This loss current is an effective value to include the mechanical losses.

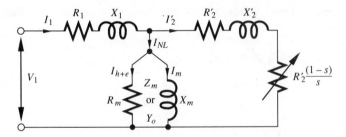

Figure 6.11 Induction motor exact equivalent circuit.

in the rotor core are a function of the rotor current, and there are stray losses such as eddy current losses in the windings themselves to name just two. However, the modeling we have developed is of such proven accuracy and convenience as to be called the exact equivalent circuit.

The results of the transformations are depicted in the exact equivalent circuit illustrated in Fig. 6.11. Except for the slip term in the fictitious electrical load, this exact equivalent circuit is similar to the circuit model of the transformer.

Exact Circuit Analysis

The general purpose of circuit analysis is to solve for the current response, given the circuit parameters and the applied voltage. For the analysis of the induction motor performance the motor speed or slip must also be known. It is further assumed that information concerning supply frequency, phases and turns and the type of connection (star or delta) is given. Unless otherwise stated the rotor-circuit terminals are short-circuited.

The first thing to determine is the input current per phase I_1 from the circuit shown in Fig. 6.11. This is accomplished by combining the parallel arrangement of the magnetizing impedance Z_m and the referred rotor impedance Z_2' with the series leakage impedance Z_1. Thus for $\mathbf{I}_1$

$$\mathbf{V}_1 = \mathbf{Z}\mathbf{I}_1 \qquad (6.5.28)$$

where

$$\mathbf{Z} = \mathbf{Z}_1 + \frac{\mathbf{Z}_m \mathbf{Z}_2'}{\mathbf{Z}_m + \mathbf{Z}_2'}, \quad \mathbf{Z}_1 = R_1 + jX_1, \quad \mathbf{Z}_m = \frac{1}{\mathbf{Y}_0}, \quad \mathbf{Z}_2' = \frac{R_2'}{s} + jX_2',$$

$$R_2' = \frac{m_1}{m_2} a^2 R_2, \quad X_2' = \frac{m_1}{m_2} a^2 X_2, \quad \text{and} \quad a = \frac{N_1}{N_2}.$$

The no-load current $\mathbf{I}_{NL}$ is obtained from

$$\mathbf{V}_1 - \mathbf{I}_1 \mathbf{Z}_1 = \mathbf{I}_{NL} \mathbf{Z}_m. \qquad (6.5.29)$$

But $\mathbf{V}_1 = -\mathbf{E}_1 + \mathbf{I}_1 \mathbf{Z}_1$. The minus sign in front of $\mathbf{E}_1$ arises from the fact that $\mathbf{V}_1$ is the sum of the voltage drops.[6] Therefore, $\mathbf{E}_1$ can be determined and $\mathbf{I}_{NL}$ follows from

$$\mathbf{I}_{NL} = -\mathbf{E}_1/\mathbf{Z}_m. \tag{6.5.30}$$

Finally the equivalent rotor circuit current $\mathbf{I}_2$ is given by

$$\mathbf{I}_2 = \frac{s\mathbf{E}_2}{R_2 + jsX_2} \tag{6.5.31}$$

where $\mathbf{E}_2 = \dfrac{1}{a}\,\mathbf{E}_1 = \dfrac{1}{a}\,(\mathbf{I}_1\mathbf{Z}_1 - \mathbf{V}_1)$. The current $\mathbf{I}_2'$ is found from

$$\mathbf{I}_2' = \mathbf{I}_1 - \mathbf{I}_{NL} \quad \text{and} \quad \mathbf{I}_2' = -\frac{1}{a}\frac{m_2}{m_1}\mathbf{I}_2. \tag{6.5.32}$$

The solution of the currents allows the component powers, the torque and the efficiency of the motor to be calculated. This will be discussed in section 6.6.

B. APPROXIMATE EQUIVALENT CIRCUIT

The magnetizing impedance Z_m of the induction motor equivalent circuit, shown in Fig. 6.11, divides the exact equivalent circuit into two loops. This complication can be eliminated, if the approximation can be made that this magnetizing impedance is placed across the supply terminals. We did this to the transformer equivalent circuit. The error will be greater than for the transformer, because the induction motor has an air gap that makes the no-load current substantially larger. Figure 6.12 depicts the approximate equivalent circuit.

If the error cannot be tolerated, a voltage correction can be made without taking away the simplicity of the approximate equivalent circuit. Compensation can be made for the neglect of the no-load leakage impedance drop $\mathbf{I}_{NL}\mathbf{Z}_1$ and the higher value of flux that results from placing the magnetizing admittance across the terminals. The applied voltage can be reduced to a value equal to

$$\mathbf{V}_1' = \mathbf{V}_1 - \mathbf{I}_{NL}\mathbf{Z}_1 = \mathbf{I}_2'(\mathbf{Z}_1 + \mathbf{Z}_2'). \tag{6.5.33}$$

From the exact equivalent circuit

$$\mathbf{I}_{NL}\mathbf{Z}_1 = -\frac{\mathbf{E}_1}{\mathbf{Z}_m}\,\mathbf{Z}_1. \tag{6.5.34}$$

We can use the approximations

$$\mathbf{E}_1 \approx -\mathbf{V}_1 \quad \text{and} \quad \frac{\mathbf{Z}_1}{\mathbf{Z}_m} \approx \frac{X_1}{X_m} \tag{6.5.35}$$

---

[6] Faraday's law is $e_1 = -N_1\dfrac{d\phi}{dt} = -\dfrac{d(Li)}{dt}$. In the steady state this becomes $\mathbf{E}_1 = -jX_m\mathbf{I}_m$, where $jX_m\mathbf{I}_m$ is the equivalent voltage drop which equals $-\mathbf{E}_1$.

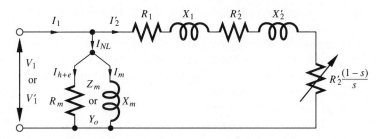

Figure 6.12 Approximate equivalent circuit.

which are acceptable approximations in this calculation. On substitution, the reduced supply voltage is

$$\mathbf{V}_1' = \mathbf{V}_1(1 - X_1/X_m) = \mathbf{I}_2'(\mathbf{Z}_1 + \mathbf{Z}_2'). \tag{6.5.36}$$

In eq. (6.5.33), $\mathbf{V}_1$ and $\mathbf{I}_{NL}\mathbf{Z}_1$ are virtually in phase, so we could use the scalar equation $V_1' \approx V_1 - I_{NL}Z_1$. This could be considered a more practical correction.

The approximate equivalent circuit, shown in Fig. 6.12, is much easier to analyze than the exact circuit. The no-load current $\mathbf{I}_{NL}$ may be obtained directly from

$$\mathbf{I}_{NL} = \mathbf{V}_1\mathbf{Y}_0 \tag{6.5.37}$$

and the referred rotor current $\mathbf{I}_2'$ is given by

$$\mathbf{I}_2' = \frac{\mathbf{V}_1}{(R_1 + R_2'/s) + \mathrm{j}(X_1 + X_2')} \tag{6.5.38}$$

so that the input current $\mathbf{I}_1$ can be calculated from

$$\mathbf{I}_1 = \mathbf{I}_{NL} + \mathbf{I}_2'. \tag{6.5.39}$$

From the computation of winding currents the motor performance in terms of power and torque is readily determined. This is discussed in section 6.6.

6.5.2. Equivalent Circuit Parameters

Similarities between the equivalent circuits of the transformer and the induction motor lead to the suggestion that an open-circuit test and a short-circuit test, adapted to the rotating machine, would enable the circuit parameters to be established. By inspection of any of the equivalent circuits, an open-circuit test on an induction motor can be simulated by having the condition of largest rotor *resistance R_2/s*. This occurs at normal motoring operation if the slip is a minimum. The nearest to zero slip occurs when there is no mechanical load and the machine is said to be *running light*.

A short-circuit test is simulated by locking the rotor in position and running a test at standstill. For this condition the slip is unity and is a maximum value for motoring action, so that the *resistance R_2/s* is a minimum. This is as close to a short circuit of the rotor windings as can be arranged, and the machine actually represents a short-circuited static transformer.

In order to obtain any of the equivalent circuits, one of the main assumptions has been that the flux pattern is sinusoidal, so that losses are taken to be proportional to the main flux, and saturation has been ignored. Analysis becomes complicated, if the variation of the effective rotor resistance R_2 is taken into account. The frequency of the rotor currents changes from the supply frequency to nearly zero, so that the skin effect can alter the value of the resistance by as much as 50% over the full range of speed. Consequently, if the values of $R_1 X_1 R_m, X_m, R_2$, and X_2 are assumed constant, even the exact equivalent circuit is an approximation. This reasoning is sufficient for choosing the approximate equivalent circuit to model the machine.

A. LIGHT LOAD TEST

When running light (no load), the induction motor speed is close to synchronism. As $s \rightarrow 0$, so $R'_2/s \rightarrow \infty$. A negligibly small current flows in the rotor circuit, and nearly all the input power to the motor is consumed as core loss and stator winding ohmic loss.

With no mechanical load connected to the rotor and normal voltage applied to the terminals, the phase value of voltage V_1, input current I_1 and input power P_i are recorded from the test. From an inspection of Fig. 6.13,

$$Y_0 = \frac{I_1}{V_1}, \quad R_m = \frac{V_1^2}{P_i}, \quad \text{and} \quad X_m = (Y_0^2 - 1/R_m^2)^{-1/2}. \quad (6.5.40)$$

In the above calculations R_m accounts for the windage and friction losses as well as the core losses.

Separation of the windage and friction losses from the core loss is accomplished by doing a simple test. By maintaining no load, the applied voltage to the motor is reduced in steps. As long as the speed remains sensibly constant the rotational losses will be constant, but the core loss diminishes, because the flux density reduces in proportion to the voltage. At zero voltage the core loss would be zero, so the only losses would be those of windage and friction. If the results of no-load power loss against applied voltage are plotted, extrapolation to the zero voltage coordinate

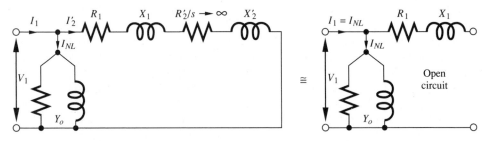

Figure 6.13 Approximate equivalent circuit for $s \approx 0$.

provides the required value of the mechanical loss P_{WF}. The resistance R_m is reduced to

$$R_m = \frac{V_1^2}{P_i - P_{WF}} = \frac{V_1^2}{P_0} \tag{6.5.41}$$

and now R_m accounts only for the hysteresis and eddy-current losses in the core. Winding resistance losses are ignored in the same way as in the no-load test of the transformer.

The magnetizing current, given by $I_1 \sin\theta_0$, is much higher than in a transformer of the same rating because of the necessary existence of an air gap. It is not unreasonable to find the no-load current to be as much as 40% of the full-load input current. This means that the power factor of an induction motor is inherently low, and can have a range from 0.2 at no load to 0.9 at full load.

B. LOCKED ROTOR TEST

If the rotor is locked

$$s = 1 \quad \text{and} \quad R_2'/s = \text{minimum value.} \tag{6.5.42}$$

Therefore the applied voltage, that is needed to allow full-load current to flow, is small (see eq. (6.5.38)). For this condition the no-load current I_{NL} is small, the flux density is low, and the core loss is small. The input current is nearly all reflected in the rotor circuit, so that the magnetizing admittance Y_0 can be neglected. Figure 6.14 illustrates this approximation.

In a test the low voltage V_1, the full-load current I_1, and the input power P_i are recorded on a per-phase basis to allow the following calculations for parameter determination.

$$R_1 + R_2' = P_i/I_1^2, \quad Z_{12} = |Z_1 + Z_2'| = V_1/I_1 \tag{6.5.43}$$

so

$$X_1 + X_2' = (Z_{12}^2 - (R_1 + R_2')^2)^{1/2}. \tag{6.5.44}$$

A separate test can be performed to measure the dc resistance of a stator phase winding. The ac resistance can be anywhere from 20% to 60% higher due to skin

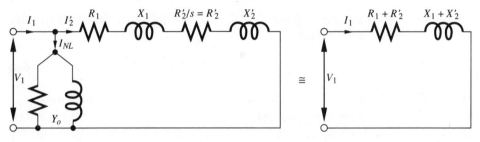

Figure 6.14 Approximate equivalent circuit for locked rotor ($s = 1$).

effect. In general, $R_1 = kR_{1DC}$. Thus

$$R_2' = (R_1 + R_2') - kR_{1DC}. \qquad (6.5.45)$$

This value of R_2' is approximate and is at its highest value, because the rotor currents are at the supply frequency while the machine is at standstill.

There are no simple means to separate the leakage reactances of the stator and rotor. When it is necessary to perform a separation for some analysis, all one can do is to assume that $X_1 = X_2'$.

6.6. PERFORMANCE OF INDUCTION MOTORS

Now that we know how to find the parameter values of the inductor motor equivalent circuit, we can use the equivalent circuit to determine motor performance in terms of power, torque, and efficiency.

6.6.1. Rotor Power

The equivalent rotor circuit, illustrated in Fig. 6.8, can be employed to obtain useful relationships, that are independent of the approximations applied to the stator circuit. The total input power P_1 to the rotor winding is

$$P_1 = m_1 E_1 I_2' \cos\theta_2 = m_2 E_2 I_2 \cos\theta_2$$

$$= m_2 I_2 ((R_2/s)^2 + X_2^2)^{1/2} \times I_2 \frac{R_2/s}{((R_2/s)^2 + X_2^2)^{1/2}}. \qquad (6.6.1)$$

That is

$$P_1 = m_2 I_2^2 \frac{R_2}{s}. \qquad (6.6.2)$$

This is the power transferred from the stator to the rotor and is called the *gap power*. The only loss accounted for by the rotor equivalent circuit is the total ohmic loss P_2, which is the loss due to the actual winding resistance, and is

$$P_2 = m_2 I_2^2 R_2. \qquad (6.6.3)$$

Thus the mechanical output power P_m, which is the difference between the input power to the rotor and the rotor-circuit losses, is

$$P_m = P_1 - P_2 = m_2 I_2^2 R_2 \frac{(1-s)}{s}. \qquad (6.6.4)$$

Whether this is gross or net mechanical power depends on how the value of the circuit parameter R_m is chosen. If R_m accounts for core loss and windage and friction loss, then P_m is the shaft mechanical power delivered to the load. Normally, R_m is only allowed to involve the core loss, so P_m is the gross mechanical power converted from the net electrical power. The latter approach is taken if the speed of the machine varies.

Three further points of interest can be noted. The slip is zero at synchronous speed and the rotor current I_2 is zero, so there is no mechanical power output. At supersynchronous speeds the slip is negative. This creates a negative value of mechanical power which is interpreted as an input power. So the induction machine can act as a generator above synchronous speed. Conversion from mechanical to electrical energy can only take place in this way if the machine stator winding is connected to an existing polyphase supply, because the generator relies upon an external source for its excitation. The third point is that, when the slip is greater than unity, the power P_m is again negative, but the motor has reversed its direction of rotation, so that it is acting as an electric brake.

All the components of rotor power have been described in terms of the rotor circuit current I_2, which can be calculated from an analysis of either the exact or the approximate equivalent circuit.

In general, the ratios of the component powers can be expressed in terms of the slip s from eqs. (6.6.2) to (6.6.4) to give

$$P_1 : P_2 : P_m = 1 : s : 1 - s. \tag{6.6.5}$$

Once one component power has been calculated the others are readily found. Note that from this relationship we obtain the slip

$$s = \frac{P_2}{P_1} = \frac{\text{Rotor ohmic loss}}{\text{Gap power}}. \tag{6.6.6}$$

The above power ratios are for motoring, whereas for negative slip

$$P_1 : P_2 : P_m = 1 : -|s| : (1 + |s|) \tag{6.6.7}$$

or, for example,

$$P_m = P_1(1 + s) \tag{6.6.8}$$

which means that the input mechanical power is greater than the output gap power by an amount sP_1, the rotor ohmic loss. This indicates that the machine is generating.

EXAMPLE 6.3

A 3-phase induction motor, with a wound-rotor winding connected in Y, has a rotor-circuit resistance of 0.5 ohm per phase and standstill leakage reactance of 1.5 ohms per phase. The open-circuit voltage between slip rings of the rotor winding is 346.4 V at standstill. Determine the gross mechanical power and the motor input power for operation at a slip of 6% with the slip rings short-circuited. For this condition, the total stator losses (stator winding loss, windage and friction losses and iron loss) are 653 W.

Solution

$E_2 = 346.4/\sqrt{3} = 200$ V.

$I_2 = sE_2/(R_2^2 + s^2X_2^2)^{1/2}$.

Numerically $I_2 = \dfrac{0.06 \times 200}{(0.5^2 + 0.06^2 \times 1.5^2)^{1/2}} = 23.62$ A.

The gross mechanical power $P_m = P_2(1 - s)/s = 3I_2^2R_2(1 - s)/s$.

Hence $P_m = 3 \times 23.62^2 \times 0.5 \times (1 - 0.06)/0.06 = 13{,}110$ W.

The gap power $P_1 = \dfrac{P_m}{1 - s} = \dfrac{13{,}110}{1 - 0.06} = 13{,}947$ W.

The input power $P = P_1 + \text{stator losses} = 13{,}947 + 653 = 14{,}600$ W.

6.6.2. Torque

In the general case, it is assumed that the rotational losses are not accounted for by the parameter R_m of the magnetizing admittance Y_0. Thus the total electromagnetic torque T_e can be expressed in terms of gross mechanical power P_m and the rotor speed ω_m from the definition

$$T_e \triangleq \frac{P_m}{\omega_m}. \tag{6.6.9}$$

The torque can be put in a different form by using the relationships

$$P_1 : P_2 : P_m = 1 : s : 1 - s \qquad \text{and} \qquad \omega_m = \omega_s(1 - s) \tag{6.6.10}$$

where ω_s = the synchronous speed (mech. rad/s) and $\omega = 2\pi f = p\omega_s$. Accordingly

$$T_e = \frac{P_1}{\omega_s} = \frac{p}{\omega} P_1. \tag{6.6.11}$$

Thus the electromagnetic torque is directly proportional to the gap power P_1. If the number of pole pairs p is unknown, the gap power is given by the expression *torque in synchronous watts*.

The electromagnetic torque minus the torque due to the mechanical losses is called the *shaft torque*.

EXAMPLE 6.4

A 4-pole, Y-connected induction motor is operated from a 440-V, 3-phase, 60-Hz supply at a slip of 0.05 per unit. The machine parameters are

$$R_1 = 0.3 \text{ ohms} \qquad R_2' = 0.35 \text{ ohms}$$
$$X_1 = X_2' = 0.6 \text{ ohms} \qquad X_m = 30 \text{ ohms}$$

If the core loss is neglected and the exact equivalent circuit is considered, determine on a per-phase basis the (a) input current, (b) input power factor, (c) input power, (d) air-gap power, (e) stator-winding ohmic loss, (f) rotor winding ohmic loss, and (g) electromagnetic torque.

Solution

(a) X_m in parallel with X_2' and R_2'/s is an impedance $\mathbf{Z}$ given by

$$\mathbf{Z} = \frac{1}{\dfrac{1}{jX_m} + \dfrac{1}{R_2'/s + jX_2'}} = \frac{1}{\dfrac{1}{j30} + \dfrac{1}{0.35/0.05 + j0.6}}$$

that is, $\mathbf{Z} = (6.387 + j2.05)$ ohms. Total input impedance $\mathbf{Z}_{IN} = \mathbf{Z} + \mathbf{Z}_1 = (6.687 + j2.65)$ ohms. The input current $\mathbf{I}_1$ is

$$\mathbf{I}_1 = \frac{\mathbf{V}_1}{\mathbf{Z}_{IN}} = \frac{254 + j0}{6.687 + j2.65} = (32.83 - j13.01) = 35.31 \angle -21.6° \text{ A.}$$

(b) Input power factor $= \cos 21.6° = 0.93$ lagging.

(c) Input power $= V_1(I_1 \cos\theta) = 254 \times 32.83 = 8340$ W/ph.

(d) Air-gap power $P_1 = I_1^2 \times \text{Re}(\mathbf{Z}) = 1247 \times 6.387 = 7964$ W/ph.

(e) Stator ohmic loss $= I_1^2 R_1 = 1247 \times 0.3 = 374$ W/ph.

(f) Rotor ohmic loss $P_2 = sP_1 = 0.05 \times 7964 = 387$ W/ph.

(g) Torque $T_e = \dfrac{P_m}{\omega_m} = \dfrac{pP_1}{\omega} = \dfrac{2}{2\pi \times 60} \times 7964 = 42.3$ N · m/ph.

A. TORQUE AND THE IDEAL EQUIVALENT CIRCUIT

The *ideal motor* is a machine with no stator circuit leakage impedance Z_1. Circuits of use are depicted in Figs. 6.7 to 6.9. By inspection of these circuits we can develop the torque expression in terms of network parameters. From eq. (6.6.11)

$$T_e = \frac{p}{\omega} P_1 = \frac{p}{\omega} m_2 E_2 I_2 \cos\theta_2 = \frac{p}{\omega} m_1 E_1 I_2' \cos\theta_2. \tag{6.6.12}$$

Since the applied voltage V_1 is constant, then, from $V_1 = E_1 = aE_2$, both the induced emfs per phase E_1 and E_2 are constants. Accordingly

$$T_e \propto I_2 \cos\theta_2. \tag{6.6.13}$$

This implies that the torque is directly proportional to the in-phase component of the rotor circuit current, and this only exists if there is rotor resistance. Now it becomes apparent why the rotor leakage impedance cannot be ignored when an ideal motor is being defined. A high-resistance rotor must signify good torque properties and high power factor, but high resistance also means high losses and thus low efficiency.

In precise terms the total torque developed by an ideal machine is

$$T_e = \frac{p}{\omega} m_1 E_1 I_2' \cos\theta_2$$

$$= \frac{p}{\omega} m_1 V_1 \frac{V_1}{((R_2'/s)^2 + X_2'^2)^{1/2}} \frac{R_2'/s}{((R_2'/s)^2 + X_2'^2)^{1/2}}$$

or

$$T_e = \frac{p}{\omega} m_1 V_1^2 \frac{R_2' s}{R_2'^2 + s^2 X_2'^2} = \frac{p}{\omega} m_1 E_1^2 \frac{R_2' s}{R_2'^2 + s^2 X_2'^2}$$

$$= \frac{p}{\omega} m_2 E_2^2 \frac{R_2 s}{R_2^2 + s^2 X_2^2}. \tag{6.6.14}$$

Equation (6.6.14) gives a functional form of torque in terms of slip. All other quantities in the equation are constant. The form of this function is illustrated in Fig. 6.15a. It is divided into three operating regions that are governed by the value of the slip. For comparison, the gross mechanical power is plotted as a function of slip and shown in Fig. 6.15b, and the regions are tabulated with respect to the torque and power in the table below.

REGION	VALUE OF SLIP s	SIGN OF TORQUE T_e	SIGN OF POWER P_m	ACTION
1	$s > 1$	Positive	Negative	Braking
2	$s < 0$	Negative	Negative	Generating
3	$0 < s < 1$	Positive	Positive	Motoring

Region 1 (s > 1)
For the slip to be greater than unity, the direction of mechanical rotation must be opposite to that of the rotating field. The electromagnetic torque is in the same direction as the rotating field, so the developed torque is trying to reduce the rotor mechanical speed.

Until the motor comes to rest the machine is acting like an electric brake. The negative sign of the mechanical output power P_m indicates an input rather than an output. The way this is interpreted is that the kinetic energy of the mechanical system is being converted to create regenerative braking, wherein electrical energy is returned to the supply.

Braking in this manner is carried out in applications such as traction, hoists, and cranes. The machine is normally in the motoring mode with the speed between standstill and synchronism. The braking action is achieved by switching two lines of the three-phase supply at the motor terminals, because, in this way, the direction of the traveling wave pattern of the field is reversed. This action is called *plugging*. Too frequent plugging causes overheating, because, with the slip greater than unity, the equivalent circuit rotor impedance is low and the current is high.

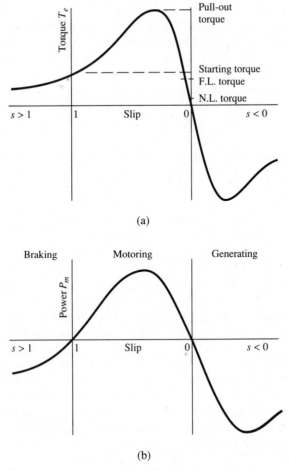

Figure 6.15 Induction motor characteristics. (a)
Torque-slip curve, (b) power-slip curve.

Region 2 (s < 0)

The rotor has to be driven above synchronous speed for the slip to be negative. In this case both the torque T_e and the mechanical power P_m become negative. This indicates that P_m is an input power and the machine is generating. An induction generator has special applications. For the cases where the induction machine drive has varying speed and the induction machine can be connected to a utility, the application is well suited. Such applications are wind generation and sea generation (tidal or current). The utility provides the excitation for the machine, so it does not matter at what speed the machine runs. Field rotation is always synchronous and the frequency of the stator winding currents is always the same as the utility.

Region 3 ($0 < s < 1$)

This is the important region over which the machine motors. The speed of rotation is between standstill ($s = 1$) and synchronous speed ($s = 0$), and is in the same direction as the rotating magnetic field. Motoring action is indicated by the positive sign of the mechanical output power. At low values of slip s, the squared term in s in eq. (6.6.14) can be ignored and the torque is approximately

$$T_e \approx \frac{p}{\omega} m_1 \frac{V_1^2}{R_2'} s. \tag{6.6.15}$$

That is, the torque-slip curve is essentially linear. This is the main operating region. At high values of slip, the squared term in s dominates, so that from standstill the torque increases with speed to give the characteristic torque-slip curve illustrated in Fig. 6.15.

Between standstill and the speed at which maximum torque occurs, the motor is unstable if the load characteristic has a smaller slope than the motor characteristic. Where the load characteristic intersects the motor torque-slip curve, there is a possible operating point. If a small disturbance reduces the shaft speed, the load torque is greater than the torque developed by the motor, so there results a further decrease in speed until the load comes to rest. If a disturbance increases the speed slightly away from the operating point, the electromagnetic torque is greater than the load torque. Consequently, there is a net torque to accelerate the load to some new operating point.

From the torque-slip curve any point that lies between the speed at maximum motor torque and synchronous speed, operation is stable. Any disturbance is such that the net torque tends to return the load speed to the equilibrium value. However, this does not mean that mechanical oscillations (hunting) do not occur.

There are two values of torque that need special mention. They are the starting torque and the maximum torque.

Starting Torque

At standstill, the slip is unity and the mechanical power is zero, but the starting torque T_{eST} is finite and is given by putting $s = 1$ in eq. (6.6.14). That is

$$T_{eST} = \frac{p}{\omega} m_1 V_1^2 \frac{R_2'}{R_2'^2 + X_2'^2} \tag{6.6.16}$$

which is some positive value that is proportional to the square of the applied voltage.

In practice, the leakage reactance dominates the circuit resistance. Therefore, the standstill rotor impedance does not vary significantly if the resistance is altered, and the starting torque becomes approximately proportional to the rotor resistance. That is, if the rotor resistance is increased, the starting torque will be greater. This has the advantage for high-inertia loads that there can be rapid acceleration to full speed. But the motor efficiency is reduced.

Maximum Torque

The expression for torque, in circuit terms, can be written in the form

$$T_e = \frac{p}{\omega} p_1 = \frac{p}{\omega} m_2 E_2 I_2 \cos\theta_2$$

$$= \frac{p}{\omega} m_2 E_2 \frac{sE_2}{(R_2^2 + s^2 X_2^2)^{1/2}} \frac{R_2}{(R_2^2 + s^2 X_2^2)^{1/2}}. \tag{6.6.17}$$

That is

$$T_e = \frac{p}{\omega} m_2 \frac{sE_2^2 R_2}{R_2^2 + s^2 X_2^2}. \tag{6.6.18}$$

It is convenient to define the ratio of the resistance to standstill leakage reaction in the manner

$$\frac{R_2}{X_2} \triangleq \alpha \tag{6.6.19}$$

so that

$$T_e = \frac{p}{\omega} m_2 \frac{E_2^2}{X_2} \frac{\alpha s}{\alpha^2 + s^2}. \tag{6.6.20}$$

Maximum torque $\hat{T}_e$ occurs when the slope $\frac{dT_e}{ds} = 0$. That is, when $s = \alpha$, so that

$$\hat{T}_e = \frac{1}{2} \frac{p}{\omega} m_2 \frac{E_2^2}{X_2}. \tag{6.6.21}$$

This result indicates that the maximum torque (or pull-out torque) is a constant and independent of the rotor circuit resistance. However, the slip, at which the maximum torque occurs, is directly proportional to the value of the resistance. Resistance modulation is one of the simplest means of altering the torque-speed characteristic, both for starting torque and load-speed control.

   Consider eqs. (6.6.20), (6.6.21), and (6.6.14). Two important ratios are the general torque to maximum torque ratio,

$$\frac{T_e}{\hat{T}_e} = \frac{2\alpha s}{\alpha^2 + s^2} \tag{6.6.22}$$

and the specific starting torque to maximum torque ratio

$$\frac{T_{eST}}{\hat{T}_e} = \frac{2\alpha}{\alpha^2 + 1}. \tag{6.6.23}$$

Maximum torque occurs at standstill if R_2 equals X_2. This follows from eq. (6.6.23).

B. TORQUE AND THE APPROXIMATE EQUIVALENT CIRCUIT

Because the stator leakage impedance Z_1 is taken into account in the approximate equivalent circuit, an estimate of the torque from the approximate equivalent circuit will be more realistic than that from the ideal equivalent circuit. This is particularly true if the terminal-voltage correction is made. The division of power remains as indicated in eq. (6.6.5), but the induced emf E_1 is no longer equal to the applied voltage V_1. The torque can be expressed in the form

$$T_e = \frac{P_m}{\omega_m} = \frac{p}{\omega} P_1 = \frac{p}{\omega} \frac{P_2}{s} = m_1 \frac{p}{\omega} I_2'^2 \frac{R_2'}{s}. \qquad (6.6.24)$$

From an inspection of Fig. 6.12, substitution can be made for the referred current I_2' to give

$$T_e = m_1 \frac{p}{\omega} \frac{V_1^2 R_2'/s}{(R_1 + R_2'/s)^2 + (X_1 + X_2')^2}. \qquad (6.6.25)$$

This torque-slip relation has the same form as that obtained from the ideal equivalent circuit. However the terms in the denominator are now greater, so the torque values are less for the same slip.

For example, the starting torque

$$T_{eST} = m_1 \frac{p}{\omega} V_1^2 \frac{R_2'}{(R_1 + R_2')^2 + (X_1 + X_2')^2} \qquad (6.6.26)$$

shows that the practical machine has a lower torque than the ideal motor. Also, at standstill the current is nearly halved by the presence of the leakage impedance Z_1, as seen from

$$\mathbf{I}_2' = \frac{\mathbf{V}_1}{(R_1 + R_2') + j(X_1 + X_2')}. \qquad (6.6.27)$$

The torque is a maximum when $\dfrac{dT_e}{ds} = 0$, and this occurs when

$$s = \frac{R'_2}{(R_1^2 + (X_1 + X_2')^2)^{1/2}}. \qquad (6.6.28)$$

If we use the practical knowledge that the leakage reactance is greater than the stator circuit resistance, we can write the slip at maximum torque to be

$$s \approx \frac{R_2'}{X_1 + X_2'}. \qquad (6.6.29)$$

Therefore

$$\hat{T}_e \approx \frac{m_1}{2} \frac{p}{\omega} \frac{V_1^2}{X_1 + X_2'}. \qquad (6.6.30)$$

This is an acceptable approximation.

An inspection of the general torque eq. (6.6.25) shows that the torques in the generating region are, in reality, higher than those in the motoring region for the same absolute value of slip. This is because, in the generating region, the slip is negative and R_1 reduces the value of the denominator. In the motoring region the addition of R_1 increases the value of the denominator.

C. HARMONIC TORQUES

The development of the torque-slip characteristic has relied on an air-gap field distribution that is sinusoidal in space and time. Time harmonic mmfs can be produced by nonsinusoidal voltages or saturation effects, and, in turn, harmonic torques are produced. Usually these harmonics are small in magnitude and may be neglected. However, mmf space harmonics need more attention, because their effects can produce the undesirable characteristic of *crawling* at low speed.

In Appendix G, the mmf analysis of a three-phase winding showed that the fifth and seventh harmonic amplitudes were much less than the fundamental amplitude, but were relatively high compared with other harmonics. The harmonic torques will have the same functional form as the fundamental torque, but the synchronous speeds for higher harmonics are different. All the harmonics are assumed to be caused by the same currents of fundamental frequency f, but the fifth harmonic flux rotates backward at one fifth of the fundamental synchronous speed, and the seventh harmonic flux rotates in the same direction as the fundamental but at one seventh of the speed. If the rotor were revolving at the fundamental synchronous speed, the fundamental slip s is

$$s = \frac{\omega/p - \omega_m}{\omega/p} = \frac{\omega/p - \omega/p}{\omega/p} = 0 \qquad (6.6.31)$$

$$\text{the seventh harmonic slip} = \frac{\omega/p - \omega/7p}{\omega/p} = \frac{6}{7} \qquad (6.6.32)$$

$$\text{and the fifth harmonic slip} = \frac{\omega/p + \omega/5p}{\omega/p} = \frac{6}{5}. \qquad (6.6.33)$$

The component torques and their resultant can be drawn as shown in Fig. 6.16. In the motoring region, the section aa' is a stable operating range. Consequently, if the load characteristic should intersect the section aa', then, on starting, the load would not rise above this low speed. This is called *crawling*.

Although we cannot analyze all harmonic effects, the reader can be aware that (i) there are possibilities of crawling at low speeds, (ii) there is a possibility of failure to start, called *cogging,* if the number of rotor and stator slots have a common factor, (iii) there are possibilities of excessive noise and shaft deflection, and (iv) there is a possibility of subsynchronous locking, if rotor harmonic fields are in synchronism with other stator harmonic fields, which lock together at a critical rotor speed, so that the machine cannot pull out of synchronism at this speed. The designer not only has to know these precise effects due to the winding mmf harmonics, magnetic satura-

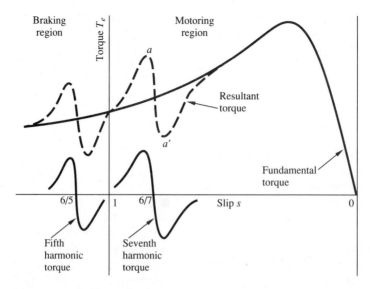

Figure 6.16 Harmonic torques.

tion, and tooth ripple, but he must also know how to compensate for them to prevent trouble.

6.6.3. Losses and Efficiency

The induction motor losses can be divided conveniently into approximately constant losses and variable losses. If the applied voltage and frequency are constant, the approximate equivalent circuit of the machine indicates that the flux is reasonably constant. Accordingly the core loss is constant. Since the speed of an induction motor changes little with load, the rotational loss is also constant. So the fixed losses comprise the core and rotational losses.

The resistance losses of the stator and rotor windings are the *variable losses*, since the ohmic loss is a function of current and current is a function of load. There are *stray losses*, which consist of the rotor core loss and additional losses due to the leakage fluxes inducing eddy currents in iron parts such as the stator frame. Stray losses may be lumped with the rotational losses.

A power distribution diagram is shown in Fig. 6.17. This illustrates the power balance.

$$
\begin{aligned}
\text{Stator input power } P_i = \text{ } &\text{Core loss } P_{NL} + \text{Stator ohmic loss } P_c \\
&+ \text{Rotor ohmic loss } P_2 \\
&+ \text{Gross mechanical power } P_m
\end{aligned}
\tag{6.6.34}
$$

where $P_i = m_1 V_1 I_1 \cos\theta_1$.

From a consideration of the approximate equivalent circuit the core loss P_{NL} is

$$P_{NL} = m_1 \frac{V_1^2}{R_m}. \tag{6.6.35}$$

The stator ohmic loss P_c is

$$P_c = m_1 I_1^2 R_1 \approx m_1 I_2'^2 R_1. \tag{6.6.36}$$

The rotor ohmic loss P_2 is

$$P_2 = m_2 I_2^2 R_2 = m_1 I_2'^2 R_2' \tag{6.6.37}$$

where $I_2' \approx V_1 / [(R_1 + R_2'/s) + j(X_1 + X_2')]$. The gross mechanical output P_m is

$$P_m = m_1 I_2'^2 R_2' \frac{(1-s)}{s} = m_2 I_2^2 R_2 \frac{(1-s)}{s} = T_e \omega_m. \tag{6.6.38}$$

Per unit efficiency η is defined as

$$\eta \triangleq \frac{\text{shaft output power}}{\text{input power}}. \tag{6.6.39}$$

As a first approximation, if the stator losses are ignored completely, the efficiency of the motor becomes

$$\eta \approx \frac{P_m}{P_1} = 1 - s \tag{6.6.40}$$

and from this we can see that for a small slip the efficiency is high.

If we take into account all the losses

$$\eta = \frac{\text{output}}{\text{output} + \text{losses}} = \frac{P_o}{P_o + P_{NL} + P_c + P_2 + P_{WF} + P_s}. \tag{6.6.41}$$

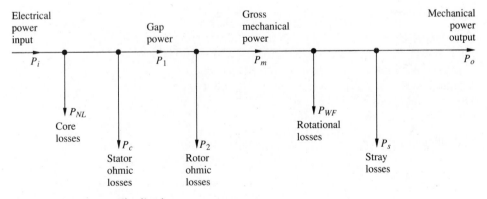

Figure 6.17 Power distribution.

The core, rotational, and stray losses ($P_{NL} + P_{WF} + P_s$) may be considered constant, and the net mechanical power can, with a little reflection, be said to be roughly proportional to the rotor-circuit current. That is

$$P_o = kI_2' \tag{6.6.42}$$

where k is a constant. In terms of the approximate equivalent circuit, the efficiency becomes

$$\eta = \frac{kI_2'}{kI_2' + (P_{NL} + P_{WF} + P_s) + m_1 I_2'^2(R_1 + R_2')}. \tag{6.6.43}$$

For maximum efficiency $d\eta/dI_2' = 0$, which occurs if

$$(P_{NL} + P_{WF} + P_s) = m_1 I_2'^2(R_1 + R_2') = P_c + P_2. \tag{6.6.44}$$

Thus, if the variable losses (ohmic losses) are equal to the fixed losses (core and rotational losses), the motor operates at maximum efficiency.

EXAMPLE 6.5

A 3-phase, 550-V, Y-connected induction motor has the following parameters.

$$R_1 = R_2' = 0.08 \text{ ohm} \quad \text{and} \quad X_1 = X_2' = 0.2 \text{ ohm}.$$

Measurements show that the magnetizing current is 40 A, the core loss is 1700 W, and the windage and friction losses are 1570 W. Determine the shaft output power, the motor efficiency, and the input power factor at a slip of 0.03 per unit. Use the approximate equivalent circuit.

Solution

$$I_2' = \frac{V_1}{((R_1 + R_2'/s)^2 + (X_1 + X_2')^2)^{1/2}}.$$

That is, $I_2' = \dfrac{550/\sqrt{3}}{[(0.08 + 0.08/0.03)^2 + (0.2 + 0.2)^2]^{1/2}} = 114 \text{ A}.$

$P_m = 3I_2'^2 R_2'(1 - s)/s = 3 \times 114^2 \times 0.08 \times 0.97/0.03 = 101{,}570 \text{ W}.$

Shaft power $P_o = P_m - P_{WF} = 101{,}570 - 1570 = 100{,}000 \text{ W}.$

Losses = WF losses + core losses + ohmic losses.

Ohmic losses = $3I_2'^2(R_1 + R_2') = 3 \times 114^2(0.08 + 0.08) = 6283 \text{ W}.$

Hence, total losses = $1570 + 1700 + 6283 = 9553 \text{ W}.$

Therefore input power = $100{,}000 + 9553 = 109{,}553 \text{ W}$

and efficiency $\eta = \dfrac{\text{output}}{\text{input}} = \dfrac{100{,}000}{109{,}553} = 0.913$ per unit.

$$\mathbf{I}_2' = I_2' \cos\theta_2 - jI_2' \sin\theta_2.$$

$$\theta_2 = \tan^{-1} \frac{X_1 + X_2'}{R_1 + R_2'/s} = \tan^{-1} \frac{0.2 + 0.2}{0.08 + 0.08/0.03} = 8.29°.$$

Thus $I_2' = (113.3 - j16.49)$ A.

Core loss $= 1700$ W, Core loss/ph $= 567$ W, $V_1 = 550/\sqrt{3} = 318$ V.

Therefore $I_{h+e} =$ core loss$/V_1 = 576/318 = 1.78$ A. Hence $I_{NL} = (1.78 - j40)$ A.

Input current $I_1 = I_{NL} + I_2' = (115.08 - j56.49) = 128.2 \angle -26.1°$ A.

Therefore, power factor $\cos\theta_1 = \cos 26.1° = 0.9$.

EXAMPLE 6.6

A 440-V, 3-phase, Y/Y-connected, 60-Hz, 6-pole, wound rotor induction motor has an effective turns ratio of $3:1$. The six circuit parameters are as follows.

$$R_1 = 0.5 \text{ ohm}, \qquad R_2 = 0.1 \text{ ohm}, \qquad R_m = 400 \text{ ohms},$$

$$X_1 = 1.8 \text{ ohms}, \qquad X_2 = 0.2 \text{ ohm}, \qquad X_m = 50 \text{ ohms}.$$

At a slip of 4% the windage, friction, and stray-load losses amount to 400 W. Compare the values of I_1, I_2, input power factor, shaft torque and motor efficiency obtained from (a) the ideal equivalent circuit, (b) the approximate equivalent circuit, and (c) the exact equivalent circuit.

Solution

(a) Figure 6.9 illustrates the circuit diagram of the ideal equivalent circuit. The applied phase voltage is

$$V_1 = 440/\sqrt{3} = 254 \text{ V}.$$

This voltage is taken as reference.

$$I_2' = \frac{V_1}{Z_2'} = \frac{V_1}{R_2'/s + jX_2'} = \frac{254}{0.9/0.04 + j1.8} = 11.218 - j0.8967$$

$$= 11.254 \angle -4.57° \text{ A}.$$

$$I_2 = 3 \times 11.254 = 33.76 \text{ A}.$$

$$Z_m = jX_m = j50 = 50 \angle 90° \text{ ohms}.$$

$$I_{NL} = \frac{V_1}{Z_m} = \frac{254}{50 \angle 90°} = 5.08 \angle -90° = 0 - j5.08 \text{ A}.$$

$$I_1 = I_2' + I_{NL} = 11.218 - j5.9767 = 12.71 \angle -28° \text{ A}.$$

Input power factor $= \cos\theta_1 = \cos 28° = 0.88$.

$$P_m = 3I_2'^2 R_2'(1-s)/s = 3 \times 11.254^2 \times 0.9 \times 0.96/0.04 = 8207 \text{ W}.$$

Shaft power output $= P_m - WF$ losses $= 8207 - 400 = 7807$ W.

$$\text{Shaft torque} = \frac{\text{shaft power}}{\omega_m} = \frac{7807}{\dfrac{\omega}{p}(1-s)} = \frac{7807}{\dfrac{2\pi 60}{3} \times 0.96} = 64.7 \text{ N} \cdot \text{m}.$$

Power input $= P_i = 3V_1(I_1 \cos\theta_1) = 3 \times 254 \times 11.218 = 8548 \text{ W}.$

Therefore efficiency $= \dfrac{\text{output}}{\text{input}} = \dfrac{7807}{8548} = 0.91$ per unit.

(b) Accounting for stator leakage impedance drop in the approximate equivalent circuit

$$V = V_1(1 - X_1/X_m) = 245 \text{ V}.$$

$$R_2' = a^2 R_2 = 0.9 \ \Omega, \quad X_2' = a^2 X_2 = 1.8 \ \Omega, \qquad \text{where} \qquad a = 3:1.$$

Now $I_2' = \dfrac{V}{Z_1 + Z_2'} = \dfrac{245}{(0.5 + j1.8) + (0.9/0.04 + j1.8)}$.

That is, $I_2' = 10.52 \ \angle -8.9° = (10.4 - j1.6)$ A.

$$I_2 = aI_2' = 31.5 \text{ A}.$$

$$Z_m = \dfrac{1}{\dfrac{1}{jX_m} + \dfrac{1}{R_m}} = \dfrac{j50 \times 400}{400 + j50} = 6.15 + j49.23 = 49.6 \ \angle 82.9° \ \Omega.$$

$$I_{NL} = \dfrac{V}{Z_m} = \dfrac{245}{49.6 \ \angle 82.9°} = 0.6 - j4.9 = 4.93 \ \angle -82.9° \text{ A}.$$

$$I_1 = I_2' + I_{NL} = (11.0 - j6.5) = 12.8 \ \angle -30.6° \text{ A}.$$

Input power factor $\cos\theta_1 = \cos 30.6° = 0.86$.

$$P_m = 3I_2'^2 R_2'(1 - s)/s = 3 \times 10.52^2 \times 0.9 \times 0.96/0.04 = 7171 \text{ W}.$$

Shaft power output $= P_m - \text{WF losses} = 7171 - 400 = 6771$ W.

Shaft torque $= \dfrac{\text{shaft power}}{\omega_m} = \dfrac{6771}{\dfrac{\omega}{p}(1 - s)} = \dfrac{6771}{\dfrac{2\pi 60}{3} \times 0.96} = 56.1 \text{ N·m}.$

Power input $P_i = 3V_1 I_1 \cos\theta_1 = 3 \times 254 \times 12.8 \cos 30.6° = 8395$ W.

Therefore efficiency $\eta = \dfrac{\text{output}}{\text{input}} = \dfrac{6771}{8395} = 0.807$ per unit.

(c) For the exact equivalent circuit let $Z = Z_2'$ in parallel with Z_m.

That is, $Z = \dfrac{1}{\dfrac{1}{Z_2'} + \dfrac{1}{Z_m}}$.

So $Z = \dfrac{1}{\dfrac{1}{22.5 + j1.8} + \left(\dfrac{1}{400} - j\dfrac{1}{50}\right)} = (17.03 + j8.51) \ \Omega.$

Let $Z_{in} = Z_1 + Z = (0.5 + j1.8) + (17.03 + j8.51) = (17.53 + j10.3) \ \Omega.$

$$I_1 = \dfrac{V_1}{Z_{in}} = \dfrac{254}{17.53 + j10.3} = 10.77 - j6.3 = 12.5 \ \angle -30.4° \text{ A}.$$

Input power factor $\cos\theta_1 = \cos 30.4° = 0.86$.

$$I_2' = \frac{E_2'}{Z_2'} = \frac{I_1 Z}{Z_2'} = \frac{(10.77 - j6.3)(17.03 + j8.51)}{22.5 + j1.8} = 10.54 \angle -8.3° \text{ A}.$$

Hence $I_2 = I_2' a = 10.54 \times 3 = 31.62$ A.

$$P_m = 3 I_2^2 R_2 (1 - s)/s = 3 \times 31.62^2 \times 0.1 \times 0.96/0.04 = 7199 \text{ W}.$$

Shaft power $= P_m - \text{WF losses} = 7199 - 400 = 6799$ W.

$$\text{Shaft torque} = \frac{\text{Shaft power}}{\omega_m} = \frac{6799}{\dfrac{2\pi \times 60}{3} \times 0.96} = 56.4 \text{ N} \cdot \text{m}.$$

Input power $P_i = 3 V_1 I_1 \cos\theta_1 = 3 \times 254 \times 12.5 \cos 30.4° = 8215$ W.

Efficiency $\eta = \dfrac{\text{output}}{\text{input}} = \dfrac{6799}{8215} = 0.83$ per unit.

The results are combined in the table below for comparison.

QUANTITY	IDEAL EQUIVALENT CIRCUIT	APPROXIMATE EQUIVALENT CIRCUIT	EXACT EQUIVALENT CIRCUIT
I_1(A)	12.7	12.8	12.5
I_2(A)	33.8	31.5	31.6
Power factor	0.88	0.86	0.86
Shaft torque (N · m)	64.7	56.1	56.4
Efficiency	0.91	0.81	0.83

It is seen from the table that the ideal equivalent circuit gives optimistic results for torque and efficiency. The results from the approximate and exact equivalent circuits are close enough to allow the approximate equivalent circuit to be used in general.

6.6.4. Output Equation

It is not the aim of this text to describe the design procedures of electrical machines. The goal is to develop the theory of machines. From the theory we can develop models that are useful to determine performance characteristics. In this section we can get some idea of the performance of a machine (that is, power and torque) in terms of speed and geometry by considering the basic design parameters called loading factors.

Loading factors were defined for the dc machine.[7] This provided a useful base, from which an approximate but useful relation was established between the main

---

[7] See Chapter 5, section 5.9.

dimensions of a machine and its power rating. The same can be done for an ac machine, whose input rating is defined by

$$VA = m_1 V_1 I_1 \approx m_1 E_1 I_1 = m_1 \times 4.44 \Phi f N_1 I_1. \tag{6.6.45}$$

The total flux per pole is given by

$$\Phi = \frac{\pi D}{2p} LB_{mean} \tag{6.6.46}$$

and the current can be expressed in the form

$$I_1 = \frac{\pi D}{m_1} \frac{ac}{2N_1} \tag{6.6.47}$$

where $\quad ac$ = specific electric loading (ampere conductors/meter)

$\quad D$ = rotor diameter (m)

$\quad L$ = rotor length (m)

and $\quad B_{mean}$ = specific magnetic loading (tesla).

Therefore

$$VA = 1.11 \pi^2 D^2 L n_s B_{mean} ac. \tag{6.6.48}$$

Since the specific loading factors B_{mean} and ac have upper limits in practice, this equation can be interpreted to mean that, for a given power rating VA, the higher the synchronous speed n_s the smaller will be the machine volume (proportional to $D^2 L$). Since power is torque times speed, eq. (6.6.48) implies that the volume of a machine is proportional to its full-load torque. This is useful information for electric drive applications.

6.7. INDUCTION MOTOR CONTROL

6.7.1. Induction Motor Starting

Good starting characteristics of an induction motor are low currents, so that the supply is not overloaded, and high torque, so that the motor will accelerate from rest to rated speed in a short time. These characteristics are not always possible to have, as we shall see.

In order to see that there can be a problem associated with starting an induction motor by applying the supply voltage directly across the motor terminals, we can consider an equivalent circuit of the machine and find the current in terms of the slip and impedances. Consider Fig. 6.12. The magnetizing admittance will be neglected, because the magnetizing current is approximately constant and is a fraction of the full-load current. The current I_2' can then be considered to be the supply current per phase in this qualitative analysis.

The current per phase for any slip s is

$$I_1 \approx I_2' = \frac{V_1}{[(R_1 + R_2'/s)^2 + (X_1 + X_2')^2]^{1/2}} \tag{6.7.1}$$

The mechanical output per phase is

$$P_m = I_2'^2 R_2' \frac{1-s}{s} \tag{6.7.2}$$

so that the electromagnetic torque is

$$T_e = \frac{p}{\omega} I_2'^2 \frac{R_2'}{s}. \tag{6.7.3}$$

Inspection of the above equations for machine starting ($s = 1$) shows that the current is a maximum. Although the high value of current tends to increase the starting torque, the maximum value of s in the denominator of the torque equation tends to reduce the value of the starting torque. These are not ideal starting characteristics, but, for small motors of only a few horsepower rating, they can be tolerated. It must be noted that these poor characteristics present themselves only when the supply is switched directly across the terminals of the motor, that is, with *direct on-line* starting. An approach to meet the two ideal conditions can be made by adding an external circuit that can give varying degrees of control.

Starting methods are

1. direct-on-line starting
2. stator-resistance starting
3. autotransformer starter
4. star-delta starter
5. special rotor cage winding
6. rotor-resistance starting.

A. DIRECT-ON-LINE STARTING

Direct-on-line starting of small motors has advantages. It is an inexpensive method, and the high starting current means that there will be a minimum delay running up to speed.

Direct-on-line starting for large motors could have a number of undesirable effects. The extremely high currents drawn from the supply could produce a supply voltage drop and the associated light dimming or flicker might not be tolerable to local domestic and industrial consumers. If a number of motors were started together, the current might be too great for the rated capacity of the supply transformer and cable, so that the protection equipment would operate to trip the supply.

The direct-on-line current ratio (DOCR) and the direct-on-torque ratio (DOTR) are defined as the ratio of the starting value to the full-load value, and DO and FL are the subscripts referring to starting and full-load quantities, respectively. As the interest is to obtain a simple and approximate estimate of the characteristics in ratio form, we can neglect the primary leakage impedance and no-load admittance, that is, we can use the ideal model of the induction motor.

Because the stator applied voltage is assumed to be constant throughout the run-up period, the required ratios can be obtained in terms independent of the voltage as follows.

$$\text{DOCR} = \frac{I_{DO}}{I_{FL}} = \frac{V_1}{Z_{DO}}\frac{Z_{FL}}{V_1} = \left[\frac{(R_2'/s_{FL})^2 + X_2'^2}{R_2'^2 + X_2'^2}\right]^{1/2}. \tag{6.7.4}$$

For example, typical parameter values may be $R_2' = 0.5\ \Omega$, $X_2' = 4\ \Omega$, and $s_{FL} = 0.05$ so that $\text{DOCR} > 7$.

$$\text{DOTR} = \frac{T_{eST}}{T_{FL}} = \frac{T_{DO}}{T_{FL}} = \frac{I_{DO}^2}{I_{FL}^2}\frac{R_2'}{R_2'/s_{FL}} = s_{FL}(\text{DOCR})^2. \tag{6.7.5}$$

For the same example $\text{DOTR} \approx 2.5$.

From this simple analysis, it can be gathered that there is a large unwanted surge of current, a starting torque which is not high and a poor power factor. In favor of this method is that it is cheap, because no auxiliary equipment is used, and the current surge occurs only once.

B. STATOR-RESISTANCE STARTER

The addition of a starter resistance in the stator circuit is suitable for induction motors with cage rotors. The implication is that the added resistance causes a voltage drop and reduces the voltage applied to the stator winding terminals. An approximate torque equation is

$$T_e = \frac{p}{\omega}I_2'^2\frac{R_2'}{s} = \frac{p}{\omega}\frac{V_1^2 R_2'/s}{[(R_2'/s)^2 + X_2'^2]^{1/2}} \tag{6.7.6}$$

and shows that the starting torque is directly proportional to the square of the terminal voltage. So a reduction of stator voltage produces a greater reduction of torque. Because the starting current is less than the direct-on value, poor starting torque is indicated.

Although this is not the best method of starting, it does offer an economic alternative to reduce the voltage, especially in those circumstances where stator impedance is used for motor speed adjustment.

C. AUTOTRANSFORMER STARTING

If starting time is important, the adjustable output autotransformer is of general use. By raising the autotransformer output voltage gradually, it is ensured that there is no large current surge. Manual operation of the transformer is possible in the laboratory, but is not suitable in industrial and remote applications. However, a fixed ratio autotransformer can be substituted. The autotransformer is relatively expensive and this method of starting can be justified only for starting large induction motors.

D. STAR-DELTA STARTER

As its name suggests this method employs the stator winding in a star connection during the starting period, and then the winding is switched to a delta connection when the motor is running normally. This means that, for the initial connection, the voltage across each phase winding is less than when the motor is in the steady-state running condition. Accordingly the method can be described as a voltage reducer with its inherent current and torque reduction. This is an inexpensive method of starting motors, whose output does not exceed about 30 bhp, and which have small, initial loads such as generators, pumps, or fans.

E. CAGE ROTORS

There have been some ingenious designs of rotor bars to get both high torque and low current at starting without affecting the full-load running characteristics.

Single-cage Rotor

If the rotor-winding resistance is increased, the speed at which maximum torque occurs is lowered and so the starting torque is increased. Also, a high resistance limits current. Resistance cannot be added to a cage rotor for starting purposes and then removed for steady-state running. However, by geometrical design, the resistance of the bars can change considerably with change in frequency.

One design is to use deep, narrow slots. At starting, the induced currents in the bars are alternating at supply frequency and are crowded to the outer surface by the skin effect. The rotor resistance is then effectively high and good starting conditions are achieved. The torque is high, the current is low, and the power factor is high. Reduced efficiency over the short period of start-up is a cheap price to pay for enhanced torque characteristics.

When the motor runs at its steady speed, the slip is low, and the frequency of the rotor currents is low also. Accordingly, the whole cross section of the rotor bar is utilized to give a more uniform distribution of current and the resistance is lower. Thus the desired running conditions of small speed variation with load and good efficiencies are achieved.

Double-cage Rotor

The outer cage of a double cage is designed to have high resistance and the inner cage is embedded in the iron to have a high leakage inductance with a low resistance. At starting, the inner cage has high reactance, because the rotor-current frequency is the same as stator current frequency. The over-all high impedance of the double-cage rotor limits the input current at starting. In order to form an idea of the torque value, the two cages can be considered separately and the results added.

The outer cage has a high ratio of resistance to reactance, so that the starting torque is high, while the same ratio for the inner cage is small and produces only a small torque initially. As the rotor speed increases, the inner cage reactance reduces in proportion to the rotor frequency and so the torque, due to this cage, increases. When the motor is rotating at the operating speed, the inner cage, with both low resistance and reactance, and in parallel with the outer cage, effectively reduces the resistance and impedance of the rotor circuit to allow efficient running. However, the double-cage motor has a lower maximum torque than the conventional single-cage machine, because the inner and outer cages do not have their respective maximum torques coinciding at the same speed.

F. ROTOR-RESISTANCE STARTING

Resistance can only be added to the rotor circuit, if the winding has its end connections brought out to slip rings on the shaft.

The principle of rotor-resistance starting is the same as that employed with special cage rotors, except that the resistance is part of an external circuit and can be varied at will. Resistance is added to the rotor circuit for starting. The increased ratio of rotor resistance to rotor leakage reactance reduces the designed speed at which the maximum torque occurs and so the starting torque is increased. As the speed builds up, the external resistance is reduced until finally the only resistance in the rotor circuit is the winding resistance.

EXAMPLE 6.7

A polyphase induction motor has a maximum (pull-out) torque equal to three times the gross full-load value at rated voltage. The ratio of the rotor standstill leakage reactance to resistance is 3. What is the full-load slip? Calculate the ratio of starting torque to full-load torque for the cases of direct-on starting, star-delta starting, and autotransformer starting with a 60% tapping.

Solution

$R_2/X_2 = \alpha = 1/3$.

Referring to eqs. (6.6.22) and (6.6.23)

$$\frac{T_e}{\hat{T}_e} = \frac{2\alpha s}{\alpha^2 + s^2}. \quad \text{That is} \quad \frac{1}{3} = \frac{2 \times (1/3) \times s}{1/9 + s^2}.$$

Thus $s^2 - 2s + 1/9 = 0$, from which $s = 0.057$. For direct-on starting, $s = 1$ and the ratio of starting torque to maximum torque is

$$\frac{T_{DO}}{\hat{T}_e} = \frac{2 \times (1/3) \times 1}{1/9 \times 1} = 0.6.$$

Therefore $\dfrac{T_{DO}}{T_{FL}} = \dfrac{T_{DO}}{\hat{T}_e} \dfrac{\hat{T}_e}{T_{FL}} = 0.6 \times 3 = 1.8$.

For star-delta starting the starting torque $T_{eST} \propto V^2$. The starting voltage is $1/\sqrt{3}$

times the normal voltage.

Therefore $\dfrac{T_{eST}}{T_{FL}} = \dfrac{1.8}{3} = 0.6$.

For autotransformer starting, $\dfrac{T_{eST}}{T_{FL}} = 1.8 \times 0.6^2 = 0.648$.

6.7.2. Induction Motor Speed Control

The torque-speed characteristic of a high-efficiency machine is shown in Fig. 6.18. If such a motor is supplied by a constant voltage, constant frequency source, the speed range from no load to a load, which would produce stalling, is within about 10% of synchronous speed. So approximate constant speed is an inherent characteristic of the induction motor.

Both because of its simple construction and because of special applications, numerous ways have been found to enable the speed of the induction motor to be adjusted. The principles associated with speed control are described in this section. Speed control methods are

1. Supply frequency adjustment
2. Pole changing
3. Supply voltage variation
4. Rotor injected voltage
5. Rotor resistance adjustment.

A. SUPPLY FREQUENCY ADJUSTMENT

The synchronous speed of an induction motor is the speed of the mmf traveling wave in the air gap. That is, $n_s = f/p$, where f is the supply frequency (hertz) and p is the number of pole pairs. The actual speed of a motor with a characteristic like that shown in Fig. 6.18 is just less than the synchronous speed and does not alter much with variation of load. Accordingly for a fixed number of poles, a change of the supply frequency would bring about a proportional change in the synchronous speed and the actual rotor speed would follow in roughly the same manner.

An induction motor is designed to work at a particular flux density, and as the electromagnetic torque is proportional to the magnetic flux, it is necessary to have a high value of flux density without going too far into the saturation region. It is usual to work at the knee of the magnetization curve. To maintain rated torque the flux amplitude should be kept constant. If the applied voltage can be said to be almost equal to the winding induced emf, as in the ideal case, then from the emf equation

$$E_1 = 4.44\Phi f N_1 \tag{6.7.7}$$

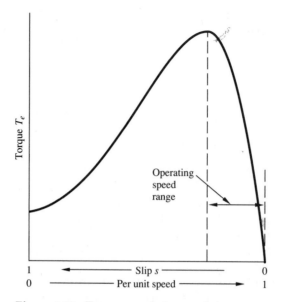

Figure 6.18   Torque-speed characteristic.

we can write

$$V_1 \approx k\Phi f \tag{6.7.8}$$

where $k = \text{constant} = 4.44N_1$. Equation (6.7.8) indicates that the ratio of applied voltage to frequency should be kept the same, or more appropriately, the dependent variable V_1 should be altered in proportion to the independent variable f for constant flux and rated torque.

It is a disadvantage to have to alter two variables (V_1 and f) separately to get efficient speed control. For small motors, or where the speed range is small, V_1 is maintained constant. But for large drives it is imperative that the voltage be altered in accordance with the frequency.

At one time the variable frequency supply was obtained from a variable-speed motor-generator set. Today the most common method of frequency adjustment is from power-electronic inverters.

We found in subsection C under section 6.6.2 that the relation for the maximum torque per phase is

$$\hat{T}_e = \frac{p}{\omega} \frac{E_2^2}{2X_2} \tag{6.7.9}$$

at a slip $s = R_2/X_2 = \alpha$.

If the voltage drop across the stator leakage impedance can be ignored (ideal case), and this becomes more acceptable the larger the motor, then for constant total flux,

$$V_1 \propto E_1 \propto E_2 \propto \omega \qquad (6.7.10)$$

and, since $X_2 \propto \omega$ very nearly,

$$\hat{T}_e \propto \frac{p}{\omega} \frac{\omega^2}{2\omega} = \text{constant.} \qquad (6.7.11)$$

The maximum torque is a constant and independent of frequency as long as the applied voltage is proportional to frequency. Also, in general, the torque at any slip s is

$$T_e = \hat{T}_e \frac{2\alpha s}{\alpha^2 + s^2} \qquad (6.7.12)$$

and

$$\alpha = \frac{R_2}{X_2}, \qquad \text{so} \qquad \alpha \propto \frac{1}{\omega} \qquad (6.7.13)$$

if the resistance can be assumed to be independent of frequency and if the leakage inductance is constant. At starting, the slip is unity, so that the starting torque is

$$T_{eST} = \hat{T}_e \frac{2\alpha}{\alpha^2 + 1} \propto \frac{\omega}{b^2 + \omega^2} \qquad (6.7.14)$$

where b is a constant. As the supply frequency is increased, the denominator in the starting torque equation increases at a greater rate than the numerator. It can be understood that the higher is the operating frequency the lower is the starting torque. This information, together with the knowledge that the maximum torque $\hat{T}_e$ is a constant and occurs at a slip inversely proportional to frequency, enables the characteristics to be plotted as shown in Fig. 6.19.

It is an advantage in traction applications that the induction motor has a high starting torque for the lower frequencies, especially if the applied voltage is proportionately lower. Other points of note are that, for the same value of a low working slip, the torque is greater at a high frequency than at a low frequency and the torque-slip relation is almost linear. For a particular load torque the power output is directly proportional to frequency, and as the frequency increases so does the efficiency. This last statement is clarified by an inspection of the efficiency expression for the ideal model. That is

$$\eta \approx \frac{P_m}{P_1} = (1 - s). \qquad (6.7.15)$$

From the information portrayed in Fig. 6.19b it can be seen that, for a constant load torque, the value of slip decreases as the frequency is increased. Accordingly the efficiency increases with frequency. It is desirable to start the motor at a low frequency to get a high torque with its associated rapid acceleration, but it is a necessary

$\omega_1 : \omega_2 : \omega_3 : \omega_4 = 1 : 2 : 3 : 4$
$V_1 \propto f$

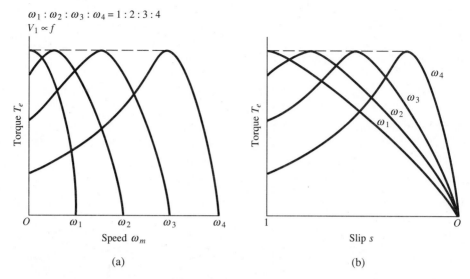

Figure 6.19 Frequency modulation for speed control. (a) Torque-speed curves, (b) torque-slip curves.

economic measure to operate at the highest possible frequency at the required operating speed.

By having the applied voltage vary in direct proportion to the frequency the horsepower will increase with the speed. This may not be required nor will it be always possible. At very low frequencies, when the reactance is small, the current limitation is produced by the winding resistance. Consequently a higher voltage at low frequency is desirable if a high torque is to be made available.

Often, as in the case of traction, constant power is required above a certain speed. In this case, while the frequency is increased the voltage should be proportional to frequency. In practice the voltage frequency characteristic could be programmed within the framework of optimum performance for a particular application.

B. POLE CHANGING

The effective speed of an induction motor is a little below the synchronous speed, which is given by the relation $n_s = f/p$, where the frequency of the supply is f hertz, and the number of pole pairs created by the winding arrangement is p.

For a constant supply frequency the maximum possible rotor speed is when $p = 1$, so that only lower speeds than f r/s can be obtained by changing the number of poles. When the supply frequency is 60 Hz, the maximum rotor speed is just below 3600 r/min. If the number of poles were equal to 6, then the speed would be just less than 1200 r/min. This method affords a stepped change of speed analogous to mechanical gear changing.

Methods of obtaining different pole configurations in the same stator frame are few, and it was not until 1957 that a completely general and acceptable approach was developed by Rawcliffe. Of historical note are two methods, coil regrouping which created complicated switching arrangements because of the large number of connections brought out to terminals, and, second, multiple windings of different pole numbers placed in the same slots but used separately and independently. The latter arrangement meant large slots to house the multiple windings, increased size of motor, and low efficiency when compared with the equivalent single-winding motor. Multiple windings need no further discussion here, but coil regrouping development is of interest.

Consequent Poles

A method of speed adjustment widely used is the consequent-pole scheme. Figure 6.20 shows schematically a single phase winding connection to give a single pole pair. Reverse the connections of half the winding (in this case one coil) and the polarity changes. Now there are two north poles, so that by consequence there must be south poles between them, giving two pole pairs, and a speed range of 2 : 1 by a change of connection.

This specific case of pole changing was devised by Dahlander in 1897. It was not until 1957 that the method of reversing half the winding was generalized by Rawcliffe to bring about both wide and close speed ratios, such as 8 : 2 or 10 : 12, respectively. Because of the similarity of the logic involved with electromagnetic waves and amplitude modulation, Rawcliffe named his method *pole-amplitude modulation.*

Pole-amplitude Modulation

The flux density wave generated by current in one phase of a 3-phase winding can be represented ideally by the space sine wave,

$$B = \hat{B} \sin p\omega t. \tag{6.7.16}$$

In particular, a single-phase mmf diagram for a normal 8-pole stator winding is depicted in Fig. 6.21a. It is shown to be rectilinear only for the convenience of simplifying the explanation of the modulation process. If the maximum flux density $\hat{B}$ has different peak values in the space distribution, and has values, which are changing periodically, then the amplitude of the magnetic field can be said to modulate. In the case of sinusoidal modulation involving k cycles around the air gap, the maximum flux density can be expressed as

$$\hat{B} = B_{\max} \sin k\omega t \tag{6.7.17}$$

where $B_{\max}$ is the maximum value of the time-varying amplitude. Figure 6.21b shows a modulating wave with one cycle, that is, $k = 1$. In practice this is obtained by reversing the connections of the windings, as shown in Fig. 6.22.

Now the equation for the general flux wave in space becomes

$$B = B_{\max} \sin k\omega t \, \sin p\omega t \tag{6.7.18}$$

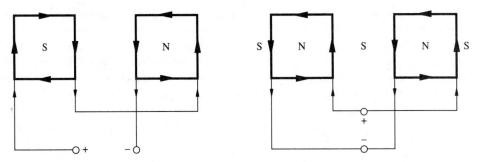

Figure 6.20 A 2 : 1 speed change for consequent-pole system. (a) 2 coils, $p = 1$, (b) 2 coils, $p = 2$.

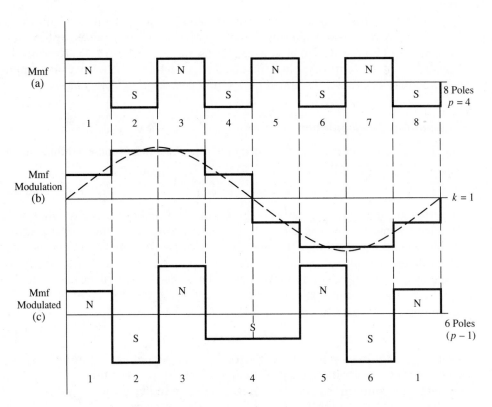

Figure 6.21 Pole-amplitude modulation for close ratio speed change (8 : 6). (a) 8-pole mmf wave, (b) modulating wave, (c) 6-pole modulated mmf wave.

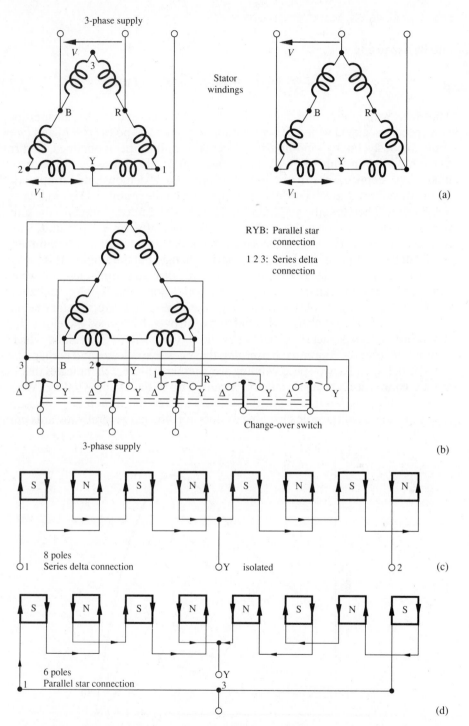

Figure 6.22 Pole-amplitude modulation. (a) Pole changing, (b) pole-changing connections, (c) pattern for 8-pole connection, (d) pattern for 6-pole connection.

and by expansion

$$B = \frac{B_{max}}{2} [\cos(p - k)\omega t - \cos(p + k)\omega t]. \qquad (6.7.19)$$

Therefore, with modulation, there are $(p \pm k)$ pole pairs and there is an appropriate new speed associated with it. There cannot be two sets of poles $(p + k)$ and $(p - k)$ simultaneously, but by winding arrangement one or the other is obtained. Returning to the modulation with $k = 1$, as shown in Fig. 6.21, the sine wave can be approximated by a stepped wave in order that the principle is demonstrated simply. In the 8-pole waveform, the coil producing the first north pole (number 1) is modulated by $+1$ per unit. Therefore the amplitude and polarity of number 1 remains the same in the modulated waveform. Pole number 2 is modulated by $+2$ per unit, so that, although the polarity remains the same, the amplitude doubles. Pole number 5 is modulated by -1 per unit. Hence the polarity changes but the amplitude is the same.

Overall, the modulated waveform, with two adjacent south poles and two adjacent north poles, gives an irregular waveform with 6 poles; that is, the pole change has been from p to $(p - 1)$ pole pairs. Although the waveform is irregular for one phase, the effect of all the three phases is to tend to give an overall balanced waveform. In fact it is the art of the designer such that both winding connections give results, which are almost free from the unwanted harmonics that cause extra losses and perhaps crawling. For the case of a two-speed winding arrangement, Fig. 6.22b indicates that there are six connections brought out to terminals from the stator winding. With 12 connections brought out, it is possible to have a three-speed motor.

Applications of the pole-changing motors are directed generally towards pumps

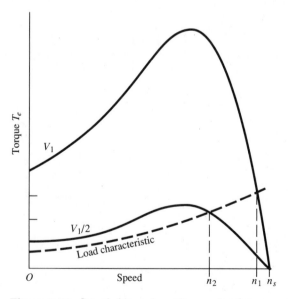

Figure 6.23   Speed change by voltage reduction.

and fans up to several thousand horsepower. One particularly suitable application is for driving cooling tower fans at a power generating station. High speed, high power is used in the summer and the low speed, low power is used in the winter when the difference between condensate and ambient temperatures is greater.

C. SUPPLY VOLTAGE VARIATION (OR ADDITIONAL STATOR IMPEDANCE)

Both supply voltage adjustment and added impedance to the stator circuit provide variable voltage at the induction motor terminals.

From eq. (6.6.25) we have seen that the torque is proportional to the square of the applied voltage. Torque-speed curves are sketched in Fig. 6.23 for a full supply voltage across the motor terminals and for half the supply voltage. With the addition of a typical load characteristic, it is seen that, if the voltage is reduced to one half the rated value, the starting torque reduces by a factor of 4 and the steady-state load speed falls from n_1 to n_2. Accordingly, by altering the applied voltage, the speed is changed. However the speed range is not great.

D. ROTOR INJECTED VOLTAGE

If the wound rotor of an induction motor has its winding connections brought out to slip rings, and if a voltage from an external source were applied to the rotor windings across the slip rings, there would be an interaction between this voltage and the induced emf of the rotor winding. A constant electromagnetic torque would drive the load at some steady speed provided that the injected voltage always has the same frequency as the induced emf. Altering the magnitude or phase of the injected voltage would alter the current, the current would affect the torque and a new steady speed would result.

This form of speed adjustment is most clearly seen by considering the circuit representation of the induction motor. Figure 6.24 shows one phase of the primitive form of the equivalent circuit with an injected voltage V_k at the rotor terminals. The loop equation of the secondary circuit is

$$sE_2 \pm V_k = I_2(R_2 + jsX_2) = I_2Z_2. \qquad (6.7.20)$$

Since the induction motor is inherently a constant speed machine, or nearly so, the

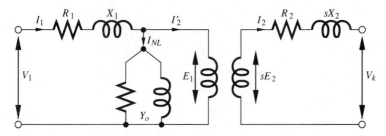

Figure 6.24 Control by injected voltage.

case of no load simplifies the interpretation of the above equation. At no load there is only need for the motor torque to overcome friction, so that the rotor current I_2 is almost zero. Hence, for no load,

$$s\mathbf{E}_2 \pm \mathbf{V}_k \approx 0. \tag{6.7.21}$$

Therefore the slip is given by

$$s = \pm |V_k/E_2| \triangleq \pm k. \tag{6.7.22}$$

The rotor induced emf per phase measured at standstill is constant. So the slip is directly proportional to the value of the injected voltage, and, what is important, the sign of the slip can be negative. This means that speeds above synchronism can be attained. The no-load speed is

$$n_0 = (1 \pm k)n_s \tag{6.7.23}$$

where n_s is the synchronous speed. For induction motors, especially those with a low resistance to reactance ratio, the speed variation between no load and full load is small. So the speed at any loading is

$$n \approx n_0 = (1 \pm k)n_s. \tag{6.7.24}$$

With an injected voltage which allows both subsynchronous and supersynchronous speeds, it is reasonable to assume that actual synchronous speed can be obtained. At this speed the slip is zero and the rotor currents have zero frequency.

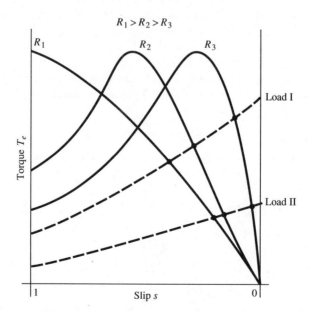

Figure 6.25 Induction motor speed control by external rotor resistance.

Thus the injected voltage is a dc one, and the machine becomes a synchronous induction motor whose speed is invariant.

The ways of employing the injected voltage principle to get adjustable speed are many. The methods vary from the simple but wasteful method of connecting resistance across the slip rings to the use of commutator machines or power electronic converters in order to draw from or return the slip power to the mains supply for maximum efficiency.

Speed Control by Rotor Resistance

This method of speed control is a special case of the control by injected voltage. That is, this voltage is the voltage drop across a resistor. Adding resistance to any circuit will increase the power wasted, but the simplicity of the method can often outweigh the disadvantage of the reduced efficiency.

The principle involved is the same as that used for starting induction motors by adding resistance to the rotor circuit. An increased rotor-winding resistance reduces the speed at which the maximum torque occurs. Thus the speed is reduced for any given and normal load torque characteristic. Figure 6.25 demonstrates this and shows that the greater speed variation is obtained from the largest possible load torque, *Load I* in this case.

Using the theory of constant flux, the ratio of the torque at any slip s to the maximum torque is

$$\frac{T_e}{\hat{T}_e} = \frac{2\alpha s}{\alpha^2 + s^2} \tag{6.7.25}$$

where α is the ratio of the rotor parameters R_2/X_2. The resistance R_2 includes the rotor winding resistance and the added resistance across the slip rings to adjust the speed. If the load torque is allowed to be constant,

$$\frac{2\alpha s}{\alpha^2 + s^2} = \text{constant} \triangleq K \tag{6.7.26}$$

since $\hat{T}_e$ is a constant no matter what R_2 is. This provides the quadratic equation

$$\left(\frac{s}{\alpha}\right)^2 - \frac{2}{K}\left(\frac{s}{\alpha}\right) + 1 = 0 \tag{6.7.27}$$

whose solution is

$$\frac{s}{\alpha} = -\frac{1}{K} \pm \left(\frac{1}{K^2} - 1\right)^{1/2} = \text{constant.} \tag{6.7.28}$$

This means that $s \propto \alpha$, that is, $s \propto R_2$, since the rotor leakage flux, inductance, and hence the leakage reactance at supply frequency are assumed to be constant. However it is not often that the load torque is constant and independent of speed, but this is the only case where a linear relationship exists.

Speed regulation is very much a function of the load torque and therefore accurate speed control is impossible without a feedback loop in the circuit to sense the speed, compare it with a given reference value and alter the resistance accordingly in a continuous manner. Figure 6.25 demonstrates this disadvantage. For a low value of rotor resistance, the torque-speed curve has a steep slope over the operating range. The speed is near the synchronous value and even if the load were to vary from zero to near-maximum torque (stalling torque or pull-out torque), the speed would not alter very much. As soon as resistance is added to the rotor circuit, the slope of the torque-speed curve decreases until it comes to the point when $R_2 \geq X_2$, where the same variation of load torque would produce such regulation that the speed could vary from the synchronous value down to zero speed. Not only is resistance used in the rotor circuit for speed control. Combinations of resistance, reactance, and saturable reactance have been used.

EXAMPLE 6.8

A 3-phase, 60-Hz, 6-pole, slip-ring motor has a rotor-circuit resistance of 0.1 ohm per phase and rotates at 1152 r/min at rated torque. Find the value of added rotor-circuit resistance to reduce the speed to 960 r/min, again at rated torque.

Solution
At 1152 r/min, slip $s_1 = (n_s - n_m)/n_s = (1200 - 1152)/1200 = 0.04$. At 960 r/min, slip $s_2 = (1200 - 960)/1200 = 0.2$. The torque equation is

$$T_e = \frac{I_2^2}{\omega_s} \frac{R_2}{s} = \frac{E_2^2}{\omega_s X_2} \frac{\alpha s}{\alpha^2 + s^2}.$$

Therefore, $\dfrac{T_1}{T_2} = \dfrac{\alpha_1 s_1}{\alpha_1^2 + s_1^2} \dfrac{\alpha_2^2 + s_2^2}{\alpha_2 s_2}$

since E_2, ω_s, and X_2 are constants. But $T_1 = T_2$.
Therefore, $\alpha_2 s_2(\alpha_1^2 + s_1^2) = \alpha_1 s_1(\alpha_2^2 + s_2^2)$.
That is, $5\alpha_2(\alpha_1^2 + 0.0016) = \alpha_1(\alpha_2^2 + 0.04)$.
Rearranging, $(5\alpha_1 - \alpha_2)(\alpha_1 \alpha_2 - 0.008) = 0$.
Consequently, $\alpha_2 = 5\alpha_1$, or $\alpha_1 \alpha_2 = 0.008$.
Take $\alpha_2 = 5\alpha_1$, and since X_2 is constant, then $(R_2 + R) = 5R_2$.
That is, $R = 5R_2 - R_2 = 4R_2 = 4 \times 0.1 = 0.4$ ohm/ph.

6.7.3. Induction Motor Braking

Especially in mills or for cranes it is often required to stop or reverse a motor quickly, that is, in a time shorter than if the supply were disconnected and the motor were free to come to rest through the frictional forces. Electromagnetic forces can be brought into play to oppose the mechanical inertia forces and slow the machine down quickly.

   Plugging is one way of braking and is also for speed reversal. This incorporates changing over two-phase connections to the supply to reverse the direction of rotation of the primary field in the air gap. The electromagnetic torque is reversed, brakes the motor until the speed is reduced to zero, and then drives the load in the opposite direction until a new equilibrium is reached. If only braking is required, then the supply to the motor has to be switched off before reversal can take place.

Slowing down a motor by regeneration can only take place in the special cases where pole changing or supply frequency changing can occur. Generation takes place only above synchronous speed and this is achieved by increasing the number of poles or decreasing the frequency. Once the new synchronous speed is reached there is no further electromagnetic braking.

To have braking over the complete speed range requires dynamic braking. A direct current is injected into the primary winding in the same way that the secondary is excited in a synchronous induction motor. The field no longer rotates in the air gap but has now a static sinusoidal pattern. The rotor acts as a generator just like a dc machine with the conductors rotating in a fixed field. The winding is shorted as in normal motoring operation, so that all the power generated is dissipated as I^2R loss in the rotor. The high current means a high retarding torque, but the slower the speed becomes, the smaller is the motional induced emf in the rotor, the smaller the current, and the less the braking effect. It is zero when the speed is zero. Resistance is usually added to the rotor circuit to limit the current to acceptable levels, so that there can be no excessive heating. This is the case of both the electrical and the mechanical power both being inputs to the machine. Electrical power is from the dc supply and the mechanical power is from the stored energy of the rotating parts. All this energy is converted to heat and special care must be taken in the design to cater for this.

6.7.4. Special Motors

Of all motors an induction machine is one of the least expensive to manufacture. It is not surprising, then, that there has been much attention given to the induction motor for special applications. There are translational motors (linear induction motors), Schrage motors (induction motors with a third winding and a commutator), synchros (for position measurement), servomotors (drag-cup and two-phase motors), and tachogenerators to name a few. In the following sections we will describe some of these special motors.

A. TWO-PHASE MOTOR

Two-phase motors are used as servo motors. They are used to control the speed of a load, whose value may be up to a few hundred watts. Their rotor windings have high resistance to give a wide speed range with change of torque (see subsection "Speed Control by Rotor Resistance," above) and high starting torque. One of the stator phases is connected to a fixed voltage, ac source, while the second phase is connected to an adjustable voltage source (controller). The frequency of the two sources is the same, but the voltages are out of phase by 90°.

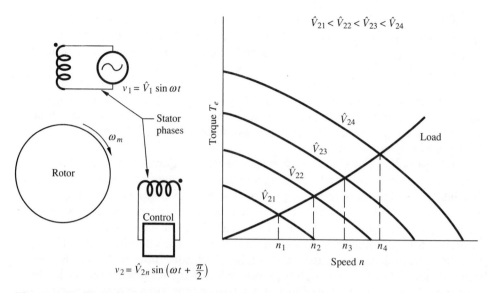

Figure 6.26   Two-phase induction motor and characteristics.

The two-phase supply enables the servo motor to have a traveling field in the air gap, so that a unidirectional driving torque is developed. By adjusting the phase-2 voltage, the torque-speed characteristic is changed. See Fig. 6.26. If the value of the phase-2 voltage is zero, the motor is designed to be at standstill. As the voltage (V_{21} to V_{24}) increases, the torque increases and the speed increases.

Speed reversal is obtained by reversing the phase of the motor's adjustable voltage from $90°$ lagging the phase-one voltage to $90°$ leading, or vice versa. Because of the high rotor winding resistance the speed range is large, but an associated disadvantage with high slip is low efficiency.

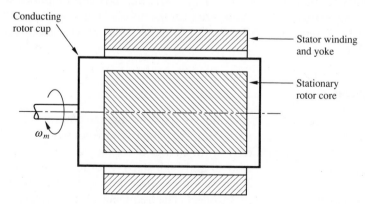

Figure 6.27   Cross section of a drag-cup motor.

To make it operate with low inertia and better mechanical response a refinement to the 2-phase servomotor has been to make only the rotor winding movable. The rotor magnetic core is made stationary and the winding is a single sheet made in the form of a cup, as depicted in Fig. 6.27. This machine is known as the *drag-cup motor*.

A machine can be made to either motor or generate as long as there is a magnetic field present. We have described the 2-phase induction machine as a motor. It can act equally well as a generator. If phase one of the two-phase stator winding is excited from a fixed-voltage, fixed-frequency supply and the rotor is driven mechanically, then the phase-two winding has a generated voltage appear at its terminals. This voltage alternates at the same frequency as the phase-one supply, and its magnitude is proportional to the speed of the rotor. Thus the machine can be used as an ac tachogenerator of fixed frequency.

B. LINEAR INDUCTION MOTOR

In Chapter 2 we read about translational-motion devices, but, in general, these were for incremental displacement. The linear induction motor is an example of a continuous motion translational machine that is suitable for such applications as conveyors, escalators, and liquid metal pumps.

There are many variations of the linear induction motor, but the principle is the same for each and is also the same as that of the conventional induction motor. Currents in a polyphase stator winding produce a magnetic traveling wave in the air gap of the machine. This magnetic field interacts with the rotor and, by induction, a torque is developed to drive the rotor in the direction of the traveling wave.

Figure 6.28 shows the cross section of a conventional induction motor, whose field and motion are cyclic. If we could open up this machine and lay it out flat, we arrive at what is called a single-sided, short-stator, linear induction motor, also shown

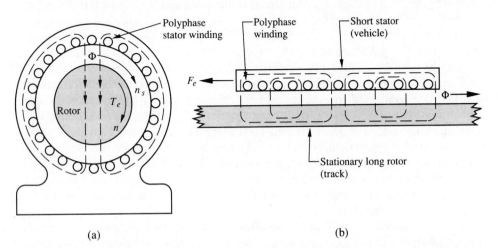

(a)   (b)

Figure 6.28 Induction motors, (a) rotational machine, (b) linear or translational machine.

in the figure. For convenience, the rotor is assumed to be solid iron. In the conventional machine the rotor moves. In the linear motor, the long rotor can be a stationary track and the short stator can move.

The linear induction motor depicted in Fig. 6.28b could be interpreted as a liquid metal pump if the stator is considered stationary and the rotor is considered to be the liquid metal (perhaps mercury or sodium) that is suitably encased. The force of electromagnetic origin will drive the liquid past the stator at some speed less than synchronous speed, depending on the resistance of the metal.

What can be done to make a linear induction motor can be done also to make a linear synchronous motor. In place of the induction motor rotor winding there is a dc-excited winding to produce a speed that is the constant synchronous speed.

C. SYNCHROS

Synchros can be classified as generators, motors, and transformers, although it is more common to think of them as transmitters and receivers, because their power handling capability is small. Synchros have other names such as selsyns and induction potentiometers. The terms *synchro* and *selsyn* (self-synchronous) are derived from the action of one machine following the action of another machine in synchronism. Their applications are found in machines such as hoists, elevators and airplanes, and their purpose is to transmit information about position to an indicator of an instrument some distance away, or to actuate a servosystem to bring about the motion of an object to a prescribed position.

These machines are not synchronous machines as described in Chapter 8. Neither are they induction machines as described in this chapter. They are more like actuators operating by induction as a function of position, and it is for this reason that they are sometimes called *induction potentiometers.*

Figure 6.29 shows the circuit diagram of two synchros. Information about the position of the shaft of the work machine is transmitted electrically by the generator and received by the motor, which develops a torque so that the motor shaft takes up the same angular position as the generator. In this way the motor can be some distance from the generator and work machine, and turn light loads such as dials to indicate the position of the work machine.

The two synchros are almost identical. There are 3-phase windings on the stators and these are connected together. There are salient poles on the rotors and their windings are connected to the same single-phase supply. This is to produce the magnetization for torque development. The difference is that a work machine positions the rotor of the generator and any electromagnetic torque developed in the generator is too small to move its rotor and the work machine, whereas any electromagnetic torque that is developed in the motor by the interaction of stator and rotor fields causes the motor shaft to move. The motor has a mechanical damper to minimize oscillations of the rotor.

For any particular position of the rotor of the synchro generator, the rotor alternating field will induce particular voltages in its stator windings. These same voltages will appear across the corresponding stator windings of the motor (Kirch-

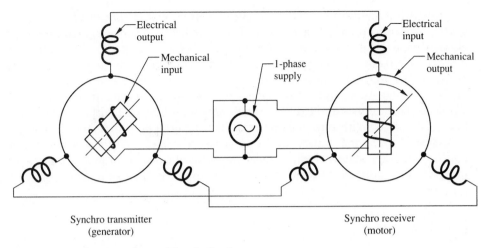

Figure 6.29 Synchros for position indication.

hoff's voltage law). If the motor rotor is in the same relative position as the generator rotor, the stator winding voltages of the motor balance the generator's voltages exactly, no currents are in the stator windings and no torque is developed to move the motor rotor. That is, the motor rotor is in the equilibrium position, and there is balance in the two machines by symmetry.

However, if the rotor of the motor is not in this equilibrium position, the stator-winding voltages that are induced in the corresponding phases of the generator and the motor are not the same. The result is the existence of stator winding currents that interact with the rotor fields to produce a small torque. The torque tends to align the fields of stator and rotor. Since the generator rotor will not move, the motor rotor must move until there is alignment. At this point the two rotors have the same angular position, there is no current in the stator windings and there is no torque to produce further movement. So an equilibrium position has been reached, whereby the synchro motor indicates the same position as the synchro generator. When the generator rotor is moved by the work machine the rotor of the motor follows.

There are other members of the synchro family that are described in detail in a reference in the bibliography at the end of this chapter. They are the differential synchro and the synchro transformer. The differential synchro generates an electrical output signal that is the sum or difference of two input signals (mechanical and electrical signals). Three-phase windings are employed on both the stator and the rotor. For example, a differential synchro could be placed between a synchro transmitter and a synchro receiver. The differential synchro receives an electrical signal from the transmitter to its stator windings plus an external mechanical signal in the form of rotor position. An electrical signal output from its rotor windings to the receiver is then the sum or difference of the two signals, which can be considered to be a position signal plus a correction signal. The receiver responds to the sum or difference of the two.

Synchro transformers are like synchro receivers except that there is no phase source connected to the rotor winding. The transformer has two inputs. One is the electrical signal from a synchro transmitter to indicate a required position of a load, and the other is a mechanical feedback signal corresponding to the actual load position. The induced voltage (transformer action) in the rotor winding is a function of the difference between the two signals and represents an error signal. This error signal can be amplified and fed to a 2-phase servomotor, whose purpose is to reduce the error to zero. With zero position error, the required position is the same as the actual position, so the transformer rotor position is such that there is no mutual coupling with the stator winding. This is the equilibrium position and there is no induced voltage in the rotor to excite the servomotor. The load remains in this position until the synchro transmitter has its input changed.

D. SCHRAGE MOTOR

The Schrage motor, whose name is taken after the inventor, was patented in 1911. It is primarily of historic interest to describe the machine here. Adjustable-speed induction motors are usually driven by variable frequency semiconductor inverters today. However, a wide range of speed (both sub- and supersynchronous speeds) can be obtained from a conventional constant frequency supply and a Schrage motor. The principle involved is speed adjustment by rotor injected voltage. See subsection D under section 6.7.2 and Fig. 6.24.

The Schrage motor is a special case of the induction motor with an ac commutator machine combined to produce the injected voltage at the correct frequency and phase to regulate the speed. There is also the possibility of adjusting the power factor as well. The primary winding is wound on the rotor and the secondary winding is on the stator. This is an inversion of the conventional induction motor and requires the supply to be connected to the primary (rotor) winding through slip rings. If the secondary winding had its terminals short-circuited, the machine would function as an ordinary induction motor. The only difference would be that the magnetic field in the air gap rotates at slip frequency, although it travels at synchronous speed relative to the primary conductors.

In addition to the primary and secondary windings, there is a third and separate winding which is wound on the rotor in the same slots as the primary winding. This is the regulator winding, whose coil connections are brought out to a commutator in order that some fraction of the voltage that is induced in the regulator winding may be injected into the secondary (stator) winding. Because the regulator winding is in the same slots as the primary winding, transformer action between the primary and regulator winding induces an emf in the conductors of the latter at the supply frequency. The value of the induced emf depends on the effective turns ratio of the windings. Although the emf of the regulator winding is alternating at the supply frequency, the voltage picked off at the commutator by the stationary brushes is at slip frequency. The secondary winding induced emf is also at slip frequency. Therefore, the commutator winding can be connected to the secondary winding so that a voltage from the former winding can be injected into the latter. The brushes that are connected to the terminals of each phase of the secondary winding can slide past each other on the commutator. Thus the voltage that is injected into the secondary winding can be made to aid or oppose the voltage generated in that winding. If the voltage aids, the motor speeds up, and if the voltage opposes, the motor slows down.

The simple equation

$$n_o = (1 \pm k)n_s \qquad (6.7.23)$$

approximately gives the no-load speed in terms of the synchronous speed and the ratio k of the regulator voltage to the secondary induced emf at standstill.

6.8. SUMMARY

The induction motor is the most common electric drive that is to be found. This is because it is relatively inexpensive to manufacture and it runs at virtually constant speed.

The operation of the machine depends on the rotating magnetic field that is set up by the currents in the stator winding. Induced rotor-winding currents interact with the air-gap field to produce a constant driving torque while steady-state conditions prevail. There is a starting torque, but there is no torque of electromagnetic origin at synchronous speed. Synchronous speed n_s is defined as the speed of the rotating magnetic field (f/p).

Slip is an expression to describe by how much the motor speed n is below synchronous speed. Slip $s = (n_s - n)/n_s$.

The induced emf E_1 in the stator phase winding is the same as the induced emf in a winding of a transformer, that is

$E_1 = 4.44\Phi f N_1$.

The induced emf E_{2s} in the rotor phase winding differs by the number of turns and the slip. That is

$E_{2s} = 4.44\Phi s f N_2$.

The rotor winding currents alternate at a frequency $f_2 = sf$.

Equivalent circuits of the induction motor have been developed. These circuits account for leakage flux, winding resistances, core and stray losses, magnetization, speed or slip, and the mechanical load. The use of these models provides a relatively easy way by which the performance of an induction motor can be determined. For given values of speed, applied voltage and supply frequency it is possible to calculate the load power, the losses, the input power, efficiency, current, and power factor.

The parameter values of the equivalent circuit are determined from the No-load Test and the Locked-rotor Test. These tests are similar to the Open-circuit Test and Short-circuit Test of the transformer. Parameters R_1, X_1, R_m, X_m, R_2, and X_2 constitute the main values that are determined by these tests. They are assumed to be constant. The leakage reactances X_1 and X_2 relate to the supply frequency f.

A useful relationship for the gap power P_1, the rotor winding ohmic loss P_2 and the output power P_m (on a per-phase basis) is $P_1 : P_2 : P_m = 1 : s : (1 - s)$.

The torque expression is $T_e = P_m / \omega_m$.

The induction machine can be used as a motor and a brake. If the speed is above synchronous speed, the machine can run as a generator, but it must be connected to an external ac supply. The starting torque is best improved if the supply frequency is reduced or resistance is added to the rotor circuit. An induction motor draws high current from the supply at starting. Means to limit this current are to lower the terminal voltage or add resistance to the rotor circuit during the starting period.

Speed control of induction motor drives is an important subject in continual development. There are relatively simple methods of speed adjustment by terminal voltage control and adding resistance to the rotor circuit. There are less simple methods such as pole changing and injecting voltage in the rotor circuit. And there are sophisticated methods using power-electronic controllers that provide both frequency and voltage adjustment, often by the technique of Pulse Width Modulation.

6.9. PROBLEMS

Section 6.2

6.1. A 4-pole induction motor has a 2-mm air gap across which all the magnetic potential drop occurs. Idealizing the windings to be represented by current sheets that have sinusoidal current distributions on the air-gap surfaces, the maximum density of the stator current is 1400 A per mech. radian, while for the rotor current it is 1000 A per mech. radian. The rotor current sheet is displaced by $\pi/12$ mech. radians from the position of total opposition to the stator current sheet in producing mmf across the air gap. Calculate the maximum value of flux density in the air gap.
   (*Answer:* 0.23 tesla.)

6.2. An m-phase induction motor has the axes of the m balanced stator windings displaced in space by $2\pi/m$ electrical radians. The balanced currents in the m windings are phase displaced in time by $2\pi/m$ electrical degrees and vary sinusoidally with time. Show that the resultant mmf pattern is a traveling field of constant amplitude $m/2$ times the maximum mmf of one phase and rotating with a speed of ω electrical radians per second.

Section 6.3

6.3. A 6-pole, polyphase, wound rotor induction motor is driven mechanically by a synchronous motor which is connected to the same 60-Hz supply and runs synchronously. If 120-Hz voltages are available at the slip rings of the induc-

tion motor, find **(a)** the rotor speed, **(b)** the direction of rotor rotation and **(c)** the number of pole pairs of the synchronous motor.
(*Answer:* (a) 1200 r/min, (b) backward, (c) 3.)

6.4. A polyphase induction motor has the stator winding arranged to accommodate 12 poles. If the motor is loaded so that the rotor speed is 59.66 mech. rad/s, estimate the frequency of the rotor winding current. The supply frequency is 60 Hz. For this same case calculate the slip.
(*Answer:* 3 Hz, 0.05 per unit.)

6.5. A 3-phase induction motor is supplied with power from a 60-Hz source. At full load the motor speed is 1728 r/min while at no load the speed is nearly 1800 r/min. Determine **(a)** the number of pole pairs for which the motor was wound, **(b)** the full-load slip, **(c)** the frequency of the rotor voltages at full load, **(d)** the speed of the rotor field with respect to the rotor at full load, **(e)** the actual speed of the rotor field in space at full load, **(f)** the speed of the rotor field with respect to the stator field at full load, and **(g)** the speed of the rotor at a slip of 0.08 per unit.
(*Answer:* (a) 2, (b) 0.04, (c) 2.4 Hz, (d) 72 r/min, (e) 1800 r/min, (f) 0, (g) 1656 r/min.)

Section 6.4

6.6. A 3-phase, 550-V, 60-Hz, 6-pole, Y-connected induction motor has a 3-phase wound rotor winding brought out to slip rings. The effective turns ratio is 1 : 3. Consider the rotor circuit open. **(a)** Calculate the voltage and its frequency across each pair of slip rings if the rotor shaft is driven at 1100 r/min in the direction of the air-gap field. **(b)** At what speed must the rotor be driven for the rotor frequency to be 90 Hz? **(c)** If the rotor is driven at 1200 r/min in the direction of the air-gap field, determine the magnitude and frequency of the rotor-circuit line voltage. **(d)** If the rotor is driven at 1200 r/min in the direction opposite to the air-gap field, determine the magnitude and frequency of the rotor circuit line voltage.
(*Answer:* (a) 138 V, 5 Hz, (b) 3000 or −600 r/min, (c) 0, 0, (d) 3300 V, 120 Hz.)

Section 6.6

6.7. A 3-phase, 550-V, 60-Hz, 8-pole induction motor develops a rated output power of 40 kW when rotating at 888 r/min, the input power factor being 0.86. Determine **(a)** the slip, **(b)** the ohmic losses of the rotor circuit, **(c)** the total input power, if the total stator losses are 2 kW, **(d)** the input line current, and **(e)** the rotor voltage frequency.
(*Answer:* (a) 1.3%, (b) 540 W, (c) 43.6 kW, (d) 53.2 A, (e) 0.78 Hz.)

6.8. The rotor of a polyphase, 440-V, 60-Hz, 8-pole induction motor can be driven at an adjustable speed to provide a variable frequency at the slip rings. **(i)(a)** At what speed must the rotor be driven to give a frequency of 240 Hz at the slip rings? **(b)** The stator and rotor have the same number of phases and the effective turns ratio is 4 : 1. What is the value of the rotor-winding voltage on open circuit at a frequency of 240 Hz? **(c)** At 240 Hz what are the per unit powers supplied by the 60-Hz source at the stator and by the shaft drive, if all losses are ignored? **(ii)** Repeat (i) if the frequency of the voltage at the slip rings is 30 instead of 240 Hz.

(*Answer:* (i)(a) 75 r/s forward or 45 r/s backward, (b) 440 V, (c) −0.25 p.u. and 1.25 p.u., or 0.25 p.u. and 0.75 p.u. (ii)(a) 22.5 r/s forward or 7.5 r/s backward, (b) 55 V, (c) −2 p.u. and 3 p.u., or 2 p.u., and −1 p.u.)

6.9. A 3-phase, 440-V, Y-connected, 60-Hz, 6-pole induction motor has a stator-winding standstill leakage impedance of $(0.6 + j1.6)$ ohms per phase and a rotor-winding leakage impedance of $(0.5 + j1.6)$ ohms per phase referred to the stator. Neglect the magnetizing impedance and calculate the starting torque.

(*Answer:* 67.2 N·m.)

6.10. A 3-phase, 400-V, 60-Hz, 4-pole, star-connected induction motor has a slip of 5% at full load. Induction motor tests provided the following data.
$R_1 = 0.64$ Ω, $R_2' = 0.68$ Ω, $X_1 + X_2' = 2.3$ Ω, and $Y_0 = (0.005 − j0.02)$S.
The motor has a wound-rotor winding and is to start with maximum torque. What are the referred values of the external rotor starting resistance, the starting current, and its ratio to full-load current? Establish the value of the maximum torque. If the minimum torque during run-up is not to fall below 45 N·m, at what speed does this occur naturally, and what alteration should be made to the external rotor resistance at this speed?

(*Answer:* 1.72 ohms, 64.1 A, 3.47, 139 N·m, 1548 r/min.)

6.11. A polyphase induction motor develops its rated full-load output of 400 kW at a slip of 1.5%, if operated at rated voltage and frequency with its rotor windings short-circuited. The effective turns ratio is 1 : 1. The rotor-circuit resistance is increased to five times the rotor-winding resistance by connecting noninductive resistors in series with each rotor slip ring. **(a)** Establish the slip at which the motor will develop the same full-load torque. **(b)** Estimate the increase in total rotor-circuit ohmic losses at full-load torque. **(c)** Calculate the power output at full-load torque. **(d)** With the aid of answers to parts (b) and (c), derive the ohmic loss of the short-circuited rotor winding at full-load torque. **(e)** Find the slip at which maximum torque occurs in terms of the slip at maximum torque with the rotor winding short-circuited. **(f)** What is the rotor current, in terms of the rotor current with a short-circuited rotor winding, at maximum torque? **(g)** At what slip will the short-circuited rotor winding develop a torque equal to the starting torque for a total rotor-circuit resistance of $5R_2$?

(*Answer:* (a) 7.5%, (b) five-fold increase, (c) 375.6 kW, (d) 6.1 kW, (e) five times, (f) same, (g) 0.2.)

6.12. An induction motor is reconnected from a 3-phase, 440-V, 60-Hz supply to a 3-phase, 440-V, 50-Hz supply. For the same full-load torque operation determine **(a)** the change of flux per pole, **(b)** the change of rotor current, and **(c)** the change of synchronous speed.

(*Answer:* (a) 20%, (b) 17%, (c) 17%.)

6.13. A polyphase induction motor has a total leakage reactance per phase $(X_1 + X_2')$ equal to twice the total resistance per phase $(R_1 + R_2')$. The rotor resistance per phase referred to the stator is equal to the stator resistance per phase. At what slip does maximum torque occur **(a)** accounting for the stator winding leakage impedance and **(b)** neglecting the stator leakage impedance?
(*Answer:* (a) 0.24, (b) 0.5.)

6.14. A polyphase induction motor operates at a slip of 0.03 per unit on full load. The slip at *pull–out* is 0.1 per unit. Determine the torque and rotor-circuit current at starting as a ratio of the full-load values, if the maximum torque is 3 per unit.
(*Answer:* 0.36, 3.5.)

6.15. A polyphase, 60-Hz, 6-pole induction motor develops a maximum torque of 100 N·m at 1080 r/min, the resistance of the rotor circuit being 0.2 ohm per phase. With the information given calculate an approximate value of the torque at a slip of 0.04 per unit. What added rotor-circuit resistance is needed to give two thirds of maximum torque at starting?
(*Answer:* 69 N·m, 5.04 or 0.56 ohm.)

6.16. Determine the ohmic loss in the rotor circuit of an induction motor, rotating at 60% of synchronous speed with a useful output of 40 kW and mechanical losses that amount to 1.5 kW. If the total stator losses amount to 3 kW, estimate the motor efficiency at this operating speed.
(*Answer:* 27.7 kW, 55.4%.)

6.17. An induction motor has an efficiency of 0.9 per unit when the load is 4000 kW. The stator and rotor ohmic losses each equals the core loss. The mechanical losses are one-third of the no-load losses. Estimate the slip.
(*Answer:* 0.03.)

6.18. A 440-V, 60-Hz, 3-phase, Y-connected induction motor has 6 poles. The locked-rotor and no-load tests give the following data. No-load current $I_{NL} = $ 20 A at 0.1 lagging power factor. Rotational losses plus stray load losses equal 2.85 kW. $R_1 = 0.1\ \Omega$, $R_2' = 0.1\ \Omega$, $X_1 = X_2' = 0.3\ \Omega$. At full load the slip is 0.03 per unit. Determine **(a)** the line current and power factor, **(b)** the torque, **(c)** the rated power output, and **(d)** the total losses, the ohmic losses, the core losses, and the efficiency of the machine at this slip. Find also the starting current and starting torque.
(*Answer:* (a) 69 A, 0.91, (b) 360 N·m, (c) 41 kW, (d) 6.75 kW, 2.48 kW, 1.41 kW, 0.96 per unit.)

6.19. A 3-phase, 440-V, 60-Hz, Y-connected, 4-pole induction motor has the following parameters, that are referred to the stator on a per phase base.
$R_1 = 0.1$ ohm, $R_2' = 0.15$ ohm, $X_1 = 0.4$ ohm, $X_2' = 0.4$ ohm.
The stator core loss is 1500 W, the rotational losses amount to 1000 W, and the no-load current is 20 A at a power factor 0.1 lagging. At a slip of 0.05 per unit and, using the approximate equivalent circuit, determine **(a)** the input current, **(b)** the input power factor, **(c)** the total electromagnetic torque, and **(d)** the motor efficiency.
(*Answer:* (a) 88.3 A, (b) 0.89 lagging, (c) 300 N·m, (d) 87%.)

6.20. A 3-phase, 4-pole, 220-V, 60-Hz, Y-connected induction motor has the following parameters referred to the stator.
$R_1 = 0.3$ ohm, $R_2' = 0.2$ ohm, $X_1 = 0.5$ ohm, $X_2' = 0.5$ ohm, $X_m = 15$ ohms. For a slip of 4% calculate **(a)** the motor speed, **(b)** the input current, **(c)** the input power factor, **(d)** the shaft torque, and **(e)** the motor efficiency. The total rotational and core loss is 500 W. For the calculations use the approximate equivalent circuit of the induction motor and neglect the core-loss circuit parameter R_m.
(*Answer:* (a) 1728 r/min, (b) 24.9 A, (c) 0.89, (d) 39 N·m, (e) 83.7%.)

6.21. A regenerative system is used to load test an induction machine. The induction motor is mechanically coupled via a gear box to an induction generator so that the generator always runs above synchronous speed. The output of the generator is connected directly to the same supply as the motor. At the beginning of the test, the source supplied 1000 kW, while the motor was running at 96% efficiency with a shaft load of 4000 kW. What is **(a)** the slip of the motor, **(b)** the efficiency of the generator, **(c)** the gear box ratio, **(d)** the ratio of rotor circuit resistances of the two machines? At the end of the test the temperature rise has increased the motor resistance of the rotor circuit by 20%. In order that the motor shaft power remains the same what are **(e)** the motor slip, **(f)** the generator efficiency, **(g)** the losses provided by the supply, and **(h)** the necessary decrease of resistance in the generator rotor circuit? Assumptions that can be made are that the gear box efficiency is constant at 0.97 per unit, all losses may be ignored except those ohmic losses in the rotor windings, and $R_2' \gg (X_1 + X_2')s$.
(*Answer:* (a) 0.04 per unit, (b) 0.819 per unit, (c) 1.27, (d) 7.25, (e) 0.048 per unit, (f) 0.826 per unit, (g) 997 kW, (h) 5% reduction.)

6.22. A 3-phase, Y-connected induction motor is operating at full load. If one line of the supply is opened to give unbalanced voltages, would the motor continue to operate? Can the induction motor start with one supply line open?
(*Answer:* yes, no.)

6.23. A polyphase, 24-pole induction motor has a cage rotor. There are 151 rotor bars each carrying an effective current of 500 A. Calculate the total loss in the end rings of the cage, if an end ring has a mean diameter of 0.5 m and a cross-sectional area of 3.142×10^{-4} m². The resistivity of the ring material is $10^{-1} \mu\Omega$-m.
(*Answer:* 1 kW.)

6.24. A 3-phase, 220-V, 60-Hz, 6-pole induction motor develops a full-load output of 11.94 kW. The stator circuit is delta-connected and has a winding factor of 0.96. Assume a mean flux density of 0.6 tesla and specific electric loading of 18.33 kA/m. Calculate the main dimensions of the rotor and the number of stator conductors, if the full-load line current is 30 A and the core length is to be the same as the pole pitch.
(*Answer:* $D = 0.21$ m, $L = 0.11$ m, $Z = 716$.)

6.25. A 3-phase, V-volt, f-Hz induction motor is connected direct-on to a 3-phase, V'-volt, f'-Hz supply. Calculate **(a)** the ratio of starting currents, starting torques, and pull-out torques for the two frequencies f and f' if V' equals V and **(b)** the ratios of V' to V to give the same values of starting current and torque at frequencies f' and f. Assume that the stator circuit impedance can be ignored.

$$\left(Answer:\ (a)\ \frac{f'}{f},\ \left(\frac{f'}{f}\right)^3,\ \left(\frac{f'}{f}\right)^2,\ (b)\ \frac{f'}{f},\ \left(\frac{f'}{f}\right)^{3/2}. \right)$$

6.26. A 3-phase, 25-Hz, 2400-V, Y-connected induction motor has a slip of 2% when delivering full-load output of 400 kW. **(a)** Compute the full-load torque. **(b)** If this motor develops three quarters of full-load torque when starting at full voltage, determine the starting torque if the motor is started with an autotransformer whose effective turns ratio is $2:1$. **(c)** If the rotor winding is Y-connected and the rotor-winding resistance is 0.2 ohms per phase referred to the stator, what resistance must be added externally to the rotor circuit in order to limit the starting current to its rated value? The effective turns ratio is $1:1$. **(d)** Determine the value of the starting torque if the total resistance of part **(c)** were used. **(e)** What is the slip at full load if half the resistance of part (c) is left in the rotor circuit?
(*Answer:*   (a) 1563 N·m,   (b) 293 N·m,   (c) 9.8 Ω/ph,   (d) 78,200 N·m, (e) 51%.)

6.27. A 3-phase, 60-Hz, 8-pole, 440-V, Y-connected, wound-rotor induction motor has the following circuit parameters per phase referred to the stator: $R'_2 = 0.4$ ohm and $X'_2 = 1.1$ ohms. The effective turns ratio is $1:1$ and the rotor winding has three phases. The stator-circuit leakage impedance can be neglected and the core and windage and friction losses can be ignored. **(a)** In order to produce maximum torque at starting, what is the value of resistance to be added to each rotor phase? **(b)** If the load has a linear characteristic of 12 N·m per rad/s and the added rotor resistance remains in the circuit, what are the motor speed and efficiency?
(*Answer:* (a) 0.7 ohm, (b) 527 r/min, 58.6%.)

6.28. A 10-pole, 3-phase, wound-rotor induction motor has a 50-Hz supply connected to the stator terminals and a 20-Hz supply connected to the slip-ring terminals. Determine the two rotor speeds for steady operation to be possible.
(*Answer:* 6 r/s, 14 r/s.)

6.29. A 3-phase induction motor has resistance connected to the slip rings in order to adjust the speed. Show that

$$R^2 + \left(2R_1 - \frac{sV_1^2}{P_2 + P_r} \right) sR + s^2(R_1^2 + X^2) = 0$$

is the equation from which the value of the regulating resistance can be found for any slip s. R is the rotor-circuit and regulator resistance referred to the stator winding. P_r is the regulator ohmic loss at slip s. $X = X_1 + X'_2$. Phase values are used throughout. The approximate equivalent circuit may be used.

6.10. BIBLIOGRAPHY

Adkins, Bernard, and W. J. Gibbs. *Polyphase Commutator Machines.* Cambridge University Press, 1951.

Alger, P. L. *Induction Machines.* 2nd ed. New York: Gordon and Breach, 1970.

Bewley, L. V. *Alternating Current Machinery.* New York: Macmillan Co., 1949.

Bose, B. K. *Power Electronics and AC Drives.* Englewood Cliffs, N.J.: Prentice-Hall, 1986.

Bose, B. K. *Microcomputer Control of Power Electronics and Drives.* New York: IEEE Press, 1987.

Carr, Laurence. *The Testing of Electrical Machines.* London: MacDonald Co., Ltd., 1960.

Chalmers, B. J. *Electromagnetic Problems of AC Machines.* London: Chapman and Hall, Ltd., 1965.

Dewan, S. B., G. Slemon, and A. Straughen. *Power Semiconductor Drives.* New York: Wiley-Interscience, 1984.

Fransua, A., and R. Magureanu. *Electrical Machines and Drive Systems.* Oxford: Technical Press, 1984.

Garik, M. L., and C. C. Whipple. *Alternating-Current Machines.* 2nd ed. New York: D. Van Nostrand, 1961.

Hindmarsh, J. *Electrical Machines and Drives: Worked Examples.* 2nd ed. Oxford: Pergamon Press, 1985.

Laithwaite, E. R., (Ed.). *Transport Without Wheels.* Boulder, Colo.: Westview Press, Inc., 1977.

Laithwaite, E. R. *A History of Linear Electric Motors.* London: Macmillan Educational, Ltd., 1987.

Lawrence, R. R., and H. E. Richard. *Principles of Alternating-Current Machinery.* 4th ed. New York: McGraw-Hill Book Co., 1953.

Leonhard, W. *Control of Electric Drives.* Berlin: Springer-Verlag, 1985.

Nasar, S. A., and Boldea, I. *Linear Motion Electric Machines.* New York: Wiley-Interscience, 1976.

Nasar, S. A. *Electric Machines and Electromechanics.* Schaum's Outline Series in Engineering. New York: McGraw-Hill Book Company, 1981.

Still, Alfred, and Charles S. Siskind. *Elements of Electrical Machine Design.* Tokyo: McGraw-Hill Book Co., 1954.

Upson, A. R., and J. H. Batchelor. *Synchro Engineering Handbook.* London: Hutchinson, 1965.

Van Valkenburgh, Nooger, and Neville, Inc. *Basic Synchros and Servomechanisms.* New York: John F. Rider Publisher, Inc., 1955.

―――― *Test Procedures for Polyphase Induction Motors and Generators.* IEEE Standard 112, 1984.

CHAPTER SEVEN

Single-Phase Induction Motors

7.1. INTRODUCTION

The single-phase induction motor usually has a distributed stator winding and a squirrel-cage rotor. The simplicity of construction results in a reliable and inexpensive motor, eminently suitable for the many fractional-horsepower drive requirements in the home, plant, office, farm, etc. In many cases, because of mass production techniques, the single-phase motors are of the three-phase type with one winding omitted or left idle. In other cases, such as for dual-speed operation, they are designed specially with two stator windings. Such is the thought given to minimizing the cost of production, that the performance ratings are very precise. There is no latitude for overloading.

Since single-phase motors are used in great numbers, it is important to describe their operation. There are two main theories. One is the cross-field theory, which is briefly described here. The other is the revolving-field theory. This latter theory is explained in detail here, because the theory of polyphase machines can be employed to develop an equivalent circuit, from which the machine performance is predicted.

Single-phase induction motors suffer the disadvantage that they are not inherently self-starting. A number of solutions to this problem are described in the following sections.

7.2. CROSS-FIELD THEORY

As its name implies, the single-phase induction motor has one winding on the stator. The rotor usually has a cage winding, as shown in Fig. 7.1a. A circuit representation

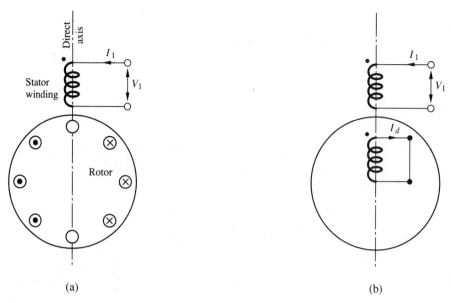

Figure 7.1 Single-phase motor. (a) Cage motor, (b) circuit representation at standstill.

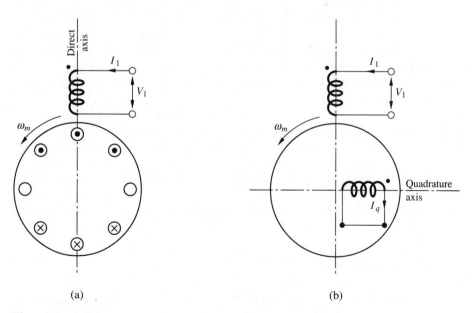

Figure 7.2 Motional emf. (a) Cage motor, (b) circuit representation.

of the motor is depicted in Fig. 7.1b and it is from this that the theory of cross fields will be explained.

Consider the rotor to be stationary. The stator-winding current I_1 produces a flux pattern, whose axis is fixed in space on the direct axis, and whose magnitude varies periodically with time. Consequently, the machine appears to be like a transformer, whose secondary winding is short-circuited. The rotor-winding currents produce a magnetic field, which is also fixed in space, and pulsates with time along the direct axis. The axes of the fields of the two windings are aligned. Thus there is no starting torque.

Now consider the rotor to be moving as illustrated in Fig. 7.2a. The rotor conductors will be moving in the stator field, so that there are motional emfs generated. Flux produced by the resulting rotor-circuit currents will be directed in the quadrature axis, just like the armature reaction of the dc machine, except that the flux pulsates in this case. Although the rotor conductors are moving, the rotor winding can be represented as being fixed on the quadrature axis, as shown in Fig. 7.2b. Since the stator and rotor-winding fields are in quadrature, maximum torque in the direction of rotation is the result. Once started, the single-phase induction motor continues to rotate.

Even though there may be rotation, the rotor winding continues to have emfs induced by transformer action. Thus the single-phase induction motor can be properly represented by three coils whose axes are stationary. This is shown in Fig. 7.3. It can be seen that, in addition to the torque produced by the fields of the stator coil and the quadrature axis rotor coil, there is torque due to the interaction of the rotor

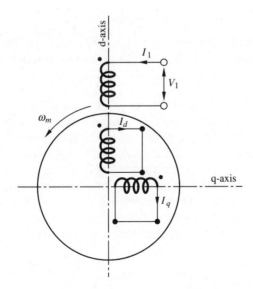

Figure 7.3 Cross-field circuit representation.

conductor *transformer* currents and the rotor-circuit quadrature-axis flux. This concept is developed further from basic principles in Chapter 10.

7.3. DOUBLE REVOLVING FIELD THEORY

The polyphase induction motor theory was developed from the concept of a magnetic field, whose pattern could be described as a rotating wave. A general expression[1] for a rotating wave of flux ϕ_f varying sinusoidally in space and time is

$$\phi_f = \frac{\hat{\Phi}}{2} \sin(\theta + \omega t). \tag{7.3.1}$$

Let this be a clockwise traveling wave. The same wave rotating in the opposite direction is

$$\phi_b = \frac{\hat{\Phi}}{2} \sin(\theta - \omega t). \tag{7.3.2}$$

If both the forward (ϕ_f) and backward (ϕ_b) waves were present in the air gap of an induction machine, assuming no saturation, the resultant field pattern would be expressed by

$$\phi = \phi_f + \phi_b = \frac{\hat{\Phi}}{2} [\sin(\theta + \omega t) + \sin(\theta - \omega t)]. \tag{7.3.3}$$

That is

$$\phi = \hat{\Phi} \sin\theta \cos\omega t. \tag{7.3.4}$$

Since the variables of space and time have been separated, the resultant flux ϕ is a standing wave. That is, there is a fixed distribution in space, but the amplitude is varying with time. This is the flux distribution set up by the stator circuit mmf of a one-phase machine. Consequently, the single-phase induction motor can be analyzed as a polyphase machine with a forward-rotating field ϕ_f, then as a polyphase machine with a backward-rotating field ϕ_b, and the results can be superimposed. The pioneers of this theory of single-phase induction motors were Ferraris and Behrend.

Since the stator circuit current does pulsate for all conditions, it would be correct to consider two equal but oppositely revolving mmf waves. This leads to the concept of replacing the single-phase machine by two identical, but opposing polyphase machines on the same shaft. This is shown in Fig. 7.4. For the currents to be the same, and for the mmf waves to rotate in opposite directions, the stator windings are in series in phase one, and in series opposition in phase two.

---

[1] See Appendix F, eq. (F4.14).

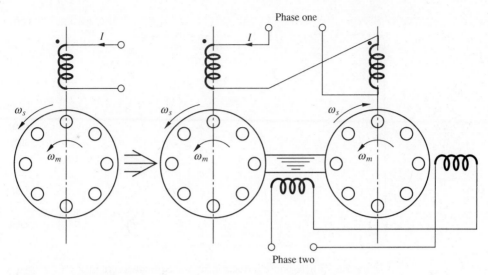

Figure 7.4 A single-phase motor equivalent to two polyphase motors.

7.3.1. Equivalent Circuit

With the polyphase motor equivalent circuit kept in mind, there is the applied voltage V_1, the stator-circuit leakage impedance $(R_1 + jX_1)$, and the input current I_1. Let I_{NLf} produce the forward and I_{NLb} produce the backward-rotating fields of amplitude $\hat{\Phi}/2$. The induced emf in the stator winding due to the forward field is $E_{1f} = E_1/2$, where $E_1 = 4.44\Phi f N_1$, and N_1 is the effective number of turns of the stator winding. The stator winding emf due to the backward field is $E_{1b} = E_1/2$. At standstill, the rotor circuit–induced emfs per phase are

$$E_{2f} = E_{2b} = 4.44 \frac{\Phi}{2} f N_2 = \frac{E_2}{2}. \tag{7.3.5}$$

Note that Φ is the total flux per pole in the air gap due to the magnetizing component of the input current in the actual machine.

It is assumed that the rotor has a speed of n_m and that the synchronous speed is n_s. Hence the forward slip is

$$s = \frac{n_s - n_m}{n_s} = 1 - \frac{n_m}{n_s} \tag{7.3.6}$$

and the backward slip is

$$s_b = \frac{n_s + n_m}{n_s} = (2 - s) \triangleq \text{the plugging slip.} \tag{7.3.7}$$

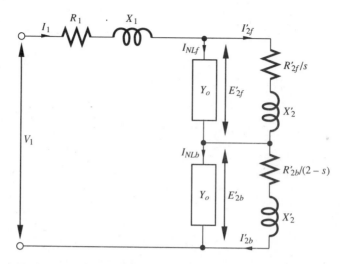

Figure 7.5   Single-phase induction motor equivalent circuit.

The rotor-circuit current set up by the forward rotating field can be expressed by

$$\mathbf{I}_{2f} = \frac{s\mathbf{E}_2}{2} \frac{1}{R_2 + jsX_2}. \tag{7.3.8}$$

This is a low-frequency (sf) current. Similarly, the rotor circuit current due to the backward-rotating field is

$$\mathbf{I}_{2b} = \frac{(2-s)\mathbf{E}_2}{2} \frac{1}{R_2 + j(2-s)X_2}. \tag{7.3.9}$$

and has a frequency almost double that of the applied voltage V_1. The rotor circuit voltage equations referred to the stator, using the approximate turns ratio, are

$$\mathbf{E}'_{2f} = (R'_{2f}/s + jX'_2)\mathbf{I}'_{2f} \tag{7.3.10}$$

and

$$\mathbf{E}'_{2b} = (R'_{2b}/(2-s) + jX'_2)\mathbf{I}'_{2b}. \tag{7.3.11}$$

The rotor circuit resistance R_{2b} is greater than the resistance R_{2f} because of the higher-frequency skin effect.

The above equations are satisfied by the circuit configuration shown in Fig. 7.5, and it is from such a circuit that the motor performance can be estimated.

7.3.2. Parameter Measurement

To determine the parameters of the equivalent circuit the same tests that are done on the polyphase machine can be carried out on the single-phase motor. These tests are the locked-rotor and the light-load tests. The parameters are calculated from voltage, current, and power measurements.

From the locked-rotor test ($s = 1$) and with reference to Fig. 7.5

$$\mathbf{V}_1 \approx [R_1 + (R'_{2f} + R'_{2b}) + j(X_1 + 2X'_2)]\mathbf{I}_1 \tag{7.3.12}$$

since the rotor circuit impedance is low and the magnetizing impedances in the equivalent circuit can be ignored. The value of R_1 can be obtained from a separate measurement. Then R'_{2f} and R'_{2b} follow. However these values are at the supply frequency f, so some correction must be made for the different frequencies. The actual value of R_{2b} may be about 50% greater than R_{2f}. We cannot separate X_1 and X'_2 from the tests, so they are assumed equal.

For the light-load test ($s \approx 0$) the equivalent circuit is simplified since R'_{2f}/s is infinite and $(R'_{2b}/2 + jX'_2)$ is small compared with Y_0. Consequently

$$\mathbf{V}_1 = (R_1 + jX_1 + 1/Y_0 + R'_{2b}/2 + jX'_2)\mathbf{I}_1. \tag{7.3.13}$$

This equation is further simplified to give

$$\mathbf{V}_1 \approx \frac{\mathbf{I}_1}{\mathbf{Y}_0}. \tag{7.3.14}$$

Hence the magnetizing admittance and its components can be determined.

7.3.3. Motor Performance

The equivalent circuit depicted in Fig. 7.5 has to be solved for the currents in order to predict the motor performance. The gap power or power input to the rotor is

$$P_1 = I'^2_{2f}\frac{R'_{2f}}{s} + I'^2_{2b}\frac{R'_{2b}}{2 - s}. \tag{7.3.15}$$

The ohmic loss in the rotor circuit is

$$P_2 = I'^2_{2f}R'_{2f} + I'^2_{2b}R'_{2b}. \tag{7.3.16}$$

Thus the gross mechanical power output is

$$P_m = P_1 - P_2 = I'^2_{2f}R'_{2f}\frac{1 - s}{s} + I'^2_{2b}R'_{2b}\frac{s - 1}{2 - s} \tag{7.3.17}$$

and the torque is

$$T_e = \frac{P_m}{\omega_m} = \frac{P_m}{\omega_s(1 - s)} = \frac{I'^2_{2f}}{\omega_s}\frac{R'_{2f}}{s} - \frac{I'^2_{2b}}{\omega_s}\frac{R'_{2b}}{2 - s} = T_{ef} - T_{eb}. \tag{7.3.18}$$

At standstill $s = 1$, $R'_{2f} = R'_{2b}$, and $I'_{2f} = I'_{2b}$, so that $T_{ef} = T_{eb}$ and there is no starting

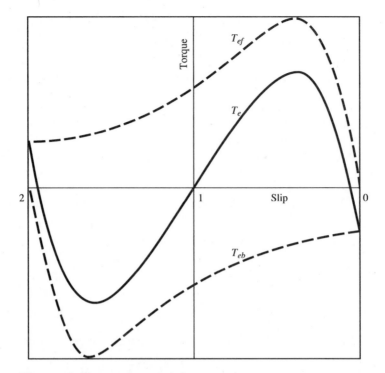

Figure 7.6 Torque characteristics.

torque. At any slip s the torque-speed characteristic, due to the forward field, is similar to that for a polyphase induction motor. The variation of T_{ef} with slip is illustrated in Fig. 7.6. At a slip of $(2 - s)$ the backward torque T_{eb} is the reverse of T_{ef}, as shown, and the difference represents the actual net torque developed by the motor. It will be noticed that the torque passes through zero before $s = 0$, so the full-load slip is usually greater than that for a polyphase machine.

The equivalent circuit, shown in Fig. 7.5, can represent the circuits of two series-connected polyphase motors driving the same load but with their phase sequence arranged for opposite rotation of fields. The two stator-circuit leakage impedances are lumped together. By inspection of this figure it will be noticed that, as the slip reduces, the motor with the forward field develops a higher impedance. Therefore this motor captures a higher proportion of the applied voltage, thereby having the higher flux amplitude. This point exemplifies the need to model the single-phase motor by two polyphase motors with their stator windings in series as shown in Fig. 7.4. The erroneous concept of a double revolving field with equal and constant fluxes demands the stator windings to be in parallel across a constant voltage source.

The usefulness of the equivalent circuit is demonstrated in the following example.

EXAMPLE 7.1

A single-phase, 240-V, 4-pole, 60-Hz, induction motor has the following parameters.

$R_1 = 2.5$ ohms $\qquad$ $R'_{2f} \approx R'_{2b} = 1.0$ ohm

$X_1 = 2X'_2 = 3$ ohms $\qquad$ $\mathbf{Y}_0 = (0.00147 - j0.00978)$ siemens.

Calculate the electromagnetic torque at a slip of 0.05.

Solution

The torque and synchronous speed expressions are

$$T_e = \frac{I'^2_{2f}}{\omega_s} \frac{R'_{2f}}{s} - \frac{I'^2_{2b}}{\omega_s} \frac{R'_{2b}}{2-s}$$

and $\omega_s = \dfrac{2\pi f}{p} = \dfrac{2\pi \times 60}{2} = \dfrac{377}{2} = 188.5$ rad/s.

It is required to find I'_{2f} and I'_{2b}.

$\mathbf{Z}_0 = 15 + j100 = 101.1 \angle 81.5° \ \Omega.$

$\mathbf{Z}'_{2f} = 1/0.05 + j1.5 = 20 + j1.5 = 20.06 \angle 4.3° \ \Omega.$

$\mathbf{Z}'_{2b} = 1/(2 - 0.05) + j1.5 = 0.513 + j1.5 = 1.59 \angle 71.1° \ \Omega.$

$\mathbf{Z}_f = \dfrac{\mathbf{Z}_0 \mathbf{Z}'_{2f}}{\mathbf{Z}_0 + \mathbf{Z}'_{2f}} = \dfrac{101.1 \angle 81.5° \times 20.06 \angle 4.3°}{15 + j100 + 20 + j1.5} = 18.9 \angle 14.8° \ \Omega.$

$\mathbf{Z}_b = \dfrac{\mathbf{Z}_0 \mathbf{Z}'_{2b}}{\mathbf{Z}_0 + \mathbf{Z}'_{2b}} = \dfrac{101.1 \angle 81.5° \times 1.59 \angle 71.1°}{15.513 + j101.5} = 1.6 \angle 71.3° \ \Omega.$

$\mathbf{Z}_T = \mathbf{Z}_1 + \mathbf{Z}_f + \mathbf{Z}_b = (2.5 + j3) + (18.3 + j4.8) + (0.5 + j1.5) =$
$\quad 23.2 \angle 23.6° \ \Omega.$

$\mathbf{I}_1 = \dfrac{V}{\mathbf{Z}_T} = \dfrac{240 \angle 0°}{23.3 \angle 23.6°} = 10.3 \angle -23.6° \ \text{A}.$

$\mathbf{E}'_{2f} = \mathbf{Z}_f \mathbf{I}_1 = 18.9 \angle 14.8° \times 10.3 \angle -23.6° = 194.7 \angle -8.8° \ \text{V}.$

$\mathbf{I}'_{2f} = \dfrac{\mathbf{E}'_{2f}}{\mathbf{Z}'_{2f}} = \dfrac{194.7 \angle -8.8°}{20.06 \angle 4.3°} = 9.7 \angle -13.1° \ \text{A}.$

$\mathbf{E}_{2b} = \mathbf{Z}_b \mathbf{I}_1 = 1.6 \angle 71.3° \times 10.3 \angle -23.6° = 16.5 \angle 47.4° \ \text{V}.$

$\mathbf{I}'_{2b} = \dfrac{\mathbf{E}'_{2b}}{\mathbf{Z}'_{2b}} = \dfrac{16.5 \angle 47.7°}{1.59 \angle 71.1°} = 10.4 \angle -23.4° \ \text{A}.$

Therefore $T_e = \dfrac{9.7^2 \times 1}{188.5 \times 0.05} - \dfrac{10.4^2 \times 1}{188.5 \times 1.95} = 9.7 \ \text{N} \cdot \text{m}.$

7.4. MOTOR STARTING

It can be seen from Fig. 7.6 that the single-phase induction motor is not self starting. There is need for some additional means to provide torque at zero speed. Once started in either direction the motor develops its own torque to accelerate the load up to the rated speed. A revolving magnetic field results from a two-phase system and this will produce the starting torque. So this is the basis for making an induction motor self-starting, if it is connected to a single-phase supply. Single-phase induction motors are grouped according to the method of starting.

7.4.1. Resistance-start Split-phase Motor

If the single-phase motor had two stator windings (a main winding and an auxiliary winding) placed in space quadrature, and if the ratio of leakage reactance to resistance were different for the two windings, the respective currents would have a time phase difference. The result would be a quasi-two-phase system, even though both windings are connected to the same single-phase supply.

A rotating field is set up in the air gap by such a configuration and starting torque is established despite the fact that the flux amplitude is not constant. The nonuniform field means low power factor and low efficiency, but this may not be too important for motors with low power ratings.

Figure 7.7a represents the symbolic form of the motor showing the auxiliary-winding axis to be 90 electrical degrees displaced in space from the main winding axis. About 45° time displacement of the currents in the two stator windings is usually

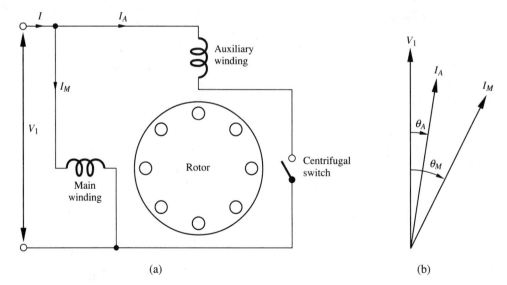

(a)   (b)

Figure 7.7   Split phase for starting. (a) Auxiliary winding connection, (b) phasor diagram.

achieved by designing the auxiliary winding to have a high resistance and low leakage reactance compared with the main winding. A centrifugal switch cuts out the auxiliary winding as the motor runs up to speed, because this winding with its high resistance is designed for a short-time power rating.

This form of starting, called the resistance-start, split-phase method, exemplifies the general principle, but there is a number of variations depending on the application of the drive. If resistance or reactance is externally connected in series with the auxiliary winding, it is possible to increase the starting torque. In general it is capacitive reactance that is added to improve the performance.

7.4.2. Capacitor Motor

Figure 7.8 shows a combination of ways a capacitor can be connected to improve both starting and running performance. For general purpose applications only capacitor C_2 and the switch S would be used to create a capacitor-start, split-phase motor. If better running performance is required, then, in addition to C_2 and S for high starting torque, there is a capacitor C_1 connected in such a way that the auxiliary winding and C_1 are permanently in the circuit. Such capacitor-start, capacitor-run motors have a better operating efficiency and power factor. Since the machine runs as a two-phase machine, noise is reduced.

Economies can be made if good running performance is required but the starting torque is not important. In this case C_2 and S are dispensed with and the machine becomes a permanent split-phase capacitor motor. Although the source is a single-phase supply, this type of motor has the running characteristics of a two-phase motor, whose speed can be readily adjusted. Just what value of capacitor to have in the

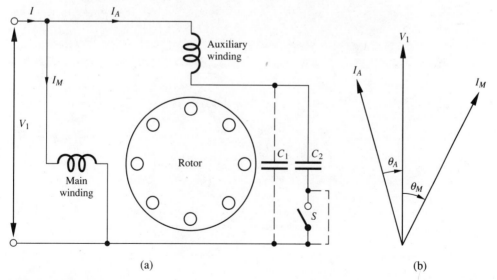

(a)   (b)

Figure 7.8 Capacitor-start motor. (a) Circuit, (b) phasor diagram.

auxiliary circuit depends both on the resistance and leakage reactance of the windings and on the performance requirements. If it were necessary to optimize the starting torque, the following analysis may be used.

The net forward starting torque can be written in the form

$$T_e = k I_M I_A \sin\theta \sin\rho \qquad (7.4.1)$$

where θ is the phase angle between the main and auxiliary winding currents I_M and I_A, and ρ is the angle between the coil axes. k is a constant of proportionality. With respect to the applied voltage as reference, the torque expression can be written

$$T_e = \frac{k V_1^2}{Z_M Z_A} (\sin\theta_M \cos\theta_A - \cos\theta_M \sin\theta_A)\sin\rho =$$

$$k \frac{V_1^2}{Z_M^2 Z_A^2} (X_M R_A - R_M X_A) \sin\rho. \qquad (7.4.2)$$

Photo 6 Capacitor-start single-phase induction motor. (Courtesy of Electric Machinery—Dresser Rand.)

The only variable is Z_A and it depends whether the machine is a split-phase motor or a capacitor-start motor for R_A or X_A to be the component that is variable.

For the case of the capacitor-start motor, maximum torque occurs at starting if $dT_e/dX_A = 0$, that is

$$R_M X_A^2 - 2X_M R_A X_A - R_A^2 R_M = 0. \qquad (7.4.3)$$

This is a quadratic in X_A and its solution is found if the machine parameters are known. It has been assumed that X_A includes the auxiliary winding leakage reactance and the externally connected capacitive reactance. A similar expression can be found for the optimum starting torque if resistance is used instead of capacitance.

7.4.3. Shaded-pole Motor

The shaded-pole motor is a single-phase, cage induction machine that has an output of less than 40 watts. It is made in great numbers for applications such as inexpensive turntable drives for record players, hair dryers and small fans. Figure 7.9 shows that the laminated stator core has salient poles and a concentrated main winding. The auxiliary winding is a short-circuited single turn of copper, called a *shading coil*. This single turn surrounds a small portion of each pole, which is displaced from the pole axis.

At the instant the main-winding current is increasing most rapidly the main flux does likewise. However, the current induced in the shading coil produces an opposing flux, according to Lenz's law, so that the net flux in the shaded pole is less compared with that in the main part of the pole. When the main current is at its peak, the change of the main flux is zero. Accordingly no current is induced in the shaded coil and the

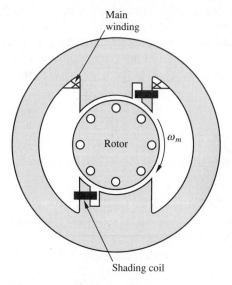

Figure 7.9 Shaded-pole motor.

shading pole gets its even share of the main flux. A short time later the main current and flux decrease at their greatest rate. Current, induced in the shading coil by transformer action, causes the flux in the shaded portion of the pole to lag the flux in the main portion of the pole. That is, flux first reaches its peak in the main pole and a short time later it reaches its peak in the shaded pole. Now we have the idea of a flux moving from the main pole to the shaded pole. This is enough of a *rotating flux* to produce a greater torque in the clockwise direction than the anticlockwise direction. The motor will start if the load is low, and this is all that is required, since a single-phase motor will run once it is started. However, the shading coil is permanent and provides the beneficial effect of a slightly increased torque due to the two-phase effect. It is seen that this is a unidirectional machine. The speed is not reversible.

A variation of the shaded-pole motor is provided by creating a nonuniform air gap under the pole piece. Flux *sweeping* occurs and a small starting torque is produced. The sweeping action occurs because the tip of the pole with the small air gap easily becomes saturated, while the other tip never becomes saturated. In a saturated region of iron the flux density B lags the excitation H, whereas in an unsaturated region B and H are in phase. Thus the flux in the small air gap lags the flux in the large air gap and this is sufficient for motion to be initiated. This type of motor is called a *reluctance-start machine*.

7.5. SUMMARY

The single-phase induction motor's operation can be explained by the two-revolving field theory. A pulsating field is resolved into forward- and backward-rotating components. The theory shows there is a net forward torque except at starting. Because there is no starting torque for a single-phase machine, means are found to change this situation by introducing a quasi–two-phase stator winding, either a split-phase, capacitor-start, or shaded-pole arrangement.

The revolving field theory leads to an equivalent circuit, from which the motor performance can be determined. This type of motor finds application wherever reasonably constant speed at a power less than 1 kW is required.

7.6. PROBLEMS

7.1. A single-phase induction motor has the following test results.

Locked-rotor test:	60 V	8 A	288 W
No-load test:	240 V	2.37 A	84.5 W

The stator winding resistance is 1.5 ohms. Determine the equivalent circuit parameters. Neglect skin effect.
(*Answer:* $X_1 = 3$ ohms, $X_2' = 1.5$ ohms, $R_{2f}' = R_{2b}' = 1.5$ ohms, $\mathbf{Y}_0 = (0.00147 - j0.00978)$ siemens.)

7.2. A 3-phase, 440-V, 4-pole, 60-Hz, Y-connected induction motor has a rated output of 10 kW. The machine phase parameters are
$\mathbf{Y}_0 = (0.001 - j0.01)$ siemens, $X_1 = X_2' = 1.5$ ohms, $R_1 = R_2' = 1.0$ ohms.
Windage and friction amounts to a loss of 100 W. This motor is to operate as a single-phase motor at 440 V by having one phase unconnected. If the machine runs at a slip of 0.05, calculate (a) the shaft torque and (b) the motor efficiency. Assume the 60-Hz skin effect increases resistance by 50% and the 120-Hz skin effect increases resistance by 70%.
(*Answer:* (a) 43.4 N·m, (b) 74.4%.)

7.3. A single-phase, 60-Hz induction motor has main and auxiliary windings whose impedances are $(50 - j50)$ ohms and $(100 + j50)$ ohms, respectively. What value of capacitance must be placed in series with the auxiliary winding in order that the two winding currents are in quadrature at starting?
(*Answer:* 21 μF.)

7.4. A single-phase induction motor has two windings whose standstill impedances are $(50 + j100)$ ohms and $(50 + j50)$ ohms. Show that the starting torque can be increased by at least four times by including a suitable reactance in series with the starting winding. What is the optimum value and nature of the added reactance?
(*Answer:* 6.2 ohms capacitive.)

7.7. BIBLIOGRAPHY

Alger, P. L. *Induction Machines.* 2nd ed. New York: Gordon and Breach, 1970.

Bewley, L. V. *Alternating Current Machinery.* New York: Macmillan Co., 1949.

Garik, M. L., and C. C. Whipple. *Alternating-Current Machines.* 2nd ed. New York: D. Van Nostrand, 1961.

Hancock, N. N. *Matrix Analysis of Electrical Machinery.* 2nd ed. Oxford: Pergamon Press, 1974.

Lawrence, R. R., and H. E. Richard. *Principles of Alternating-Current Machinery.* 4th ed. New York: McGraw-Hill Book Co., 1953.

Nasar, S. A. *Electric Machines and Electromechanics.* Schaum's Outline Series in Engineering. New York: McGraw-Hill Book Co., 1981.

Veinott, C. G., and J. E. Martin. *Fractional- and Subfractional-Horsepower Electric Motors.* 4th ed. New York: McGraw-Hill Book Co., 1986.

Veinott, C. G. *Theory and Design of Small Induction Motors.* New York: McGraw-Hill Book Co., 1959.

CHAPTER EIGHT

Synchronous Machines

8.1. INTRODUCTION

The largest machines manufactured are polyphase synchronous generators and almost all the electric energy that we consume is derived from such machines. In their present form, individual units can be constructed to give outputs up to 1000 MW. Although they are not made in such large numbers as induction machines, much attention has been given to the theory of synchronous machines, because of their roles in the balanced operation, fault behavior, and stability of power systems.

The action of synchronous machines is reversible. That is, the machines can generate or motor. Synchronous motors find application where constant and precise speed is required. A typical specification of one of the largest motors is a 13.8 kV, three-phase, 60-Hz, solid pole, synchronous motor delivering 29.5 MW at unity power factor for driving air compressors. This specification divulges one of the reasons why ac machines have developed in preference to dc machines. It is because the ac machine can operate at a much higher voltage level.

In this chapter, the theory of the synchronous machine is developed so that the performance for steady-state operation can be predicted if the machine parameters are known.

8.2. BASIC STRUCTURE AND OPERATION

In terms of currents in the windings, the synchronous machine is a doubly-excited system. In the slots of a laminated stator there is a uniformly distributed polyphase winding similar to that found in the induction motor. This winding is excited by alternating currents. Excitation of the rotor winding is by direct current. The rotor

342

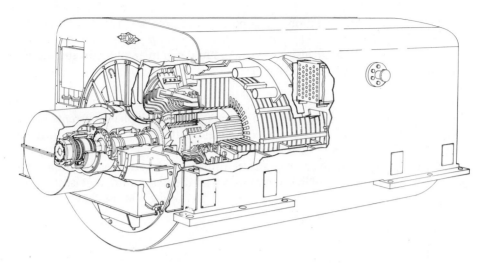

Three-phase turbo generator, 3600 rpm, 13,800 volts, 20,000 kW capacity. (Courtesy of Electric Machinery—Dresser Rand.)

can be round as found in high-speed machines, or it can have saliency as found in slow-speed machines. The dc winding produces the magnetic field that acts as the catalyst in the energy conversion process. The energy that is converted is associated with the power in the ac winding.

A third winding exists, but it is not directly associated with energy conversion. The third winding is a cage embedded in the pole shoes of the field structure, or it could be the pole shoe itself if it were solid. Its purpose is twofold. It provides a motor-starting mechanism in conjunction with the polyphase winding, and provides a stabilizing influence by damping any mechanical oscillations of the rotor.

Stator and rotor functions are interchangeable. The armature winding could be on the rotor with the field winding on the stator. This is not common practice except for synchronous converters, because the high armature power and voltage are more easily handled on the stator than through slip rings on the rotor. The power dissipated in the excitation winding may be 5% or less or the rated power of the machine. This power dissipation, rather than the armature power, is more conveniently handled by means of slip rings. In this chapter stator or rotor dc excitation is chosen by the convenience of the analysis.

There are common features between synchronous machines and other machines. For example, the polyphase windings are found in induction motors. So the treatment of induction, harmonics, leakage flux, and winding factors is the same. Also, if a commutator were connected to the armature winding on the rotor, the synchronous machine would be converted to a dc machine. However, there is one major difference between synchronous machines and all other machines, and this difference is implied in the name. This machine operates only at synchronous speed, a constant speed that can be determined by the number of poles and the fundamental frequency of alternation of the armature-winding voltage.

For generator action, rotation of a N pole and a S pole past a given point on the stator constitutes a cycle of field variation. A rotor with p pole pairs moving at n_s revolutions per second would give rise to an induced emf in a stator coil and the emf would have a fundamental frequency of

$$f = n_s p \text{ hertz.} \tag{8.2.1}$$

This equation applies equally well to motoring action. Polyphase currents of frequency f in the stator winding give rise to a field pattern of p pole pairs traveling at n_s revolutions per second in the air gap. The stator winding mmf pattern interacts with the rotor field pattern to produce an electromagnetic torque. For the torque to be constant the rotor must travel in synchronism with the stator mmf, that is, at n_s also. At any other speed the instantaneous value of the torque will oscillate at slip frequency and the average torque would be zero.

8.3. ELEMENTARY SYNCHRONOUS MACHINES

The operation of synchronous machines is readily understood by stripping the machine down to its basic elements, which are the ac winding and the main field poles (produced by the dc excitation). Polyphase machines have an advantage over single-phase machines. Ideally, the former have constant power and torque, whereas the latter have pulsating power and torque. This becomes apparent if we consider the single-phase machine first.

8.3.1. Single-phase Generator

An elementary single-phase generator is shown in Fig. 8.1. A two-pole rotor, excited by direct current, rotates at a constant angular speed ω. The total flux per pole that crosses the air gap and enters the armature is Φ. The single-phase armature winding is considered to be concentrated and to have N turns.

Let the coil axis and the pole axis be coincident at time $t = 0$ and assume the flux to be sinusoidally distributed in space. Thus, at any instant t, the flux linkage with the stator coil is

$$\lambda = N\Phi \cos\alpha = N\Phi \cos\omega t \tag{8.3.1}$$

and the emf induced in the coil is, by Faraday's law,

$$e = -\frac{d\lambda}{dt} = -N\Phi \frac{d(\cos\omega t)}{dt} = \omega N\Phi \sin\omega t = \hat{E} \sin\omega t \tag{8.3.2}$$

where $\hat{E} = \omega N\Phi = 2\pi\Phi fN$. Hence the effective (rms) value of the induced emf is

$$E = \frac{\hat{E}}{\sqrt{2}} = 4.44\Phi fN. \tag{8.3.3}$$

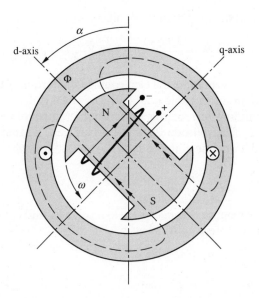

Figure 8.1 Elementary single-phase generator.

This induced emf, whose frequency is f, can only be measured if the coil is open-circuited. The emf for a distributed winding would have the same form, but N would be the effective number of turns to account for the winding factor.[1]

If the generator's armature winding is connected to a load, whose impedance together with that of the coil is $\mathbf{Z} = R + jX$, the effective load current is

$$\mathbf{I} = \frac{\mathbf{E}}{\mathbf{Z}} = \frac{E}{Z} \angle -\theta \tag{8.3.4}$$

where $\mathbf{I}$ lags $\mathbf{E}$ by the angle θ, given by $\theta = \tan^{-1} X/R$. If $e = \hat{E} \sin\omega t$, then $i = \hat{I} \sin(\omega t - \theta)$ and the instantaneous value of the power generated is

$$p(t) = ei = \hat{E} \sin\omega t \times \hat{I} \sin(\omega t - \theta)$$
$$= \frac{\hat{E}\hat{I}}{2} [\cos\theta - \cos(2\omega t - \theta)] \tag{8.3.5}$$
$$= EI[\cos\theta - \cos(2\omega t - \theta)].$$

Thus the average value of power is

$$P = EI \cos\theta. \tag{8.3.6}$$

The instantaneous electric power is derived from the instantaneous mechanical

---

[1] The winding factor is described in Appendix G, sections G2 and G4.1.

power ($p(t) = T_e\omega$), so the instantaneous value of the torque is

$$T_e = \frac{p}{\omega} = \frac{EI\cos\theta}{\omega} - \frac{EI\cos(2\omega t - \theta)}{\omega}. \tag{8.3.7}$$

Hence, in a single-phase generator the electromagnetic torque comprises a constant term $EI\cos\theta/\omega$ and an undesirable double-frequency term.

The armature winding current, together with the N turns of the coils, gives rise to a square mmf wave of amplitude $Ni/2$ in the air gap, so that the fundamental component directed along the coil axis is

$$F_A = \frac{2}{\pi} N\hat{I}\sin(\omega t - \theta). \tag{8.3.8}$$

To note the armature reaction it is necessary to resolve F_A into components along the pole axis and in quadrature with it. That is

$$\mathbf{F}_A = F_d + jF_q = F_A\cos\alpha - jF_A\sin\alpha$$
$$= \frac{2}{\pi} N\hat{I}\sin(\omega t - \theta)(\cos\omega t - j\sin\omega t)$$

or

$$\mathbf{F}_A = \frac{N\hat{I}}{\pi}\{[-\sin\theta + \sin(2\omega t - \theta)] + j[-\cos\theta + \cos(2\omega t - \theta)]\}. \tag{8.3.9}$$

The two components of armature reaction comprise constant- and double-frequency terms. The constant mmf term in the direct axis opposes the main field if the power factor is lagging, but assists if the power factor is leading and might put the iron parts into saturation and so alter the induced emf. The constant mmf term in the quadrature axis does not change direction. It is this cross field that gives rise to the electromagnetic torque that balances the driving torque.

8.3.2. Three-phase Generator

A 3-phase elementary generator has three identical, concentrated coils on the stator. This is shown in Fig. 8.2a in which the three individual coils, aa', bb', and cc', are shown to be displaced in space by $120°$.

On open circuit the instantaneous induced emfs of each phase are shown in Fig. 8.2b and are described by

$$e_a = \hat{E}\sin\omega t$$
$$e_b = \hat{E}\sin(\omega t - 2\pi/3) \tag{8.3.10}$$
$$e_c = \hat{E}\sin(\omega t - 4\pi/3) = \hat{E}\sin(\omega t + 2\pi/3)$$

if $\alpha = 0$ at $t = 0$. Consequently, if the armature coils are connected to a balanced

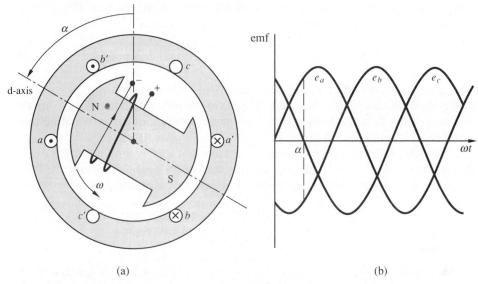

(a)   (b)

Figure 8.2 Elementary three-phase generator. (a) Basic structure, (b) induced emfs.

three-phase load, the individual currents would be

$$i_a = \hat{I} \sin(\omega t - \theta)$$
$$i_b = \hat{I} \sin(\omega t - 2\pi/3 - \theta) \qquad (8.3.11)$$
$$i_c = \hat{I} \sin(\omega t + 2\pi/3 - \theta)$$

where θ is the time angle[2] of lag of **I** behind **E**.

The total instantaneous power $p(t)$ generated[3] is the sum of the powers generated in each phase.

$$p(t) = p_a + p_b + p_c = 3\,\frac{\hat{E}\hat{I}}{2}\cos\theta = 3EI\cos\theta = P. \qquad (8.3.12)$$

The total instantaneous power is constant and equal to the sum of the average power in each of the phases. The instantaneous torque is also constant, and is

$$T_e = \frac{P}{\omega} = \frac{3EI\cos\theta}{\omega}. \qquad (8.3.13)$$

The polyphase machine has the advantage over the single-phase machine in that

---

[2] The angle is measured in radians to suit the sinusoidal function. To transform to true time (in seconds) the angle (in radians) is divided by the angular frequency of the supply.

[3] The reader may care to expand the instantaneous phase powers in terms of the currents and voltages in order to prove that eq. (8.3.12) is correct.

there is no double-frequency term. The fundamental component of the resultant armature-reaction mmf is developed from eq. (8.3.9) and is

$$\mathbf{F}_A = F_d + jF_q = \mathbf{F}_a + \mathbf{F}_b + \mathbf{F}_c = -\frac{3N\hat{I}}{\pi}(\sin\theta + j\cos\theta). \qquad (8.3.14)$$

This mmf is a constant and is equal to three times the constant mmf value of the single-phase winding and is fixed with respect to the pole axis. The pole axis rotates synchronously. This reaffirms that the polyphase currents set up an armature reaction field that rotates synchronously.

Although we have used the generator as an example, this simple analysis can be applied equally well to the motor, because the same form is taken by the induced emf, the armature currents, the armature reaction, and the armature power for conversion.

A. TORQUE

A round rotor (uniform air gap) machine is shown in Fig. 8.3a. The space vector of the field mmf $\mathbf{F}_F$ in the air gap is on the rotor magnetic axis (d-axis) and rotates at synchronous speed. The currents in the three-phase windings produce an mmf vector $\mathbf{F}_A$, which rotates at synchronous speed and which lags $\mathbf{F}_F$ by an angle δ_{af} that depends on the load. If these vectors are aligned there is no torque. If these vectors are displaced by 90 electrical degrees the torque is a maximum. Since the distribution of the rotor and stator winding fields can be assumed to be sinusoidal in space, the

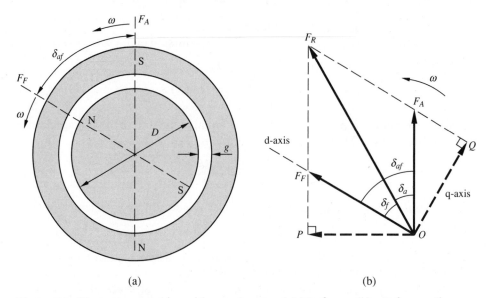

(a)   (b)

Figure 8.3 Elementary machine with round motor. (a) Mmf axes, (b) mmf space diagram.

torque at any angle δ_{af} is

$$T_e = kF_FF_A \sin\delta_{af} \tag{8.3.15}$$

where F_F and F_A are peak values.

The resultant peak mmf F_R in the air gap is given by

$$F_R^2 = F_F^2 + F_A^2 + 2F_FF_A \cos\delta_{af} \tag{8.3.16}$$

and if the phase angles between $\mathbf{F}_R$ and $\mathbf{F}_A$, and $\mathbf{F}_R$ and $\mathbf{F}_F$ are δ_a and δ_f, respectively, a diagram showing the mmf space components can be drawn as illustrated in Fig. 8.3b. There is now a number of torque expressions available by considering the interaction between the resultant field and the field mmf. Thus

$$T_e = kF_RF_A \sin\delta_a = kF_RF_F \sin\delta_f. \tag{8.3.17}$$

This can be checked by noting the perpendicular projections.

$$OP = F_R \sin\delta_a = F_F \sin\delta_{af} \quad \text{and} \quad OQ = F_R \sin\delta_f = F_A \sin\delta_{af}. \tag{8.3.18}$$

The constant of proportionality k can be evaluated from a consideration of the coenergy stored in the air gap. The coenergy W'_s in the air gap is

$$W'_s = \tfrac{1}{2}\mu_0(H^2)_{av} \times \text{volume} = \tfrac{1}{2}\mu_0(H^2)_{av} \times \pi DLg \tag{8.3.19}$$

where L is the effective length of the machine. Since the resultant field distribution is considered to be sinusoidal, the average value of the magnetic field intensity squared is

$$(H^2)_{av} = \frac{1}{\pi} \int_0^\pi \hat{H}^2 \sin^2\theta \, d\theta = \frac{\hat{H}^2}{2} \tag{8.3.20}$$

so the coenergy is given by

$$W'_s = \mu_0 \frac{\pi}{4} DLg\hat{H}^2. \tag{8.3.21}$$

Since the peak value of the resultant mmf is $F_R = \hat{H}g$, the coenergy is

$$W'_s = \mu_0 \frac{\pi}{4} \frac{DL}{g} F_R^2 = \mu_0 \frac{\pi}{4} \frac{DL}{g} (F_F^2 + F_A^2 + 2F_FF_A \cos\delta_{af}). \tag{8.3.22}$$

Hence the torque is

$$T_e = \frac{\partial W'_s}{\partial \delta_{af}} = -\frac{\pi\mu_0}{2} \frac{DL}{g} F_FF_A \sin\delta_{af} \tag{8.3.23}$$

and so

$$k = -\frac{\pi\mu_0}{2} \frac{DL}{g}. \tag{8.3.24}$$

The significance of the negative sign is that the electromagnetic torque tends to reduce the angle δ_{af} and so opposes the mechanical driving torque. We note that the torque is proportional to the product of the main field and the component of the armature reaction field in the quadrature axis. This torque expression is for each pair of poles so that, in general, for a p pole-pair machine

$$T_e = -p \times \frac{\pi\mu_0}{2} \frac{DL}{g} F_F F_A \sin\delta_{af}. \qquad (8.3.25)$$

The outputs of a dc machine and an induction motor were obtained in terms of the resultant flux density and the ampere conductors. A similar expression can be obtained for the synchronous machine. Since the resultant flux density $\hat{B}_R$ is given by

$$\hat{B}_R = \frac{\mu_0 F_R}{g} \qquad (8.3.26)$$

and the resultant flux per pole is given by

$$\Phi_R = B_{av} \times \text{pole area} = \frac{2}{\pi} \hat{B}_R \frac{\pi DL}{2p} \qquad (8.3.27)$$

then

$$T_e = -\frac{\pi p}{2} DL\hat{B}_R F_F \sin\delta_f = -\frac{\pi}{2} p^2 \Phi_R F_F \sin\delta_f. \qquad (8.3.28)$$

B. PHASOR DIAGRAMS

Time and space angles (in electrical radians) are identical in synchronous machines and this is convenient when drawing phasor diagrams. Consider the round rotor generator to be on open circuit. There is no armature reaction. The rotating field pattern induces an emf E in each phase winding with an effective magnitude

$$E = 4.44\Phi fN \qquad (8.3.29)$$

lagging the flux Φ by 90°. On open circuit the resultant air-gap mmf F_R is the actual main field mmf F_F, so the phasor diagram is as in Fig. 8.4a.

If the generator is on load we can make assumptions, so that certain machine features become clear. Consider the load to be an infinite bus (constant voltage V and constant frequency f). The field flux Φ would induce an emf E in the armature windings, as shown in Fig. 8.4b. However, the armature current I sets up an armature reaction mmf[4] F_A, which together with the field mmf F_F produces a resultant mmf F_R and a resultant flux Φ_R. There is an induced emf E_r given by

$$E_r = 4.44\Phi_R fN \qquad (8.3.30)$$

and lagging the flux Φ_R by 90°. If the leakage flux and armature winding resistance

---

[4] The main flux axis is the d-axis. If we take the components of the armature-reaction mmf, given in eq. (8.3.14), and plot them on Fig. 8.4b, we see that the armature reaction mmf is in phase with the phase current.

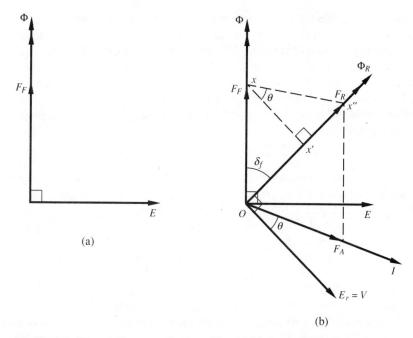

Figure 8.4 Phasor diagram of a generator. (a) No load, (b) connected to an infinite bus.

are neglected, this equation indicates that Φ_R is constant since

$$E_r = V = \text{constant.} \tag{8.3.31}$$

A relation between mmfs is given by the perpendicular projection xx', that is

$$F_F \sin\delta_f = F_A \cos\theta \tag{8.3.32}$$

where θ is the phase angle between load current and voltage. The term $F_F \sin\delta_f$ is proportional to torque, since Φ_R has been found to be constant, and the term $F_A \cos\theta$ is proportional to the power converted since V is constant (since the speed is constant the torque is proportional to power).

Also, from Fig. 8.4b, the component mmfs are related using the fact that $Ox'' = Ox' + x'x''$ or $F_R = F_F \cos\delta_f + F_A \sin\theta$. Thus, for an infinite-bus load, with Φ_R and hence F_R constant, the reactive component of power $VI \sin\theta$ (proportional to $F_A \sin\theta$) can be controlled by the field excitation, $F_F \cos\delta_f$. This is an important mode of control in power system operation.

8.4. PRACTICAL SYNCHRONOUS MACHINES

In the previous section the discussion of elementary machines gives an insight into the operation of a synchronous generator or motor. Actual machines differ from the elementary machines not in principle but in detail of construction. Consequently we need to know something about the actual flux distribution in a real machine, so that we can modify the ideal equations. For convenience consider the generator.

The field structure of a generator is on the rotor and is driven at synchronous speed. There are salient structures and round structures, as shown schematically in Figs. 8.5a and b. Direct current in the field coils excite the poles and the shape of the resulting flux density distribution in the air gap the distribution can be represented by a Fourier series

$$B = \sum \hat{B}_n \cos(2n - 1)\theta, \, n = 1, 2, 3 \, \ldots \tag{8.4.1}$$

The actual distribution is usually flat-topped and contains appreciable harmonics. A pure sinusoid is theoretically possible by shaping the salient-pole face or by having a sinusoidally-distributed field winding for the round rotor.

Another source of flux density distortion is the armature winding slots. The difference in reluctance between the tooth and the slot gives rise to a ripple superimposed on the flux density distribution. In the analysis to predict the performance of the machine, the harmonic effects will be considered small enough to be neglected. Accordingly, the flux distribution will be assumed to be sinusoidal.

A traveling wave of flux produced by the field structure will cause an emf to be induced in the armature windings of the stator. Both the armature construction and armature windings resemble closely those of a polyphase induction motor. Thus, the topics on the rotating magnetic field, induced emf, and ac windings in Chapter 6 apply equally well to the synchronous generator armature windings. We can make the same analysis and conclusions. In brief, when an electric load is connected to the terminals of the armature, the induced emf causes armature winding currents to flow. Currents in a polyphase winding produce an mmf pattern whose fundamental com-

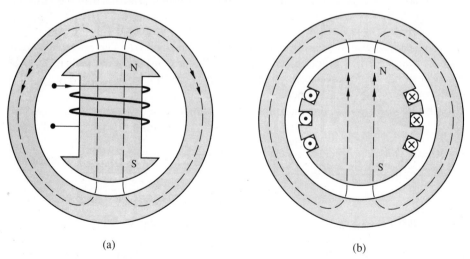

(a)   (b)

Figure 8.5 Field structure. (a) Salient poles, (b) round rotor.

ponent rotates synchronously in the air gap. This constitutes armature reaction, whose flux amplitude position in space depends on the phase angle of the armature-winding current. Since the main field and the armature field rotate at the same speed, the main field may be enhanced or effectively reduced, depending on the phase of the armature winding currents. Accordingly, there is a resultant flux density in the air gap and this modifies the induced emf in the armature windings. The modified emf and the armature winding current constitute the components of the electric output power that is converted from the mechanical power driving the rotor.

Finally, consideration must be given to leakage flux. Field leakage flux is that flux that does not cross the air gap to link with the armature winding. It has no effect except perhaps to saturate the field poles, so it is ignored in the analysis. However, an effective value of the field flux must be used in calculations. Not all the flux produced by the armature winding currents crosses the air gap. This is common to all kinds of machines. This leakage flux can be divided into its components, slot leakage flux, end-turn leakage flux, and air-gap (or differential) leakage flux. The last accounts for all the space harmonics of armature reaction excluding the space fundamental. In synchronous machine analysis these three components are lumped together and their effect is accounted for by a fictitious leakage reactance.

8.5. SYNCHRONOUS GENERATOR (ROUND ROTOR)

8.5.1. Induced Emf on No Load

The configuration of the round-rotor generator is shown in Fig. 8.5b. If there is no load connected to the terminals, the expression for the induced emf in the armature winding can be developed in the same way as that described in Chapter 6.

The set of field poles revolving at synchronous speed produces a flux density traveling wave of the form

$$B = \hat{B} \sin(\theta - \omega t). \tag{8.5.1}$$

The flux linked with an arbitrary concentrated coil of N turns on the armature is

$$\phi = \Phi \sin(\theta - \omega t) \tag{8.5.2}$$

where Φ is the total flux per pole. Thus the coil-induced emf is

$$e = -N \frac{d\phi}{dt} = N\Phi\omega \cos(\theta - \omega t). \tag{8.5.3}$$

Let the actual number of series turns per phase be N'. To account for the fact that the coil is not concentrated but forms part of a fractional-pitch, distributed, phase winding, we must use a winding factor k_ω. Therefore the instantaneous value of the induced emf per phase is

$$e = k_\omega \Phi N' \omega \cos(\theta - \omega t) = \Phi\omega N \cos(\theta - \omega t). \tag{8.5.4}$$

N' and k_ω can be combined to give an effective number of turns N per phase. The equation can be written in the form

$$e = \hat{E} \cos(\theta - \omega t) \tag{8.5.5}$$

where $\hat{E} = \Phi \omega N$. Hence the effective value of the emf is

$$E = \frac{\hat{E}}{\sqrt{2}} = \frac{2\pi}{\sqrt{2}} \Phi f N = 4.44 \Phi f N. \tag{8.5.6}$$

The frequency of voltage alternation f depends on the rotor speed and the number of pole pairs, and is given by eq. (8.2.1).

EXAMPLE 8.1

A 3-phase, 6.6-kV, 24-pole, Y-connected generator is driven at 300 r/min. The armature winding has 210 turns in series per phase and the winding factor is 0.95. Estimate the effective number of turns per phase and the no-load terminal voltage, if the field winding is excited to produce a sinusoidal air-gap flux per pole of 75 mWb.

Solution
Effective turns/phase $N = k_\omega N' = 0.95 \times 210 = 200$.

Frequency $f = n_s p = \dfrac{300}{60} \times \dfrac{24}{2} = 60$ Hz.

The induced emf per phase on no load is

$$E = 4.44 \Phi f N = 4.44 \times 75 \times 10^{-3} \times 60 \times 200 = 3996 \text{ V}.$$

Therefore the terminal voltage on no load is $V = \sqrt{3} \times 3996 = 6920$ V.

8.5.2. Armature Reaction

If a load is connected to the armature terminals, the armature winding emf E, induced by the main field, causes armature winding currents to flow. Because the armature winding is polyphase and distributed, we can use the results of Appendix F, eq. (F4.13) to write the expression for the mmf distribution in the air gap as

$$F = \frac{9N\hat{I}}{\pi^2} \left[\sin(\theta - \omega t) + \frac{1}{25} \sin(5\theta + \omega t) - \frac{1}{49} \sin(7\theta - \omega t) + \cdots \right]. \tag{8.5.7}$$

The component of significance is the fundamental one, which we can call F_A, so that

$$F_A = \frac{9N\hat{I}}{\pi^2} \sin(\theta - \omega t) \tag{8.5.8}$$

where $\hat{I}$ is the maximum value of the phase current. The other harmonic terms will be lumped with leakage later, as they do not contribute directly in the useful energy conversion process. This mmf F_A, which is called armature reaction, produces a wave pattern of flux in the air gap traveling at the same speed as the field structure. There is interaction between the main flux and the armature reaction flux to produce a resultant flux.

Consider a particular instant during the rotation of the field structure when the pole axis is ahead of the phase a axis by 90 electrical degrees. This is shown in Fig. 8.6, in which the rotor and stator have been separated for clarity. In Fig. 8.6a the field winding is represented symbolically by two coils carrying a current I_F and the main field distribution in the air gap is shown. The pole axis (d-axis) acts as a reference for the other figures. In Fig. 8.6b only phase a of the armature winding is shown and it is depicted as a concentrated winding.

Since time and space angles are identical in the synchronous machine and since at this instant of concern, the main field is leading phase a by 90°, the instantaneous value of the induced emf of phase a is a maximum. Let the generator load be such that the phase current I is in phase with the emf E. Therefore the current is a maximum when the emf is a maximum. The polyphase currents set up a synchronously rotating field pattern, whose fundamental amplitude is coincident with the phase a axis, when phase a current is a maximum.[5] This is shown in Fig. 8.6b. Hence the actual distribution of the field in the air gap is the sum of two sine waves, both rotating synchronously. The resultant is the synchronously rotating sine wave, illustrated in Fig. 8.6c. The time and space phasor diagrams to represent this particular condition are shown in Fig. 8.7.

The condition that the current I is in phase with the emf E is a special case. In general the current will lag behind the emf. Superimposing the time and space angles, a general phasor diagram is illustrated in Fig. 8.8. The emf E lags behind the flux Φ (which produces it) by 90 electrical degrees. Lagging behind E is the current I by some angle β. F_A, the armature reaction mmf, is in phase with I, and the resultant, F_R, of the field and armature mmf can be drawn at an angle δ_f to the field mmf F_F, which is in phase with the field flux Φ. In terms of the two field quantities, the electromagnetic torque expression is

$$T_e = -\frac{\pi}{2} p^2 \Phi_R F_F \sin\delta_f. \tag{8.5.9}$$

If the field current and the resultant air-gap flux are constant, the generator adjusts to a change in torque by a change in the torque angle δ_f. This case is found in practice if the armature winding is connected to a constant-voltage bus of V volts. If the resultant flux Φ_R causes an emf E_r to be induced in each phase of the armature winding and if we permit the approximation that V and E_r are nearly equal, then

$$V \approx E_r = 4.44\Phi_R fN. \tag{8.5.10}$$

This provides the information that Φ_R is nearly constant.

The resultant air-gap flux Φ_R comprises the components Φ due to the main field and Φ_A due to the armature reaction field, such that, in the absence of magnetic

---

[5] Refer to Appendix F, sections F4 and F5.

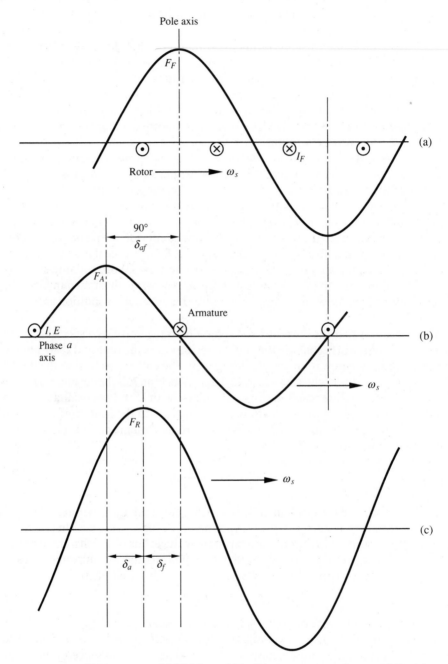

Figure 8.6 Field patterns. (a) Field mmf F_F, (b) armature reaction mmf F_A, (c) resultant mmf F_R.

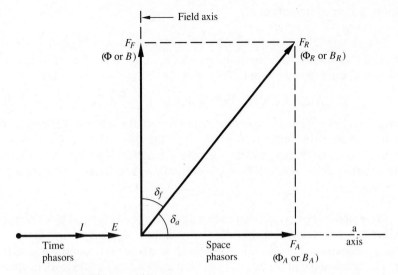

Figure 8.7 Phasor diagrams.

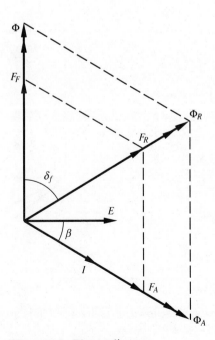

Figure 8.8 Phasor diagram.

saturation, superposition gives us

$$\Phi_R = \Phi + \Phi_A. \tag{8.5.11}$$

Similarly, the armature winding induced emf $\mathbf{E}_r$, due to the flux Φ_R, has components $\mathbf{E}$ due to the field flux Φ and $\mathbf{E}_{ar}$ due to the armature reaction flux Φ_A, so that

$$\mathbf{E}_r = \mathbf{E} + \mathbf{E}_{ar} \tag{8.5.12}$$

where $E_{ar} = 4.44\Phi_A fN$. In a simple equivalent circuit, the self-induced emf E_{ar} can be modeled by a reactive voltage drop Ix_a since the flux Φ_A is proportional to the armature winding current I, and the armature reaction reactance x_a indicates the inductive nature of the voltage. Consequently the voltage equation becomes

$$\mathbf{E}_r = \mathbf{E} - j\mathbf{I}x_a. \tag{8.5.13}$$

The negative sign is used because an induced emf has been replaced by a voltage drop.

Figure 8.9 shows a partial equivalent circuit and its phasor diagram that satisfy this voltage equation. The circuit represents one phase of a round rotor, armature circuit. The generated voltage equals the actual induced emf on no load, in series with a reactance that accounts for armature reaction. Consequently, a generator can be modeled by a voltage behind a reactance.

In this foregoing description, the individual fluxes due to the field and the armature reaction have been added to obtain the resultant air-gap flux. Thus, it has been assumed that there was no saturation of the magnetic circuit.

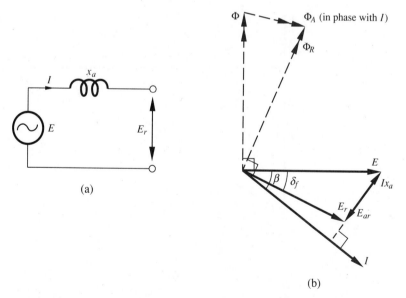

(a)

(b)

Figure 8.9 Armature reaction and resultant emf. (a) Partial equivalent circuit, (b) phasor diagram.

8.5.3. Synchronous Impedance

The terminal voltage of an armature winding is different from the induced emf E_r by an amount equal to the resistance and leakage reactance drops in the winding. Thus, the terminal voltage is written as

$$\mathbf{V} = \mathbf{E}_r - \mathbf{I}R_a - j\mathbf{I}x_l. \tag{8.5.14}$$

R_a is the resistance of the armature winding per phase. In large machines R_a is usually small compared with the reactances, so it is often ignored. The leakage reactance x_l accounts for all the fluxes due to the armature current except the fundamental field of armature reaction. The leakage fluxes, which incorporate slot and end-winding leakage and all the space harmonic fields, cause an emf to be induced in the winding. This emf is proportional to the leakage fluxes, which in turn are proportional to the armature winding current. Accordingly, the emf induced by leakage fluxes has the same effect as the emf induced by the fundamental armature reaction flux.

The detailed equation for the terminal voltage becomes

$$\mathbf{V} = \mathbf{E} - \mathbf{I}R_a - j\mathbf{I}x_l - j\mathbf{I}x_a = \mathbf{E} - \mathbf{I}R_a - j\mathbf{I}(x_l + x_a). \tag{8.5.15}$$

The sum of the two reactances is defined as the synchronous reactance X_s, where

$$X_s \triangleq x_l + x_a \tag{8.5.16}$$

and accounts for all the armature current fluxes. In addition, it is usual to define the synchronous impedance $\mathbf{Z}_s$ such that $\mathbf{Z}_s \triangleq R_a + jX_s$. Hence

$$\mathbf{V} = \mathbf{E} - \mathbf{I}(R_a + jX_s) = \mathbf{E} - \mathbf{I}\mathbf{Z}_s. \tag{8.5.17}$$

This equation provides the basis for synchronous machine analysis.

EXAMPLE 8.2

A 3-phase, 6.6-kV, 12-MVA, 60-Hz, Y-connected generator has an armature winding with 210 turns per phase, a winding factor of 0.95, and a synchronous impedance of $(0.02 + j1.0)$ ohms per phase. Determine the flux per pole produced by the field winding for the generator to deliver full load at rated voltage and 0.8 power factor lagging.

Solution
On a per-phase basis the relation between the terminal voltage V and the no-load induced emf E is given by

$$\mathbf{E} = \mathbf{V} + \mathbf{I}(R_a + jX_s).$$

$V = 6600/\sqrt{3} = 3810$ volts. (This is taken as reference.)

$$I = \frac{12 \times 10^6}{\sqrt{3} \times 6600} = 1050 \text{ amperes.}$$

Consequently $\mathbf{E} = 3810 + 1050(0.8 - j0.6)(0.02 + j1.0)$.

That is $\mathbf{E} = 4533 \angle 10.5°$ volts

The no-load emf equation is $E = 4.44\Phi fN$.

Hence the flux per pole $\Phi = \dfrac{4533}{4.44 \times 60 \times 0.95 \times 210} = 85 \times 10^{-3}$ webers.

8.5.4. Equivalent Circuit and Phasor Diagrams

Figure 8.10 shows an equivalent circuit that satisfies the voltage eq. (8.5.17).

Construction of the phasor diagrams, illustrated in Fig. 8.11, begins with drawing the terminal voltage $\mathbf{V}$ as the reference. The armature phase current $\mathbf{I}$ is drawn at an angle θ to $\mathbf{V}$, where $\cos\theta$ is the power factor of the load. The voltage drop due to the armature winding resistance and the voltage drop due to the synchronous reactance are respectively in phase with and 90° leading the armature winding current. These voltage drops $\mathbf{I}R$ and $j\mathbf{I}X_s$ are added to the terminal voltage $\mathbf{V}$ to obtain the no-load induced emf per phase $\mathbf{E}$.

What is of note is that, as the load becomes more and more capacitive, the power factor changes from lagging to leading and the terminal voltage magnitude eventually becomes greater than the induced emf. This change of magnitude implies a problem associated with the terminal voltage regulation. Another point to note is asociated with the angle δ between the terminal voltage $\mathbf{V}$ and the no-load induced emf $\mathbf{E}$. This angle δ is defined as the *load angle* and is important in connection with stability and power transfer. Its similarity with the torque angle δ_f is evident from a study of the phasor diagrams for minimum and maximum power output. For example, δ is zero for the case of no load because the only way the phasors $\mathbf{V}$ and $\mathbf{E}$ can be coincident is if the load current $\mathbf{I}$ is zero. When the load angle makes $\mathbf{E}$ lead $\mathbf{V}$ the machine is generating, but when $\mathbf{E}$ lags $\mathbf{V}$ the machine is motoring.

In practice, the armature winding resistance has a range between the per unit limits $0.05 \geq R_a \geq 0.01$, decreasing as the machine rating increases, whereas the synchronous reactance value may be between the per unit limits $0.5 \leq X_s \leq 1.0$, increasing as the machine rating increases. The ratio of reactance to resistance can

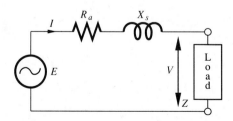

Figure 8.10 Equivalent circuit of a round rotor generator.

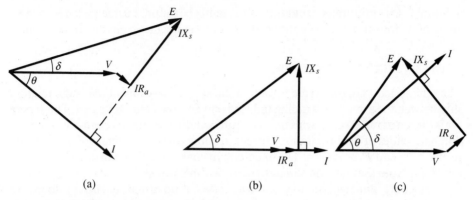

(a)   (b)   (c)

Figure 8.11 Phasor diagrams of a round rotor generator. (a) Lagging power factor, (b) unity power factor, (c) leading power factor.

increase from 10 to 100 as the rating increases. So resistance is frequently omitted from calculations. If the voltage drop due to the resistance can be ignored, the synchronous machine is modeled by a voltage behind a synchronous reactance.

EXAMPLE 8.3

A 3-phase, 6.6-kV, 12-MVA, 24-pole, Y-connected generator has a synchronous impedance of $(0.02 + j1.0)$ ohms per phase. Estimate the load angle δ in mechanical degrees, if the generator delivers full load at rated voltage and with unity power factor.

Solution
On a per phase basis the induced emf on no load is given by eq. (8.5.17). Take **V** as the reference phasor, where
$$\mathbf{V} = \frac{6600}{\sqrt{3}} = 3810 \angle 0° \text{ volts per phase and } \mathbf{I} = \frac{12 \times 10^6}{\sqrt{3} \times 6600} = 1050 \angle 0° \text{ amperes.}$$
Thus $\mathbf{E} = 3810 + 1050(0.02 + j1.0) = 3974 \angle 15.3°$ volts.
The load angle δ is the phase angle difference between **E** and **V**. Hence $\delta = 15.3$ electrical degrees. In mechanical degrees $\delta = 15.3/12 = 1.28°$.

8.5.5. Parameter and Loss Separation

There are some simple tests that give the values of the parameters R_a and X_s used in the equivalent circuit and that aid separation of the different losses. There is no precise way to determine the synchronous reactance X_s. The saturation of the iron parts of the magnetic circuit influences the effects of armature reaction. This means that X_s varies according to the field excitation and the load conditions. However a value of X_s for unsaturated conditions can readily be found. Further, an approximate value of X_s for saturated conditions of parts of the magnetic circuit can be found.

A. OPEN-CIRCUIT CHARACTERISTIC

The open-circuit characteristic (OCC) of the synchronous machine is the relationship betwen the armature-winding induced emf and the field excitation. The generator is driven at rated speed with the armature-winding terminals open-circuited. For different values of field-excitation current, readings of the terminal voltage, which is in effect the emf E, are taken. The open-circuit characteristic is illustrated as OCC in Fig. 8.12. Note that the coordinates are in per unit values.

Losses are windage, friction, and core losses, if no armature current flows. The core loss is a function of the air-gap flux, which in turn is proportional to the open-circuit induced emf. Measurement of the input power gives all these losses, but plotting them against the induced emf allows separation of the core loss. At zero induced emf the only losses are windage and friction losses. This is depicted in Fig. 8.13.

B. SHORT-CIRCUIT CHARACTERISTIC

The short-circuit characteristic (SCC) is a plot of the armature-winding current against the field current or mmf, when the armature winding terminals are short-

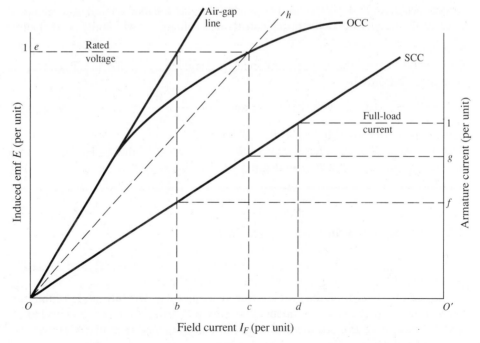

Figure 8.12   Open-circuit characteristic (OCC) and short-circuit characteristic (SCC).

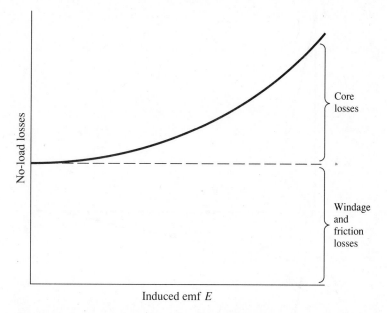

Figure 8.13 Separation of the windage and friction loss.

circuited. This is illustrated in Fig. 8.12 as the curve SCC. It is a linear curve, because there is no saturation. If the terminal voltage V is zero, more than 90% of the voltage drop appears across the synchronous reactance.

Inspection of the generator phasor diagram in Fig. 8.14 for the short-circuit condition shows that the armature winding current $\mathbf{I}$ lags the induced emf $\mathbf{E}$ by nearly 90°. Accordingly, the armature reaction flux produced by $\mathbf{I}$ is almost directly opposed to the main field flux $\mathbf{\Phi}$, which produces $\mathbf{E}$. The two opposing fluxes keep the resultant air-gap flux to a low level, so that saturation in the teeth or pole shoes is never reached.

Measurement of the mechanical input power during the short-circuit test gives the windage and friction losses, which are constant, plus the electric loss known as the short-circuit load loss. The windage and friction losses are constant since the speed is constant and can be separated from the other losses by having no current in the field coils and hence no armature winding current. The form of the loss curve as a function of armature winding current is shown in Fig. 8.15. It is usual to take the loss per phase P_{sc} at full-load armature winding current and determine an effective value of resistance per phase R_a given by

$$R_a = \frac{P_{sc}}{I_{FL}^2}. \tag{8.5.18}$$

This value is assumed constant for the purposes of calculating the machine performance. It is different from a dc test value in that it includes the effect of stray load losses such as iron losses caused by the armature leakage flux and by the resulting air-gap flux, eddy current losses in the windings themselves, and the skin effect.

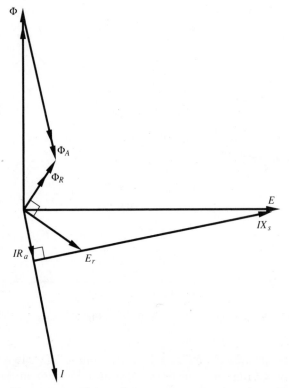

Figure 8.14 Phasor diagram of generator on short circuit.

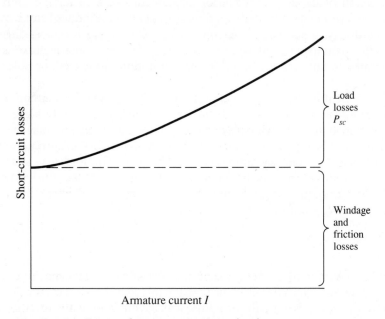

Figure 8.15 Losses of generator on short circuit.

C. SYNCHRONOUS REACTANCE

If the load at the terminals of a generator is a short circuit, Fig. 8.10 indicates that the synchronous impedance is given by $Z_s = E/I$. That is, the synchronous impedance is the induced emf divided by the short-circuit current, all in phase values, for a particular field mmf. The induced emf is the open-circuit voltage for the same mmf. Accordingly, from the OCC and SCC for a particular field current Ob, as shown in Fig. 8.12,

$$Z_s = \frac{Oe}{O'f} \text{ per unit.} \qquad (8.5.19)$$

The air-gap line was used to establish Oe, the induced emf, since the armature reaction of the short-circuit current opposes the main field and saturation does not prevail. Hence the value of Z_s obtained this way is called the unsaturated value of the synchronous impedance. Since it is common to neglect the armature-winding resistance R_a for large machines, the unsaturated value of the synchronous reactance becomes equal to the synchronous impedance.

In general, it might be expected that some parts of the magnetic circuit are saturated under normal running conditions. Thus saturation is accounted for in some way if the magnetization curve is approximated by the straight line Oh in Fig. 8.12, where Oh intersects the OCC at the rated voltage. The saturated value of the synchronous reactance at rated voltage is defined as

$$X_s \approx \frac{Oe}{O'g} \text{ per unit.} \qquad (8.5.20)$$

The line Oh is representative of a saturated condition. Results from measurements cannot have great accuracy, but for many cases they can be satisfactory. For example, it may be necessary to find the excitation current for given terminal conditions V, I, and $\cos\theta$. Use is made of the saturated value of synchronous reactance to calculate the induced emf E from $\mathbf{E} = \mathbf{V} + \mathbf{j}\mathbf{I}X_s$, and the field current I_F is obtained from E on the line Oh. This is because this emf E is the value under saturated conditions. The value of the excitation current I_F, that is needed to obtain the given terminal conditions, can be read from the OCC.

A more accurate way to determine the saturated value of the synchronous reactance is to use a saturation factor k, which is applied only to the armature reaction x_a. The leakage reactance x_l is independent of saturation conditions in the iron and remains sensibly constant. In this method we find first the unsaturated value of the armature reaction reactance from

$$x_a(\text{unsat.}) = X_s(\text{unsat.}) - x_l \qquad (8.5.21)$$

where $X_s(\text{unsat.}) = Oe/O'f$ from the OCC and SCC in Fig. 8.12. (The leakage reactance x_l is found from the Potier triangle presented in subsection E under section 8.5.5.) The saturated value of the armature reactance is given by

$$x_a(\text{sat.}) = x_a(\text{unsat.})/k \qquad (8.5.22)$$

where the saturation factor k is defined as the ratio of the field current required for an armature induced emf on the OCC to the field current required for the same voltage on the air-gap line. So, from Fig. 8.12,

$$k \triangleq Oc/Ob. \tag{8.5.23}$$

Consequently the new value of the saturated synchronous reactance is

$$X_s(\text{sat.}) = x_a(\text{sat.}) + x_l. \tag{8.5.24}$$

EXAMPLE 8.4

A 3-phase, 6.6-kV, 12-MVA, Y-connected generator has the following test results.

Open-circuit test (OCC)	EMF E (volts/phase)		1000	2000	3000	3500	4000
	Field current I_F (amperes)	24		48	75	98	160
Short-circuit test (SCC)	$V = 0$, $I = 1050$ A, $I_F = 27$ A.						

Estimate the unsaturated and saturated values of the synchronous reactance from the OCC and SCC, if the armature-circuit resistance R_a can be neglected and if the leakage reactance can be assumed to have a value of 0.2 ohm per phase.

Solution
From the air-gap line of the OCC a field winding current of 92 A is required to produce the rated open-circuit voltage of 3810 V per phase. From the SCC a short-circuit current of 3810 A flows in the armature winding, if the field winding current is 92 A. Hence the unsaturated value of the synchronous reactance is given by

$$X_s = E/I = 3810/3810 = 1.0 \text{ ohm.}$$

From the saturated part of the OCC a field-winding current of 127 A is required to produce the rated voltage of 3810 V. The corresponding short-circuit current is 5400 A. Hence the saturated value of the synchronous reactance at this particular field current is $X_s(\text{sat.}) = 3810/5400 = 0.706$ ohm.

The saturation factor k is the ratio of field currents to produce rated voltage on the OCC and the air-gap line. That is, $k = 127/92 = 1.4$.
Since $x_a(\text{unsat.}) = X_s - x_l = 1.0 - 0.2 = 0.8$ ohm
$x_a(\text{sat.}) = x_a(\text{unsat.})/k = 0.8/1.4 = 0.59$ ohm
and $X_s(\text{sat.}) = x_a(\text{sat.}) + x_l = 0.59 + 0.2 = 0.79$ ohm.

D. ZERO POWER FACTOR CHARACTERISTIC

An additional test for the purpose of determining leakage reactance and armature reaction is the zero power factor load test. This test is one in which the generator is loaded either by means of reactors or an underexcited synchronous motor. While the armature winding current I is maintained at its full-load value, readings of terminal voltage V are taken as the field mmf F_F is varied. Plotting V against F_F gives the zero power factor characteristic (ZPFC), which, as shown in Fig. 8.16, is similar to but

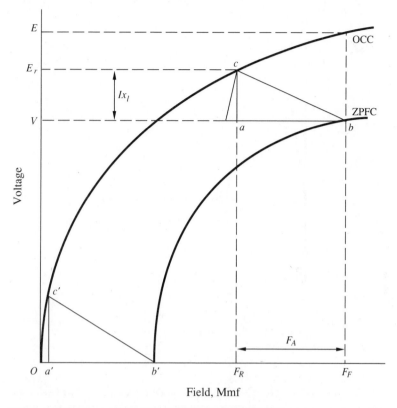

Figure 8.16 Zero power factor characteristic (ZPFC).

displaced from the OCC. The similarity can be explained by a simple theory which employs the assumptions that both the armature winding resistance and the field pole leakage can be ignored.

We can start with the phasor diagram for the condition of zero power factor load, as shown in Fig. 8.17. The armature winding current **I** lags the terminal voltage **V** (reference phasor) by 90° so that the armature leakage-reactance drop $j\mathbf{I}x_l$ is in phase with **V**. The resistance drop $\mathbf{I}R_a$ is shown added to **V** and $j\mathbf{I}x_l$ to require an induced emf $\mathbf{E}_r$ to produce them. Accordingly, armature reaction is accounted for already and the emf E_r is induced by the resultant flux Φ_R in the air gap. That is, the field mmf $\mathbf{F}_F$ was sufficient to neutralize the armature reaction mmf $\mathbf{F}_A$ to provide the resultant mmf $\mathbf{F}_R$, as shown in Fig. 8.17.

If the resistance R_a is ignored, scalars can be employed to give

$$V = E_r - Ix_l \quad \text{and} \quad F_F = F_R + F_A. \tag{8.5.25}$$

Both of these equations are satisfied by the relations of the triangle *abc* in Fig. 8.16.

Since the armature winding current I is maintained constant in the ZPFC, both the leakage reactance drop *ac* and the armature reaction *ab* must be constant for any

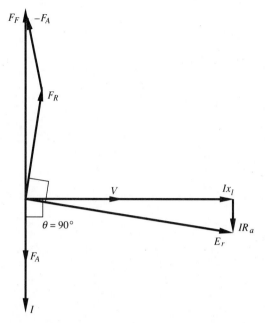

Figure 8.17 Phasor diagram for zero power factor load.

field excitation. Therefore, the two characteristics OCC and ZPFC are parallel, because, if triangle *abc* is moved parallel to itself with *c* on the OCC, then *b* must be on the ZPFC. If the triangle *abc* is taken to its limit ($V = 0$, and $b = b'$, which is the mmf condition to produce full-load current on short circuit), the triangle *abc* becomes part of the larger triangle $Ob'c'$, called the Potier triangle.

E. LEAKAGE REACTANCE

Due to the construction of the Potier triangle between the OCC and ZPFC, we are able to find both the leakage reactance and the armature reaction. A closer approximation to the saturated value of synchronous reactance follows from this.

The graphical determination of the Potier triangle is shown in Fig. 8.18. Information needed to start the construction is the OCC and two points on the ZPFC. Common points are the field current or mmf to produce rated current on short circuit ($V = 0$) and the field current or mmf to produce rated voltage across a zero power factor load for the same full-load current. These are the points b' and b, respectively. The ZPFC could be drawn in because it is parallel to the OCC, but it is sufficient to

1. draw bO' produced parallel to the abscissa axis
2. measure bO' equal to $b'O$
3. draw $O'c$ parallel to the air-gap line to meet the OCC in c
4. draw ca normal to bO' to meet bO' in a.

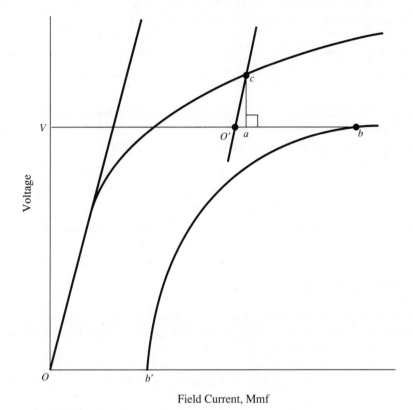

Figure 8.18 Graphical construction of the Potier triangle.

Thus the Potier triangle can be drawn. The leakage reactance is obtained from $Ix_l = ca$ and the armature reaction is obtained from $F_A = ab$, which may be in terms of field quantities. The value of the leakage reactance is likely to be a little high because the main field-pole leakage was not subtracted from the resultant flux Φ_R to produce the emf E_R. This value of leakage reactance also decreases with increasing saturation, but this is ignored for calculations if the voltage is about the normal rated value and steady-state conditions prevail. The leakage reactance obtained in this fashion is known also as the Potier reactance, named after the French engineer who devised the ingenious process of separating the leakage reactance from the armature reaction.

EXAMPLE 8.5

A 3-phase, 6.6-kV, 12-MVA, Y-connected synchronous generator has the following test results.

Open-circuit test (OCC)						
	E volts/phase	1000	2000	3000	3500	4000
	I_F amperes	24	48	75	98	160

Short-circuit test (SCC) $V = 0$, $I = 1050$ A, $I_F = 27$ A.
Zero power factor test $V = 3810$ volts/phase, $I = 1050$ A, $I_F = 182$ A.
  (ZPFC)

Determine the value of the leakage reactance from the Potier triangle.

Solution
From the nomenclature in Fig. 8.18 and the SCC results from value of Ob' is 27 A.
For $O'b$ equal to 27 A on the ZPFC at 3810 V, the line $O'c$ is drawn parallel to the
air-gap line to meet the OCC in c. The line ac is drawn normal to bO' to meet bO'
in a. By measurement

$$Ix_l = ca = 210 \text{ V}.$$

Since $I = 1050$ A, $x_l = 210/1050 = 0.2$ ohm.

8.5.6. Voltage Regulation

Regulation is associated with the change of terminal voltage as the load alters. It is
defined as the percentage rise of voltage, when the load is removed and the speed and
excitation are maintained constant. That is

$$\text{Regulation} \triangleq REG \triangleq \frac{E - V}{V} \times 100\% \tag{8.5.26}$$

where V is the rated voltage at a given load and power factor, and E is the no-load
value of the terminal voltage.

From some given data of the generator load, V, I, and $\cos\theta$, it is straight forward
to predict regulation, if the equivalent circuit model is used. However, the result may
not be too accurate, because it depends on how the synchronous reactance takes into
account the complicated saturation effects. It is simple to write down the equation for
the emf E from eq. (8.5.15), but it takes the next four sections to describe some of the
ways to solve for E and obtain the regulation.

A. EMF METHOD

This method of calculating the voltage regulation of the round-rotor generator relies
upon estimating the synchronous reactance from the OCC and SCC curves. For any
particular field current, neglecting the resistance R_a, the synchronous reactance is

$$X_s \approx \frac{\text{open-circuit voltage } E \text{ (from OCC)}}{\text{short-circuit current } I \text{ (from SCC)}}. \tag{8.5.27}$$

For high values of field excitation the magnetic circuit becomes saturated and X_s
decreases. However, the SCC was obtained under conditions of no saturation so the
value of X_s from eq. (8.5.27) will be high.

A general phasor diagram that is appropriate is seen in Fig. 8.11a. The voltage equation

$$\mathbf{E} = \mathbf{V} + \mathbf{I}(R_A + jX_s) \tag{8.5.28}$$

is expanded to real and imaginary parts with the current $\mathbf{I}$ taken as reference. That is

$$\begin{aligned}
\mathbf{E} &= V(\cos\theta + j\sin\theta) + I(R_a + jX_s) \\
&= (V\cos\theta + IR_a) + j(V\sin\theta + IX_s)
\end{aligned} \tag{8.5.29}$$

so

$$E = ((V\cos\theta + IR_a)^2 + (V\sin\theta + IX_s)^2)^{1/2}. \tag{8.5.30}$$

All values on the right-hand side are known so E can be calculated and the percentage regulation can be found from eq. (8.5.26). Since the value of X_s is too high, the calculation of E is too high and so the estimate of the regulation is high and pessimistic.

A less pessimistic value of the regulation is predicted if the generated emf $\mathbf{E}$ were first calculated using one of the saturated values of the synchronous reactance obtained from the methods presented in subsection C under section 8.5.5. The magnitude of $\mathbf{E}$ on the line oh in Fig. 8.12 determines the field current I_F, which in turn produces the emf on the OCC. It is the open-circuit emf from the OCC, which is used to calculate the machine's regulation.

B. MMF METHOD

To calculate regulation this method uses the OCC and SCC just as in the previous section but utilizes the phasor relations more. The steps to be taken to calculate the induced emf E are described with the aid of Fig. 8.19. From the equivalent circuit, shown in Fig. 8.19a, the voltage equation can be written in the form

$$\mathbf{E} - j\mathbf{I}X_s = \mathbf{V} + \mathbf{I}R_a \triangleq \mathbf{V}_1. \tag{8.5.31}$$

The last two expressions can be represented on the phasor diagram in Fig. 8.19c. V_1 can be considered to be an induced emf due to the main flux, armature reaction, and leakage. So for V_1 the equivalent mmf F_1 can be obtained from the OCC as depicted in Fig. 8.19b. $\mathbf{F}_1$ can be added to the phasor diagram to lead $\mathbf{V}_1$ by $90°$.

The effects of leakage and armature reaction are included in the armature mmf F_A, which is in phase with the current $\mathbf{I}$, as shown in the phasor diagram. The value of F_A is obtained from the SCC for the appropriate value of current I.

$$\mathbf{F}_F = \mathbf{F}_1 - \mathbf{F}_A \tag{8.5.32}$$

is the equation for the field excitation and its value is derived from the phasor diagram. Having the value of F_F, the value of the open-circuit voltage E is obtained from the OCC, and the regulation can be estimated. The armature mmf F_A is obtained from the SCC for unsaturated conditions. Therefore, F_A is too small for saturated conditions. Then, in turn, F_F and E are too small. Thus the value of

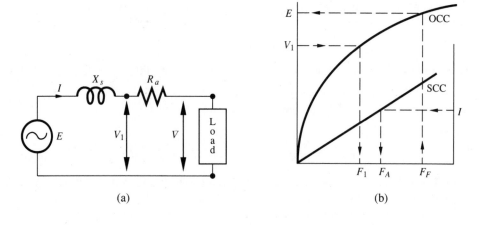

(a)   (b)

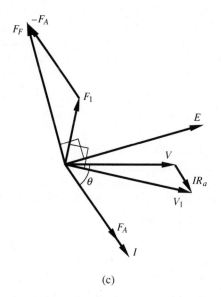

(c)

Figure 8.19 Regulation by mmf method. (a) Equivalent circuit, (b) OCC and SCC, (c) phasor diagram.

regulation becomes optimistic. This method is of interest because it forms the basis for the accepted standard method [American Standards Association (ASA) method].

C. GENERAL METHOD

This method is based on the generator model of a voltage E_r behind a leakage impedance, as shown in Fig. 8.20a, and relies on the OCC and ZPFC.

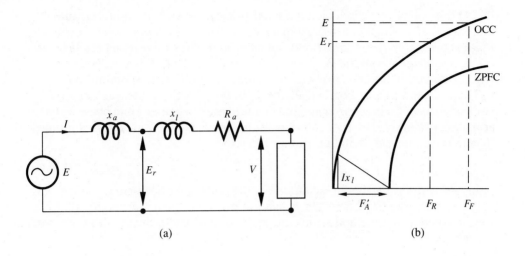

(a)   (b)

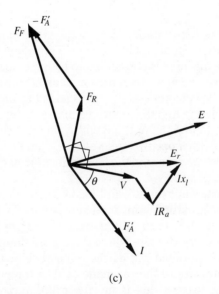

(c)

Figure 8.20 Regulation by the general method. (a) Equivalent circuit, (b) OCC, ZPFC, and Potier triangle, (c) phasor diagram.

The value of E_r is either determined from the phasor diagram of Fig. 8.20c or from the equation

$$\mathbf{E}_r = V + I(\cos\theta - j\,\sin\theta)(R_a + jx_l) \tag{8.5.33}$$

where in both cases the leakage reactance x_l is obtained from the Potier triangle in Fig. 8.20b.

At this point in the discussion it is appropriate to separate the armature mmf F_A that is associated with the current I and the mmf of armature reaction that we can call F'_A. F_A covers both the leakage (modeled by x_l) and armature reaction (modeled by x_a). F'_A is obtained from the Potier triangle, which is shown in Figure 8.20b.

Use is made of the OCC to obtain the resultant mmf F_R corresponding to E_r, and $\mathbf{F}_R$ is added to the phasor diagram, leading $\mathbf{E}_r$ by $90°$. The armature reaction mmf F'_A is obtained from the Potier triangle, and on the phasor diagram $\mathbf{F}'_A$ is in phase with the armature winding current $\mathbf{I}$. As shown in the phasor diagram $\mathbf{F}'_A$ and $\mathbf{F}_R$ are combined to determine the field excitation $\mathbf{F}_F$ according to

$$\mathbf{F}_F = \mathbf{F}_R - \mathbf{F}'_A. \tag{8.5.34}$$

Equipped with a value for the excitation F_F, the value of the open-circuit voltage E is obtained from the OCC, and the regulation can be estimated using eq. (8.5.22). Results based on this method are in close agreement with tests but the ZPFC must be available.

D. ASA METHOD

The American Standards Association (ASA) modified the *mmf method* of obtaining the voltage regulation to give more reliable results. The modified method is called the *ASA method.* First, the machine is considered to be in an unsaturated state and the field mmf is determined. Then an incremental value of mmf is found in order to correct for saturation. This incremental mmf is added to the unsaturated value of mmf and the result is used to find the induced emf E. A value for regulation follows.

Figure 8.21 is used to find the induced emf E on open circuit. The mmf F that would induce an *unsaturated* open-circuit voltage V can be found from the air-gap line. See Fig. 8.21b. $\mathbf{F}$ can be drawn on the phasor diagram at right angles to $\mathbf{V}$, which is used as the reference phasor.

From the appropriate value (usually full-load value) of load current I on the SCC a value is found for the armature mmf F_A that accounts for leakage and armature reaction. $\mathbf{F}_A$ is in phase with $\mathbf{I}$. The resultant mmf $(\mathbf{F} - \mathbf{F}_A = \mathbf{F}'_F)$ corresponds to an open-circuit induced emf for the case where there is no saturation.

See Fig. 8.21a. It is considered that $\mathbf{E}_r$, the voltage behind the leakage impedance $(R_a + jx_l)$, is a good measure of the magnetic field in the poles, where saturation occurs. This is because leakage flux is mostly in unsaturated regions. $\mathbf{E}_r$ can be determined from the phasor diagram (Fig. 8.21b), or from eq. (8.5.33), since the values of V, I, θ, R_a, and x_l are assumed to be known. At this emf E_r, ΔF (see Fig. 8.21b) gives the difference between the *unsaturated* mmf (air-gap line) and the *saturated* mmf (OCC). The amount ΔF is added to the *unsaturated* mmf F'_F to give the total required field mmf F_F. The value of the mmf F_F on the OCC gives the open-circuit emf E, as shown in Fig. 8.21b. The value of regulation follows.

This method is standard practice in the United States. Ignoring the voltage drop IR_a makes little difference to the results.

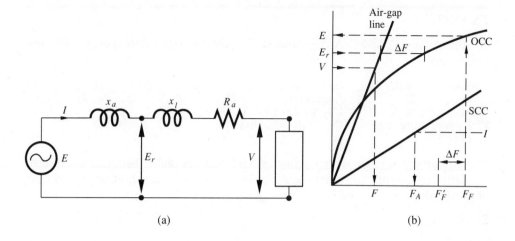

(a)

(b)

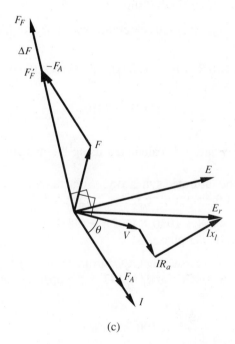

(c)

Figure 8.21   Regulation by ASA method. (a) Equivalent circuit, (b) OCC and SCC, (c) phasor diagram.

EXAMPLE 8.6

A 3-phase, 6.6-kV, 12-MVA, Y-connected, synchronous generator has the following test results.

Open-circuit test (OCC)	E volts/phase	1000	2000	3000	3500	4000
	I_F amperes	24	48	75	98	160
Short-circuit test (SCC)	$V = 0$, $I = 1050$ A, $I_F = 27$ A.					
Zero power factor test (ZPFC)	$V = 3810$ volts/phase, $I = 1050$ A, $I_F = 182$ A.					

The armature winding has a resistance of 0.02 ohm per phase. Estimate the voltage regulation by the emf, mmf, general, and ASA methods for the condition of a full load at rated voltage and 0.8 power factor lagging.

Solution
Emf method
(a) From Fig. 8.11a the generated emf E is given by

$$E = ((V \cos\theta + IR_a)^2 + (V \sin\theta + IX_s)^2)^{1/2}.$$

The unsaturated synchronous reactance was found in Example 8.4 to be 1.0 ohm. Therefore

$$E = ((3810 \times 0.8 + 1050 \times 0.02)^2 + (3810 \times 0.6 + 1050 \times 1.0)^2)^{1/2}$$
$$= 5433 \text{ volts per phase.}$$

$$REG = \frac{E - V}{V} \times 100 = \frac{4533 - 3810}{3810} \times 100 = 19\%.$$

This value of regulation is calculated using the unsaturated value of synchronous reactance.
(b) A value of the saturated synchronous reactance was found in Example 8.4 to be 0.79 ohm, when the saturation factor k was used. Thus

$$E = ((3810 \times 0.8 + 1050 \times 0.02)^2 + (3810 \times 0.6 + 1050 \times 0.79)^2)^{1/2}$$
$$= 4377 \text{ volts per phase.}$$

From the *Oh* line (intersecting OCC at rated voltage) $I_F = 147$ A for $E = 4373$ V. Therefore from the OCC for $I_F = 147$ A the open-circuit emf is 3940 V. The new value of regulation is

$$REG = \frac{3940 - 3810}{3810} \times 100 = 3.4\%.$$

Mmf method
The reference diagram is Fig. 8.19. Let **V** be the reference phasor $3810 \angle 0°$ V.
$\mathbf{V}_1 = \mathbf{V} + \mathbf{I}R_a = 3810 + 1050(0.8 - j0.6)0.02 = 3827 \angle -0.19°$ V.
From the OCC for $V_1 = 3827$ V, $F_1 = 130$ A (equivalent field amperes).
That is $\mathbf{F}_1 = 130 \angle(-0.19° + 90°) = 130 \angle 89.8°$ A.
From the SCC for $I = 1050$ A, $F_A = 27$ A (equivalent field amperes).

Since $\mathbf{F}_A$ is in phase with $\mathbf{I}$, $\mathbf{F}_A = 27\angle -\cos^{-1}0.8 = 27\angle -36.9°$ A. Accordingly the field-winding excitation is

$$\mathbf{F}_F = \mathbf{F}_1 - \mathbf{F}_A = 130\angle 89.8° - 27\angle -36.9° = 148\angle 98.2°\ \text{A}.$$

The corresponding open-circuit emf E for a field-winding current of 148 A from the OCC is $E = 3950$ volts per phase.

$$REG = \frac{E-V}{V}\times 100 = \frac{3950-3810}{3810}\times 100 = 3.7\%.$$

This value of regulation is higher than that obtained by the emf method, as expected, because saturation has been taken into consideration to some extent.

General method
The reference diagram is Fig. 8.20. From the Potier triangle the leakage reactance x_l is 0.2 ohm. The voltage $\mathbf{E}_r$ behind the leakage impedance is

$$\mathbf{E}_r = V + I(\cos\theta - j\sin\theta)(R_a + jx_l).$$

That is, $\mathbf{E}_r = 3810 + 1050(0.8 - j0.6)(0.02 + j0.2) = 3956\angle 2.3°$ V.
The resultant mmf F_R corresponding to E_r on the OCC is $F_R = 150$ A (equivalent field amperes).
$\mathbf{F}_R$ leads $\mathbf{E}_r$ by 90°, so $\mathbf{F}_R = 150\angle 92.3°$ A.
From the Potier triangle the armature reaction mmf is
$F'_A = 19$ A (equivalent field amperes).
$\mathbf{F}'_A$ is in phase with $\mathbf{I}$, so $\mathbf{F}'_A = 19\angle -36.9°$ A.
Consequently the field winding excitation is given by

$$\mathbf{F}_F = \mathbf{F}_R - \mathbf{F}'_A = 150\angle 92.3° - 19\angle -36.9° = 163\angle 97.4°\ \text{A}.$$

From the OCC for a field winding current of 163 A the open-circuit emf E is 4020 V per phase.

$$REG = \frac{E-V}{V}\times 100 = \frac{4020-3810}{3810}\times 100 = 5.5\%.$$

This result is considered to be reliable, since saturation is taken into consideration more fully.

ASA method
The reference diagram is Fig. 8.21.
From the air-gap line for $V = 3810$ V, $F = 91.5$ A (equivalent field amperes).
That is, $\mathbf{F} = 91.5\angle 90°$ A.
From the SCC for $I = 1050$ A, $F_A = 27$ A (equivalent field amperes).
That is, $\mathbf{F}_A = 27\angle -36.9°$ A.
Accordingly the *unsaturated* value of the field-winding excitation is

$$\mathbf{F}'_F = \mathbf{F} - \mathbf{F}_A = 91.5\angle 90° - 27\angle -36.9°\ \text{A}.$$

That is, $F'_F = 108.9$ A (equivalent field amperes).

From the calculation in the *General method* $E_r = 3956$ V.

From the air-gap line and the OCC for $E_r = 3956$ V, $\Delta F = 53$ A (equivalent field amperes).

Therefore the *saturated* value of the field-winding excitation is $F_F = F'_F + \Delta F = 161.9$ A.

From the OCC for $F_F = 161.9$ A, $E = 4020$ V. By chance this is the same value obtained from the *General method,* so the value of the regulation is the same, that is, 5.5%.

8.5.7. Power and Load Angle

The voltage eq. (8.5.17) for the round rotor generator provides a means to express the current **I** in the form

$$\mathbf{I} = \frac{\mathbf{E} - \mathbf{V}}{R_a + jX_s}. \tag{8.5.35}$$

Making use of the geometry provided by the phasor diagram in Fig. 8.22 and rearranging real and imaginary parts, the current becomes

$$\mathbf{I} = \frac{(E\cos\delta - V) + jE\sin\delta}{R_a + jX_s} \tag{8.5.36}$$

where it will be noticed that **V** is the reference phasor $V \angle 0°$. The aim of this section is to find an expression for the generator power output on a per phase basis as a function of the load angle δ and independent of the current. Expressions for the complex power are

$$\mathbf{S} = P + jQ = \mathbf{VI}^* \tag{8.5.37}$$

where **I*** is the conjugate of **I**. On substitution for the current

$$P + jQ = \frac{(VE\cos\delta - V^2)R_a + VEX_s\sin\delta}{Z_s^2}$$
$$+ \frac{j(VE\cos\delta - V^2)X_s - VER_a\sin\delta}{Z_s^2}. \tag{8.5.38}$$

The real power is

$$P = \frac{(VE\cos\delta - V^2)R_a + VEX_s\sin\delta}{Z_s^2} \tag{8.5.39}$$

and the reactive power is

$$Q = \frac{(VE\cos\delta - V^2)X_s - VER_a\sin\delta}{Z_s^2}. \tag{8.5.40}$$

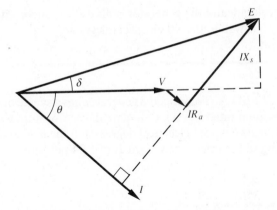

Figure 8.22 Phasor diagram for a lagging power factor load.

It is quite practical to make the assumptions that $R_a \approx 0$ and $X_s \approx Z_s$, so that

$$P \approx \frac{VE}{X_s} \sin\delta \quad \text{and} \quad Q \approx \frac{VE \cos\delta - V^2}{X_s}. \tag{8.5.41}$$

The power angle characteristic is drawn in Fig. 8.23. For negative δ, with **E** lagging **V**, the power output is negative. This means that the power has reversed and the synchronous machine has become a motor. The maximum power output of the generator occurs when $\delta = \pi/2$. Any further increase of mechanical power input causes a decrease in the electrical output power. The difference between the two powers causes

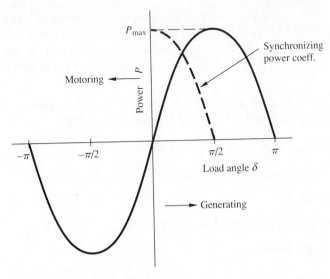

Figure 8.23 Power and load angle.

the machine to accelerate and the generator pulls out of synchronism. The normal operating load angle is well under 90 electrical degrees.

EXAMPLE 8.7

A 3-phase, 6.6-kV, 12-MVA, Y-connected generator has a synchronous reactance of 0.275 per unit (based on rated MVA). Calculate the MVA output and power factor assuming the load angle δ is 15.3 electrical degrees and the field excitation is set at a value corresponding to the open-circuit emf E of 1.04 per unit.

Solution

$$X_s \text{ per unit} = \frac{X_s}{X_{sb}} = \frac{X_s I_b}{V_b} = X_s \frac{V_b I_b}{V_b^2}.$$

The subscript b refers to a base value. Therefore, $X_s = 0.275 \times 6.6^2/12 = 1.0$ ohm. E per unit $= E/V_b$. Hence $E = 1.04 \times 3810 = 3974$ V per phase.

Total power $P = 3 \dfrac{VE}{X_s} \sin\delta = 3810 \times 3974 \times \sin 15.3° = 12 \times 10^6$ W.

Total reactive power

$$Q = 3 \frac{VE \cos\delta - V^2}{X_s} = 3(3810 \times 3974 \times \cos 15.3° - 3810^2) = 0 \text{ VAr}.$$

Total volt-amperes $\mathbf{S} = P - jQ = 12 \times 10^6 \angle 0°$ VA.

EXAMPLE 8.8

A 3-phase, 6.6-kV, 12-MVA, Y-connected generator is connected to an infinite bus whose voltage is 3.81 kV per phase. Calculate the two possible values of load angle δ if the generator delivers full-load current. Assume the input power to the generator is 10 MW, the synchronous reactance is 1.0 ohm and the losses can be neglected.

Solution

Total power $P = 3\,VI \cos\theta$. P, V, and I are constant, therefore $\cos\theta$ is constant, either leading or lagging. $\cos\theta = 10/12 = 0.833$, so $\sin\theta = \pm0.553$.

Full-load current $I = \dfrac{12 \times 10^6}{\sqrt{3} \times 6.6 \times 10^3} = 1050$ A.

The load angles can be determined graphically from a phasor diagram or analytically.

Lagging power factor

$$\mathbf{E} = \mathbf{V} + j\mathbf{I}X_s = 3810 + j1050(0.833 - j0.533) \times 1.0$$

or $E = 4470$ V per phase.

$$P = 3 \frac{VE}{X_s} \sin\delta.$$

Therefore $\sin\delta = \dfrac{10 \times 10^6}{3 \times 3810 \times 4470}$

that is $\delta = 11.3°$ electrical.

Leading power factor

$$E = 3810 + j1050(0.833 + j0.553) \times 1.0$$

or $E = 3346$ V per phase. Therefore $\sin\delta = \dfrac{10 \times 10^6}{3 \times 3810 \times 3346}$

or $\delta = 15.2$ electrical degrees.

A. SYNCHRONIZING POWER

The *synchronizing power coefficient* is defined as the rate at which the power varies with the load angle, and has other names such as stability factor, stiffness of coupling, and rigidity factor.

Consider the generator to have fixed excitation and to be connected to an infinite bus (constant voltage V and frequency f). The stability of the generator depends on the fact that, if the load angle δ is given an incremental increase due to some small disturbance, the additional electric power output causes the electromagnetic torque to be greater than the driving torque. Retardation is produced by the torque imbalance and a return to the original rotor position should ensue. This excess torque is known as the *synchronizing torque* and the power associated with it is the *synchronizing power*.

For steady-state conditions the generator rate of change of power for constant V and E is

$$\frac{dP}{d\delta} = \frac{d}{d\delta}\left(\frac{VE}{X_s}\sin\delta\right) = \frac{VE}{X_s}\cos\delta. \qquad \textbf{(8.5.42)}$$

Thus the synchronizing power ΔP due to an increment of load angle $\Delta\delta$ is

$$\Delta P = \frac{dP}{d\delta}\Delta\delta = \frac{VE}{X_s}\cos\delta\,\Delta\delta. \qquad \textbf{(8.5.43)}$$

As shown in Fig. 8.23, the synchronizing power coefficient $dP/d\delta$ is greatest at no load, when $\delta = 0$, and decreases to zero at maximum power, when $\delta = \pi/2$. The latter case is termed the limit of stability. An increase of excitation increases the synchronizing power coefficient.

8.6. SYNCHRONOUS GENERATOR (SALIENT POLES)

Due to the field excitation, a synchronous machine has a fundamental flux distribution in the air gap of the form $\phi = \Phi \sin(\theta - \omega t)$. This distribution is the same for both a round rotor and a salient-pole structure. The steady-state emf per phase, generated in the armature windings on no load, is given by

$$E = \frac{\hat{E}}{\sqrt{2}} = 4.44\Phi f N. \tag{8.3.3}$$

A generator on load will have sinusoidal currents flowing in the armature windings and these currents give rise to an air-gap mmf F_A. In a round rotor machine without saturation the armature winding mmf and the field excitation mmf could be summed to find the resultant flux in the air gap, since the reluctance is constant around the air gap. In a salient-pole machine the mmfs cannot be summed to find the resultant flux, because the reluctance of the air gap varies with position. In the pole axis the air gap is small so the value of reluctance is low, whereas the air gap in the quadrature axis is relatively large and the reluctance is high.

A method of two reactances is used to account for armature reaction and to facilitate the analysis of salient-pole, synchronous machines.

8.6.1. Armature Reaction

In section 8.3.2 the fundamental armature reaction mmf was resolved into two components, one in the direct axis and the other in the quadrature axis, with the form

$$\mathbf{F}_A = F_d + jF_q = \frac{-3N\hat{I}}{\pi}(\sin\theta + j\cos\theta). \tag{8.6.1}$$

If position α is measured from the pole axis, the armature reaction mmf can be expressed as

$$F_A(\alpha) = F_d\cos\alpha + F_q\sin\alpha. \tag{8.6.2}$$

This component distribution in space is shown in Fig. 8.24a.

The flux due to armature reaction is obtained from the general equation

$$\phi_A = \frac{F_A}{\mathcal{R}_A} = P_A F_A \tag{8.6.3}$$

where the reluctance $\mathcal{R}_A$ or its inverse, the permeance P_A, is a function of position α. As indicated in Fig. 8.24a, the permeance can be approximately expressed by the first two terms of the Fourier series

$$P_A = P_0 + P_2\cos2\alpha \tag{8.6.4}$$

that is, a constant term plus a second harmonic term. Thus the flux distribution is given by

$$\begin{aligned} \phi_A &= (P_0 + P_2\cos2\alpha)(F_d\cos\alpha + F_q\sin\alpha) \\ &= \left(P_0 + \frac{P_2}{2}\right)F_d\cos\alpha + \left(P_0 - \frac{P_2}{2}\right)F_q\sin\alpha \\ &\quad + \frac{P_2}{2}(F_d\cos3\alpha + F_q\sin3\alpha). \end{aligned} \tag{8.6.5}$$

Hence, there is a third-harmonic component of flux due to the second harmonic in the permeance. This third harmonic will be incorporated into the leakage flux components, while the fundamental components of flux

$$\phi_{Ad} = \left(P_0 + \frac{P_2}{2} \right) F_d \quad \text{and} \quad \phi_{Aq} = \left(P_0 - \frac{P_2}{2} \right) F_q \quad \text{(8.6.6)}$$

will be considered as armature reaction effects. As would be expected, the direct axis component ϕ_{Ad} is greater than the quadrature axis component ϕ_{Aq}, because of the difference in reluctances or permeances. For a round rotor machine P_2 becomes zero and the third harmonic of flux vanishes. Only in an unsaturated machine can the armature reaction flux be added to the field flux. There is little saturation in the quadrature axis. However, if there is saturation in the direct axis, the field mmf must be added to the direct axis component of the armature reaction mmf and the resultant flux is found from the magnetization curve.

A space phasor diagram is shown in Fig. 8.24b to illustrate the positions of the fundamental field quantities. It can be assumed that there is a fictitious space component of current I_d to account for the armature reaction in the direct axis and another fictitious component I_q to account for the armature reaction in the quadrature axis. The components are related to actual quantities by the expressions

$$\Phi_A^2 = \Phi_{Ad}^2 + \Phi_{Aq}^2 \quad \text{and} \quad I^2 = I_d^2 + I_q^2. \quad \text{(8.6.7)}$$

Although the pairs of components Φ_{Ad}, I_d, and Φ_{Aq}, I_q are cophasors, the actual space phasor Φ_A is not in phase with the armature winding current per phase **I** because the reluctances on the direct and quadrature axes differ.

8.6.2. Two-reactance Theory

For the round rotor machine analysis the effect of the armature flux was modeled by a synchronous reactance. A similar means can be employed in the analysis of salient-pole machines. The difference in this case is that two reactances have to be introduced.

The field excitation produces a flux per pole Φ. When the prime mover drives the generator at some speed, the open-circuit armature-winding induced emf is **E**, which is in quadrature with the space phasor Φ. On load the armature winding current per phase is some value **I**. If the current **I** is in phase with the emf **E**, then the armature reaction flux is centered on the quadrature axis. But if the current **I** lags the emf **E** by 90°, the armature-reaction flux is centered on the direct axis and has a demagnetizing effect. In general the current **I** may lag the emf **E** by some angle less than 90°, in which case there will be components of armature-reaction flux on both the direct axis and the quadrature axis.

Let the armature winding current **I** be resolved into the two direct and quadrature axis space components I_d and I_q, respectively. The current I_d is responsible for the flux Φ_{Ad} on the direct axis and the current I_q is responsible for the quadrature axis flux Φ_{Aq}. Associated with these current components are the voltage drops $I_d X_d$ and

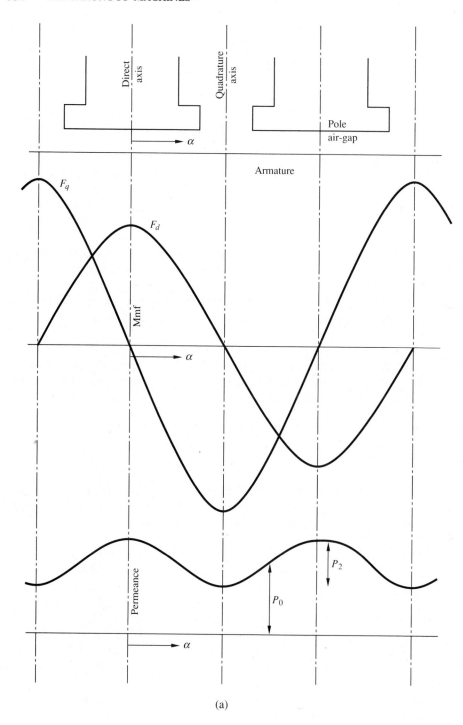

(a)

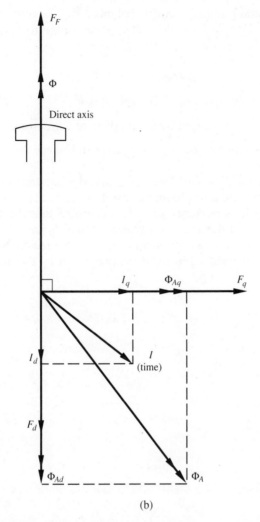

(b)

Figure 8.24 Armature reaction of a salient-pole generator. (a) Component mmfs and permeance, (b) space phasor diagram.

$I_q X_q$, where by definition

$X_d \triangleq$ synchronous reactance on the direct axis

$X_q \triangleq$ synchronous reactance on the quadrature axis.

In this way the inductive voltage drops account for armature reaction and leakage flux on each axis in the same way that the voltage drop IX_s modeled the armature reaction and leakage in the round rotor machine.

The concept of the component synchronous reactances can be developed in greater detail from the armature circuit voltage equations based on phase values. The total armature-winding induced emf is balanced by the voltage drop due to the armature winding resistance and the voltage drop across the load at the terminals. Thus

$$\mathbf{E} + \mathbf{E}_A + \mathbf{E}_l = \mathbf{V} + \mathbf{I}R_a \qquad (8.6.8)$$

where E = emf induced by the field flux Φ

                E_A = emf induced by the fundamental armature reaction flux Φ_A

and E_l = emf induced by the equivalent leakage flux Φ_l.

The excitation emf E is given by eq. (8.3.3) and $\mathbf{E}$ lags the flux Φ by 90°. This is the basis for the space and time phasor diagrams in Fig. 8.25, in which the quadrature axis and $\mathbf{E}$ are taken as reference axes. It is assumed that the armature winding current $\mathbf{I}$ lags $\mathbf{E}$, so that the space components I_q and I_d have directions as shown. These currents give rise to the components of armature reaction flux Φ_{Aq} and Φ_{Ad}. Emfs E_{Aq} and E_{Ad} are induced in the armature circuit by the fluxes Φ_{Aq} and Φ_{Ad},

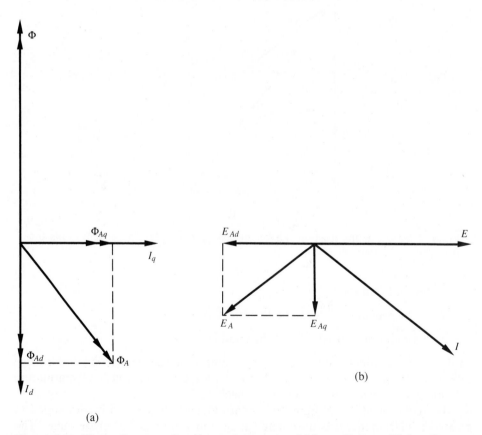

(a)

(b)

Figure 8.25 Phasor diagrams of a salient-pole generator. (a) Space diagram, (b) time diagram.

respectively, and the relations between them are

$$E_{Ad} = 4.44\Phi_{Ad}fN \quad \text{and} \quad E_{Aq} = 4.44\Phi_{Aq}fN \tag{8.6.9}$$

where the emfs lag their respective fluxes by 90°, as shown in the figure. In an unsaturated machine

$$\Phi_{Ad} \propto I_d, \quad \text{therefore} \quad E_{Ad} \propto I_d.$$

That is

$$E_{Ad} = I_d x_d \tag{8.6.10}$$

where x_d is the proportionality constant and is an equivalent reactance of armature reaction on the direct axis, since E_{Ad} is an induced emf. Thus the emf E_{Ad} can be replaced by an inductive voltage drop. The interpolar path is unlikely to be saturated because of the high reluctance, so in a similar fashion the emf E_{Aq} can be represented by a voltage drop given by

$$E_{Aq} = I_q x_q. \tag{8.6.11}$$

From the phasor diagram in Fig. 8.25b the resultant emf $\mathbf{E}_A$ due to the armature reaction is

$$\mathbf{E}_A = -E_{Ad} - jE_{Aq} = -I_d x_d - jI_q x_q. \tag{8.6.12}$$

Thus the emf E_A has been transformed into equivalent voltage drops.

The equivalent leakage flux Φ_l gives rise to the induced emf E_l. Resolving into direct and quadrature axis components in the same way as for armature reaction, we arrive at the similar expressions

$$\mathbf{E}_l = -E_{ld} - jE_{lq} = -I_d x_l - jI_q x_l. \tag{8.6.13}$$

The only difference is that the equivalent leakage reactances in the direct and quadrature axis paths differ very little, so they assume the same value x_l.

If the induced emfs $\mathbf{E}_A$ and $\mathbf{E}_l$ are replaced by their equivalent voltage drops, the voltage eq. (8.6.8) becomes

$$\mathbf{E} - I_d x_d - jI_q x_q - I_d x_l - jI_q x_l = \mathbf{V} + \mathbf{I}R_a. \tag{8.6.14}$$

But $\mathbf{I} = I_q - jI_d$, so

$$\mathbf{E} = \mathbf{V} + I_d X_d + jI_q X_q + I_q R_a - jI_d R_a \tag{8.6.15}$$

where $X_d \triangleq x_d + x_l$ and $X_q \triangleq x_q + x_l$.

This voltage equation can be compared with that obtained for the round rotor machine by putting

$$\mathbf{I} = I_q - jI_d \quad \text{and} \quad X_d = X_q = X_s. \tag{8.6.16}$$

Figure 8.26 depicts the phasor diagrams for lagging, unity, and leading power factors. The diagrams are constructed from the voltage equation above. All three diagrams are drawn to illustrate how the power factor changes from lagging to leading as the excitation, and hence E, is reduced when the generator is connected to an infinite bus (V is constant).

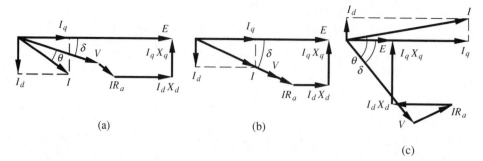

Figure 8.26 Phasor diagrams of a salient-pole generator. (a) Lagging power factor, (b) unity power factor, (c) leading power factor.

A. PHASOR DIAGRAM CONSTRUCTION

In Fig. 8.26 the emf **E** was used as a reference to draw the phasor diagrams, but the phase relationship between **E** and the armature winding current per phase **I** is not usually known initially. In practice the terminal voltage V, the current I, the power factor $\cos\theta$, and the synchronous reactances X_d and X_q are known. The load angle δ and the fictitious current components I_d and I_q are unknown, but can be found, together with the emf E, from the following construction of the phasor diagram, which is illustrated in Fig. 8.27.

1. Draw **V** as the reference.
2. Draw **I** at an angle θ to **V**.
3. Add to **V** the resistive voltage drop IR_a in phase with **I**.
4. Draw the line $0'a'$ perpendicular to **I** and equal to IX_q.
5. Extend $0a'$ to c'. It will be proved that $0c'$ equals E (in scaled length).

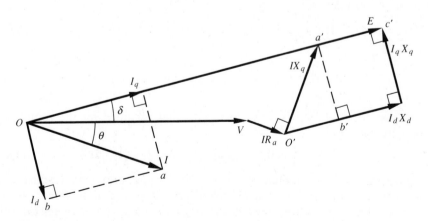

Figure 8.27 Construction of the phasor diagram.

6. Resolve the current **I** into the components I_q along $0a'$ and I_d perpendicular to $0a'$.
7. Now I_dX_d can be added to $\mathbf{I}R_a$ at right angles to I_d, and I_qX_q can be added to I_dX_d at right angles to I_q to meet $0a'$ extended to c'.

The phasor diagram has been drawn.

Proof of construction

Let the phasor diagram in Fig. 8.27 represent the voltage equation for one phase of the armature circuit. For this to be the case we have to prove that the line $0'a'$, drawn at right angles to $\mathbf{I}R_a$ to meet the phasor **E** in a', is equal to IX_q.

From the notation used in the figure, the triangles $0ab$ and $0'a'b'$ are similar, because corresponding sides are perpendicular. Therefore

$$\frac{0'a'}{0a} = \frac{b'a'}{ba} \qquad \text{or} \qquad 0'a' = \frac{b'a'}{ba}\, 0a.$$

Hence

$$0'a' \equiv \frac{I_qX_q}{I_q}\, I = IX_q$$

This proves the construction.

8.6.3. Power and Load Angle

In terms of terminal quantities the output power per phase of a salient-pole generator is

$$P = VI \cos\theta. \tag{8.6.17}$$

In order to find the power as a function of the load angle δ, the angle between the phasors **V** and **E**, it is a reasonable approximation to assume that the effect of the armature winding resistance R_a is negligible. If the armature winding resistance is ignored, the phasor diagram in Fig. 8.28 can be considered. From an inspection of this figure

$$I \cos\theta = I_q \cos\delta + I_d \sin\delta. \tag{8.6.18}$$

Thus

$$P = V(I_q \cos\delta + I_d \sin\delta). \tag{8.6.19}$$

Also from the figure, $V\cos\delta = E - I_dX_d$ and $V\sin\delta = I_qX_q$.
Hence $I_d = (E - V\cos\delta)/X_d$ and $I_q = V\sin\delta/X_q$.
On substitution for I_d and I_q in the power equation

$$P = \frac{VE}{X_d}\sin\delta + \frac{V^2(X_d - X_q)}{2X_dX_q}\sin 2\delta. \tag{8.6.20}$$

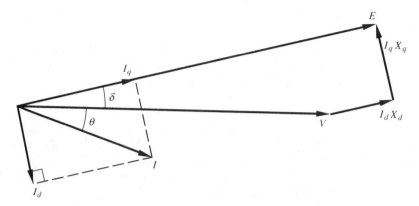

Figure 8.28 Phasor diagram neglecting armature winding resistance.

This equation has two terms, one of which is independent of the excitation emf E and relies on a difference in the values of the direct and quadrature axis synchronous reactances. If $X_d = X_q = X_s$, there is no saliency and the above power equation reduces to that obtained for the round rotor machine in section 8.5.7.

The two component terms of power, together with the resultant sum, are plotted as a function of the load angle in Fig. 8.29. The information obtained from this curve

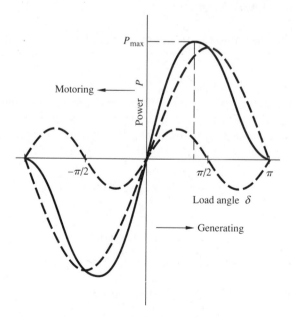

Figure 8.29 Power and load angle for saliency.

is that the maximum theoretical power output that can be achieved occurs at a load angle less than 90° for the salient-pole generator and is greater than the maximum power of the round rotor machine because of the saliency term. Further, the synchronizing power coefficient, which is the slope of the power and load-angle curve, is greater for the salient-pole generator, and consequently the machine with saliency is said to be stiffer than the machine with a round rotor.

For a salient-pole machine the synchronizing power coefficient is

$$\frac{dP}{d\delta} = \frac{VE}{X_d}\cos\delta + \frac{V^2(X_d - X_q)}{X_d X_q}\cos 2\delta \qquad \textbf{(8.6.21)}$$

and the synchronizing power ΔP for a small change of load angle $\Delta\delta$ is

$$\Delta P = \frac{dP}{d\delta}\Delta\delta. \qquad \textbf{(8.6.22)}$$

In Fig. 8.29 the first-quadrant curve represents the relationship between the power and the load angle for the generator. The curve in the third quadrant, in which the power and load angle are both negative, is for the salient-pole machine acting as a motor. The convention for the positive and negative signs must be explained.

If the electrical power output of a generator is taken by convention to be positive, then an electrical power input can be considered to be negative. For the machine to act as a motor it must have an electrical power input. Therefore for motoring action the electrical power P is deemed to be negative.

Load angle δ is the phase difference between the excitation emf $\mathbf{E}$ and the terminal voltage $\mathbf{V}$ and has no physical meaning. However, a physical picture can be constructed in order to give significance to the positive and negative signs attached to δ. The field poles of the machine rotate synchronously and the armature mmf pattern rotates synchronously. Interaction between the excitation field and the armature field is the cause of the electromagnetic torque, which balances the mechanical torque for steady-state operation. When the machine generates, the rotor is mechanically driven, and the electromagnetic torque opposes the mechanical torque tending to pull the field poles back into alignment with the armature poles. Thus the excitation field leads the armature reaction field. Conceptually, the excitation and armature reaction phase angle for the fields can be associated with the way the excitation emf leads the terminal voltage by the angle δ for generating action. By convention, a leading angle δ is called positive. On the other hand, when the machine motors, it can be considered that the armature field is pulling the field poles around. The electromagnetic torque tends to align the armature and field poles, but the mechanical load torque makes the field poles lag the synchronously rotating armature poles by an amount so that the torques balance.

Again, the angle between the field phasors can be associated with the fact that the excitation emf lags the terminal voltage by the angle δ for motoring action. It is the lagging δ that is termed negative with respect to a leading δ. Accordingly, we place the power and load-angle curve in the first quadrant for generator action and in the third quadrant for motor action.

A. REACTIVE POWER

The reactive power Q for the salient-pole generator can be found as a function of load angle δ in the same manner as the real power P was determined, if the armature resistance R_a is neglected. The output reactive power per phase is given by

$$Q = VI \sin\theta. \qquad (8.6.23)$$

From Fig. 8.28 we obtain the relations

$$I \sin\theta = I_d \cos\delta - I_q \sin\delta$$

$$V \cos\theta = E - I_d X_d \quad \text{and} \quad V \sin\delta = I_q X_q.$$

Substitution into eq. (8.6.23) yields

$$Q = \frac{VE}{X_d} \cos\delta - \frac{V^2}{X_d} \cos^2\delta - \frac{V^2}{X_q} \sin^2\delta. \qquad (8.6.24)$$

8.6.4. Determination of Synchronous Reactances

For the steady-state analysis of the salient-pole machine the synchronous reactances X_d and X_q must be known. They may be determined experimentally with the aid of a method called the slip test.

The slip test is accomplished by mechanically driving the rotor slightly above or below synchronous speed, leaving the field circuit open, applying a low voltage at normal frequency to the armature circuit, and recording the applied voltage, current and the induced field-winding emf by means of oscillograms. For example, see Fig. 8.30. The applied voltage and the slip are kept as low as possible to minimize the errors introduced by eddy currents in pole faces or damper windings. Another reason for the low voltage is to prevent the rotor being pulled into synchronism by the armature field.

Since the armature field rotates synchronously, it experiences in turn the direct and quadrature axis magnetic circuits of the rotor that is driven sub- or super-synchronously. Consequently, the armature winding currents are influenced by changing reactances. When the armature field is centered on the direct axis, the reactance is maximum, so the armature-winding current is a minimum. The current is a maximum when the armature field is centered on the quadrature axis, and the reactance is a minimum. The current and reactance alternate between these two extremes as the armature field rotates at slip frequency with respect to the rotor. Due to a high source impedance, the terminal voltage V may be influenced by current and vary as a result. Thus, when the current is maximum, the voltage will be minimum and vice versa. The induced field winding voltage serves as a check, since it is zero when the armature field is in the direct axis and a maximum when the armature field is in the quadrature axis. If the armature winding resistance is neglected, direct

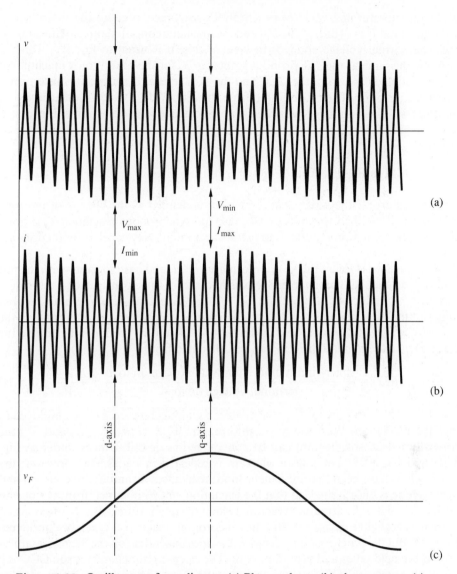

Figure 8.30 Oscillograms from slip test. (a) Phase voltage, (b) phase current, (c) field-induced emf.

measurement from the oscillograms gives

$$X_d = \frac{V_{max}}{I_{min}} \quad \text{and} \quad X_q = \frac{V_{min}}{I_{max}}. \tag{8.6.25}$$

The slip tends to pulsate, because of the fluctuation of reluctance torque with relative pole position, so that X_d can be underestimated. Calculated by this means, X_d

is also the unsaturated value of synchronous reactance, because the main field is unexcited (and $E = 0$) during the test and the magnetic conditions are not the same as with the normal voltage applied. In view of this, it is usual to find X_d, which is virtually the same as the synchronous reactance X_s for the round rotor machine for the unsaturated case, from the OCC and the SCC tests as described in subsection C under section 8.5.5. Then X_q is found from the ratio X_d/X_q, which is determined by the slip test. The saturated value of X_d can be calculated as shown in subsection C under section 8.5.5.

8.6.5. Voltage Regulation

The regulation for the salient-pole generator is defined as the change of terminal voltage when the load is removed. Because the synchronous reactance X_d is not a constant the regulation is difficult to predict accurately. Nevertheless predictions are made.

One method is to use the given information V, I, and $\cos\theta$ together with the test values of X_d and X_q. This enables the phasor diagram to be constructed as shown in subsection A under section 8.6.2 and Fig. 8.27. A value for the emf E is obtained, but this is for the case of an unsaturated magnetic circuit. However, this value of E on the line Oh of Fig. 8.12 gives the field current I_F, which in turn produces the emf on the OCC. Hence the regulation REG can be calculated from

$$\text{pu}REG = \frac{\text{open-circuit emf} - \text{terminal voltage on load}}{\text{terminal voltage on load}} = \frac{E - V}{V}. \quad (8.6.26)$$

If the ZPFC and the OCC are available, a graphical means can be employed to find the regulation. With the given information V, I, $\cos\theta$, R_a, X_d, and X_q, the conventional phasor diagram can be constructed as described in A under section 8.6.2 and Fig. 8.27. The phasor diagram is repeated in Fig. 8.31a. However, the assumptions have been that the synchronous reactances X_d and X_q are constant and the machine is unsaturated, so that the excitation emf E, obtained from the phasor diagram, is high. The armature reaction field in the quadrature axis is of low density, so there is negligible saturation and the synchronous reactance X_q can be considered constant. But the synchronous reactance X_d varies depending on the degree of saturation in the field poles and armature core. The open-circuit characteristic OCC is a reasonable measure of this saturation and may be used to find a more accurate value of the excitation emf E.

First we find the Potier reactance from the zero power factor characteristic ZPFC as described in subsection E under section 8.5.5. The value obtained is used as the leakage reactance x_l. This value is slightly higher than the true leakage reactance, because the armature opposes the field mmf to increase the pole leakage flux and the true magnetization curve lies slightly lower than the OCC from which the Potier reactance was estimated. However, satisfactory results are obtained using this value of x_l.

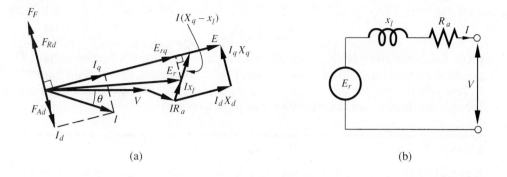

(a)   (b)

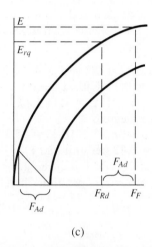

(c)

Figure 8.31 Regulation with saliency. (a) Phasor diagram, (b) equivalent circuit, (c) OCC and ZPFC.

If we add to the terminal voltage $\mathbf{V}$, the phasor voltage drops due to the armature-winding resistance R_a and leakage reactance x_l, as shown in Fig. 8.31a, the result is the voltage E_r behind the leakage reactance as illustrated in the equivalent circuit diagram of Fig. 8.31b, that is,

$$\mathbf{E}_r = \mathbf{V} + \mathbf{I}R_a + j\mathbf{I}x_l. \qquad (8.6.27)$$

The quadrature component $\mathbf{E}_{rq}$ of this voltage, as shown in the phasor diagram, is measured. It is also given by

$$\mathbf{E}_{rq} = \mathbf{V} + \mathbf{I}R_a + \mathbf{I}x_l + j(X_q - x_l)\mathbf{I}_q. \qquad (8.6.28)$$

The magnitude of the induced emf E_{rq} determines the net field mmf F_{Rd}, whose flux saturates the poles. F_{Rd} is obtained from the OCC as shown in Fig. 8.31c. The net field mmf is the phasor sum of the true field mmf and the direct axis armature mmf, and is given by

$$F_{Rd} = F_F + F_{Ad}. \qquad (8.6.29)$$

From the Potier triangle the armature reaction is F_{Ad}. Since the lagging current I produces an mmf on the direct axis opposing the true field mmf, the values of F_{Rd} and F_{Ad} are added to give the mmf F_F. The mmf F_F produces the true excitation emf E on the OCC as shown, and the regulation can be determined.

EXAMPLE 8.9

A synchronous generator has d-axis and q-axis synchronous reactances of 1.0 and 0.55 per unit, respectively. For a 0.8 lagging power factor load find the per unit excitation emf E and the load angle δ neglecting saturation, and calculate the regulation. Compare these values for the case when saliency is neglected.

Solution
Let V be the reference phasor. $I = 0.8 - j0.6 = 1.0 \angle -36.9°$ per unit.
$jIX_q = j(0.8 - j0.6) \times 0.55 = 0.33 + j0.44$ per unit.
Hence Oa' in Figure 8.27 is $V + jIX_q = 1.33 + j0.44 = 1.44 \angle 18.3°$.
Therefore $\delta = 18.3$ electrical degrees.
$\theta = \cos^{-1}0.8 = 36.9°$. $\theta + \delta = 36.9° + 18.3° = 55.2°$.
Hence $I_d = 1.0 \sin(55.2°) = 0.82$ per unit and $I_q = 1.0 \cos(55.2°) = 0.57$ per unit.
From Fig. 8.27 $E \equiv Oa' + a'c' = |V + jIX_q| + I_d(X_d - X_q)$.
So $E = 1.4 + 0.82(1.0 - 0.55) = 1.77$ per unit. Therefore $E = 1.77 \angle 18.3°$ per unit.
Regulation $REG = \dfrac{E - V}{V} \times 100 = \dfrac{1.77 - 1.0}{1.0} \times 100 = 77\%$.

Neglect saliency.
$X_d = X_q = X_s = 1.0$ per unit.
$E = V + jIX_s = 1.0 + j1.0(0.8 - j0.6) \times 1.0 = 1.79 \angle 26.6°$ per unit.
Hence $\delta = 26.6°$.
Regulation $REG = \dfrac{E - V}{V} \times 100 = \dfrac{1.79 - 1.0}{1.0} \times 100 = 79\%$.

The results indicate that, for a normal excitation, the salient-pole machine can be represented by a cylindrical machine without appreciable error.

8.7. SYNCHRONOUS MOTOR

The synchronous machine is reversible. If the rotor is driven by a prime mover, the machine is a generator, converting the mechanical energy input to electrical energy output. If the machine draws electrical energy from the supply, it will motor to drive a mechanical load connected to its shaft.

For a constant electromagnetic torque to be developed, the rotor poles have to lock into the synchronously rotating armature field, which is produced by the poly-

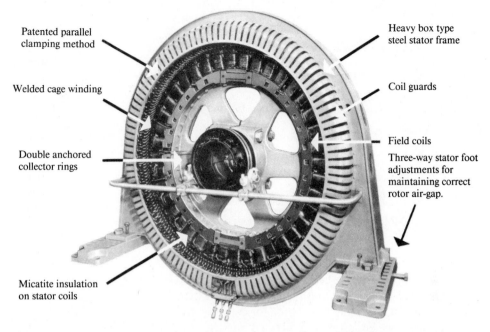

Patented parallel
clamping method

Welded cage winding

Double anchored
collector rings

Micatite insulation
on stator coils

Heavy box type
steel stator frame

Coil guards

Field coils

Three-way stator foot
adjustments for
maintaining correct
rotor air-gap.

Photo 7 Low-speed, synchronous motor. (Courtesy of Electric Machinery—Dresser Rand.)

phase currents. Therefore, the motor is a constant-speed machine. It rotates at the synchronous speed n_s, which is given in terms of the frequency f and the number of pole pairs p in eq. (8.2.1). Thus, the same theory that was developed for the synchronous generator can be applied to the synchronous motor.

It was noted that the magnitude of the excitation of a generator connected to an infinite bus affected the power factor. A similar effect occurs for the motor. Therefore, synchronous motors find application where constant speed is required and there is a need for power factor correction. One disadvantage of the synchronous motor is that it is not self-starting. Some additional means must be used to bring the machine up to its operational speed. Normally the applications are for low-speed drives, so the machine may have 12 poles or more and the motor is usually of salient-pole construction.

8.7.1. Motor Analysis

The applied voltage V must be balanced by the excitation emf E and the voltage drops due to the armature winding resistance R_a and the synchronous reactances X_d and X_q. For the salient-pole motor the armature-circuit voltage equation on a per phase basis is

$$\mathbf{V} = \mathbf{E} + \mathbf{I}R_a + \mathrm{j}\mathbf{I}_d X_d + \mathrm{j}\mathbf{I}_q X_q. \tag{8.7.1}$$

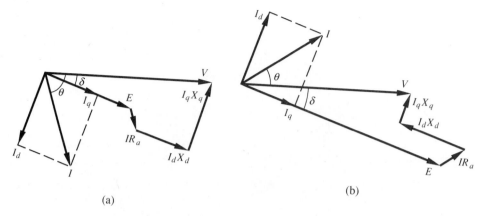

Figure 8.32 Motor phasor diagrams for saliency. (a) Lagging power factor, (b) leading power factor.

This equation could have been written down from the generator voltage eq. (8.6.15) by maintaining the convention from the circuit diagrams that the current $+I$ was directed out of the machine. Thus, for a motor the current would be $-I$ in the generator voltage equation and directed into the machine.

Just as for the generator, the motor phasor diagram follows from the voltage equation. Figure 8.32 depicts the phasor diagrams for both lagging and leading power factors. The differences between the generator and the motor become apparent from an inspection of the phasor diagrams in Figs. 8.26 and 8.32. For the motor, low values of excitation (and hence low E) are accompanied by a lagging power factor $\cos\theta$, and high values of excitation produce a leading power factor. For the generator connected to an infinite bus, the reverse occurs. Further, the generator has the emf $\mathbf{E}$ leading the terminal voltage $\mathbf{V}$ by the load angle δ, but the motor has the emf $\mathbf{E}$ lagging the voltage $\mathbf{V}$. In the motor, the pole field lags the armature field by an amount such that the

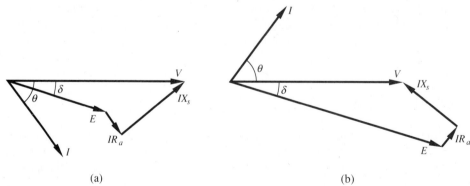

Figure 8.33 Phasor diagrams for round-rotor motor. (a) Lagging power factor, (b) leading power factor.

developed electromagnetic torque balances the gross mechanical torque exerted by the load and the windage and friction. On no load the emf $\mathbf{E}$ is in phase with the voltage $\mathbf{V}$, the load angle δ is zero and the pole field is aligned with the armature field, if the effect of windage and friction is neglected.

A round-rotor motor has no difference between the direct and quadrature axis reluctances. Accordingly $X_d = X_q = X_s$, and since $\mathbf{I} = \mathbf{I}_d + \mathbf{I}_q$, the voltage equation for the round rotor motor simplifies to

$$\mathbf{V} = \mathbf{E} + \mathbf{I}(R_a + jX_s). \tag{8.7.2}$$

The phasor diagrams for this condition are illustrated in Fig. 8.33 with lagging and leading power factors produced by under and overexcitation, respectively. Further simplifications can be made if the armature winding resistance R_a is ignored.

EXAMPLE 8.10

A 3-phase, 11-kV, Y-connected synchronous motor has a synchronous impedance of $(0.95 + j29)$ ohms per phase. Calculate the power supplied to the motor and the induced emf E if the current input is 50 A at a power factor 0.8 leading.

Solution
Power input $P = 3VI \cos\theta = \sqrt{3} \times 11000 \times 50 \times 0.8 = 762 \times 10^3$ W.
The voltage equation is $\mathbf{V} = \mathbf{E} + \mathbf{I}Z_s$, or from the phasor diagram of Fig. 8.33b

$$E^2 = (V - IR_a \cos\theta + IX_s \sin\theta)^2 + (IR_a \sin\theta + IX_s \cos\theta)^2.$$

That is, $E^2 = (6352 - 38 + 870)^2 + (28.5 + 1160)^2$. Hence $E = 7281$ volts per phase. The line value of the induced emf is $\sqrt{3} \times 7281$ or $12,610$ V.

8.7.2. Power Torque and Load Angle

The power input expression for the salient-pole motor is the same as the power output expression for the generator, so that if the armature winding resistance R_a is ignored, the power input per phase (from section 8.6.3) is given by eq. (8.6.20) and the reactive power per phase is given by eq. (8.6.24).

On no load the real power is zero and so δ is zero. Therefore, the reactive power becomes

$$Q_{NL} = \frac{V}{X_d}(E - V). \tag{8.7.3}$$

Since V and X_d are assumed constant, the magnitude of the excitation determines whether the reactive power is leading (positive) or lagging (negative). This suggests the use of an overexcited synchronous motor without a mechanical load connected to the power system for the purpose of power factor correction. An unloaded motor for such an application is called a *synchronous compensator*. An overexcited motor on load will also provide power factor correction. At 0.8 power factor leading, the reactive power may be about 70% of the real rated power of the motor.

The torque per phase developed by the salient-pole motor is given by

$$T_e = \frac{P}{\omega_s} = \frac{VE}{\omega_s X_d} \sin\delta + \frac{V^2(X_d - X_q)}{2\omega_s X_d X_q} \sin 2\delta. \qquad (8.7.4)$$

This shows that even without excitation ($E = 0$) there is a torque due to the difference between the direct axis and quadrature axis reluctances. It is called the reluctance torque. If the motor is overexcited to an extent that $E = 2V$, for example, and if the load angle δ is small so that $\sin\delta \approx \delta$ and $\sin 2\delta \approx 2\delta$, then, since the ratio X_q/X_d is approximately 0.6, the electromagnetic torque is

$$T_e \approx \frac{2V^2}{\omega_s X_d} (1 + 1/3)\delta. \qquad (8.7.5)$$

This shows the considerable contribution made by the reluctance torque (in this case, one third of the excitation torque).

The expressions for power and torque developed in the round rotor motor can be derived from the above equations by putting $X_d = X_q = X_s$. Accordingly

$$P = \frac{VE}{X_s} \sin\delta, \quad Q = \frac{VE}{X_s} \cos\delta - \frac{V^2}{X_s}, \quad \text{and} \quad T_e = \frac{VE}{\omega_s X_s} \sin\delta. \quad (8.7.6)$$

The torque has no saliency term and is zero if there is no excitation. Actually, there is a difference between the direct and quadrature axis reluctances caused by the slots for the field winding, but the effect is negligible. The reactive power can be controlled in the same way as for the salient-pole motor, and power factor correction can be made. It is seen from a comparison of Figs. 8.23 and 8.29 that the peak power and the synchronizing power coefficient $dP/d\delta$ are less for the round rotor motor, making it less stable under the influence of disturbances.

For both the round rotor and salient pole motors, once the maximum power has been reached, any increase in load torque is accompanied by an increase in load angle δ. The power input then decreases, the electromagnetic torque decreases, and the difference between the load torque and developed torque tends to reduce the speed. As a result, the motor drops out of synchronism and tends to come to rest. The maximum torque is sometimes called the *pull-out* torque because of the loss of synchronism ensuing from any incremental change in load torque.

An inspection of either of the power equations yields the information that the lower the synchronous reactance the higher the pull-out and synchronizing power. The same trend occurs from an increased excitation, which also decreases the reactance on account of saturation.

EXAMPLE 8.11

A 2300-V, 3-phase, Y-connected, cylindrical-rotor, synchronous motor has a synchronous reactance X_s of 11 ohms per phase. If the load is 149.2 kW, the efficiency of the motor is found to be 90% exclusive of field loss, and the load angle δ is 15 electrical degrees. Find **(a)** the induced voltage per phase E, **(b)** the line current I, and **(c)** the power factor. Neglect the armature winding resistance.

Solution

(a) By neglecting the armature winding resistance R_a, the power input and power converted are the same.

Thus $P = 3 \dfrac{VE}{X_s} \sin\delta = \text{shaft power}/\eta$

and $E = \dfrac{149,200}{0.9} \times \dfrac{11}{3} \times \dfrac{\sqrt{3}}{2300} \times \dfrac{1}{\sin 15°} = 1790$ V per phase.

(b) $\mathbf{V} = \mathbf{E} + j\mathbf{I}X_s$. Let $\mathbf{V}$ be the reference phasor.

Therefore $\mathbf{I} = \dfrac{1330 \angle 0° - 1790 \angle -15°}{11 \angle 90°} = 55.4 \angle 41.4°$ A.

(c) Since the angle of $\mathbf{I}$ is positive, the power factor is leading.
Power factor is $\cos\theta = \cos 41.4° = 0.755$ leading.

8.7.3. Motor Starting

Self-starting of the synchronous motor is not possible. At rest, with the armature winding connected to a polyphase supply, the stationary poles experience a rapid alternation of armature field N and S poles. Because of the inertia, the rotor cannot respond quickly enough to pull into step with the armature field. Consequently, the field poles have exerted on them a fluctuating electromagnetic torque whose average value is zero and the rotor remains at rest.

The induction motor is self-starting and, as the induction and synchronous motors have similar armature windings, this suggests a means to bring the synchronous motor up to speed. A partial or complete cage winding can be placed in the pole shoes of the synchronous motor to provide induction torque at starting. In fact, any of the induction motor rotor windings may be employed. The one chosen depends on the starting specification for the particular load application. Once the motor is at synchronous speed, the starting winding becomes inactive because it experiences no change of field.

During the starting process the field circuit is short-circuited, otherwise dangerously high voltages could be induced by the armature field. Induction torque accelerates the rotor close to synchronous speed, at which time the field circuit is opened and excited with direct current. If the total load torque is lower than the pull-in torque developed by the armature and excitation fields, the motor will attain synchronism. This simple statement belies the complex nature of pull-in, which depends on the induction torque characteristic, the load characteristic, the inertia and natural frequency of the rotor and the synchronous motor torque, the load angle, and the motor speed at the time the field circuit is energized. These problems are much reduced for the case of an adjustable-speed motor, which is connected to a variable frequency supply ($n_s = f/p$). Starting is achieved at low frequency and the frequency is increased until the required speed is attained.

The starting winding has another useful purpose for both motors and generators. A machine may experience a momentary disturbance such as a load change, a fault, or a switching action. The load angle change is produced by an acceleration or deceleration such that the speed becomes subsynchronous or supersynchronous or both in an oscillatory manner.[6] As soon as the rotor loses synchronism the cage winding is activated by the changing field. The resulting induction torque tends to accelerate the rotor at subsynchronous speeds by motor action, and to decelerate the rotor at supersynchronous speed by generator action. In both cases the tendency is to damp mechanical oscillations and restore stability. Induction motor windings on the rotor of a synchronous machine are called *damper windings* or *amortisseur windings*.

8.7.4. Special Motors

There are several special motors that come within the category of synchronous machines, that is, those motors whose speed is determined precisely by the frequency of the supply. Three are described here. They are the reluctance motor, the hysteresis motor, and the stepper motor.

A. RELUCTANCE MOTOR

A *reluctance motor* is a salient-pole synchronous machine with no field excitation. In section 2.5.3 a primitive reluctance machine was studied in order to demonstrate the application of the principles of electromechanical energy conversion. Here, we compare the reluctance motor with a synchronous machine that has field excitation.

In section 8.7.2 the torque per phase for a salient-pole synchronous motor was found to be

$$T_e = \frac{P}{\omega_s} = \frac{VE}{\omega_s X_d} \sin\delta + \frac{V^2(X_d - X_q)}{2\omega_s X_d X_q} \sin2\delta. \qquad (8.7.4)$$

If there is no excitation, the induced emf E is zero and the torque expression becomes

$$T_e = \frac{V^2(X_d - X_q)}{2\omega_s X_d X_q} \sin2\delta. \qquad (8.7.7)$$

This torque exists only because the synchronous reactances are different in value. The difference is due to the long air gap in the quadrature axis magnetic circuit, so that the reluctance of the quadrature axis path is greater than the reluctance of the direct axis path. The resulting reluctance torque gives this type of motor its name.

---

[6] This oscillatory mode is known as *hunting*.

By inspection of the above two equations the reluctance motor develops less torque than the excited synchronous motor of the same size. Since the volume of a machine is roughly proportional to the torque, the example described in section 8.7.2 and eq. (8.7.5) indicates that for the same torque and speed the reluctance motor has a frame about three times the size of an excited synchronous motor. Nevertheless, in such hazardous environments, where the likely arcing and sparking at the sliding contacts of an exciter system cannot be tolerated, the constant-speed reluctance motor finds application.

The excited synchronous motor is more stable than the reluctance motor. Figure 8.29 depicts the power and load angle characteristics of the synchronous machines. The full-line curve is for the salient-pole, excited machine and the double-frequency, broken curve is for the reluctance motor. The load angle, at which the maximum power, or limit of stability, is reached, is almost half for the unexcited machine. The slope of the power curve (representing the synchronizing power coefficient) is higher for the excited machine, which is thus more capable of handling disturbances and remain operating stably.

Starting reluctance motors require the same methods as all synchronous machines and employ an induction motor starting technique.

Reluctance motors can operate at synchronous speed from a single-phase supply and they utilize the same methods of starting as single-phase induction motors. One of the most common applications of the single-phase reluctance motor is for electric clock drives.

B. HYSTERESIS MOTOR

While operating at steady state the *hysteresis machine* is a permanent-magnet, synchronous motor. It is during the starting period (or subsynchronous operation) that the torque is developed because of magnetic hysteresis, from which the motor derives its name.

Consider a magnetic material, whose *B-H* characteristic has the form as shown in Fig. 8.34. This is the static hysteresis loop. For a particular excitation H (sinusoidal in this case), the resulting induction B can be derived from the loop instant by instant. The flux density waveform is shown in the figure to lag the magnetic field intensity H by the angle α, which is called the *hysteretic angle.* The wider is the hysteresis loop, that is, the greater the magnetic material coercivity, the greater is the hysteretic angle. Hard magnetic materials, used for permanent magnets, have large hysteretic angles.

The hysteresis motor comprises a polyphase winding on the stator and a round rotor of hard magnetic material. As long as the rotor is not running at synchronous speed, each element of rotor volume experiences a hysteresis cycle due to the synchronous rotation of the stator mmf pattern. Because of the hysteresis, the induced flux density in the rotor lags behind the air-gap mmf produced by the stator winding currents by the angle α. This is shown in Fig. 8.35. There is a tendency for the field of the rotor to align itself with the field of the stator. Accordingly, there is a torque due to hysteresis to accelerate the rotor. The torque is expressed by

$$T_e \propto HB \sin\alpha. \tag{8.7.8}$$

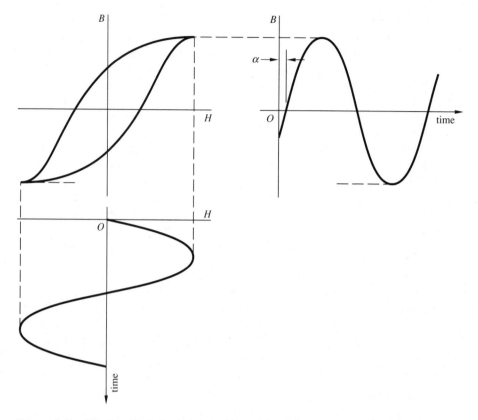

Figure 8.34   The hysteresis loop and the hysteretic angle σ.

This torque exists as long as the rotor material experiences the hysteresis cycle. This is the case as long as the relative speed between the traveling wave and the rotor is not zero. Thus the hysteresis motor is self-starting. Further, the torque is constant and independent of subsynchronous speed for constant flux, since the hysteretic angle is independent of the rate of hysteresis cycling.

At synchronous speed the rotor travels at the same speed as the air-gap field, so there is no hysteresis torque. However, the magnetic domains of the rotor are aligned by the magnetic field intensity H as the rotor arrives at synchronous speed, so that the rotor has permanent magnetic poles as long as the rotor speed remains synchronous. At this speed the machine operates as a permanent magnet synchronous motor and the load angle δ adjusts automatically so that the load torque and electromagnetic torque balance. Consequently, a torque exists from zero up to and including synchronous speed. This feature does not exist in the synchronous induction motor or the reluctance motor which relies on low inertia and which must snap into synchronism as the induction motor–type torque is reducing to zero.

A hysteresis motor, whose rotor is round and not laminated, has induction torque added to the hysteresis torque until synchronous speed is reached. This is characterized in Fig. 8.36. The linear induction torque arises from the high value of

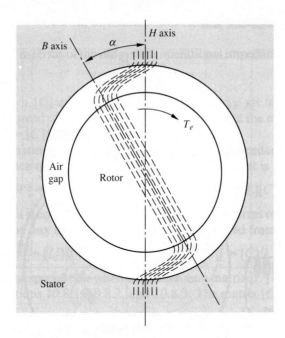

Figure 8.35 Hysteresis effect.

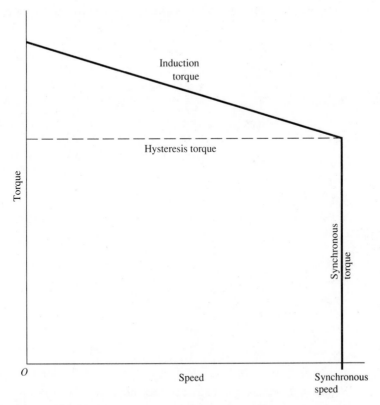

Figure 8.36 Torque-speed curve of the hysteresis motor.

the rotor material resistance. See Fig. 6.25 for the high value of rotor winding resistance R_1.

In single-phase hysteresis motors the stator winding is usually the permanent-split-capacitor type in order to approximate 2-phase conditions. If low torques are tolerated, shading coils can be employed instead.

Motors are built up to about 200 W for precise-speed drives. Clocks, record player motors, and servomechanisms are typical examples where hysteresis motors find application.

C. STEPPER MOTOR

The *stepper motor* is another form of the synchronous reluctance machine with a special supply and a special controller. A source supplies a train of pulses and the controller directs pulses sequentially to the motor windings. In this way the axis of the motor's magnetic field *steps* around the air gap at a speed that is related to the frequency of pulses. The rotor tends to align itself to the axis of the magnetic field. Consequently the rotor *steps* in synchronism with the motion of the magnetic field. For this reason the motor is called a *stepper motor*. It can be used for precise speed control or for accurate position control.

A primitive form of stepper motor is depicted in Fig. 8.37. The rotor is a permanent magnet. It will align itself with the axis of any stator coil that is excited. If the

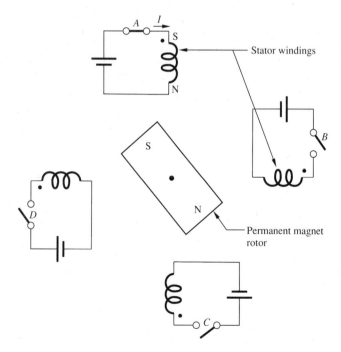

Figure 8.37 Permanent-magnet stepper motor.

switches *A*, *B*, *C*, and *D* are opened and closed sequentially in order, the rotor will move (or *step*) through 90 mechanical degrees in a clockwise direction at every switching action. By this sequential switching the rotor speed would be controlled to be the same as the frequency of pulsing any one of the switches. Alternatively, it is possible to position the rotor at any one of the stator-coil axes. A variation of the switching is *A*, *D*, *C*, *B*, *A*, *D*, *C*, ... This gives rotation in an anticlockwise direction. Another variation, *A*; *A* and *B*; *B*; *B* and *C*; *C*; etc., gives 45° steps of rotation.

Figure 8.38 illustrates how the variable-reluctance stepper motor works. The stator has three pairs of poles designated *A*, *B*, and *C*, which are excited one at a time by direct current in their windings (called phases). If phase *A* is excited, the axis of the magnetic field is along the pole A axis, as shown in the figure. The rotor has two pairs of teeth, 1 and 2, so that around the air gap there is variable reluctance. Hence, the rotor tends to align itself so that there is minimum reluctance in the magnetic path. As seen in the figure, the axis of tooth pair 1 is already aligned with pole *A* to offer minimum reluctance, and this is a stable operating position.

Next, if phase *B* is excited alone, rotor tooth pair 2 tends to align along the axis of pole pair *B* to a new equilibrium position. The rotor has moved anticlockwise through an angle of 30 mechanical degrees. Exciting phase *C* alone provides another 30° step of rotation in an anticlockwise direction, so that tooth pair 1 aligns along the axis of pole pair *C*, which is the new equilibrium position.

Under proper control, a continuous train of pulses, sequenced in turn, to *A*, *B*, *C*, *A*, *B*, *C*, *A*, . . . produces motion in an anticlockwise direction at a rate that depends on the frequency of the pulses. Equally well, a train of pulses in the sequence *A*, *C*, *B*, *A*, *C*, *B*, *A*, . . . causes motion in the reverse direction.

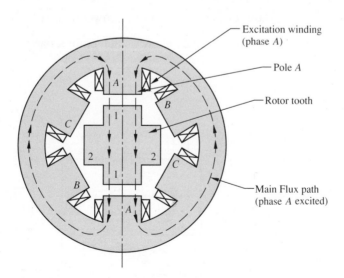

Figure 8.38 Variable-reluctance stepper motor.

A particular configuration of a reluctance stepper motor has been described. In general, the number of stator phases and rotor teeth can be designed to provide different angular steps, typically $2°$ or more. If $2p$ is the number of phases and N is the number of teeth, the rotor angular motion per pulse is a *step* of π/Np radians. Figure 8.39 shows a control implementation, in which a train of f pulses per second is supplied to the digital driver circuit. The input of the controller is divided so that the output is directed in sequence to one phase at a time. Consequently, the rotor makes f steps per second. That is, the angular speed is precisely $\pi f/Np$ rad/s.

Hybrid stepper motors are perhaps more common than either the permanent magnet motors or the reluctance motors. In the hybrid motor there are both permanent magnets and teeth on the rotor, which provide both excitation torque and reluctance torque. Often there is a number of stators and rotors *stacked*. Adjacent rotors (on the same shaft) are phase displaced. With this arrangement smaller steps become possible. However, the basic principles remain the same.

The application of stepper motors is found in open-loop control equipment such as tape drives, floppy disk drives, and printers, in numerically-controlled machine tools, and in industrial robots. The power levels can be about 1 kW, but, usually they are a fraction of this.

8.8. EFFICIENCY

A knowledge of the values of the power losses for a given power output enable the efficiency η of a synchronous machine to be obtained from

$$\text{pu } \eta = \frac{\text{output power}}{\text{input power}} = 1 - \frac{\text{losses}}{\text{output power} + \text{losses}}. \qquad (8.8.1)$$

For the case of a generator, a power chart, showing the major separation of losses, is

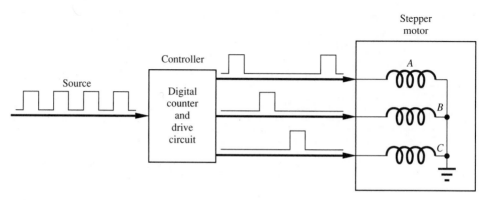

Figure 8.39 Driver for stepper motor.

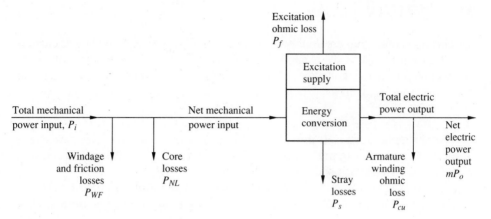

Figure 8.40 Generator power chart.

depicted in Fig. 8.40, so that

$$\text{pu } \eta = 1 - \frac{P_{cu} + P_f + P_s + P_{NL} + P_{WF}}{mP_o + P_{cu} + P_f + P_s + P_{NL} + P_{WF}} \tag{8.8.2}$$

where m = number of armature winding phases

P_o = power output per phase ($VI \cos\theta$)

P_{cu} = armature winding ohmic loss (mI^2R_a)

P_f = field winding ohmic loss ($I_F^2 R_f$)

P_s = stray load losses

P_{NL} = no-load core losses due to hysteresis and eddy currents

and P_{WF} = windage and friction losses.

The no-load core losses, due to the field flux, and the windage and friction losses are found from the open-circuit test, which was described in subsection A under section 8.5.5. The armature winding resistance is measured so that the ohmic loss can be calculated for a particular load current. From the short-circuit test, described in subsection B under section 8.5.5, the measured loss has subtracted from it the windage, friction, core, and armature winding ohmic losses to give the stray losses, which comprise the additional core losses and conductor eddy current loss due to the armature winding current. The measured input power into the field winding is the excitation ohmic loss.

For the case of a synchronous motor, the efficiency can be calculated from

$$\text{pu } \eta = \frac{\text{input} - \text{losses}}{\text{input}} = 1 - \frac{P_{cu} + P_f + P_s + P_{NL} + P_{WF}}{P_i} \tag{8.8.3}$$

where P_i is the total mechanical input power.

8.9. SUMMARY

Synchronous machines are doubly-excited. The field system is excited by dc currents and the armature winding is excited by polyphase currents. The interaction of the two excitations produces a constant electromagnetic torque at synchronous speed. Synchronous speed is given by

$n_s = f/p$ r/s.

Synchronous generators provide us with the greatest amount of electrical energy in the form of an ac supply at 50 or 60 Hz. Synchronous motors find application where precisely constant-speed drives are required.

Salient-pole machines are used for slow speed (large number of poles) and round rotor machines are used for high-speed machines (2-pole, 60-Hz machines rotate at 3600 r/min).

A 3-phase synchronous machine converts energy in the armature at the rate

$P = 3EI \cos\theta$ watts

where θ is the angle between **E** and **I** in this case. The electromagnetic torque is

$T_e = P/\omega_s$ newton-meters

where ω_s is the synchronous speed in mechanical rad/s.

The reactive component of power is controlled by the level of excitation of the field winding. An overexcited motor or an underexcited generator will enable the power factor to be leading. A synchronous machine on no load is used in power systems to improve the overall power factor by excitation control. In this mode the machine is called a *synchronous compensator*.

The induced emf in an armature phase is

$E = 4.44\Phi f N$ volts

for steady-state operation.

Round-rotor Machines

Armature reaction is accounted for by the use of a reactance x_a. Leakage fluxes and harmonic fluxes are accounted for by the use of a reactance x_l. Together, the two reactances are called the synchronous reactance X_s ($X_s = x_a + x_l$). The synchronous impedance is

$\mathbf{Z}_s = R_a + jX_s$ ohms

in round rotor machines. For a generator the terminal voltage is

$\mathbf{V} = \mathbf{E} - \mathbf{I}\mathbf{Z}_s$ volts.

From this model the generator is described as a voltage (E) behind a synchronous impedance (Z_s).

The terminal voltage regulation of a generator is defined as

$$REG = \frac{E - V}{V} \times 100\%.$$

The problem to predict the regulation with accuracy is associated with the calculation of the induced emf E. The difficulty arises because the synchronous reactance is a function of the saturation of the magnetic circuit.

If the armature winding resistance is small compared with the synchronous reactance ($R_a : X_a \approx 1 : 10$), it is reasonable to express power in terms of load angle δ as

$$P \approx 3 \frac{VE}{X_s} \sin\delta \quad \text{and} \quad Q = 3 \frac{VE \cos\delta - V^2}{X_s}.$$

Synchronizing power is a measure of resistance to unstable operation (stiffness) and is defined as $dP/d\delta$.

Salient-pole Machines

The two-reactance theory simplifies the modeling of a synchronous machine with saliency. Armature reaction is divided into components along the direct axis and the quadrature axis. This leads to the concept of the synchronous reactance X_d on the direct axis and the synchronous reactance X_q on the quadrature axis. The armature winding current is modeled by two components I_d and I_q on the two axes. If **V** is reference, the voltage equation for the generator is

$$\mathbf{E} = V + I_d X_d + jI_q X_q + I_q R_a - jI_d R_a.$$

In terms of the load angle δ, the power equation is

$$P \approx \frac{VE}{X_d} \sin\delta + \frac{V^2(X_d - X_q)}{2X_d X_q} \sin2\delta.$$

The second term on the right-hand side of the above equation is the contribution of saliency. It becomes zero for a round rotor machine ($X_d = X_q$).

Synchronous motors have a third winding for starting. The winding is like the induction motor cage winding and provides induction torque for starting. At synchronous speed this third winding is inoperative. This starting winding is also used to damp out speed oscillations.

8.10. PROBLEMS

Section 8.3

8.1. Prove that two small and identical synchronous generators on a shaft and a voltmeter can be used to obtain a direct reading of the power transmitted by the shaft.

Section 8.5

8.2. A 4-pole generator on no load develops 250 V per phase at 60 Hz, if its field current is 2.1 A. Calculate the emf if the speed is 1200 r/min and the field current is 3.5 A. Assume the magnetic circuit to be unsaturated.
(*Answer:* 278 V.)

8.3. Find the synchronous impedance and reactance of a synchronous generator, in which a given field excitation current produces an armature winding current of 250 A on short circuit and an emf of 1500 V on open circuit. The armature winding resistance is 2 ohms. Calculate the terminal voltage if a load with a current of 250 A at 6.6 kV and 0.8 power factor lagging is disconnected.
(*Answer:* 6 ohms, 5.65 ohms, 7892 V.)

8.4. A 3-phase, 60-Hz synchronous generator is rated 12,000 kVA, 13,800 V, 0.8 power factor. **(a)** What should be its kVA and voltage rating at 0.8 power factor and 50 Hz if the field and armature ohmic losses are to be the same as at 60 Hz? **(b)** If its voltage regulation at rated load and 60 Hz is 18%, what will be the value of the voltage regulation at its rated load for 50-Hz operation? (The effect of armature resistance voltage drop on regulation may be neglected.)
(*Answer:* (a) 10,000 kVA, 11,500 V, (b) 18%.)

8.5. A 3-phase, 4600-V, 1000-kVA, Y-connected synchronous generator has an armature winding resistance and synchronous reactance of $(2 + j20)$ ohms per phase. What is the voltage regulation if the power factor is **(a)** 0.75 leading, **(b)** 0.4 leading, **(c)** unity, and **(d)** 0.75 lagging?
(*Answer:* (a) -11%, (b) -50.6%, (c) 44.4%, (d) 81.3%.)

8.6. A 3-phase, 11-kV, 16-MVA, Y-connected synchronous generator has a no-load characteristic and zero power-factor data as follows.

(OCC) Terminal voltage (line to line) kV	3.2	6.4	10.9	13.3	14.6
Field current A	50	100	200	300	400
(ZPFC) $V = 0$, $I_F = 186$ A, and $V = 11$ kV, $I_F = 450$ A.					

For operation at rated voltage, full-load current and unity power factor determine **(a)** the leakage reactance x_l, **(b)** the saturated synchronous reactance X_s, **(c)** the field current I_F, and **(d)** the regulation by the emf, mmf, and general methods. Neglect saliency and the armature winding resistance.
(*Answer:* (a) 1.52 ohms, (b) 6.1 ohms, (c) 260 A, (d) 14.5%, 16.4%, 19.1%.)

8.7. A 3-phase, 60-MVA synchronous generator has a synchronous reactance of 10 ohms per phase. The generator is connected to an infinite bus whose voltage is 13.2 kV per phase. The shaft power is 50 MW and is maintained constant. What is the ratio of maximum to minimum field current for full-load armature winding current? Neglect saturation and losses.
(*Answer:* 1.85 : 1.)

Section 8.6

8.8. A generator has a direct axis synchronous reactance of 0.8 per unit and a quadrature axis synchronous reactance of 0.5 per unit. Draw the phasor dia-

gram for full load at a lagging power factor of 0.8 and find the per unit open-circuit induced emf E per phase. Neglect saturation.
(*Answer:* 1.6 per unit.)

8.9. A 2-phase, salient-pole synchronous generator operates at full load, 0.94 lagging power factor. For this condition the excitation voltage E is 1.5 per unit leading the current by $30°$, and the per unit resistance drop is 0.1. Calculate the value of the synchronous reactances on the direct and quadrature axes.
(*Answer:* 0.86, 0.26.)

8.10. A salient-pole synchronous generator has $X_d = 1$ per unit and $X_q = 0.6$ per unit. The machine is operated at a power angle $\delta = 45°$. The output terminal voltage is the rated value and the open-circuit induced emf E is 1.0 per unit. What percentage of the total power output is contributed by the machine's saliency?
(*Answer:* 31%.)

Section 8.7

8.11. A 500-V, single-phase synchronous motor has a synchronous impedance of 3.2 ohms. The armature resistance is 0.2 ohm. To what voltage E must the motor be excited so that it may develop 30 kW at the shaft and operate at unity power factor? The mechanical losses are 3.57 kW. What is the armature-winding current?
(*Answer:* 534 V, 69 A.)

8.12. A single-phase synchronous motor has a synchronous reactance of 3.3 ohms. The exciting current I_F is adjusted to such a value that the induced emf E is 950 V. Find the power factor at which the motor would operate when taking 80 kW from a 800-V supply. Neglect armature winding resistance.
(*Answer:* 0.964 leading.)

8.13. A 3-phase, 2300-V, Y-connected motor has a synchronous impedance of $(0.021 + j0.95)$ ohms per phase. If the motor is taking the full-load current of 628 A at unity power factor the field-winding current is 4.5 A. Find the field-winding current if the motor takes 565 A at a leading power factor of 0.8. It may be assumed that operation takes place over the linear portion of the open-circuit characteristic.
(*Answer:* 5.2 A.)

8.14. A 3-phase, 20-pole, 660-V, Y-connected synchronous motor is operating at no load with its generated voltage per phase E exactly equal to the phase voltage V applied to its armature winding. At no load the power angle δ is 0.5 mechanical degrees. The synchronous reactance is 10 ohms per phase and the effective armature-winding resistance is 1.0 ohm per phase. Calculate **(a)** the power angle δ in electrical degrees, **(b)** the synchronous-impedance drop per phase, **(c)** the armature winding current I per phase, **(d)** the power per phase and the total power drawn by the motor from the bus, and **(e)** the armature winding power loss.
(*Answer:* (a) $5°$, (b) $33.2 \angle 87.3°$V, (c) $3.32 \angle 3°$ A, (d) 1265 W/ph, 3795 W, (e) 33 W.)

8.15. A 3-phase, Y-connected synchronous motor, taking 93 A at 500 V and unity power factor on full load, has an armature winding resistance and synchronous reactance per phase of 0.03 and 0.3 ohm, respectively. Calculate for the same input power, but at 0.8 power factor leading, the total mechanical power developed and the induced emf E generated by the motor. Assume an efficiency of 90%.
(*Answer:* 85.7 kW, 533.9 V.)

8.16. A synchronous motor operates at a load angle of 30° at rated voltage and frequency. If the terminal voltage and frequency both reduce by 10%, find the load angle for **(a)** constant load power and **(b)** constant load torque. Neglect losses and saliency.
(*Answer:* (a) 33.7°, (b) 30°.)

8.17. A synchronous motor has a load angle of 35 electrical degrees for normal conditions of rated voltage, frequency, and torque. Find the steady-state load angle if **(a)** the supply frequency is reduced by 5% and the load torque is increased by 10%, **(b)** the supply frequency is reduced by 5% and the load power is increased by 10%, **(c)** the supply frequency and voltage are reduced by 15% but the torque is the rated value, and **(d)** the supply frequency and voltage are reduced by 15% but the load power is the rated value. Neglect saliency, saturation, and losses and assume the field current is maintained constant.
(*Answer:* (a) 37°, (b) 39.2°, (c) 35°, (d) 42.5°.)

8.18. A 3-phase synchronous generator supplies power P to a three-phase synchronous motor. Prove that

$$P = 3 \frac{E_g E_m}{X_g + X_m} \sin(\delta_g + \delta_m)$$

where E_g, E_m are the no-load induced emfs per phase of the generator and the motor, X_g, X_m are the synchronous reactances per phase of the generator and the motor, and δ_g, δ_m are the load angles of the generator and motor. Armature winding resistance and saliency are ignored.

8.19. A 3-phase, 440-V, 60-Hz, 8-pole, Y-connected synchronous motor has a synchronous reactance of 2.4 ohms per phase. The motor operates underexcited at a generated phase voltage E of 240 V. If the load angle is 20 electrical degrees, calculate the electromagnetic torque.
(*Answer:* 278 N·m.)

8.20. A 2300-V, 60-Hz, 1492-kW, 3-phase, Δ-connected, 8-pole, unity power factor synchronous motor has a synchronous reactance X_s of 1.95 ohms per phase. **(a)** Neglecting losses, saturation, and saliency, calculate the maximum torque. It is assumed that the excitation for the unity power factor, rated load remains constant. **(b)** For the same assumptions as (a), estimate the maximum output kVA and power factor if the machine were run as a generator. **(c)** If the machine were run as a synchronous compensator (zero torque) what would be the kVAr input to the machine for **(i)** rated field current, **(ii)** 1.1 times rated field current, and **(iii)** half rated field current.

(*Answer:* (a) 105,400 N·m, (b) 1160 kVA at 0.72 leading power factor, (c) (i) 141 kVAr leading, (ii) 970 kVAr leading, (iii) 4000 kVAr lagging.)

8.21. A 60-Hz synchronous motor drives a synchronous generator, whose output frequency is required to be 400 Hz with a tolerance of ±4% variation. Determine the lowest number of poles of each machine.

(*Answer:* 4, 26.)

8.22. A mechanical load is driven by a 1865-kW, 3-phase, 4400-V induction motor having a full-load efficiency of 0.93 per unit and a power factor of 0.87. A synchronous motor, having a rating of 3000 kVA, is installed to provide additional drive and its input is found to be 1750 kW. The resistance and synchronous reactance are 0.006 per unit and 0.35 per unit, respectively, and the magnetization curve at rated speed is as follows.

Field current A	11.5	23	34.5	47.5	67	115
Generated line voltage V	1000	2000	3000	4000	5000	6000.

Determine the synchronous motor field current required to improve the power factor of the total load to unity.

(*Answer:* 70 A.)

8.23. A 3-phase, 2300-V, 2000-kVA, 60-Hz, Y-connected synchronous generator, whose synchronous reactance is 2 ohms per phase, supplies power to a 3-phase, 30-pole, 2300-V, 2000-kW, 60-Hz, Y-connected synchronous motor, whose synchronous reactance is 1.5 ohms per phase. For a particular mechanical load, whose torque varies as the square of the speed, the field currents are set for full-load power, unity power factor, at rated speed and frequency. If the speed of the generator is slowly increased find the maximum torque that can be developed by the motor without loss of synchronism. Neglect losses.

8.11. BIBLIOGRAPHY

Acarnley, P. P. *Stepping Motors.* 2nd ed. IEE Control Engineering Series 19. Peter Peregrinus, Ltd., 1984.

Bewley, L. V. *Alternating Current Machinery.* New York: Macmillan Co., 1949.

Carr, Laurence. *The Testing of Electrical Machines.* London: MacDonald Co., Ltd., 1960.

Chalmers, B. J. *Electromagnetic Problems of AC Machines.* London: Chapman and Hall, Ltd., 1965.

Concordia, Charles. *Synchronous Machines Theory and Performance.* Schenectady, N.Y.: General Electric Co., 1951.

Elgerd, Olle. *Electric Energy Systems Theory,* 2nd ed. New York: McGraw-Hill Book Co., 1982.

Garik, M. L., and C. C. Whipple. *Alternating-Current Machines,* 2nd ed. New York: D. Van Nostrand, 1961. John Wiley & Sons, Inc., 1986.

Kenjo, T. *Stepping Motors and Their Microprocessor Controls.* Oxford: Clarendon Press, 1984.

Kimbark, E. W. *Power System Stability: Synchronous Machines.* New York: Dover, 1968.

Kuo, B. C. *Step Motors.* West Publishing Co., 1974.

Lawrence, R. R., and H. E. Richard. *Principles of Alternating-Current Machinery,* 4th ed. New York: McGraw-Hill Book Co., 1953.

Nasar, S. A. *Electric Machines and Electromechanics.* Schaum's Outline Series in Engineering. New York: McGraw-Hill Book Co., 1981.

Nasar, S. A. (Ed). *Handbook of Electric Machines.* New York: McGraw-Hill Book Co., 1987.

Still, Alfred, and Charles S. Siskind. *Elements of Electrical Machine Design.* Tokyo: McGraw-Hill Book Co., 1954.

Walker, J. H. *Large Synchronous Machines.* Oxford: Oxford University Press, 1981.

———. *General Requirements for Synchronous Machines.* ANSI Standard C50.10, 1977.

———. *Test Procedures for Synchronous Machines.* IEEE Standard 115, 1983.

CHAPTER 9

Single-Phase AC Commutator Motors

9.1. INTRODUCTION

We are well acquainted with the operation of a dc machine. The commutator in the dc machine preserves the special distribution of current in the armature conductors so that the developed torque is constant and unidirectional. If the polarity of the applied dc voltage is changed such that the currents in the field and armature windings both reverse, as would happen in a dc series motor, the torque does not change direction. It follows that, if alternating current is supplied to a series motor, the torque is always in the same direction and operation is ensured.

There are two types of ac commutator motors that operate from single-phase supplies. One is the series motor and the other is the repulsion motor. They are similar in that they both have a stator winding which produces the main field, and they both have a distributed armature winding connected to a commutator. The differences are in the connections and the commutator brush positions. As its name implies, the series motor has its stator winding connected in series with the armature winding and the brushes are on the geometric neutral or quadrature axis. However, the repulsion motor has its armature winding electrically isolated by having the brushes short-circuited, and the brush axis can be shifted. See Fig. 9.1. Consequently, the electric energy for conversion is conductively transferred to the armature circuit of the series motor, whereas the energy is inductively transferred to the armature of the repulsion motor.

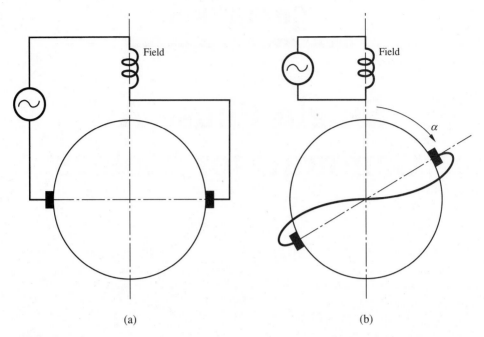

(a)   (b)

Figure 9.1   Single-phase commutator motors. (a) Simple series motor, (b) repulsion motor.

9.2. SERIES AC COMMUTATOR MOTOR

The single-phase ac series motor has torque-speed characteristics similar to the dc series motor. Consequently, ac series motors up to several thousand kilowatts output have found application for traction purposes. Because of commutation difficulties the large motors must operate at low frequencies. For ac electric railways, frequencies of $16\frac{2}{3}$ and 25 Hz have usually been used. However, power frequencies can be employed if the motor draws a small current and runs at high speeds (3000 to 16,000 r/min). Accordingly a great number of ac commutator motors with outputs not exceeding 500 W are manufactured for such purposes as household appliances.

There is a structural resemblance between the ac and dc series motors. In fact the ac series machine, in small sizes, can run equally well, if not better, on direct current as on alternating current. For this reason it is also called a *universal* motor. Alternating current causes appreciable reactance-voltage drops in the field and armature windings. Therefore, for a given applied voltage, current, and torque, the motional emf in the armature winding is less during ac operation than during dc operation. Accordingly, the speed of the ac machine tends to be lower. Also, alternating current may cause the magnetic circuit of a dc machine to be saturated at the peaks of the current wave. So the rms value of flux may be appreciably less with alternating current. Therefore the torque tends to be less and the speed higher with alternating current than with direct current in the series commutator motor. Further, there are some structural differences between the ac and dc machines. For example, the ac machine has both the stator and rotor cores laminated to minimize iron losses and it may have salient or non-salient field poles.

Reactance and commutation in the ac machine cause problems. The high reactance is due to the sum of the field and armature reactances and this results in a low power factor. Designers combat this by making both the magnetic circuit and the air gap short in order to keep the leakage flux low. The field winding is made to have only a few turns to reduce the reactance. The armature winding must have many turns, since the torque is proportional to the product of the main field and the armature ampere turns. In large machines the large armature reaction mmf can be neutralized to some extent by means of a distributed compensating winding, centered on the quadrature axis, and placed in slots in the pole face. Reactance effects can also be reduced by operating at a low frequency but this is economically feasible only for large machines.

The problem of commutation is worse in ac machines than in dc machines. This comes about because the axis of the commutating coil is in the direct axis. Consequently there is an additional transformer emf to contend with. No special precautions are taken with the small machines except that they are designed to operate at high speeds. The high speed means that the commutating coil experiences very little change in the main flux during commutation, so that the transformer emf is small, the commutation difficulties are alleviated, and the machine can operate at power frequencies. However, there is a need to keep the armature volts per turn low and to have very few turns per coil between commutator segments. This restricts ac commutator motors to supply voltages less than 400 V.

Large motors may have interpole windings with a shunt resistor. These windings aid the current reversal in the coil undergoing commutation. The resistor changes the phase angle of the current so that the interpole ampere turns can better tend to neutralize the transformer emf and the self-induced emf in the commutation coil. Improvements have also been made by using high-resistance brushes or high-resistance leads connecting the coil ends to the risers of the commutator segments. The leads are in series to the short-circuit current in the commutating coil, but are in parallel to the load current.

9.2.1. Analysis of Simple Series Motor

In a series motor a single armature coil experiences both a transformer induced emf due to the alternation of the main field and a motional induced emf due to the relative motion of the coil in the main field. If the commutator brushes are on the quadrature axis, the net transformer emf between the brushes is zero, while the net motional emf is a maximum. Since it is the motional emf that promotes energy conversion, the natural place for the brushes is on the quadrature axis.

If the supply were direct current, then the armature winding motional or back emf would be

$$E = \phi n Z \frac{p}{a}. \tag{9.2.1}$$

But in the ac motor the main flux alternates. The instantaneous values of the motional emf and the flux alternate simultaneously and the rms values are in phase. Therefore the rms value of the motional emf E_m between the brushes of the ac machine is

$$E_m = \frac{\hat{\Phi}}{\sqrt{2}} n Z \frac{p}{a} \tag{9.2.2}$$

where $\hat{\Phi}$ is the maximum value of the flux per pole. Since the motional frequency f_m and the number of series turns N_a on the armature winding are

$$f_m = np \quad \text{and} \quad N_a = \frac{Z}{4a} \tag{9.2.3}$$

$$E_m = 2\sqrt{2}\hat{\Phi}f_m N_a. \tag{9.2.4}$$

E_m is in phase with ϕ, which is in phase with I if there are no losses. For this condition the mechanical power developed is $E_m I$. If there are iron losses the flux is not in phase with the current. Account of the losses is made by resolving the main, direct-axis flux into two components. One component, Φ_m, is in phase with I and is associated with mechanical power. The second component Φ_i accounts for iron losses. Thus, $\Phi = \Phi_m - j\Phi_i$. The phase angle between Φ_m and Φ is known as the *hysteretic angle*. Now the total motional voltage V_m is written in the form

$$\mathbf{V}_m = 2\sqrt{2}(\hat{\Phi}_m - j\hat{\Phi}_i)f_m N_a = \mathbf{I}\frac{f_m}{f}\left(\frac{2\sqrt{2}\hat{\Phi}_m f N_a}{\mathbf{I}} - \frac{j2\sqrt{2}\hat{\Phi}_i f N_a}{\mathbf{I}}\right). \tag{9.2.5}$$

The two terms within parentheses have units of ohms. The first is associated with magnetization at the supply frequency and can be represented by the reactance X_m, while the second term is associated with magnetic circuit losses at the supply frequency and can be represented by the resistance R_i. If the ratio f_m/f is called S then

$$\mathbf{V}_m = \mathbf{I}S(X_m - jR_i). \tag{9.2.6}$$

In addition to the motional voltage, the current I through the field and armature windings will cause voltage drops at the supply frequency due to the field leakage impedance Z_f, the field magnetizing impedance Z_{of}, and the armature impedance Z_a. The sum of the impedance drops and the motional voltage is the voltage V applied to the motor. Thus

$$\mathbf{V} = \mathbf{I}(\mathbf{Z}_f + \mathbf{Z}_{of} + \mathbf{Z}_a) + \mathbf{V}_m \tag{9.2.7}$$

or

$$\mathbf{V} \triangleq \mathbf{I}(R + jX) + \mathbf{I}S(X_m - jR_i) = \mathbf{I}[(R + SX_m) + j(X - SR_i)]. \tag{9.2.8}$$

A phasor diagram is shown in Fig. 9.2.

The useful performance factors of a motor are power factor, efficiency, and torque. From the voltage equation the power factor is

$$\cos\theta = \frac{R + SX_m}{[(R + SX_m)^2 + (X - SR_i)^2]^{1/2}}. \tag{9.2.9}$$

If the speed is low, $X \to 0$ and $\cos\theta \to R/Z$, and if the speed is high $S \to \infty$ and $\cos\theta \to X_m/Z_m$.

Efficiency can be calculated from a knowledge of the input and output powers. The input power is

$$P_i = VI\cos\theta = \frac{V^2(R + SX_m)}{(R + SX_m)^2 + (X - SR_i)^2}. \tag{9.2.10}$$

The developed or gross mechanical power output is

$$P_m = E_m I = SX_m I^2 = \frac{V^2 SX_m}{(R + SX_m)^2 + (X - SR_i)^2}. \tag{9.2.11}$$

If the windage and friction losses are P_{WF} then the per unit efficiency is

$$\eta = \frac{P_m - P_{WF}}{P_i}. \tag{9.2.12}$$

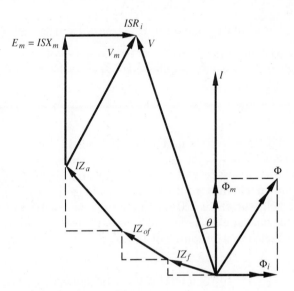

Figure 9.2 Phasor diagram.

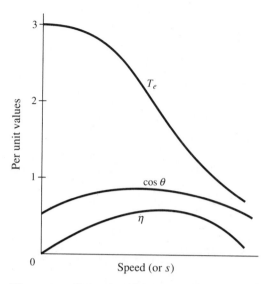

Figure 9.3   General performance curves.

The gross torque is

$$T_e = \frac{P_m}{2\pi f_m/p} = \frac{p}{\omega}\frac{P_m}{S} = \frac{p}{\omega}\frac{V^2 X_m}{(R+SX_m)^2 + (X-SR_i)^2}. \tag{9.2.13}$$

For low speeds $S \to 0$ and $T_e \to \dfrac{p}{\omega}\dfrac{V^2 X_m}{Z^2}$, and for the high speeds $S \to \infty$ and

$T_e \to \dfrac{p}{\omega}\dfrac{V^2 X_m}{S^2 Z_m^2}$. The torque is a maximum at standstill and decreases inversely as the

square of the speed at high speeds. Figure 9.3 illustrates the variation of the performance factors with speed which is proportional to the factor S.

EXAMPLE 9.1

A single-phase, 110-V, universal motor has a standstill impedance of $(20 + j40)$ ohms. If the speed is 127 rad/s, the current is 2 A, the motional voltage is 40 V, and the windage and friction losses are 2 W. Saturation and core losses can be neglected and it can be assumed that the torque loss due to windage and friction is proportional to speed. If the field current is 1.0 A estimate **(a)** the motor speed, **(b)** the shaft torque, and **(c)** the machine efficiency. **(d)** Repeat the calculations for the case when the motor is connected to a dc source of 110 V.

Solution

(a) From eq. (9.2.6) $\mathbf{V}_m = \mathbf{I}SX_m$. Thus the motional voltage $V_m = K_E I \omega_m$, where K_E is a constant.

From the data $K_E = \dfrac{40}{2 \times 127} = 0.157$. The voltage equation is $\mathbf{V} = \mathbf{IZ} + \mathbf{V}_m$.

Since the core losses are ignored it can be seen from the phasor diagram in Fig. 9.2 that $\mathbf{V}_m$ is in phase with the current $\mathbf{I}$.

Therefore, $V^2 = (IR + V_m)^2 + I^2X^2$,

or $V_m = (V^2 - I^2X^2)^{1/2} - IR = (110^2 - 40^2)^{1/2} - 20 = 82.5$ V.

The speed $\omega_m = \dfrac{V_m}{K_E I} = \dfrac{82.5}{0.157 \times 1.0} = 525$ rad/s.

(b) Loss torque $T_{WF} = D\omega_m$. Therefore windage and friction loss $P_{WF} = D\omega_m^2$.

From the data $D = \dfrac{2}{127^2} = 1.24 \times 10^{-4}$ N·m·s/rad.

At 525 rad/s, $P_{WF} = 1.24 \times 10^{-4} \times 525^2 = 34.2$ W. The developed power $P = V_m I = 82.5$ W. The output power $P_o = P - P_{WF} = 48.3$ W.

Therefore the shaft torque $T_s = P_o/\omega_m = 48.3/525 = 0.092$ N·m.

(c) Ohmic loss $P_c = I^2R = 20$ W. Therefore input power $P_i = P + P_c = 82.5 + 20 = 102.5$ W.

Efficiency $\eta = \dfrac{P_o}{P_i} \times 100 = \dfrac{48.3}{102.5} \times 100 = 47.1\%$.

(d) If the motor operates from a dc supply the reactance is zero. Consequently the motional armature emf is $E = V - IR = 110 - 1.0 \times 20 = 90$ V.

The speed $\omega_m = \dfrac{E_a}{K_E I} = \dfrac{90}{0.157 \times 1.0} = 572$ rad/s.

So we see that the speed for the ac motor is lower than the dc motor for the same operating conditions.

Developed power $P = E_a I = 90$ W.

Windage and friction loss $P_{WF} = D\omega_m^2 = 1.24 \times 10^{-4} \times 572^2 = 40.6$ W.

Output power $P_o = P - P_{WF} = 90 - 40.6 = 49.4$ W.

Therefore shaft torque $T_s = \dfrac{P_o}{\omega_m} = \dfrac{49.4}{572} = 0.086$ N·m.

Ohmic losses $P_c = I^2R = 20$ W. Therefore input power $P_i = P + P_c = 110$ W.

Efficiency $\eta = \dfrac{49.4}{110} \times 100 = 44.9\%$.

9.3. REPULSION MOTOR

The repulsion motor, depicted in Fig. 9.1b, is similar to a series ac commutator motor, except that the armature winding is inductively rather than conductively connected to the stator winding.

The principle of repulsion can be explained with the use of the diagram in Fig. 9.4. A fixed coil S is connected to a single-phase supply. The flux ϕ_s, produced by the current i_s, will tend to link the short-circuited coil R. By transformer action there will

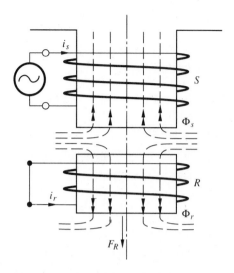

Figure 9.4   Repulsion principle.

be a current i_r that tends to oppose the change of flux linkage. The result is similar to having two like magnetic poles adjacent to one another to produce a force of repulsion F_R. For torque to exist the axis of coil R must be inclined at an angle to the axis of coil S.

Since the currents in the rotor winding of a repulsion motor are induced by transformer action, the rotor field along the brush axis is mainly opposite to the main field. These two fields will repel each other and it is from this action that the motor derives its name.

An advantage of inductive coupling is that the armature winding can be designed as a low voltage winding so that the problems of commutation can be reduced. Another advantage, when the brush shift α from the main field-winding axis is adjustable, is that the motor speed can be controlled. Figure 9.5 depicts the action for different brush settings. For this discussion it is assumed that the main field has the polarity shown and is increasing.

For the case where the brushes are on the direct axis, induction causes emf polarities and current flow as shown in Fig. 9.5a. As illustrated, the interaction between the main flux and the armature currents produces two equal bands of torque that oppose each other, so there is no net torque. This also follows from the observation that stator and rotor fields are in alignment.

Figure 9.5b shows the brush axis to be in quadrature with the main flux. Each individual armature coil will have an emf induced in it, and the pattern of emf polarities is the same as in Fig. 9.5a. The net voltage between the brushes is zero since the coil emfs neutralize one another. There is no current flowing in the armature conductors. Without current there can be no torque.

If the brush axis is shifted to an angle α with the direct axis, the induced emfs do not cancel each other completely, so currents flow in the armature conductors as

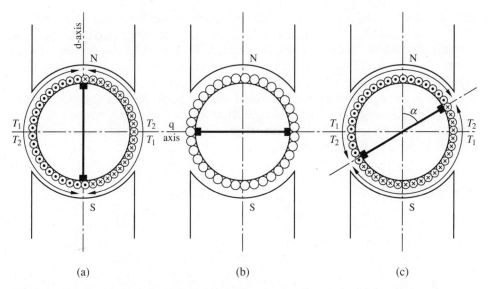

Figure 9.5 Repulsion motor action. (a) Brushes on the direct axis, (b) brushes on the quadrature axis, (c) brush axis inclined at angle α.

indicated in Fig. 9.5c. There are four torque bands shown. Motion results from the net electromagnetic torque $T_e = 2(T_1 - T_2)$.

The main field may be resolved into two components, a transformer or current producing component along the brush axis, $\phi \cos\alpha$, and a torque producing component, $\phi \sin\alpha$, perpendicular to the brush axis. The torque component can not produce any current in the rotor. As in dc machines, the torque can be expressed by

$T_e \propto$ (main flux component normal to brush axis) $\times$ (current between brushes).

But the current between brushes is proportional to that component of main flux along the brush axis. Hence

$$T_e \propto \Phi \sin\alpha \times \Phi \cos\alpha \propto \frac{\Phi^2}{2} \sin2\alpha. \qquad (9.3.1)$$

Thus the torque is a maximum when $\alpha = 45°$. If the brushes are shifted over an angle α opposite to the direction shown in Fig. 9.5c and if the main field is resolved into its components, it can be seen that the direction of the rotor currents is reversed with respect to the torque-producing field component. Hence the developed torque and the resulting direction of rotation are reversed. A repulsion motor develops a torque in the direction in which the brushes are shifted from the direct axis.

A single-phase induction motor has no starting torque. It is necessary to introduce some means of starting and accelerating it. However, it has excellent running characteristics. A repulsion motor has excellent starting torque with the additional quality of drawing low starting current with respect to torque, but in its simplest form has poor running characteristics. It is only natural that the repulsion motor principle has been incorporated into a single-phase induction motor to create a repulsion-start, induction-motor-run machine. In practice, the rotor winding is connected to a commutator whose brushes are short-circuited. This provides the starting mechanism. Once the machine is up to speed, the commutator segments are short-circuited by a conducting ring, so that the rotor winding becomes a quasi-cage winding.

The plain repulsion motor has found limited application. Examples include centrifuges, air compressors, hay hoists, and hay driers. However, the repulsion-start, induction-run motor has had a wide range of applications from use in air conditioners to use in washing machines.

9.4. SUMMARY

AC commutator motors find useful applications in the home and in industry. The ac series motor is a universal motor in that it can operate from both ac and dc sources. The repulsion motor is useful for adjustable-speed applications.

9.5. PROBLEMS

9.1. If saturation is neglected, show that the average torque of a single-phase, series commutator motor is proportional to the square of the rms current.

9.2. A single-phase, 240-V, 50-Hz, universal motor has a field winding self-inductance of 0.1 H and an armature winding self-inductance of 0.05 H. The machine constant K_E is 0.3 volts per ampere per radian per second and the total circuit resistance is 20 ohms. Determine (a) the speed, (b) the electromagnetic torque, and (c) the power factor if the current is 2 A. Plot speed and power factor as a function of current.
(*Answer:* (a) 301 rad/s, (b) 1.2 N·m, (c) 0.92.)

9.3. A single-phase, 6-pole, 25-Hz series motor has a full-load current of 500 A at the rated speed of 10 revolutions per second. The ratio of the armature winding turns in series to the total field winding turns is 3 : 1. The field magnetizing impedance, the field leakage impedance, and the armature winding impedance are (0.005 + j0.05) ohms, (0.005 + j0.01) ohms, and (0.005 + j0.03) ohms, respectively. If the windage and friction losses at full load are 2 kW, determine (a) the power factor and (b) the efficiency of the motor. *Note:*

$$\mathbf{Z}_m = -j\mathbf{Z}_{of}N_a/N_f.$$

(*Answer:* (a) 0.94, (b) 88.2%.)

9.4. A single-phase, 25-Hz, 10-pole commutator motor has three turns per field pole and the armature has a lap winding with 1000 conductors. The field leakage impedance and the magnetizing impedance amount to (0.011 + j0.062) ohms and the armature impedance is (0.004 + j0.02) ohms. A full-load current of 700 A produces 2.45-kW iron loss and 2-kW windage and friction loss at

600 r/min. The maximum flux per pole is 0.014 webers. Calculate the power factor and efficiency at full load if magnetic saturation is neglected. (*Answer:* 0.92, 87.6%.)

9.6. BIBLIOGRAPHY

Bewley, L. V. *Alternating Current Machinery.* New York: Macmillan Co., 1949.

Garik, M. L., and C. C. Whipple. *Alternating-Current Machines.* 2nd ed. New York: D. Van Nostrand, 1961.

Lawrence, R. R., and H. E. Richard. *Principles of Alternating-Current Machinery.* 4th ed. New York: McGraw-Hill Book Co., 1953.

Nasar, S. A. (Ed.). *Handbook of Electric Machines.* New York: McGraw-Hill Book Co., 1987.

Veinott, C. G., and J. E. Martin. *Fractional- and Subfractional-Horsepower Electric Motors.* 4th ed. New York: McGraw-Hill Book Co., 1986.

CHAPTER 10

Dynamic Circuit Analysis of Rotating Machines

10.1. INTRODUCTION

In Chapter 2 we applied dynamic circuit analysis to a general class of electromechanical energy converters. In this chapter we will concentrate on rotating machines, in which energy conversion is continuous. The aim is to develop models, from which the machine equations can be systematically formulated for the purpose of either transient or steady-state performance studies.

The traditional approach to the steady-state analysis of rotating machines was taken in Chapters 5 to 9 in order to obtain some physical insight into their operation. Dynamic circuit analysis is more abstract and the aids of physical interactions can be lost during the mathematical manipulations. However this latter method of analysis, which has been developed largely by Gabriel Kron, does offer advantages. It provides the basic elements of a unified or generalized theory. Also, from the sets of differential equations defining the transient behavior, it is a simple matter to arrive at the steady-state performance of an electric machine.

A rotating machine is to be represented by a *primitive*[1] model wherever possible.

---

[1] A *primitive* is similar to the elementary machines used in Chapter 4 to describe the basic operation of the various types of machine.

428

From this model the general differential equations can be found. Once the constraints of voltages and system parameters are applied, it is desired to find a solution for the variables, which are usually currents and angular displacement or speed. The solution is relatively simple, if the equations are linear and the coefficients are constants. Consequently we try to reduce the equations, by transformation of variables if necessary, to this type. Most of the discussion centers about this type. There are important cases where there is no method of reducing the equations of motion to a linear form. Then a solution has to be obtained by means of the digital computer using numerical methods.

The assumptions and limitations unfold as the unified theory is developed and the results are compared with the traditional approach for the dc, induction, and synchronous machines.

10.2. TWO COUPLED COILS

Figure 10.1 shows two windings that are both excited. They are stationary, have the same axis of symmetry, and are assumed to be concentrated windings. Air is the medium so that the permeability is constant. This leads to a linear relationship between the magnetic flux and the exciting current. The result is that, even though inductance may vary with the relative position of the two coils, the inductances remain independent of current.

The Maxwell-Lorentz voltage equations for this doubly-excited system are

$$v_1 = R_1 i_1 + \frac{d\lambda_1}{dt} = R_1 i_1 + \frac{d}{dt}(L_{11}i_1 + L_{12}i_2) \qquad \textbf{(10.2.1)}$$

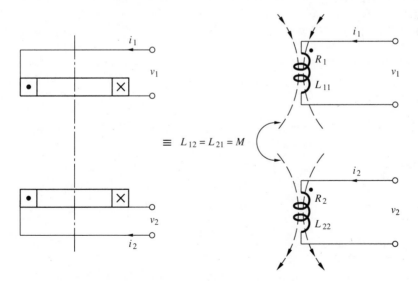

Figure 10.1 Two coupled coils.

and

$$v_2 = R_2 i_2 + \frac{d\lambda_2}{dt} = R_2 i_2 + \frac{d}{dt}(L_{21} i_1 + L_{22} i_2). \tag{10.2.2}$$

The voltage equations can be put into the matrix form

$$\underline{v} = [R + pL]\underline{i} \tag{10.2.3}$$

where $p = \dfrac{d}{dt}, \underline{v} = \begin{bmatrix} v_1 \\ v_2 \end{bmatrix}, \underline{i} = \begin{bmatrix} i_1 \\ i_2 \end{bmatrix}, [R] = \begin{bmatrix} R_1 & 0 \\ 0 & R_2 \end{bmatrix}$ and $[L] = \begin{bmatrix} L_{11} & L_{12} \\ L_{21} & L_{22} \end{bmatrix}$. This matrix form generalizes the circuit equations to include any number of circuits. Since p, the differential operator d/dt, acts on both $[L]$ and $\underline{i}$, then an expansion of the matrix equation gives

$$\underline{v} = [R]\underline{i} + [Lp]\underline{i} + [pL]\underline{i}. \tag{10.2.4}$$

These matrix vectors are described by

$\underline{v}$ = the vector of terminal voltages

$[Lp]\underline{i}$ = the set of transformer induced emfs (p operates only on i)

$[pL]\underline{i}$ = the set of motionally (or generated or speed or back) induced emfs.

The voltage equation can be written down as the simple Ohm's law expression

$$\underline{v} = [Z]\underline{i} \tag{10.2.5}$$

where $[Z]$ is known as the impedance matrix. This impedance is not that impedance associated with a single frequency sinusoidal excitation. It can be viewed as an operational impedance in the time domain. The entries of the $[Z]$ matrix in this case

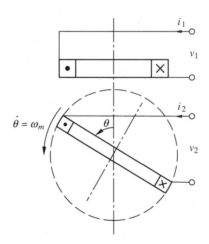

Figure 10.2   Doubly-excited system with rotational freedom.

are

$$[Z] = \begin{bmatrix} R_1 + pL_{11} & pL_{12} \\ pL_{21} & R_2 + pL_{22} \end{bmatrix} \tag{10.2.6}$$

and p operates both on the inductances and the currents.

When there is no relative coil motion, the coils of Fig. 10.1 represent a static transformer. Then the only induced emf is the transformer emf $[Lp]\underline{i}$. The motional emf $[pL]\underline{i}$ is zero because there is no change of inductance with respect to time or angular position. We have already dealt with the transformer in some detail. We wish to progress to the general formulation of the equations governing the behavior of primitive coils with cyclic changes of position such as rotary motion. The inclusion of motion is an extension of the case of the static transformer to that where one coil is given rotational freedom as shown in Fig. 10.2.

10.3. SLIP-RING PRIMITIVE

The primitive coils shown in Fig. 10.2 are arranged to have coil 1 stationary and coil 2 free to rotate about an axis. Both coils are excited. In order to deliver electric energy to a rotating coil from a stationary power supply the coil is attached to slip rings. The source supplies energy through brushes which slide on the surface of the slip rings.

This simple form of electric machine is called a *slip-ring primitive*. Consideration of the slip-ring primitive shows that

1. the current and voltage in the rotating coil are the same at every instant as the current and voltage from the power supply
2. if both coils are excited, then the resulting fields will attempt to align and motion ensues (motoring action)
3. if one coil is excited and the free coil is rotated, electric energy can be obtained from the unexcited coil (generator action).

This basic electromechanical energy conversion system can be a model for many of the complex, practical systems of motors and generators.

10.3.1. Voltage Equation of the Slip-ring Primitive

The general forms of the voltage equation are

$$\underline{v} = [Z]\underline{i} = [R]\underline{i} + p\underline{\lambda} \tag{10.3.1}$$

which can be expanded in general circuit terms to

$$\underline{v} = [R]\underline{i} + p[L]\underline{i} = [R]\underline{i} + [Lp]\underline{i} + [pL]\underline{i}. \tag{10.3.2}$$

This is the relation between terminal voltages, potential drops, transformer emfs, and motional emfs. The motional emf expression $[pL]\underline{i}$ contains inductance, which is a function of coil position, and the operator p acting on the inductance. It is convenient to put the motional emf term of eq. (10.3.2) into another form to become

$$[pL]\underline{i} = \left[\frac{d}{dt} L(\theta)\right]\underline{i} = \left[\frac{dL}{d\theta} \cdot \frac{d\theta}{dt}\right]\underline{i}$$

$$= \left[\frac{dL}{d\theta}\dot{\theta}\right]\underline{i} = \left[\frac{dL}{d\theta}\omega_m\right]\underline{i} = \omega_m\left[\frac{dL}{d\theta}\right]\underline{i}. \tag{10.3.3}$$

Hence

$$\underline{v} = [R]\underline{i} + [Lp]\underline{i} + \omega_m[G_s]\underline{i} \tag{10.3.4}$$

where $[dL/d\theta] = [G_s]$ is known as the motional inductance matrix or just the G matrix and $\dot{\theta} = \omega_m$ is the relative rotational speed of the two coils. The G matrix is present in the voltage equation if there is relative motion of coils which gives rise to a change of self- or mutual inductance with displacement.

In the two coupled coils shown in Fig. 10.2 there are transformer emfs in both coils if there is a change of current with respect to time. If coil 2 rotates there will be a motional induced emf in both coils, because the mutual inductance of each coil changes with relative position.

The entries of the matrices of eq. (10.3.4) for the two coupled coils in relative motion are

$$\begin{bmatrix} v_1 \\ v_2 \end{bmatrix} = \begin{bmatrix} R_1 & 0 \\ 0 & R_2 \end{bmatrix}\begin{bmatrix} i_1 \\ i_2 \end{bmatrix} + \begin{bmatrix} L_{11}p & L_{12}p \\ L_{21}p & L_{22}p \end{bmatrix}\begin{bmatrix} i_1 \\ i_2 \end{bmatrix}$$

$$+ \omega_m \begin{bmatrix} \dfrac{dL_{11}}{d\theta} & \dfrac{dL_{12}}{d\theta} \\[2mm] \dfrac{dL_{21}}{d\theta} & \dfrac{dL_{22}}{d\theta} \end{bmatrix}\begin{bmatrix} i_1 \\ i_2 \end{bmatrix}. \tag{10.3.5}$$

Since the coils are in air, their self-inductances are constants. The mutual inductance between the two coils is a function of the position of coil 2, defined by the angle θ. Thus

$$L_{12} = L_{21} = M(\theta). \tag{10.3.6}$$

At $\theta = \pi/2$ or $3\pi/2$ the axes of the two coils are coincident. Each coil experiences maximum flux linkage, that is, $M(\theta)$ is a maximum and the transformer emf is a maximum. At these positions the conductors of coil 2 are moving in the same direction as the field lines of coil 1, so there is no motional emf. This means that, although $M(\theta)$ is a maximum, the rate of change of $M(\theta)$ with position is zero. Therefore the motional emf in coil 1 will be zero also. (The reader may like to argue from the physical point of view the case that the motional emf in coil 1 is zero at $\theta = \pi/2$.) When the fields are aligned in this way there is no electromagnetic torque.

At $\theta = 0$ or π, the axes of the two coils are perpendicular. The mutual flux is zero and the transformer emf is zero. However, the conductors of coil 2 are sweeping at right angles through the field of coil 1. Therefore the motional emf is a maximum and the rate of change of $M(\theta)$ is a maximum. This holds true for coil 1 also. Further, when the fields of the two coils are perpendicular, the strongest torque exists to attempt to align the coil axes.

It would seem reasonable to conclude that the torque is a function of the motional induced emf and independent of the transformer emf. Since the motional emf is a function of $[G_s]$, we would like to find the relation between the electromagnetic torque acting on the coils and the G matrix.

A. SIMPLE CASE

Eq. (10.3.6) describes the mutual inductance M between the two coils shown in Fig. 10.2 as a function of the angle θ. Since the mutual inductance is zero at $\theta = 0$, and is a maximum (M_d say) at $\theta = \pi/2$, an approximation for the mutual inductance as a function of θ is $M = M_d \sin\theta$.

This simple relation is the objective of machine designers. The reason of the objective is not like ours, which is to simplify analysis. It is done to ensure that the machine will have, very nearly, a sinusoidal flux density distribution in the air gap of a slip-ring machine. This helps to optimize the efficiency. Consequently the simple relation describes the general case.

In the voltage eq. (10.3.4) the inductance matrix becomes

$$[L] = \begin{bmatrix} L_{11} & M_d \sin\theta \\ M_d \sin\theta & L_{22} \end{bmatrix} \tag{10.3.7}$$

and the motional inductance matrix becomes

$$[G_s] = \frac{d}{d\theta} \begin{bmatrix} L_{11} & M_d \sin\theta \\ M_d \sin\theta & L_{22} \end{bmatrix} = \begin{bmatrix} 0 & M_d \cos\theta \\ M_d \cos\theta & 0 \end{bmatrix}. \tag{10.3.8}$$

If we write the mutual inductance matrix as

$$[M] = \begin{bmatrix} 0 & M_d \sin\theta \\ M_d \sin\theta & 0 \end{bmatrix} = \begin{bmatrix} 0 & M_d \cos(\theta - \pi/2) \\ M_d \cos(\theta - \pi/2) & 0 \end{bmatrix} \tag{10.3.9}$$

we see that the G and M matrices are complementary and symmetric. The difference between $[G_s]$ and $[M]$ concerns an angle of $\pi/2$. It is this particular difference that allows us to formulate a rule for the G matrix entries.

B. RULE FOR G ENTRIES

An entry of $[G_s]$ will be other than zero only if there is relative motion of conductors of one coil in the field pattern of the second coil. We will find it useful in this chapter[2] to remember the following rule.

- Where an entry of $[G_s]$ is other than zero, its numerical value is the same as the mutual inductance of the two coils under consideration when the rotating coil has been advanced through $\pi/2$ in the direction of rotation.

---

[2] The rule for the G matrix entries is used in sections 10.8 onward.

We can check this rule by writing down the mutual inductance matrix of the coupled coils shown in Fig. 10.2 as

$$[M] = \begin{bmatrix} 0 & M_d \sin\theta \\ M_d \sin\theta & 0 \end{bmatrix}. \tag{10.3.10}$$

The mutual coupling, after advancing the rotating coil through $\pi/2$ in the direction of motion is

$$
\begin{aligned}
[M]_{\pi/2} &= \begin{bmatrix} 0 & M_d \sin(\theta + \pi/2) \\ M_d \sin(\theta + \pi/2) & 0 \end{bmatrix} \\
&= \begin{bmatrix} 0 & M_d \cos\theta \\ M_d \cos\theta & 0 \end{bmatrix} = [G_s].
\end{aligned}
\tag{10.3.11}
$$

The rule applies only for the conditions that (1) the system is linear and (2) the variation of mutual coupling is sinusoidal.

This rule provides a simple way to formulate the detailed equations that involve the G matrix.

10.3.2. Torque

The slip-ring primitive is an elementary electromechanical system. Electric energy is converted to mechanical energy or vice versa. If the machine is acting as a motor the electromagnetic torque is balanced by the mechanical torque of inertia, friction and load torques. The dynamic equation of motion[3] of the system shown in Fig. 10.2 is

$$T_e = J\ddot{\theta} + D\dot{\theta} + T_m \tag{10.3.12}$$

where $T_e =$ the electromagnetic torque (N·m)

           $T_m =$ the mechanical load torque (N·m)

           $J =$ the moment of inertia of the drive (kg·m²)

           $D =$ the friction factor (N·s)

           $\dot{\theta} =$ the rotor angular speed (rad/s)

and $\ddot{\theta} =$ the rotor angular acceleration (rad/s²).

This equation is quite general, so that it does not matter how many coils there are on the rotor or the stator of the electric machine. What does matter is how many magnetic poles are created. The two concentrated coupled coils create a 2-pole system and this is the basic system considered throughout this chapter. If the angular displacement is measured in electric radians, then the torque equation remains unaltered if multi–pole pair primitives are considered. The relations between the

---

[3] See Appendix B for the development of the dynamic equation of motion.

electric and mechanical quantities are

$$\theta \triangleq \text{no. pole pairs} \times \theta_p \quad \text{and} \quad \dot\theta \triangleq \omega_m = \text{no. pole pairs} \times \dot\theta_p \quad (10.3.13)$$

where θ_p is the angular displacement in mechanical radians.

The object is to solve the torque eq. (10.3.12) for θ or $\dot\theta$ in order to determine the machine performance under different load conditions T_m.

A. GENERAL TORQUE EXPRESSION FOR THE SLIP-RING PRIMITIVE

For any system, in which there is electromechanical energy conversion, the torque[4] is

$$T_e = -\frac{\partial W_s(\lambda,\theta)}{\partial\theta} = \frac{\partial W_s'(i,\theta)}{\partial\theta} \qquad (10.3.14)$$

and for any number of coils comprising a linear system

$$W_s = W_s' = \tfrac{1}{2}\underline{i}^t[L]\underline{i} \qquad (10.3.15)$$

so that

$$T_e = \tfrac{1}{2}\underline{i}^t\left[\frac{dL(\theta)}{d\theta}\right]\underline{i} \qquad (10.3.16)$$

or

$$T_e = \tfrac{1}{2}\underline{i}^t[G_s]\underline{i}. \qquad (10.3.17)$$

Accordingly, the objective stated in the last paragraph of section 10.3.1 has been met. We have found a relation between the electromagnetic torque and the G matrix.

For our simple case of $M = M_d \sin\theta$, that is applied to the two coil arrangement, shown in Fig. 10.2, the expansion of eq. (10.3.17) gives

$$T_e = i_1 i_2 M_d \cos\theta. \qquad (10.3.18)$$

Now that we have the analytic form of the electromagnetic torque it is possible to explore the dynamic equation of motion, eq. (10.3.12).

10.3.3. Power

The gross mechanical power output p_m of a slip-ring primitive motor is

$$p_m = T_e\dot\theta = \tfrac{1}{2}\underline{i}^t\left[\frac{dL}{d\theta}\right]\dot\theta\underline{i} = \tfrac{1}{2}\underline{i}^t[G_s\dot\theta]\underline{i}. \qquad (10.3.19)$$

---

[4] See Chapter 2, section 2.5 for the development of this equation.

The total instantaneous electric power input p_e is readily calculated by multiplying the row vector current by the column vector voltage, that is

$$p_e = \underline{i}^t \underline{v}. \tag{10.3.20}$$

However

$$\underline{v} = [R]\underline{i} + [Lp]\underline{i} + [pL]\underline{i}. \tag{10.3.21}$$

Accordingly

$$p_e = \underline{i}^t[R]\underline{i} + \underline{i}^t[Lp]\underline{i} + \underline{i}^t[pL]\underline{i} \tag{10.3.22}$$

and this can be written

$$p_e = \underline{i}^t[R]\underline{i} + \underline{i}^t[Lp]\underline{i} + \underline{i}^t[G_s\dot\theta]\underline{i}. \tag{10.3.23}$$

Some of the terms on the right of this equation are recognized. Others can be deduced.

$\underline{i}^t[R]\underline{i} = $ the ohmic power loss dissipated as heat.
$\frac{1}{2}\underline{i}^t[G_s\dot\theta]\underline{i} = $ gross mechanical power output. Therefore, by deduction
$\underline{i}^t[Lp]\underline{i} + \frac{1}{2}\underline{i}^t[G_s\dot\theta]\underline{i} = $ the rate of change of stored energy in the magnetic field.

All these equations represent the mathematical model of an electric machine. From them the performance should be able to be calculated. Just how useful the slip-ring primitive is can be seen from the following example.

EXAMPLE 10.1

Consider the multicoil slip-ring primitive shown in Fig. 10.3.

A four-coil arrangement is chosen as there is frequently more than one winding on either the rotor or stator of a real machine. Two concentrated windings are shown to be stationary. Their axes are in space quadrature; one axis is called the direct axis (d-axis) and the other is called the quadrature axis (q-axis). The q-axis is $\pi/2$ electric radians behind the d-axis. The rotor also has two windings in quadrature. We could have chosen any number of separate coils on the stator and rotor displaced by any angle. As with forces acting at a point in mechanics we can resolve the effects of currents in coils into components at right angles to simplify the solution. Determine the voltage equations in terms of the parameters of the primitive.

Solution

This example is quite general except that we have not chosen any axis on the stator or rotor to have more than one winding. There is no reason for this other than the aim to maintain simplicity at this stage. A real machine which could be represented by this elementary model is a two-phase induction motor.

The assumptions are as follows.

1. The rotor windings are excited through slip rings. This means that the rotor-winding flux axes rotate at the same speed as the rotor.

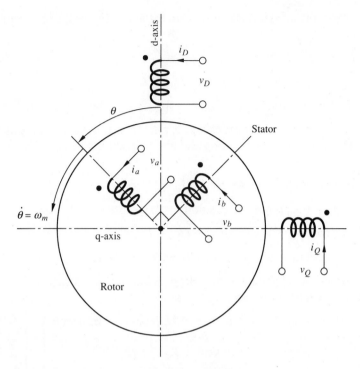

Figure 10.3 A multicoil slip-ring primitive.

2. All windings are concentrated (hence the term *primitive*).
3. As in a real machine, the rotor and stator windings are balanced.
4. The system has linear magnetic characteristics, consequently the inductances are functions of geometry and are independent of the currents.
5. The medium is air. No matter what the position of the rotor is, all self-inductances are constants.
6. The magnetic flux experienced by one coil due to current in a second coil is assumed to be a sinusoidal function of position. Therefore the mutual inductances are sinusoidal functions of position.

The voltage equations are

$$\underline{v} = [Z]\underline{i} = [R + pL]\underline{i}.$$

Use is made of the symbols depicted in Fig. 10.3 in order to write the voltage equations in full.

$$v_D = R_D i_D + p(L_D i_D) + p(L_{DQ} i_Q) + p(L_{Da} i_a) + p(L_{Db} i_b)$$

$$v_Q = p(L_{QD} i_D) + R_Q i_Q + p(L_Q i_Q) + p(L_{Qa} i_a) + p(L_{Qb} i_b)$$

$$v_a = p(L_{aD} i_D) + p(L_{aQ} i_Q) + R_a i_a + p(L_a i_a) + p(L_{ab} i_b)$$

$$v_b = p(L_{bD} i_D) + p(L_{bQ} i_Q) + p(L_{ba} i_a) + R_b i_b + p(L_b i_b).$$

From the constraints of symmetry and from the assumptions the following relations are introduced.

$$R_D = R_Q \triangleq R_1, \quad R_a = R_b \triangleq R_2, \quad L_D = L_Q \triangleq L_1, \quad \text{and} \quad L_a = L_b \triangleq L_2.$$

Further, $L_{DQ} = L_{QD} = 0$, $L_{ab} = L_{ba} = 0$, $L_{Da} = L_{aD} = M_d \cos\theta$,

$$L_{Db} = L_{bD} = M_d \sin\theta, \quad L_{Qa} = L_{aQ} = -M_d \sin\theta, \quad \text{and} \quad L_{Qb} = L_{bQ} = M_d \cos\theta.$$

All the above values are constants except θ, and M_d is the maximum mutual inductance between any one rotor winding and any one stator winding. The new voltage equations are

$$v_D = (R_1 + L_1 p)i_D + 0 + M_d p \cos\theta \, i_a + M_d p \sin\theta \, i_b$$

$$v_Q = 0 + (R_1 + L_1 p)i_Q - M_d p \sin\theta \, i_a + M_d p \cos\theta \, i_b$$

$$v_a = M_d p \cos\theta \, i_D - M_d p \sin\theta \, i_Q + (R_2 + L_2 p)i_a + 0$$

$$v_b = M_d p \sin\theta \, i_D + M_d p \cos\theta \, i_Q + 0 + (R_2 + L_2 p)i_b.$$

Since in future we will formulate the voltage equations in matrices, we can begin by writing in matrix form now. So

$$\begin{bmatrix} v_D \\ v_Q \\ v_a \\ v_b \end{bmatrix} = \begin{bmatrix} R_1 + L_1 p & 0 & M_d p \cos\theta & M_d p \sin\theta \\ 0 & R_1 + L_1 p & -M_d p \sin\theta & M_d p \cos\theta \\ M_d p \cos\theta & -M_d p \sin\theta & R_2 + L_2 p & 0 \\ M_d p \sin\theta & M_d p \cos\theta & 0 & R_2 + L_2 p \end{bmatrix} \cdot \begin{bmatrix} i_D \\ i_Q \\ i_a \\ i_b \end{bmatrix}.$$

This is the solution. We should inspect it.

Because p operates on θ and $\underline{i}$, all entries in $[Z]$ are not constant. That is, the differential equations do not have constant coefficients. Consequently the solution of the differential equations (to find $\underline{i}$ for a given $\underline{v}$) is not an easy task. Since we are looking for a simple means to solve for the currents, the torque, and the power at this elementary stage, we must develop further the idea of a primitive to see how we can obtain constant coefficients.

Our mathematical training tells us to make transformations so that the new equations are linear with constant coefficients. We could solve the new equations simply, then transform back to the real variables. Fortunately, people like Park and Kron produced transformations that we can use. We look at these transformations in section 10.8.

10.3.4. Conditions for Average Power Conversion

We would like to find the conditions that must prevail for the rotor of the primitive machine to have unidirectional and continuous motion. A net average value of the mechanical power converted satisfies the unidirectional motion requirement. If the average mechanical power were zero, the net motion must be zero.

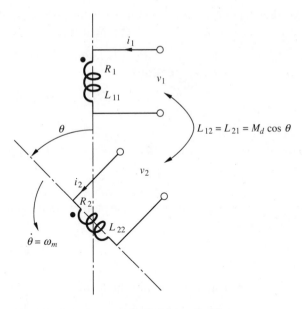

Figure 10.4 A two-coil slip-ring primitive.

Consider a slip-ring primitive machine with one winding on the stator and one winding on the rotor, as shown in Fig. 10.4. From eq. (10.3.17) the electromagnetic torque is

$$T_e = \frac{\partial W_s'}{\partial \theta} = \tfrac{1}{2}\underline{i}^t[G_s]\underline{i}.$$ (10.3.24)

That is

$$T_e = \tfrac{1}{2}[i_1 \quad i_2] \cdot \begin{bmatrix} 0 & -M_d \sin\theta \\ -M_d \sin\theta & 0 \end{bmatrix} \cdot \begin{bmatrix} i_1 \\ i_2 \end{bmatrix} = -M_d \sin\theta \, i_1 i_2.$$ (10.3.25)

From this value of torque we can calculate the instantaneous power converted to mechanical form. It is

$$p_m = T_e \omega_m = T_e \dot\theta = -M_d \sin\theta \, \dot\theta i_1 i_2.$$ (10.3.26)

We are interested in the average power, so we want to know the steady-state values of the two currents from the voltage equations. These steady-state values can be put in a general form

$$i_1(t) = \hat{I}_1 \sin\omega_1 t \quad \text{and} \quad i_2(t) = \hat{I}_2 \sin\omega_2 t$$ (10.3.27)

where ω_1 and ω_2 are the angular frequencies of the sources supplying coils 1 and 2, respectively. Either frequency could be zero, so that a source could be supplying direct current.

Since $\dot{\theta} = \omega_m$, then by integration $\theta = \omega_m t + \delta$, where $\theta = \delta$ at $t = 0$. On substitution of currents and position into eq. (10.3.26) the instantaneous power becomes

$$p_m = -M_d \sin(\omega_m t + \delta)\omega_m \hat{I}_1 \hat{I}_2 \sin\omega_1 t \sin\omega_2 t$$
$$= -\tfrac{1}{2}M_d\omega_m \hat{I}_1 \hat{I}_2 [\cos(\omega_1 t - \omega_m t - \delta)\sin\omega_2 t \qquad (10.3.28)$$
$$- \cos(\omega_1 t + \omega_m t + \delta)\sin\omega_2 t].$$

Further expansion of the two products yields

$$p_m = -\tfrac{1}{4}M_d\omega_m \hat{I}_1 \hat{I}_2 [\sin(\omega_1 t + \omega_2 t - \omega_m t - \delta)$$
$$- \sin(\omega_1 t - \omega_2 t - \omega_m t - \delta) - \sin(\omega_1 t + \omega_2 t + \omega_m t + \delta) \quad (10.3.29)$$
$$+ \sin(\omega_1 t - \omega_2 t + \omega_m t + \delta)].$$

A sine term as a function of time has zero average value. Therefore the condition that the power p_m has an average value is that the coefficient of t in the sine terms must be zero. These conditions are

$$\omega_1 + \omega_2 - \omega_m = 0, \qquad \omega_1 - \omega_2 - \omega_m = 0, \qquad \omega_1 + \omega_2 + \omega_m = 0 \qquad \text{and}$$
$$\omega_1 - \omega_2 + \omega_m = 0.$$

These frequency conditions can be written in the form

$$\omega_m = \pm\omega_1 \pm \omega_2. \qquad (10.3.30)$$

For example, if $\omega_m = \omega_1 + \omega_2$, from eq. (10.3.29) the average value of the power P_m is

$$P_m = \tfrac{1}{4}M_d\omega_m \hat{I}_1 \hat{I}_2 \sin\delta. \qquad (10.3.31)$$

However, for this case there are superfluous power and torque oscillations still present. The number of ways in which the frequency conditions are satisfied leads to different types of rotating machines. Without average conversion to mechanical power there can be no unidirectional motion of the rotor.

A specific configuration has been used to determine the test conditions for continuous energy conversion. There are often more than two coils. However, the principle is the same.

10.4. DISTRIBUTED WINDINGS

Concentrated windings are useful for formulating the equations from which to analyze the transient interaction of currents in coils. The mutual inductance relations have been assumed to be simple sinusoids. In practice, windings are more often distributed rather than concentrated. If the mutual relations between distributed windings and concentrated windings have the same sinusoidal nature, we can use the same kind of analysis as that which has been described in the foregoing sections.

The ideal case of a distributed winding with adjacent conductors touching on the surface of a cylinder is shown in Fig. 10.5. This is a rare feature of a real machine, because, with nearly all practical designs, the coil sides are in slots around the armature. However, it is a fair approximation, and it enables a simple calculation of the

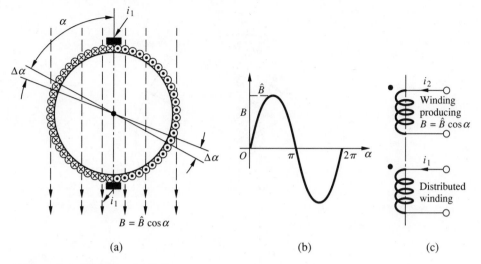

Figure 10.5 Distributed winding.

mutual inductance to be made. To get a uniformly distributed winding the conductors of each turn are diametrically opposite in the figure.

In order to calculate the mutual inductance of the whole winding it is necessary to find the flux linkage due to a current in a second winding. The second winding will remain unspecified except for its axis and the main assumption that the flux density has a sinusoidal distribution.

The maximum value of the mutual inductance is obtained when the axis of the distributed winding is made coincident with the axis of the applied field. The brush contacts for this situation give a current distribution as shown. With stationary brushes making a rubbing contact with the conductors, the current distribution in space remains fixed and the winding axis is fixed also. (Use is made of this principle in all commutator machines.)

For a single turn, as shown in Fig. 10.6, the conductors are at α and $(\alpha + \pi)$ from the common axes of both windings. The flux linking that turn is found from the integration of the flux linking the elemental area $lr\,\Delta\delta$ between the limits of $\pm\alpha$, where l is the effective length of the conductor in the field B, and r is the radius of the winding on the cylindrical armature. The flux linking the elemental area is

$$\Delta\Phi = \hat{B}\cos\delta \; lr\,\Delta\delta \tag{10.4.1}$$

so that the total flux linking the single turn is

$$\Phi = \int_{-\alpha}^{\alpha} \hat{B}lr\,\cos\delta\,d\delta. \tag{10.4.2}$$

That is

$$\Phi = 2\hat{B}lr\,\sin\alpha. \tag{10.4.3}$$

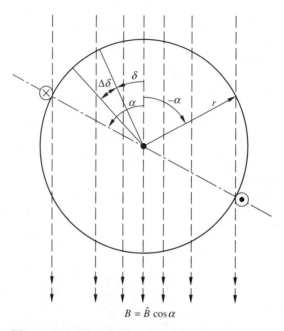

$$B = \hat{B} \cos \alpha$$

Figure 10.6 Flux linking a single turn.

In the limit, within the incremental angle $\Delta\alpha$, shown in Fig. 10.5a, the flux will be the same as that of a single turn at α. That is

$\Phi_\alpha = 2\hat{B}lr \sin\alpha.$

If Z is the total number of conductors on the cylindrical armature, there are $N_\alpha = (Z/2\pi)\Delta\alpha$ turns within the band $\Delta\alpha$ and the flux linkage is

$$\Delta\lambda_\alpha = 2\hat{B}rl \frac{Z}{2\pi} \sin\alpha \, \Delta\alpha. \tag{10.4.4}$$

Hence the total flux linkage of the winding with the coincident axes is

$$\lambda = \int_{\alpha=0}^{\pi} d\lambda_\alpha = \int_0^{\pi} 2\hat{B}rl \frac{Z}{2\pi} \sin\alpha \, d\alpha = 4\hat{B}rl \frac{Z}{2\pi}. \tag{10.4.5}$$

If the flux path is air, the relation between B and i_2 is a linear one. If the coils are wound on a ferromagnetic material, as is usual, when the exciting current is increased, saturation eventually becomes apparent. This analysis limits itself to the linear portion of the magnetization curve. Then the maximum value of the mutual inductance M_d of the winding is given by

$$M_d = \frac{\lambda}{i_2} = \frac{2\hat{B}rlZ}{\pi i_2}. \tag{10.4.6}$$

The general case is where the axis of the distributed winding is displaced by θ from the axis of the exciting winding. That is, the brushes are assumed to be at θ from the axis of the applied field. The single turn, first considered, is at α from the axis of the distributed winding, so that the flux linkage of the single turn becomes

$$\lambda_1 = 2\hat{B}lr \sin(\theta + \alpha) \tag{10.4.7}$$

and the flux linkage for an elemental band within $\Delta\alpha$ is

$$\Delta\lambda_\alpha = 2\hat{B}lr \frac{Z}{2\pi} \sin(\theta + \alpha)\Delta\alpha. \tag{10.4.8}$$

Hence the total flux linkage for the whole of the distributed winding is

$$\lambda(\theta) = \int_{\alpha=0}^{\pi} d\lambda_\alpha. \tag{10.4.9}$$

That is

$$\lambda(\theta) = \int_0^{\pi} 2\hat{B}lr \frac{Z}{2\pi} \sin(\theta + \alpha)d\alpha = 4\hat{B}lr \frac{Z}{2\pi} \cos\theta. \tag{10.4.10}$$

Therefore the mutual inductance is

$$M = \frac{\lambda(\theta)}{i_2} = \frac{2\hat{B}lrZ}{\pi i_2} \cos\theta = M_d \cos\theta. \tag{10.4.11}$$

This has the same form as that of the concentrated winding in subsection A under section 10.3.1. Analysis of the interaction between a distributed winding and other windings can thus be conducted in the manner described already in section 10.3.

10.5. COMMUTATOR PRIMITIVE

A machine with a commutator has a distributed winding. We have just shown that we can treat such a winding as a concentrated winding. Therefore the two windings shown in Fig. 10.7 are equivalent. We will use the symbol of the concentrated winding, because the axis of the winding is shown clearly.

The winding represented in Fig. 10.7b is a primitive commutator winding. Although the conductors of this winding rotate at the same speed as the rotor, the winding axis coincides at all times with the brush axis as shown. The winding axis is the axis of the magnetic field, which is produced by the winding current. For all commercially manufactured machines the brushes remain stationary with respect to the machine's stator. Therefore the commutator winding has a fixed axis. This is going to make the analysis easier than that for the slip-ring primitive, whose rotor winding axis rotates.

We will deal only with voltages and currents that can be measured at the terminals or brushes. At this stage we will consider air-cored coils for simplicity. Therefore,

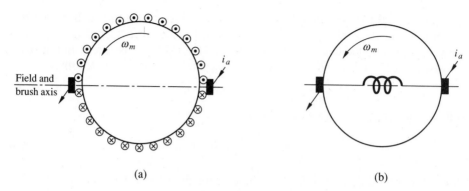

(a)   (b)

Figure 10.7 Commutator winding equivalents. (a) Distributed winding, (b) concentrated winding.

there must be at least two coils if electromechanical energy conversion is to take place. Consider a stationary coil, either distributed or concentrated. It is represented by the primitive coil which has the terminal voltage v_f in Fig. 10.8. We can call it the *field winding* or *exciting winding*. The second winding shown in Fig. 10.8 is a commutator winding which rotates at ω_m. The brushes are fixed at an angle α to the field winding axis, which is the direct axis. The resulting energy converter is called the *commutator primitive*.

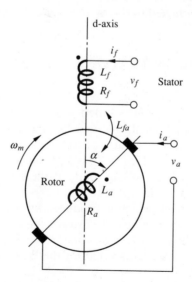

Figure 10.8 Doubly-excited commutator primitive.

10.5.1. Commutator Primitive Voltage Equations

In Fig. 10.8 the conductors of the field winding are fixed in space, and the field winding axis is fixed on the direct axis. The rotor conductors rotate at ω_m. However, the rotor or armature winding axis is fixed at an angle α to the direct axis by means of the brushes on the commutator. Hence, both winding axes are stationary with respect to each other. This is important to note because it limits motional emfs to the armature winding.

The voltage equation for the field winding is

$$v_f = R_f i_f + L_f p i_f + L_{fa} p i_a \qquad (10.5.1)$$

where L_f = the self-inductance and is a constant

L_{fa} = the mutual inductance = $M_d \cos\alpha$

and M_d = the mutual inductance for coincident axes.

It must be noted that there is no motional emf in the field winding because there is no relative motion between the field winding conductors and the magnetic field of the armature winding.

The armature winding conductors move at ω_m relative to both the stator field flux and the armature winding flux. There is a net motional emf due to the armature conductors moving in the stator field. However, the net motional emf in the armature winding due to motion in its own field is always zero. Figure 10.9 illustrates the cancellation effect, so that at the brushes there is no self-motional emf. The motional

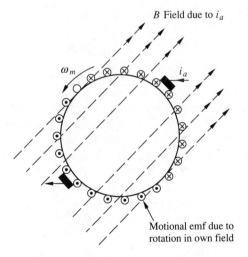

Figure 10.9 Motional emf by self-induction.

emf is given by

$$e_m = G\omega_m i_f \qquad (10.5.2)$$

where, as before, the G term is the value of the mutual inductance between the two windings when the armature winding has been rotated through $\pi/2$ in the direction of motion, that is, $G = -M \sin\alpha$.

The armature voltage equation can be written in the form

$$v_a = \{R_a i_a\} + \{L_a p i_a + M_d \cos\alpha \, p i_f\} - \{M_d \sin\alpha \, \omega_m i_f\} \qquad (10.5.3)$$

where L_a = the self-inductance of the armature winding and is a constant.

The first term in braces on the right of eq. (10.5.3) is the voltage drop across the armature winding resistance. There are two expressions in the second term in braces. The first is a transformer emf induced by the flux of the armature winding current and the second is a transformer emf induced by the flux of the field winding current. Finally, the end term is the motional emf.

The voltage equations for the field winding and the armature winding can be combined in the matrix form.

$$\underline{v} = [Z]\underline{i} \qquad (10.5.4)$$

where $[Z] = [R + Lp + G\omega_m]$. Hence the voltage equation for the commutator primitive with two windings is

$$\begin{bmatrix} v_f \\ v_a \end{bmatrix} = \begin{bmatrix} R_f + L_f p & M_d \cos\alpha p \\ M_d \cos\alpha p & R_a + L_a p \end{bmatrix} \cdot \begin{bmatrix} i_f \\ i_a \end{bmatrix}$$
$$+ \omega_m \begin{bmatrix} 0 & 0 \\ -M_d \sin\alpha & 0 \end{bmatrix} \cdot \begin{bmatrix} i_f \\ i_a \end{bmatrix}. \qquad (10.5.5)$$

The G matrix is

$$[G] = \begin{bmatrix} 0 & 0 \\ -M_d \sin\alpha & 0 \end{bmatrix}. \qquad (10.5.6)$$

The G matrix is asymmetric and contains terms only in entries pertaining to the armature. In contrast, the slip-ring primitive G matrix is always symmetric and contains terms in both the entries for stator and rotor windings.

The motional emf at the brushes of the armature of the commutator primitive is zero when $\alpha = 0$. This is similar to the case for the motional induced emf of the winding whose conductors move relative to their own field. The motional induced emf is a maximum when $\alpha = \pi/2$, at which point the transformer emf in the armature winding is a minimum. Since the energy conversion depends on the magnitude of the motional induced emf, it would be expected that the brushes should be fixed at $\alpha = \pi/2$ ideally. The axis of the armature winding is in quadrature with the field winding (or direct axis) for this case. So, if $\alpha = \pi/2$, the armature-winding axis is said to be on the quadrature axis.

Inspection of the voltage equations shows that they are linear differential equations with constant coefficients if ω_m is constant. Hence the equations are relatively simple to solve by well-known methods such as Laplace transformation. If the slip-ring primitive could be mathematically transformed to a commutator primitive then the solution of the slip-ring primitive equations would also be simplified.

Now we want to know how much torque is developed electromagnetically by the commutator primitive.

10.5.2. Commutator Primitive Torque and Power

For the commutator primitive shown in Fig. 10.8 the torque expression is

$$T_e = \frac{\partial W'_s(i,\alpha)}{\partial \alpha}. \qquad (10.5.7)$$

The system is considered to be linear so that coenergy W'_s and the stored energy W_s have the same values. Hence, at any instant

$$W_s = W'_s = \tfrac{1}{2}L_f i_f^2 + \tfrac{1}{2}L_a i_a^2 + L_{fa} i_f i_a. \qquad (10.5.8)$$

Since the coils are in air L_f and L_a are constant and independent of α. However, $L_{fa} = M_d \cos\alpha$. Therefore

$$T_e = -i_a i_f M_d \sin\alpha. \qquad (10.5.9)$$

If $\alpha = 0$, $T_e = 0$, and if $\alpha = \pi/2$, $T_e = -i_a i_f M_d$ (a maximum value). For this reason α is usually made equal to $\pi/2$.

In matrix form the general expression for torque in eq. (10.5.9) can be written as

$$T_e = [i_f \ \ i_a] \cdot \begin{bmatrix} 0 & 0 \\ -M_d \sin\alpha & 0 \end{bmatrix} \cdot \begin{bmatrix} i_f \\ i_a \end{bmatrix} = \underline{i}^t [G] \underline{i}. \qquad (10.5.10)$$

The matrix $[G]$ was defined in eqs. (10.5.4) and (10.5.6).

Note that for the commutator primitive

$$T_e = \underline{i}^t [G] \underline{i} \qquad (10.5.11)$$

whereas for the slip-ring primitive

$$T_e = \tfrac{1}{2}\underline{i}^t [G_s] \underline{i}. \qquad (10.3.17)$$

The G terms are different in the two cases, because of the differences in operation of the commutator primitive and the slip-ring primitive.

The gross mechanical power output p_m for the commutator primitive, shown in Fig. 10.8, is

$$p_m = T_e \omega_m = \underline{i}^t [G] \underline{i} \omega_m. \qquad (10.5.12)$$

EXAMPLE 10.2

Figure 10.10 illustrates a multicoil commutator primitive. There are four coils on two fixed axes at right angles. These axes are called the direct and quadrature axes and define the reference frame.

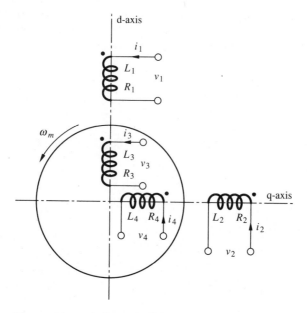

Figure 10.10   A four-winding commutator primitive.

Coils 1 and 2 are fixed; the distribution of conductors is not defined in space but the resultant magnetic flux or mmf axes are. The conductors of coils 3 and 4 rotate at a fixed angular speed, but they form a commutator winding with a pair of brushes on the direct axis and a pair of brushes on the quadrature axis, so the rotor winding axes are fixed as shown. Although the coils are supposed to be in air to give linear magnetic relations, sinusoidal flux distributions must prevail. Formulate the voltage equations in matrix form and determine the electromagnetic torque.

Solution

In the following the subscript d is associated with the direct axis and the subscript q is associated with the quadrature axis. The voltage equation is

$$\underline{v} = [Z]\underline{i} = [R]\underline{i} + [Lp]\underline{i} + [G\omega_m]\underline{i}.$$

We want to examine the entries of the impedance matrix, so we write the voltage equation as

$$
\begin{bmatrix} v_1 \\ v_2 \\ v_3 \\ v_4 \end{bmatrix}
=
\begin{bmatrix}
a_{11} & a_{12} & a_{13} & a_{14} \\
a_{21} & a_{22} & a_{23} & a_{24} \\
a_{31} & a_{32} & a_{33} & a_{34} \\
a_{41} & a_{42} & a_{43} & a_{44}
\end{bmatrix}
\cdot
\begin{bmatrix} i_1 \\ i_2 \\ i_3 \\ i_4 \end{bmatrix}.
$$

Multiplying this out

$$v_1 = a_{11}i_1 + a_{12}i_2 + a_{13}i_3 + a_{14}i_4$$

$$v_2 = a_{21}i_1 + a_{22}i_2 + a_{23}i_3 + a_{24}i_4$$

$$v_3 = a_{31}i_1 + a_{32}i_2 + a_{33}i_3 + a_{34}i_4$$
$$v_4 = a_{41}i_1 + a_{42}i_2 + a_{43}i_3 + a_{44}i_4.$$

A general voltage is $a_{ij}i_j$ where a_{ij} is the entry in the ith row and jth column. If $i = j$, a_{ij} must be resistance and self-inductance and if $i \neq j$, a_{ij} must be either a mutual inductance or a motional inductance term associated with $[G]$. a_{ij} could have zero value. Let us find the a_{ij} values.

a_{11} Voltages associated with coil 1 due to the current in coil 1 are the voltage drop $R_1 i_1$ and the self-induced emf $L_1 di_1/dt$, so that

$$a_{11} = R_1 + L_1 p.$$

a_{12} The only voltages that can be associated with coil 1 due to current in coil 2 are mutually or motionally induced emfs arising from the flux produced by i_2. Coil 1 and flux linkage λ_2 are at right angles, so there is no mutual coupling and so no induced emf. The conductors of coil 1 are stationary and the axis of λ_2 is stationary, so without relative motion there can be no motionally induced emf. Hence the voltage in coil 1 due to current in coil 2 is zero and

$$a_{12} = 0.$$

a_{13} Voltage associated with coil 1 due to the flux produced by the current in coil 3 is a mutually induced emf, because the flux of coil 3 links directly with coil 1. The coils are both on the same axis. The induced emf is $M_{d13}di_3/dt$. There can be no motionally induced emf in the stationary conductors of coil 1.

$$a_{13} = M_{d13} p.$$

a_{14} There is no mutual flux linkage so there is no mutually induced emf. Although the flux produced by i_4 is at right angles to the axis of coil 1, the conductors of coil 1 do not move in that stationary field, so the motionally induced emf is also zero. Hence the voltage associated with the coil 1 due to the flux produced by the current in coil 4 is zero.

$$a_{14} = 0.$$

The same argument can be followed for coil 2, which is stationary. We can study the action of the flux linkage produced by each of the currents in turn to arrive at

$$a_{21} = 0, \quad a_{22} = R_2 + L_2 p, \quad a_{23} = 0, \quad \text{and} \quad a_{24} = M_{q24} p.$$

For the entries of the voltage equation for coil 3 we can study the action of the currents.

a_{31} Voltage associated with coil 3 due to the flux produced by current in coil 1 is a mutual one, because there is a maximum direct linkage. With a maximum flux linkage there can be no motionally induced emf at the brushes, even though the conductors of coil 3 are moving. Hence

$$a_{31} = M_{d31} p.$$

a_{32} Voltage associated with coil 3 due to the flux produced by the current in coil 2 cannot be a mutual one because there is no mutual flux linkage associated with the two. However, the flux axis of coil 2 is at right angles to the axis of coil 3 whose conductors move in that flux. Accordingly there is a motionally induced emf, $G_{32}\omega_m i_2$. By the rule formulated in subsection B under section 10.3.1, the G term is equal to the value of the mutual inductance between coils 3 and 2 if coil 3 is advanced through $\pi/2$ in the direction of rotation of the moving member. By the convention of Fig. 10.10, this is $-M_{q32}$ and the minus sign is added because the fluxes in the coils 3 and 2 would oppose each other. Hence

$$a_{32} = -M_{q32}\omega_m.$$

a_{33} Voltages associated with coil 3 due to its own current are the resistance drop $R_3 i_3$ and the self-induced emf $L_3 di_3/dt$. Hence

$$a_{33} = R_3 + L_3 p.$$

a_{34} Voltage associated with coil 3 due to the flux produced by the current in coil 4 cannot be mutual because the coils are at right angles. However, the conductors of coil 3 rotate in the stationary flux of coil 4 which is at right angles to the coil 3 axis, so the induced emf is motional and is $G_{34}\omega_m$. Advancing coil 3 through $\pi/2$ makes the mutual inductance between coils 3 and 4 equal to $-M_{q34}$. Hence

$$a_{34} = -M_{q34}\omega_m.$$

A similar argument for coil 4 can be offered to obtain

$$a_{41} = M_{d41}\omega_m, \quad a_{42} = M_{q42}p, \quad a_{43} = M_{d43}\omega_m \quad \text{and} \quad a_{44} = R_4 + L_4 p.$$

The matrix equation can be written as

$$
\begin{bmatrix} v_1 \\ v_2 \\ v_3 \\ v_4 \end{bmatrix} =
\begin{bmatrix}
R_1 + L_1 p & 0 & M_{d13}p & 0 \\
0 & R_2 + L_2 p & 0 & M_{q24}p \\
M_{d31}p & -M_{q32}\omega_m & R_3 + L_3 p & -M_{q34}\omega_m \\
M_{d41}\omega_m & M_{q42}p & M_{d43}\omega_m & R_4 + L_4 p
\end{bmatrix}
\cdot
\begin{bmatrix} i_1 \\ i_2 \\ i_3 \\ i_4 \end{bmatrix}.
$$

If the two rotor coils are the same, and they are for most real machines, some of the impedance entries can be rewritten as follows.

$$M_{d13} = M_d, \ M_{q24} = M_q$$

$$M_{d31} = M_d, \ G_{32} = -M_{q32} = M_{q42} = -M_q$$

$$G_{34} = -M_{q34} \triangleq L_q, \ G_{41} = M_{d41} = M_{d31} = M_d$$

$$M_{q42} = M_q, \ G_{43} = M_{d43} \triangleq L_d.$$

Splitting the impedance matrix into its component forms,

$$
[R] =
\begin{bmatrix}
R_1 & 0 & 0 & 0 \\
0 & R_2 & 0 & 0 \\
0 & 0 & R_3 & 0 \\
0 & 0 & 0 & R_4
\end{bmatrix}
$$

and

$$[L] = \begin{bmatrix} L_1 p & 0 & M_d p & 0 \\ 0 & L_2 p & 0 & M_q p \\ M_d p & 0 & L_3 p & 0 \\ 0 & M_q p & 0 & L_4 p \end{bmatrix}.$$

These are both symmetrical but

$$[G] = \begin{bmatrix} 0 & 0 & 0 & 0 \\ 0 & 0 & 0 & 0 \\ 0 & -M_q & 0 & -L_q \\ M_d & 0 & L_d & 0 \end{bmatrix}$$

is asymmetrical.

The torque is $T_e = \underline{i}^t[G]\underline{i}$, which is

$$T_e = [i_1 \quad i_2 \quad i_3 \quad i_4] \cdot \begin{bmatrix} 0 & 0 & 0 & 0 \\ 0 & 0 & 0 & 0 \\ 0 & -M_q & 0 & -L_q \\ M_d & 0 & L_d & 0 \end{bmatrix} \cdot \begin{bmatrix} i_1 \\ i_2 \\ i_3 \\ i_4 \end{bmatrix}.$$

That is

$$T_e = M_d i_1 i_4 - M_q i_2 i_3 + L_d i_3 i_4 - L_q i_3 i_4.$$

If the stator coils are identical $M_d = M_q$ and $L_d = L_q$. Then

$$T_e = M_d(i_1 i_4 - i_2 i_3).$$

From the assumptions made, all commutator primitives have voltage equations that are linear differential equations with constant coefficients. The solution of the voltage equations to find the currents will therefore pose no great difficulty. The torque and power delivered by the machine can be calculated once the currents are known.

Not all machines are commutator machines, but there is no reason why we should not mathematically transform other machines to commutator primitives if the resultant equations are easy to solve. This is the basis of Kron's generalized theory of electrical machines. There is no better unifying theory yet available.

10.6. LINK BETWEEN CIRCUITS AND MACHINES

We have considered the action of circuits on each other. First, static coils were considered and then the relations for the voltage, power, energy, and torque were developed for a doubly-excited system capable of motion. By putting these equations into matrix form, the equations become general and satisfy a system with any number of interacting circuits. Examples have been given for a quadruply-excited device to clarify the principles used.

By expanding the concept, defining limitations and specifying premises, the theory can be used to analyze many of the common rotating electromagnetic machines. Not all windings are concentrated however. Except for salient poles, the windings are usually distributed and this subject is a useful introduction to the real machine.

The real machine, whether it is a motor or a generator, may have distributed windings or quasiconcentrated windings. It will have iron to define the magnetic path and an air gap between the rotor and the stator. The air gap may be uniform or there may be saliency. If all these differences between the primitive machine and the real machine can be eliminated then the analysis of the real machine is simplified.

10.6.1. The Real Machine

However a real machine is constructed, its model must be able to be reduced to the ideal elements already used in this chapter in order to apply the generalized theory. The aim has been to get the electrical equations into a set of linear differential equations with constant coefficients in order that operational methods will provide a solution. If the equations cannot be obtained in this form then computer methods of solution must be sought.

The premises used so far have been that the mmf and flux waveforms of the concentrated windings are sinusoidal so that the inductances, if not constant, vary with position in the same way. The speed of the rotating member has been required to be constant. If it were otherwise, inspection would show the entries of the Z matrix to have variables, and hence the differential equations would no longer have constant coefficients.

Many real machines have distributed as well as concentrated windings, but the former has been shown in section 10.4 to provide no limitation, as the mutual inductance can still be a sinusoidal function. Whereas the windings of the previous simple electromechanical systems have been in air, the real machine has its coils wound on ferromagnetic materials with the minimum air gap between the stationary and rotating members. This material is used to define, as well as possible, the path and distribution of the flux and also to keep the magnetizing mmf as small as possible for a required flux density in order that the efficiency of energy conversion is kept as high as possible. Up to about 1.0 tesla for magnetic material used in practice, the relation between the flux density and the magnetizing force is a linear one and the permeability can be thousands of times greater than that of air. However, at higher flux densities, the relation is nonlinear because of saturation. A nonlinear relationship between the flux linkages λ and current i means that inductance becomes a function of current as well as position. The solution of the resulting differential equations, if possible at all, might well need the use of graphical or digital computer techniques. Both can be arduous and far from easy to apply universally to machine theory.

To get the greatest possible output with the highest efficiency from a machine of given dimensions, the designer would have a machine working at the knee of the saturation curve. Up to that point on the magnetization curve, a linear analysis will not produce great errors, for the slight saturation can generally be disregarded, at least in rotating machines, because the reluctance of the air gap is usually much greater than that of the iron path. Thus the air gap, which has linear magnetic properties, dominates. There are approximate techniques to account for saturated values of inductance or reactance.

Another disadvantage with the real machine is the nonlinearity produced by the hysteresis of the magnetic materials. This has to be ignored if a simple solution is to be expected. Errors are small if it is neglected because the core materials are chosen for their low hysteresis loss. It is only in the case of the actual hysteresis motor that the hysteresis effect is useful during the run up to synchronous speed and a hard magnetic material is chosen for its pronounced hysteresis.

The conductors of windings, if distributed, are placed in slots in the armature. It could be argued that as the conductors are now magnetically screened by the armature teeth there can be no force on those conductors. The force will be acting on the teeth instead. This is true, but is of no consequence as far as the choice of analysis is concerned.

To develop the established relations of the elementary or primitive machines and apply them to the real machine, the three main types of machine will be studied. It is not intended that we formulate the theory of the steady-state operation and performance of dc, induction, and synchronous machines. The intention is to show the similarity of results obtained from the unified theory and the traditional approach. Throughout, the machines are analyzed as motors. The only difference for generation will be that the input and output powers will have negative signs. The complications of multipolar machines are eliminated by considering the machine to have only two poles throughout the analysis. During the final reckoning, account can be taken of the number of poles.

10.7. DC MACHINE AND DYNAMIC CIRCUIT ANALYSIS

The dc machine, which is considered in its simplest form here, has salient poles with quasiconcentrated windings for excitation. The field produced in this way is not sinusoidal nor need it be in practice. However, after splitting the field distribution by Fourier analysis into a fundamental sinusoidal component plus the higher harmonics, the harmonics will be neglected. Because the armature is a cylinder, and neglecting any slot effect, the reluctance path of the exciting field does not vary. Thus the self-inductance of the field winding is constant.

The armature winding is connected to a commutator, so that the current and flux pattern of the armature reaction remain the same in space even though the conductors rotate. Thus the armature winding axis remains fixed and the self-inductance of this winding is constant also. This axis is taken to be along the quadrature axis in order that the energy conversion may be a maximum. As has already been indicated in section 10.4, the distributed armature winding has mutual inductance which varies sinusoidally with the position of its axis. That axis is at right angles to the field winding axis, so that the overall mutual coupling is zero, but the G coupling is a maximum. It is stressed again that it is assumed there is no saturation to prevent the superposition theorem from being applied to the flux linkages. This gives a linear system and the machine can be considered to be modeled as a commutator primitive.

Figure 10.11 shows a diagrammatic sketch of the dc machine and its circuit representation. So far, all the premises stated have been satisfied. As a result, the voltage equation employing the instantaneous values of voltage and current should produce two linear differential equations with constant coefficients, as long as the armature speed does not vary significantly. Let us determine the voltage and torque equations.

The voltage equation is

$$\underline{v} = [Z]\underline{i} \qquad (10.7.1)$$

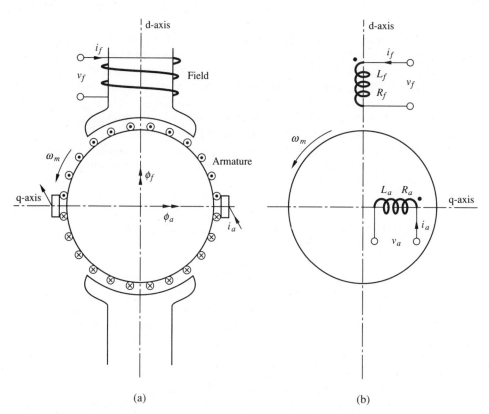

(a)   (b)

Figure 10.11   DC machine. (a) Symbolic representation, (b) circuit representation.

which expands to

$$\begin{bmatrix} v_f \\ v_a \end{bmatrix} = \begin{bmatrix} R_f + L_f p & 0 \\ G_{af}\omega_m & R_a + L_a p \end{bmatrix} \cdot \begin{bmatrix} i_f \\ i_a \end{bmatrix} \tag{10.7.2}$$

where $[Z] = [R] + [Lp] + [G]\omega_m$

$$[R] = \begin{bmatrix} R_f & 0 \\ 0 & R_a \end{bmatrix}, \quad [L] = \begin{bmatrix} L_f & 0 \\ 0 & L_a \end{bmatrix} \quad \text{and} \quad [G] = \begin{bmatrix} 0 & 0 \\ G_{af} & 0 \end{bmatrix}$$

where G_{af} is given by the maximum mutual inductance that could exist between the two windings. There are no mutual inductance terms because the two windings have their axes at right angles to each other. Application of the rule for G entries, found in subsection B under section 10.3.1, gives $G_{af} = +M_d$, where M_d is the maximum possible mutual inductance. The voltage equations in full are

$$v_f = (R_f + L_f p)i_f \quad \text{and} \quad v_a = M_d\omega_m i_f + (R_a + L_a p)i_a. \tag{10.7.3}$$

The electromagnetic torque, or gross mechanical torque, is

$$T_e = \underline{i}^t[G]\underline{i} = M_d i_f i_a. \tag{10.7.4}$$

If the field and armature currents have steady-state dc values, then the torque is a constant. The equations show that, because the driving torque T_e is unvarying for a given steady-state power input, then, if the load is constant, the speed must be constant and this is an initial condition that the analysis be acceptable for the simplest solution. The voltage equations are linear differential equations with constant coefficients. If the constraints and values of inductances are known, a standard method of say the Laplace transform would produce the solutions. If the brushes had not been on the quadrature axis, it would have been necessary to apply a linear transformation to have windings along the principal axes of symmetry. Otherwise the differential equations would not be in an acceptable form.

EXAMPLE 10.3

Consider a separately-excited dc motor, whose commutator primitive is depicted in Fig. 10.11b. Let the armature winding be supplied from a constant-current source. The problem is to find the variation of speed with time, if the motor is started against a constant load torque T_m when there is a step change of voltage applied to the field winding.

Solution
This example shows how the analytical tools of transform methods can be utilized to solve machine problems involving differential equations.

Since the armature winding is supplied with a constant current I_a, the armature voltage equation is superfluous to the requirements of finding a solution for the speed response.

From an inspection of Fig. 10.11b, the field winding voltage equation is

$$V_f = (R_f + L_f p)i_f.$$

The electromagnetic torque is

$$T_e = i^t[G]\underline{i} = M_d I_a i_f = K_a i_f$$

where $K_a = M_d I_a$ = constant, since I_a is constant. The dynamic equation of motion is

$$T_e = Jp\omega_m + D\omega_m + T_m.$$

These three equations, which define the transient or dynamic response of the motor, can be transformed from the time domain (t) to the frequency domain (s) in order to find a solution. If it is assumed that the initial conditions are zero, the equations become

$$V_f(s) = (R_f + L_f s)I_f(s) = R_f(1 + \tau_f s)I_f(s)$$

where τ_f is the time constant (L_f/R_f) of the field winding,

$$T_e(s) = k_a I_f(s) \qquad \text{and} \qquad T_e(s) = D(1 + \tau_m s)\Omega_m(s) + T_m(s)$$

where τ_m is the mechanical time constant (J/D). The electromagnetic torque $T_e(s)$ and the field current $I_f(s)$ can be eliminated to give the expression for the speed in the form

$$\Omega_m(s) = \frac{K_a}{DR_f} \frac{V_f(s)}{(1 + \tau_f s)(1 + \tau_m s)} - \frac{T_m(s)}{D(1 + \tau_m s)}.$$

This equation can be solved since the transforms $V_f(s)$ and $T_m(s)$ are known. The motor starts against a constant load torque T_m and there is a step change of voltage applied to the field winding. The initial conditions are zero and

$$V_f(s) = V_f/s \qquad \text{and} \qquad T_m(s) = T_m/s.$$

The transform of the speed becomes

$$\Omega_m(s) = \frac{K_a V_f}{DR_f \tau_f \tau_m} \frac{1}{s(s + 1/\tau_f)(s + 1/\tau_m)} - \frac{T_m}{D\tau_m} \frac{1}{s(s + 1/\tau_m)}.$$

Transforming back into the time domain the speed is

$$\omega_m(t) = \frac{K_a V_f}{DR_f} - \frac{T_m}{D} + \frac{K_a V_f \tau_f}{DR_f(\tau_m - \tau_f)} e^{(-t/\tau_m)} - \frac{K_a V_f \tau_m}{DR_f(\tau_m - \tau_f)} e^{(-t/\tau_f)} - \frac{T_m}{D} e^{(-t/\tau_m)}.$$

The speed response is shown in Fig. 10.12, where the final steady-state speed is given by

$$\omega_m(\infty) = \frac{K_a V_f}{DR_f} - \frac{T_m}{D}.$$

K_a and V_f govern the maximum speed while D, R_f, and T_m tend to reduce the speed. Because the frictional damping D is very small in practice, a small error in the value of D can result in a large error in ω_m. Similarly small changes in the damping

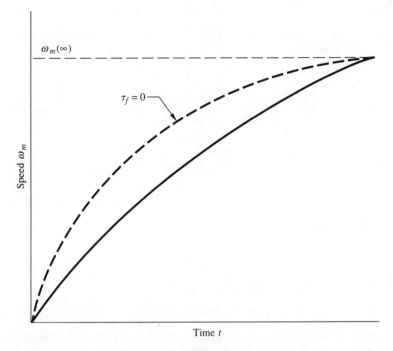

Figure 10.12   Separately-excited dc motor speed response.

during motor operation will result in fluctuations in the speed. Damping cannot be ignored in the speed expression, but simplifications can be made if the field winding time constant τ_f can be neglected. That is, if $R_f \gg L_f$ and $\tau_m \gg \tau_f$, then $\tau_f \approx 0$. The speed expression becomes

$$\omega_m(t) \approx \frac{K_a V_f}{D R_f} - \frac{T_m}{D} (1 - e^{(-t/\tau_m)}).$$

10.7.1. Power Flow

It is interesting to draw up a power flow diagram for the dc machine. The total electrical power input to the motor is

$$p_e = \underline{i}^t \underline{v} \tag{10.7.5}$$

and in detail this is

$$p_e = i_a v_a + i_f v_f. \tag{10.7.6}$$

From eq. (10.7.3) this becomes

$$p_e = i_f^2 R_f + i_a^2 R_a + i_f L_f p i_f + i_a L_a p i_a + M_d \omega_m i_a i_f. \tag{10.7.7}$$

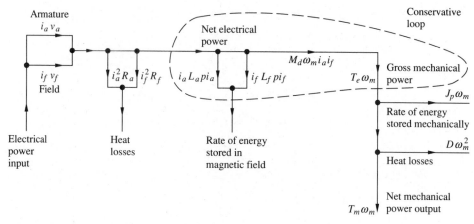

Figure 10.13 DC machine power flow.

The differential equation of motion is

$$T_e = T_m + Jp\omega_m + D\omega_m. \tag{10.7.8}$$

The power relation for the mechanical system is obtained by multiplying each term in the equation of motion by the speed ω_m. Thus

$$p_m = T_e\omega_m = T_m\omega_m + \omega_m Jp\omega_m + D\omega_m^2. \tag{10.7.9}$$

The power flow is depicted in Fig. 10.13. It will be noticed that on the electrical side it has been necessary to neglect the eddy currents induced by the tooth effect, hysteresis in the iron and stray losses as a whole. These iron losses can be accounted for when determining the efficiency of the machine.

Further inspection of the power balance demonstrates that, although the power converted from electrical to mechanical form is influenced by the value of the exciting current, it is only the input to the armature that is actually converted. The power into the field winding is used to store energy magnetically and to dissipate energy because of the winding resistance; none is converted into mechanical power.

10.7.2. Steady-state Back Emf

General investigation can be made, without defining constraints for the series, shunt, compound, or cross-field dc machines, concerning the motionally induced emf during steady-state conditions.

From the voltage eqs. (10.7.3) the motionally induced or back emf E_a in the armature circuit is

$$E_a = M_d\omega_m i_f \tag{10.7.10}$$

and from section 10.4

$$M_d = \frac{2\hat{B}rlZ}{\pi i_f}.$$ (10.7.11)

If the armature speed is n revolutions per second,

$$\omega_m = 2\pi n \quad \text{so} \quad E_a = 4\hat{B}rlnZ.$$ (10.7.12)

This value is the back emf if all the conductors were in series. In practice this is not so, because with wave and lap windings there is a number of parallel paths in the armature winding. If a is the number of pairs of parallel paths, the actual back emf between the brushes would be

$$E_a = 4\hat{B}rln\frac{Z}{2a}$$ (10.7.13)

where $Z/2a$ = the number of conductors in series per path and $4\hat{B}rln$ = average back emf per conductor. Now the mean and maximum flux densities are related by

$$\hat{B} = \frac{\pi}{2}B_{av}$$ (10.7.14)

since the field distribution has been assumed to be sinusoidal. The total flux Φ_T and the flux per pole Φ are related by

$$\Phi_T = 2p\Phi$$ (10.7.15)

where p is the number of pole pairs on the machine. Finally, the total flux crossing the air gap is

$$\Phi_T = B_{av} \times 2\pi rl$$ (10.7.16)

to give

$$\hat{B} = \frac{\Phi_T}{4rl} \quad \text{or} \quad \hat{B} = 2p\frac{\Phi}{4rl}.$$ (10.7.17)

After substitution we arrive at the expression

$$E_a = \Phi nZ\frac{p}{a}$$ (10.7.18)

which is the steady-state equation of the back emf obtained by classical methods.

10.7.3. DC Shunt Motor Operating at Steady State

The dc shunt motor has its field and armature windings connected in parallel across the same voltage source. We want to apply dynamic circuit analysis to confirm some of the steady-state characteristics that were obtained by classical theory in Chapter 5. In particular we wish to determine the torque-speed characteristic.

From an inspection of Fig. 10.11b, which represents the schematic form of the motor primitive without constraints, the voltage equation is

$$\begin{bmatrix} v_f \\ v_a \end{bmatrix} = \begin{bmatrix} R_f + L_f p & 0 \\ \omega_m M_d & R_a + L_a p \end{bmatrix} \cdot \begin{bmatrix} i_f \\ i_a \end{bmatrix}. \tag{10.7.19}$$

The constraints are that the applied voltage is constant and the current variables are constant dc values. Consequently, for steady-state conditions, terms containing p (i.e., d/dt) are put equal to zero and

$$v_a = v_f = V, \quad i_f = I_f, \quad \text{and} \quad i_a = I_a. \tag{10.7.20}$$

Hence

$$\begin{bmatrix} V \\ V \end{bmatrix} = \begin{bmatrix} R_f & 0 \\ \omega_m M_d & R_a \end{bmatrix} \cdot \begin{bmatrix} I_f \\ I_a \end{bmatrix} \tag{10.7.21}$$

and in full

$$V = I_f R_f \quad \text{and} \quad V = \omega_m M_d I_f + I_a R_a. \tag{10.7.22}$$

The electromagnetic torque is given by

$$T_e = \underline{i}^t [G] \underline{i} = [I_f \quad I_a] \cdot \begin{bmatrix} 0 & 0 \\ M_d & 0 \end{bmatrix} \cdot \begin{bmatrix} I_f \\ I_a \end{bmatrix} = M_d I_f I_a. \tag{10.7.23}$$

On substitution for I_f and I_a from eqs. (10.7.22) the torque expression becomes

$$T_e = \frac{V 2 M_d}{R_a R_f^2} (R_f - \omega_m M_d). \tag{10.7.24}$$

Thus the torque-speed characteristic is a straight line with a negative slope. The maximum torque $T_{\max}$ occurs at zero speed, that is, at starting. It is given by

$$T_{\max} = \frac{V 2 M_d}{R_a R_f}. \tag{10.7.25}$$

Rearranging the torque equation we obtain

$$\omega_m = \frac{R_f}{M_d} (1 - T_e / T_{\max}) \tag{10.7.26}$$

and this is a straight line with a negative slope. That is, as the load torque increases the speed will reduce. However, the speed variation from no load to full load is small since, in practice, $T_e / T_{\max} < 0.1$ so the shunt motor is considered to be essentially a constant-speed machine.

For a given torque it is seen that the speed ω_m is proportional to the value of the resistance R_f in the field circuit. This indicates a simple means of speed adjustment.

The armature circuit back emf E_a is

$$E_a = \omega_m M_d I_f \tag{10.7.27}$$

and the armature voltage equation becomes

$$V = \omega_m M_d I_f + I_a R_a = E_a + I_a R_a \tag{10.7.28}$$

and this is the traditional form. If the machine is driven at constant speed, if the field winding is separately-excited and if there is no load applied, a plot of E_a against I_f is the open-circuit magnetization characteristic. The slope of this curve for any field current is $\omega_m M_d$, so that this suggests one method for experimentally determining the parameter M_d. The speed-torque characteristic provides another method for finding M_d, if the characteristic is empirical. At no load the speed equation becomes

$$\omega_m = \frac{R_f}{M_d}(1-0) \tag{10.7.29}$$

so that M_d can be calculated. Although the value of M_d depends on saturation conditions and is a function of the field current, it is usual to linearize about an operating point and assume M_d to be a constant.

10.8. AC MACHINES AND LINEAR TRANSFORMATIONS

Most machines of the ac type are polyphase and more often than not they have three-phase windings. So the three-phase motor will be the point of discussion, but the method of obtaining the right kind of differential equations (linear with constant coefficients) describing the machine behavior is quite general.

It is necessary that the mmf distribution around the air gap be sinusoidal for the torque to be constant and the losses to be at a minimum. The distributed windings of the induction and synchronous machines are arranged so that this is nearly so, and if salient, the poles of the synchronous machine can be shaped to provide it. With linear magnetization, the mutual inductances vary sinusoidally with position and this satisfies another postulate of the generalized theory.

With a commutator, dc armature windings have an axis that is fixed relative to the stator. There is no commutator on the synchronous or common induction motor. Accordingly, the axes of the rotor windings rotate, but to correspond to the generalized theory, the winding axes must be fixed and preferably fixed in the axes of symmetry, that is, the direct and quadrature axes. So long as the windings are balanced, fixed-axis windings can be created mathematically by linear transformation. *Winding balance,* which means that an interchange of windings on the same member (rotor or stator) would produce no change in results, must be supplemented by the fact that energy balance must remain the same after every transformation. It must be stressed that the sole purpose of transformation is to simplify the method of solution.

The first transformation must bring the three-phase winding to an equivalent set of windings that have axes at right angles. This will be a two-phase winding, rotating in space.

10.8.1. Three-phase to Two-phase Transformation

We want to make a three-phase to two-phase transformation of currents in the slip-ring primitive in order to simplify the circuit configuration. Assuming energy

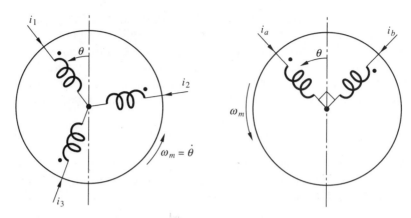

Figure 10.14 Slip-ring primitive 3-phase to 2-phase transformation.

balance (or power invariance, by which it is usually specified), simple equivalence is to be found for the two sets of balanced windings, diagrammatically represented by Fig. 10.14, distributed in practice but concentrated in representation. For equivalence, the mmf of the three phases must be the same as the mmf of the transformed set of windings in pattern, direction, and magnitude.

For the balanced three-phase system the instantaneous values of the phase currents can be described by

$$i_1 = \hat{I}\cos\omega t, \quad i_2 = \hat{I}\cos(\omega t - 2\pi/3), \quad \text{and} \quad i_3 = \hat{I}\cos(\omega t - 4\pi/3). \quad (10.8.1)$$

For the two-phase system it is convenient, although not necessary for the transformation, to have one of the currents in phase with one of the three-phase currents. In this case i_a is chosen to be in phase with i_1. Accordingly, the instantaneous values of the two-phase currents can be written

$$i_a = I_m \cos\omega t \quad \text{and} \quad i_b = I_m \cos(\omega t - \pi/2). \quad (10.8.2)$$

Then, if

$$\hat{I} = k_1 I_m \quad (10.8.3)$$

where k_1 is some constant to be determined, it follows that

$$i_1 = k_1 i_a = \hat{I}\cos\omega t. \quad (10.8.4)$$

Also

$$k_1 i_b = k_1 I_m \cos(\omega t - \pi/2) = \hat{I}\cos(\omega t - \pi/2) = \hat{I}\sin\omega t.$$

Since i_2 can be expanded from $i_2 = \hat{I}\cos(\omega t - 2\pi/3)$ to $i_2 = \hat{I}(\cos\omega t \cos 2\pi/3 + \sin\omega t \sin 2\pi/3)$,

$$i_2 = -1/2 \, \hat{I}\cos\omega t + \sqrt{3}/2 \, \hat{I}\sin\omega t. \quad (10.8.5)$$

Substitution gives

$$i_2 = k_1(-1/2 \, i_a + \sqrt{3}/2 \, i_b). \quad (10.8.6)$$

Similarly

$$i_3 = k_1(-1/2 \; i_a - \sqrt{3}/2 \; i_b).$$ (10.8.7)

Writing them together, the relations for the currents are

$$i_1 = k_1 i_a, \quad i_2 = k_1(-1/2 \; i_a + \sqrt{3}/2 \; i_b), \quad \text{and} \quad i_3 = k_1(-1/2 \; i_a - \sqrt{3}/2 \; i_b).$$ (10.8.8)

The three-phase currents produce a rotating mmf F_3 of amplitude[5] $\frac{3}{2}$ times the peak value of the mmf of one of its phases. The two-phase currents produce a rotating mmf F_2 of amplitude[6] equal to the peak value of one of its phases. Consequently, the ratio of mmfs F_3/F_2 is equal to $3\hat{I}N_3/2I_mN_2$, where N_3 and N_2 are the number of turns per phase of the three-phase and two-phase windings, respectively. Now for equivalence, $F_3 = F_2$, so it is possible to arrange either the currents or the effective number of turns, or both, to maintain the mmfs the same.

If it is arranged that $N_2 = \frac{3}{2}N_3$, then the currents are the same and $i_a = i_1$. This means that $k_1 = 1$. The same final result would occur if $k_1 = \frac{2}{3}$ by taking the turns N_3 and N_2 to be the same. An equal division between turns and current yields $k_1 = \sqrt{2/3}$. This last value of k_1 is chosen, because it simplifies the transformations for voltage, operational impedance, and flux linkage.

The transformation of currents can be put into the matrix form

$$\begin{bmatrix} i_1 \\ i_2 \\ i_3 \end{bmatrix} = k_1 \begin{bmatrix} 1 & 0 \\ -1/2 & \sqrt{3}/2 \\ -1/2 & -\sqrt{3}/2 \end{bmatrix} \cdot \begin{bmatrix} i_a \\ i_b \end{bmatrix}.$$ (10.8.9)

Transforming three variables into two has given a singular transformation matrix, that is, one without an inverse. If a matrix is not square it can never have an inverse and then the relation cannot be transformed to get i_a and i_b in terms of the three-phase currents. This situation is remedied here by using the balanced condition that

$$i_1 + i_2 + i_3 = 0.$$

This indicates that, from symmetrical components theory, the zero-sequence current is zero in the three-phase system, so the zero-sequence current in the two-phase balanced winding is zero also. This enables us to write

$$\begin{bmatrix} i_1 \\ i_2 \\ i_3 \end{bmatrix} = \sqrt{2/3} \begin{bmatrix} 1 & 0 & x \\ -1/2 & \sqrt{3}/2 & x \\ -1/2 & -\sqrt{3}/2 & x \end{bmatrix} \cdot \begin{bmatrix} i_a \\ i_b \\ 0 \end{bmatrix}.$$ (10.8.10)

The entry x in the above matrix is arbitrary. It would seem reasonable to choose the value of x so that the inverse of the matrix can be obtained most conveniently. Consequently x is chosen to be $1/\sqrt{2}$, because the matrix becomes orthogonal and a property of the orthogonal matrix is that its inverse is equal to its transpose.

We can write

$$\underline{i}_{123} = [C_1]\underline{i}_{abo}$$ (10.8.11)

---

[5] See eqs. (F4.10) and (F4.12).
[6] See Example F1.

where $[C_1]$ is known as a transformation or connection matrix and has been given the value

$$[C_1] = \sqrt{2/3} \begin{bmatrix} 1 & 0 & 1/\sqrt{2} \\ -1/2 & \sqrt{3}/2 & 1/\sqrt{2} \\ -1/2 & -\sqrt{3}/2 & 1/\sqrt{2} \end{bmatrix}. \tag{10.8.12}$$

If the *new* variables i_{abo} are multiplied by the connection matrix $[C_1]$, the values of the *old* variables i_{123} are obtained. Although it is not done in a practical problem, we can transform back to find the two-phase currents in terms of the three-phase currents. That is

$$i_{abo} = [C_1]^{-1} i_{123} \tag{10.8.13}$$

and, since $[C_1]^{-1} = [C_1]^t$ in this particular case we can write by inspection

$$\begin{bmatrix} i_a \\ i_b \\ 0 \end{bmatrix} = \sqrt{2/3} \begin{bmatrix} 1 & -1/2 & -1/2 \\ 0 & \sqrt{3}/2 & -\sqrt{3}/2 \\ 1/\sqrt{2} & 1/\sqrt{2} & 1/\sqrt{2} \end{bmatrix} \cdot \begin{bmatrix} i_1 \\ i_2 \\ i_3 \end{bmatrix}. \tag{10.8.14}$$

The ab phase currents expressed as $ab0$ components are the same as $\alpha\beta0$ components, the positive, negative and zero sequence values of symmetrical components for unbalanced conditions. That is

$$\begin{bmatrix} i_1 \\ i_2 \\ i_3 \end{bmatrix} = [C_1] \cdot \begin{bmatrix} i_\alpha \\ i_\beta \\ i_0 \end{bmatrix}. \tag{10.8.15}$$

A. VOLTAGE TRANSFORMATION

We will explore the voltage transformation from a 3-phase to a 2-phase slip-ring primitive in order to consolidate the choice of k_1 for the current transformation discussed in the previous section.

In a similar manner to that by which the current relations were found (see eq. (10.8.8)), the three-phase voltages can be expressed in terms of the two-phase voltages in the following form.

$$v_1 = k_2 v_a, \quad v_2 = k_2 \left(-\frac{1}{2} v_a + \frac{\sqrt{3}}{2} v_b \right), \quad \text{and} \quad v_3 = k_2 \left(-\frac{1}{2} v_a - \frac{\sqrt{3}}{2} v_b \right) \tag{10.8.16}$$

where k_2 is a constant to be determined.

The current transformation was derived by considering the mmfs to be the same for both primitives. If the mmfs are the same then there will be equal air-gap fluxes. By Faraday's law the emf induced in a winding is given by $e = -N d\phi/dt$, so that, for two windings, x and y, linked by the same flux ϕ, $e_x/e_y = N_x/N_y$. Phase 1 of the three-phase winding has N_3 turns and has been given the same angular position as phase a of the two-phase winding, which has N_2 turns per phase. Accordingly we can write

$$v_1/v_a = N_3/N_2 = k_2. \qquad (10.8.17)$$

For equal mmfs the requirement was $(3/2)\hat{I}N_3 = I_m N_2$. We had $\hat{I}/I_m = k_1$. So that $k_1 N_3/N_2 = \frac{2}{3}$. That is, $k_1 k_2 = \frac{2}{3}$. As a check we can transform the instantaneous power into the three-phase primitive by substituting the two-phase values into the three-phase variables.

$$
\begin{aligned}
v_1 i_1 + v_2 i_2 + v_3 i_3 &= k_2 v_a \times k_1 i_a \\
&+ k_2 k_1 \left(-\frac{1}{2} v_a + \frac{\sqrt{3}}{2} v_b\right)\left(-\frac{1}{2} i_a + \frac{\sqrt{3}}{2} i_b\right) \\
&+ k_2 k_1 \left(-\frac{1}{2} v_a - \frac{\sqrt{3}}{2} v_b\right)\left(-\frac{1}{2} i_a - \frac{\sqrt{3}}{2} i_b\right) \\
&= k_1 k_2 \frac{3}{2} (v_a i_a + v_b i_b).
\end{aligned}
\qquad (10.8.18)
$$

If the power input is to be the same for both primitives then

$$v_1 i_1 + v_2 i_2 + v_3 i_3 = v_a i_a + v_b i_b. \qquad (10.8.19)$$

and $k_1 k_2 = \frac{2}{3}$.

The arbitrary choice of one of the constants k_1 or k_2 fixes the value of the other. If the constants are conveniently chosen to be equal, that is, $k_1 = k_2 = \sqrt{2/3}$, then the connection matrices for current and voltage transformations assume the same values. By introducing a zero sequence voltage component to the two-phase system in the same way as for the currents, the voltage relation for the two systems becomes

$$
\begin{bmatrix} v_1 \\ v_2 \\ v_3 \end{bmatrix} = \sqrt{2/3}
\begin{bmatrix} 1 & 0 & 1/\sqrt{2} \\ -1/2 & \sqrt{3}/2 & 1/\sqrt{2} \\ -1/2 & -\sqrt{3}/2 & 1/\sqrt{2} \end{bmatrix} \cdot
\begin{bmatrix} v_a \\ v_b \\ v_0 \end{bmatrix}.
\qquad (10.8.20)
$$

For the assumed balanced conditions the zero sequence component is zero. In practice the three-phase voltages are usually known, so that the two-phase voltages can be found by inverting the connection matrix to give

$$
\begin{bmatrix} v_a \\ v_b \\ v_0 \end{bmatrix} = \sqrt{2/3}
\begin{bmatrix} 1 & -1/2 & -1/2 \\ 0 & \sqrt{3}/2 & -\sqrt{3}/2 \\ 1/\sqrt{2} & 1/\sqrt{2} & 1/\sqrt{2} \end{bmatrix} \cdot
\begin{bmatrix} v_1 \\ v_2 \\ v_3 \end{bmatrix}.
\qquad (10.8.21)
$$

10.8.2. Transformation from Rotating to Fixed Axes

For the transformation from a slip-ring to a commutator primitive the 2-phase rotating mmf pattern is to be resolved into components along the direct and quadrature axes, that is, along fixed axes. Consider the notation in Fig. 10.15, which illustrates a slip-ring primitive and its equivalent commutator primitive. The mmfs in

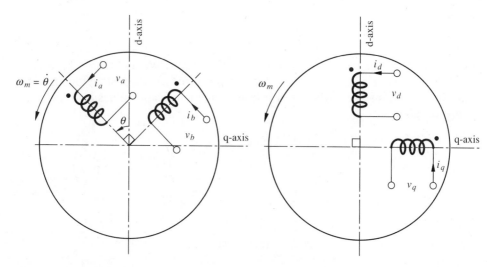

Figure 10.15 Transformation from rotating to fixed axes.

space must be the same for the two primitives. Therefore

$$i_d = i_a \cos\theta + i_b \sin\theta \quad \text{and} \quad i_q = -i_a \sin\theta + i_b \cos\theta. \quad (10.8.22)$$

In matrix form

$$\begin{bmatrix} i_d \\ i_q \end{bmatrix} = \begin{bmatrix} \cos\theta & \sin\theta \\ -\sin\theta & \cos\theta \end{bmatrix} \cdot \begin{bmatrix} i_a \\ i_b \end{bmatrix}. \quad (10.8.23)$$

Inverting the square matrix, it follows that

$$\begin{bmatrix} i_a \\ i_b \end{bmatrix} = \begin{bmatrix} \cos\theta & -\sin\theta \\ \sin\theta & \cos\theta \end{bmatrix} \cdot \begin{bmatrix} i_d \\ i_q \end{bmatrix}. \quad (10.8.24)$$

For the voltage transformation, the concept of power invariance can be used. That is

$$i_a v_a + i_b v_b = i_d v_d + i_q v_q. \quad (10.8.25)$$

Substituting for i_d and i_q the values already obtained in terms of i_a and i_b

$$i_a v_a + i_b v_b = (i_a \cos\theta + i_b \sin\theta) v_d + (-i_a \sin\theta + i_b \cos\theta) v_q. \quad (10.8.26)$$

Rearranging,

$$i_a v_a + i_b v_b = i_a (v_d \cos\theta - v_q \sin\theta) + i_b (v_d \sin\theta + v_q \cos\theta). \quad (10.8.27)$$

For equivalence the coefficients of each current must be the same, giving

$$\begin{bmatrix} v_a \\ v_b \end{bmatrix} = \begin{bmatrix} \cos\theta & -\sin\theta \\ \sin\theta & \cos\theta \end{bmatrix} \cdot \begin{bmatrix} v_d \\ v_q \end{bmatrix}. \quad (10.8.28)$$

Since v_a and v_b are usually given, the values of v_d and v_q are found from

$$\begin{bmatrix} v_d \\ v_q \end{bmatrix} = \begin{bmatrix} \cos\theta & \sin\theta \\ -\sin\theta & \cos\theta \end{bmatrix} \cdot \begin{bmatrix} v_a \\ v_b \end{bmatrix}. \tag{10.8.29}$$

It is necessary to add the zero sequence winding to the primitives in order to allow a direct transformation from a three-phase, slip-ring primitive to a two-phase commutator primitive. Thus we can write

$$\begin{bmatrix} i_a \\ i_b \\ i_0 \end{bmatrix} = \begin{bmatrix} \cos\theta & -\sin\theta & 0 \\ \sin\theta & \cos\theta & 0 \\ 0 & 0 & 1 \end{bmatrix} \cdot \begin{bmatrix} i_d \\ i_q \\ i_0 \end{bmatrix}. \tag{10.8.30}$$

or

$$\underline{i}_{ab0} = [C_2]\underline{i}_{dq0} \tag{10.8.31}$$

where $[C_2]$ is the connection matrix. So when the *new* variables $\underline{i}_{dq0}$ of the commutator primitive are multiplied by the connection matrix $[C_2]$, the values of the *old* variables $\underline{i}_{ab0}$ of the slip-ring primitive are obtained. Similarly the voltage relations for the two systems are

$$\begin{bmatrix} v_d \\ v_q \\ v_0 \end{bmatrix} = \begin{bmatrix} \cos\theta & \sin\theta & 0 \\ -\sin\theta & \cos\theta & 0 \\ 0 & 0 & 1 \end{bmatrix} \cdot \begin{bmatrix} v_a \\ v_b \\ v_0 \end{bmatrix}. \tag{10.8.32}$$

This connection matrix is the transpose of $[C_2]$, that is

$$\underline{v}_{dq0} = [C_2]^t\underline{v}_{ab0}. \tag{10.8.33}$$

10.8.3. Real Machine to Commutator Primitive Transformation

The two steps taken to first transform variables from a three-phase rotating reference frame to a two-phase rotating reference frame and then to transform variables from the two-phase rotating reference frame to a fixed reference frame can be combined. The result is the relationship between variables of the three-phase rotating reference frame and the two-phase fixed reference frame.

We have determined the two relations (from eqs. (10.8.11) and (10.8.31))

$$\underline{i}_{123} = [C_1]\underline{i}_{ab0} \quad \text{and} \quad \underline{i}_{ab0} = [C_2]\underline{i}_{dq0}.$$

Therefore

$$\underline{i}_{123} = [C_1][C_2]\underline{i}_{dq0} \tag{10.8.34}$$

where

$$[C_1] \cdot [C_2] = \sqrt{2/3} \begin{bmatrix} 1 & 0 & 1/\sqrt{2} \\ -1/2 & \sqrt{3}/2 & 1/\sqrt{2} \\ -1/2 & -\sqrt{3}/2 & 1/\sqrt{2} \end{bmatrix} \cdot \begin{bmatrix} \cos\theta & -\sin\theta & 0 \\ \sin\theta & \cos\theta & 0 \\ 0 & 0 & 1 \end{bmatrix}. \tag{10.8.35}$$

Since $\cos 2\pi/3 = \cos 4\pi/3 = -\frac{1}{2}$, $\sin 2\pi/3 = \sqrt{3}/2$, and $\sin 4\pi/3 = -\sqrt{3}/2$, then

$$[C_1] \cdot [C_2] = \sqrt{2/3} \begin{bmatrix} \cos\theta & -\sin\theta & 1/\sqrt{2} \\ \cos(\theta - 2\pi/3) & -\sin(\theta - 2\pi/3) & 1/\sqrt{2} \\ \cos(\theta - 4\pi/3) & -\sin(\theta - 4\pi/3) & 1/\sqrt{2} \end{bmatrix}. \qquad \textbf{(10.8.36)}$$

Hence

$$\begin{bmatrix} i_1 \\ i_2 \\ i_3 \end{bmatrix} = \sqrt{2/3} \begin{bmatrix} \cos\theta & -\sin\theta & 1/\sqrt{2} \\ \cos(\theta - 2\pi/3) & -\sin(\theta - 2\pi/3) & 1/\sqrt{2} \\ \cos(\theta - 4\pi/3) & -\sin(\theta - 4\pi/3) & 1/\sqrt{2} \end{bmatrix} \cdot \begin{bmatrix} i_d \\ i_q \\ i_0 \end{bmatrix}. \qquad \textbf{(10.8.37)}$$

It is to be noted that for balanced conditions i_0 is zero.

10.8.4. Transformation and Power Invariance

It is usually possible to write down the differential equations of the rotating machine's electric system in the form

$$\underline{v} = [Z]\underline{i} \qquad \textbf{(10.8.38)}$$

but it is not always possible to solve them. In this set of equations the voltage vector $\underline{v}$ and the operational impedance matrix $[Z]$ are known and the real (or *old*) current variables are to be found. For ac machines such as induction and synchronous motors the real machine is modeled by a slip-ring primitive. If it is not possible to solve the slip-ring primitive equations, we turn our attention to a transformed system, the commutator primitive, with *new* variables described by

$$\underline{v}' = [Z']\underline{i}'. \qquad \textbf{(10.8.39)}$$

At this moment $\underline{v}'$, $[Z']$, and $\underline{i}'$ are all unknown. However, when the voltage vector $\underline{v}'$ and the operational impedance matrix $[Z']$ are determined, it is hoped that a standard form can be found for the solution of the *new* current vector $\underline{i}'$.

We have an *old* system and a *new* system that are linked by mmf equivalence and power invariance. Accordingly

$$\underline{i} = [C]\underline{i}' \quad \text{and} \quad \underline{v}' = [C]^t\underline{v} \qquad \textbf{(10.8.40)}$$

where $[C]$ is the connection matrix, whose entries have been determined from a knowledge of the mmf constraints. We can make use of these relations in the *old* system eq. (10.8.38) to obtain

$$[C^t]^{-1}\underline{v}' = [Z][C]\underline{i}' \qquad \textbf{(10.8.41)}$$

or

$$\underline{v}' = [C]^t[Z][C]\underline{i}'. \qquad \textbf{(10.8.42)}$$

But the *new* system equation is eq. (10.8.39), so the values of the *new* system operational impedances in terms of the *old* system operational impedances are found from

$$[Z'] = [C]^t[Z][C]. \tag{10.8.43}$$

$[Z]$ and $\underline{v}$ are known. $[C]$ is determined and then $[Z']$ and $\underline{v}'$ are found. Accordingly, it is possible to proceed with the solution of $\underline{v}' = [Z']\underline{i}'$ and the real or *old* variables are found from $\underline{i} = [C]\underline{i}'$.

The transformation that applies to the operational impedance matrix applies also to the resistance matrix and the inductance matrix. That is

$$[R'] = [C]^t[R][C] \qquad \text{and} \qquad [L'] = [C]^t[L][C]. \tag{10.8.44}$$

In some instances it is convenient to work with voltages in terms of the rate of change of flux linkage. The flux linkage transformation is obtained from

$$\underline{\lambda}' = [L']\underline{i}' = [C]^t[L][C]\underline{i}' = [C]^t[L]\underline{i} = [C]^t\underline{\lambda}. \tag{10.8.45}$$

There is also the special case when the connection matrix $[C]$ is orthogonal, such as the cases in sections 10.8.1, 10.8.2, and 10.8.3. The matrix $[C]$ is orthogonal if

$$[C]^{-1} = [C]^t \qquad \text{or} \qquad [C^{-1}]^t = [C]$$

so that the voltage transformation from the *new* system to the *old* system becomes

$$\underline{v} = [C^{-1}]^t\underline{v}' = [C]\underline{v}'.$$

This shows that the current and voltage transformations employ the same connection matrix, if the matrix is orthogonal.

10.8.5. A Quadruply-excited Machine with Saliency

The quadruply-excited primitive in Example 10.1 concerned coils in air. The real machine uses ferromagnetic materials. If the air gap between the rotor and stator is constant, the self-inductances of the windings are constants and independent of rotor position. However, some excitation windings are wound on salient poles and the air gap is not constant around the periphery of the rotor. The self-inductances of some of the machine windings are functions of the position of the rotor. If the inductances are functions of position the voltage equation is not a linear differential equation with constant coefficients. Therefore, the solution for the currents and hence the torque is difficult. We will use a general example of saliency to show how the transformations aid solution.

The first step is to transform to a two-phase, slip-ring primitive. From this slip-ring primitive the complexity of the equations defining the performance becomes apparent. Consequently the transformation to the commutator primitive will demonstrate the resulting equations' simplicity.

The machine has salient poles which define the direct axis. Consideration is given only to a 2-pole machine. The salient poles are on the stator and there are two stator windings. One winding has its axis on the direct axis and the second winding is on the quadrature axis. A balanced polyphase winding is on the round rotor. The rotor must be round if the stator is salient, otherwise the generalized theory of dynamic circuit analysis is of no avail. A two-phase winding on the rotor is chosen. Any other polyphase winding can be reduced to a two-phase case by the transformation indicated in section 10.8.1. Rotor speed is constant, the magnetic circuit is linear, and the flux distribution in the air gap is sinusoidal.

The above description of the machine is sufficient to be able to form a quadruply-excited, slip-ring primitive as shown in Fig. 10.16. From this the performance equations are to be derived.

A. SLIP-RING PRIMITIVE

The nomenclature used in the equations to describe the behavior of the salient-pole machine follows that shown in Fig. 10.16.

It is necessary to know the change of inductance with respect to position. L_1 and L_2 are constants because the reluctance paths along the principal axes do not vary. This is not the case for the other inductances. The mutual coupling between the two stator windings is zero because the magnetic flux axes are orthogonal. The same

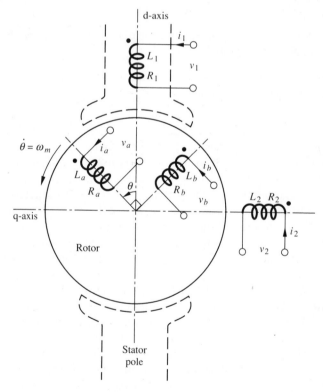

Figure 10.16 A quadruply-excited slip-ring primitive.

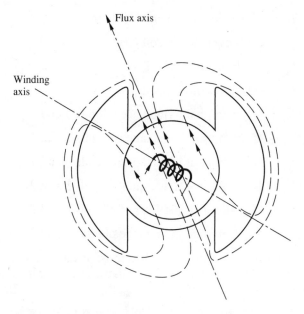

Figure 10.17   Coil axis and field axis with saliency.

cannot be said for the mutual coupling between the two rotor windings. Although the two coil axes are in quadrature the flux axes are not necessarily in quadrature. Saliency makes the coil axes and the flux axes noncoincident as shown in Fig. 10.17. Only if the rotor coil axes are on the direct and quadrature axes is there zero mutual interaction. By resolving the mmf of each coil along the quadrature and direct axes the function of the mutual coupling in terms of maximum values, minimum values and the angular displacement θ can be determined.

B. MACHINE INDUCTANCES OF THE SLIP-RING PRIMITIVE

The geometry of all machines is not the same. The only generalization that can be made about the inductances, in the first instance, is that they vary cyclically with the rotation of the rotor. For example, consider Fig. 10.16. The self-inductance of phase a winding is a maximum when it lies on the direct axis and is a minimum when it lies on the quadrature axis. A cyclic variation of the inductance with position means that the inductance coefficient can be represented by a Fourier series.

The general form of the inductance matrix can be written down by inspection. It is

$$[L] = \begin{bmatrix} L_1 & M_{12} & M_{1a} & M_{1b} \\ M_{21} & L_2 & M_{2a} & M_{2b} \\ M_{a1} & M_{a2} & L_a & M_{ab} \\ M_{b1} & M_{b2} & M_{ba} & L_b \end{bmatrix} \qquad \textbf{(10.8.46)}$$

where L_a = self-inductance of coil a (phase a)

L_b = self-inductance of coil b (phase b)

L_1 = self-inductance of coil 1

L_2 = self-inductance of coil 2

$M_{ab} = M_{ba}$ = mutual inductance of coils a and b

$M_{a1} = M_{1a}$ = mutual inductance of coils a and 1

$M_{b1} = M_{1b}$ = mutual inductance of coils b and 1

$M_{12} = M_{21}$ = mutual inductance of coils 1 and 2

$M_{a2} = M_{2a}$ = mutual inductance of coils a and 2

and $M_{b2} = M_{2b}$ = mutual inductance of coils b and 2.

Of these inductance coefficients L_1 and L_2 are constant, M_{12} is zero, and L_b has the same form as L_a but lags cyclically by $\pi/2$ radians. Therefore it is only necessary to consider the four terms L_a, M_{ab}, M_{a1}, and M_{b2}. M_{b1} and M_{a2} can be deduced from M_{a1} and M_{b2}.

L_a and L_b

The self-inductance L_a has its maximum value when θ is zero. L_a can be represented by the cosine series

$$L_a = a_0 + a_1 \cos\theta + a_2 \cos2\theta + a_3 \cos3\theta + \cdots \qquad (10.8.47)$$

Higher inductance harmonics cause harmonic currents to flow and these increase the losses without increasing the average torque or power. The winding of an ac machine is distributed around the periphery and the pole pieces are shaped to suppress the higher harmonics. Thus the series converges rapidly enough to ignore terms in which the coefficient of θ is 3 or greater. That is

$$L_a \approx a_0 + a_1 \cos\theta + a_2 \cos2\theta. \qquad (10.8.48)$$

The coefficients a_0, a_1, and a_2 have to be found in terms of constants that can be readily determined. Inspection of Fig. 10.16 shows that the reluctance path of coil a is greatest when θ is $\pi/2$ radians. At this position the inductance is a minimum. The inductance is a maximum when θ is π radians. Therefore, a_1 must be zero. Hence

$$L_a \approx a_0 + a_2 \cos2\theta. \qquad (10.8.49)$$

Let $L_a = L_d$ when $\theta = 0$, π, 2π, . . . , etc., and let $L_a = L_q$ when $\theta = \pi/2$, $3\pi/2$, $5\pi/2$ . . . , etc. L_d and L_q are constants and supposedly they can be measured from machine tests, or calculated from the known magnetic field quantities and the geometry. These two conditions lead to the equations

$$L_d = a_0 + a_2 \qquad \text{and} \qquad L_q = a_0 - a_2 \qquad (10.8.50)$$

from which

$$a_0 = \tfrac{1}{2}(L_d + L_q) \quad \text{and} \quad a_2 = \tfrac{1}{2}(L_d - L_q). \tag{10.8.51}$$

It is found in practice that $L_d \approx 2L_q$ for salient-pole machines, so that $a_0 \approx 3a_2$ and the series converges rapidly. Substitution of the values for the coefficients of the Fourier series gives

$$L_a = \tfrac{1}{2}(L_d + L_q) + \tfrac{1}{2}(L_d - L_q)\cos2\theta \tag{10.8.52}$$

or, if L_d and L_q are separated

$$L_a = L_d \cos^2\theta + L_q \sin^2\theta. \tag{10.8.53}$$

For the self-inductance of coil b, θ is replaced by $(\theta - \pi/2)$, so that

$$L_b = L_d \sin^2\theta + L_q \cos^2\theta. \tag{10.8.54}$$

M_{ab}

The mutual inductance between the two phase windings a and b is zero if θ is zero. Therefore the inductance can be represented by the series

$$M_{ab} = m_0 + m_1 \sin\theta + m_2 \sin2\theta + m_3 \sin3\theta + \cdots \tag{10.8.55}$$

or

$$M_{ab} \approx m_0 + m_1 \sin\theta + m_2 \sin2\theta \tag{10.8.56}$$

since convergence is considered to be rapid.
$M_{ab} = 0$ for $\theta = 0$ and $\theta = \pi/2$. Therefore, $m_0 = 0$ and $m_1 = 0$. Consequently the series is reduced to

$$M_{ab} = m_2 \sin2\theta = \tfrac{1}{2}(L_d - L_q) \sin2\theta \tag{10.8.57}$$

if we allow $m_2 = a_2$. (In practice it has been shown that a_2 is a smaller fraction of m_2.)

M_{a1} and M_{b1}

The value of the mutual inductance of coils a and 1 is a maximum when θ is zero. Therefore, M_{a1} can be represented by a cosine series, which is truncated to

$$M_{a1} = f_0 + f_1 \cos\theta + f_2 \cos2\theta \tag{10.8.58}$$

because of rapid convergence.
At $\theta = \pi/2$ the mutual inductance is zero so that either $f_0 = f_2 = $ constant, which does not provide convergence, or both f_0 and f_2 are zero. The latter case must be correct because

$$M_{a1}(\theta = 0) = -M_{a1}(\theta = \pi) \tag{10.8.59}$$

is only satisfied if $f_0 = f_2 = 0$. Therefore

$$M_{a1} = f_1 \cos\theta. \tag{10.8.60}$$

At $\theta = 0$ the mutual inductance is the maximum value M_d say, so that

$$M_{a1} = M_d \cos\theta. \tag{10.8.61}$$

The mutual inductance M_{b1} is obtained by replacing θ by $(\theta - \pi/2)$ in the above equation so that

$$M_{b1} = M_d \sin\theta. \tag{10.8.62}$$

M_{b2} and M_{a2}

The value of the mutual inductance of coils b and 2 is the maximum value M_q when θ is zero. So M_{b2} can be represented by the truncated cosine series

$$M_{b2} = g_0 + g_1 \cos\theta + g_2 \cos2\theta. \tag{10.8.63}$$

At $\theta = 0$, $M_{b2} = M_q = g_0 + g_1 + g_2$. At $\theta = \pi$, $M_{b2} = -M_q = g_0 - g_1 + g_2$. Therefore, $g_1 = M_q$ and $g_0 + g_2 = 0$.
At $\theta = \pi/2$, $M_{b2} = 0 = g_0 - g_2$. Therefore, $g_0 = g_2 = 0$, and

$$M_{b2} = M_q \cos\theta. \tag{10.8.64}$$

The mutual inductance M_{a2} is obtained by replacing θ by $(\theta + \pi/2)$ in the above equation for M_{b2}. Hence

$$M_{a2} = -M_q \sin\theta. \tag{10.8.65}$$

The machine inductances have been reduced to constant inductance coefficients and sinusoidal functions of the rotor position θ. The inductance matrix of eq. (10.8.46) becomes

$$[L] = \begin{bmatrix} L_1 & 0 & M_d \cos\theta & M_d \sin\theta \\ 0 & L_2 & -M_q \sin\theta & M_q \cos\theta \\ M_d \cos\theta & -M_q \sin\theta & L_d \cos^2\theta + L_q \sin^2\theta & \frac{1}{2}(L_d - L_q) \sin2\theta \\ M_d \sin\theta & M_q \cos\theta & \frac{1}{2}(L_d - L_q) \sin2\theta & L_d \sin^2\theta + L_q \cos^2\theta \end{bmatrix}. \tag{10.8.66}$$

Inspection of the inductance matrix shows that many of the entries are functions of the angular displacement θ. Consequently, the machine's voltage equation does not have constant coefficients since p acts on $[L]$ as well as $\underline{i}$. Therefore it is not practicable to expand the equations to find the torque in terms of currents and inductance coefficients.

The only way to make the coefficients constant is to transform to the commutator primitive. First, some points must be noted.

Pole Pairs

In this section we have assumed that there is only one pole pair on the machine, or else the angle θ is measured in electric radians. If there are p pole pairs and the angle θ is measured in mechanical radians, then $p\theta_{mech} = \theta_{elec}$ has to be substituted in the expressions for L_a, M_{ab}, etc.

Conditions

The relations that have been obtained for the inductances are subject to certain conditions. L_d, L_q, and M_{ab} are only constants if hysteresis effects are not present and saturation of the iron parts is neglected. Further, the currents and flux are considered to vary sinusoidally in time and space, respectively. In a linear system the inductance is only a function of geometry and is independent of the current. However, the conclusion that $a_2 = m_2$ requires the current to be sinusoidal. Finally, the slot permeances are ignored and the polyphase windings are balanced.

Special Case

For the condition that both the stator and rotor members of a machine are round, that is, if the air gap is constant around the periphery, then $L_a = L_b$, $L_d = L_q$ and remain the same constant values. In addition, $M_{ab} = M_{ba} = 0$, $M_{a1} = M_d \cos\theta$, and $M_{b2} = M_q \cos\theta$. Consequently, in this case the only variables are the mutual inductances between the stator windings and the rotor windings.

C. TRANSFORMATION TO THE COMMUTATOR PRIMITIVE

The transformation to be used converts the slip-ring primitive (or real machine) to the commutator primitive (or generalized machine). We follow the process outlined in sections 10.8.2 and 10.8.4. The slip-ring primitive is analyzed by the equations

$$v = [Z]\underline{i} \quad \text{and} \quad T_e = \tfrac{1}{2}\underline{i}^t[dL/d\theta]\underline{i} = \tfrac{1}{2}\underline{i}^t[G_s]\underline{i}. \tag{10.8.67}$$

The commutator primitive, as illustrated in Fig. 10.18, is analyzed by the new equations

$$\underline{v}' = [Z']\underline{i}' \quad \text{and} \quad T'_e = \underline{i}'^t[G_c]\underline{i}' \tag{10.8.68}$$

where $[G_c]$ is the G matrix for the commutator primitive.

The relationship between the slip-ring and commutator primitive currents is defined by

$$\underline{i} = [C]\underline{i}'. \tag{10.8.69}$$

If power invariance is accepted for the two primitives $T'_e = T_e$ since the speeds for the two primitives are assumed to be the same. Also $[Z'] = [C]^t[Z][C]$.

From an inspection of Figs. 10.18 and 10.16 the connection between the *old* and *new* current is

$$\begin{bmatrix} i_1 \\ i_2 \\ i_a \\ i_b \end{bmatrix} = \begin{bmatrix} 1 & 0 & 0 & 0 \\ 0 & 1 & 0 & 0 \\ 0 & 0 & -\cos\theta & -\sin\theta \\ 0 & 0 & \sin\theta & \cos\theta \end{bmatrix} \cdot \begin{bmatrix} i_1 \\ i_2 \\ i_d \\ i_q \end{bmatrix}. \tag{10.8.70}$$

If the equation

$$\underline{v}' = [C]^t[Z][C]\underline{i}' \tag{10.8.71}$$

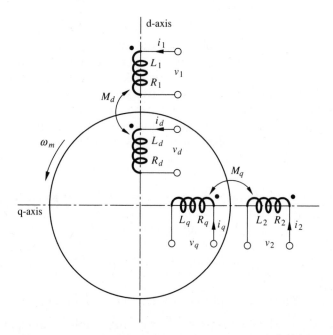

Figure 10.18 A quadruply-excited commutator primitive.

is worked out in detail, the product $[Z] \cdot [C]$ is calculated first. Since $[Z] \cdot [C]$ contains the operator p which acts on $\underline{i}'$ as well as θ, then the next product calculated is $[Z][C]\underline{i}'$. Finally the answer will be multiplied by the transpose of $[C]$. The reader may like to work this out in detail as a check for the voltage equation, which is written down from an inspection of Fig. 10.18 as

$$
\begin{bmatrix} v_1 \\ v_2 \\ v_d \\ v_q \end{bmatrix} = \begin{bmatrix} R_1 + L_1 p & 0 & M_d p & 0 \\ 0 & R_2 + L_2 p & 0 & M_q p \\ M_d p & -M_q \omega_m & R_d + L_d p & -L_q \omega_m \\ M_d \omega_m & M_q p & L_d \omega_m & R_q + L_q p \end{bmatrix} \cdot \begin{bmatrix} i_1 \\ i_2 \\ i_d \\ i_q \end{bmatrix}. \quad \textbf{(10.8.72)}
$$

This has constant coefficients so that $\underline{i}'$ can be found readily using operational methods, if the speed ω_m is constant and the voltages and parameters are known.

D. ELECTROMAGNETIC TORQUE

For the commutator primitive, the currents $\underline{i}'$ can be found in terms of known voltages, inductances, and speed, and the torque can be determined. For completeness the torque will be calculated for the real machine, for the real machine transformed and for the generalized machine. The three results should be the same.

The Real Machine

The electromagnetic torque for the real machine of Fig. 10.16 is

$$T_e = \tfrac{1}{2}\underline{i}^t[dL/d\theta]\underline{i} = \tfrac{1}{2}\underline{i}^t[G_s]\underline{i}. \tag{10.8.73}$$

Applying the transformation to the current enables comparison of the *impedance* coefficients. So

$$T_e = \tfrac{1}{2}\underline{i}''^t[C]^t[dL/d\theta]\underline{i}' = \tfrac{1}{2}\underline{i}''^t[G_s']\underline{i}'. \tag{10.8.74}$$

where $[G_s']$ is the transformed G matrix of the slip-ring primitive. From the inductance matrix $[L]$ of eq. (10.8.66)

$$\left[\frac{dL}{d\theta}\right] = \begin{bmatrix} 0 & 0 & -M_d\sin\theta & M_d\cos\theta \\ 0 & 0 & -M_q\cos\theta & -M_q\sin\theta \\ -M_d\sin\theta & -M_q\cos\theta & -(L_d-L_q)\sin2\theta & (L_d-L_q)\cos2\theta \\ M_d\cos\theta & -M_q\sin\theta & (L_d-L_q)\cos2\theta & (L_d-L_q)\sin2\theta \end{bmatrix} \tag{10.8.75}$$

and

$$[G_s'] = [C]^t \cdot \left[\frac{dL}{d\theta}\right] \cdot [C] = \begin{bmatrix} 0 & 0 & 0 & M_d \\ 0 & 0 & -M_q & 0 \\ 0 & -M_q & 0 & L_d-L_q \\ M_d & 0 & L_d-L_q & 0 \end{bmatrix}. \tag{10.8.76}$$

So that

$$T_e = \tfrac{1}{2}[i_1 \; i_2 \; i_d \; i_q] \cdot \begin{bmatrix} 0 & 0 & 0 & M_d \\ 0 & 0 & -M_q & 0 \\ 0 & -M_q & 0 & L_d-L_q \\ M_d & 0 & L_d-L_q & 0 \end{bmatrix} \cdot \begin{bmatrix} i_1 \\ i_2 \\ i_d \\ i_q \end{bmatrix} \tag{10.8.77}$$

or

$$T_e = i_1 i_q M_d - i_2 i_d M_q + i_d i_q L_d - i_d i_q L_q. \tag{10.8.78}$$

The Generalized Machine

The torque for the commutator primitive, shown in Fig. 10.18, is expressed by

$$T_e' = \underline{i}''^t[G_c]\underline{i}'. \tag{10.8.79}$$

The motional inductance coefficients of $[G_c]$ can only be associated with the rotor windings that move past the stationary stator winding field. The rotor mmf is fixed in space, the stator windings are fixed in space, there is no relative motion and so there is no motionally induced emf in the stator windings. Motional induced emfs are in quadrature with transformer emfs; that is, the $[G_c]$ terms, where they exist, are complementary to the mutual terms. Accordingly, the $[G_c]$ terms are equal to the mutual inductance coefficients obtained when the rotor windings are advanced by $\pi/2$ electric radians. *Advanced* implies the same direction as the direction of rotor motion. Hence

$$[G_c] = \begin{bmatrix} 0 & 0 & 0 & 0 \\ 0 & 0 & 0 & 0 \\ 0 & -M_q & 0 & -L_q \\ M_d & 0 & L_d & 0 \end{bmatrix}. \tag{10.8.80}$$

For example, the entry a_{32} is finite because the d-winding conductors are moving past the field of the winding 2, there is a motional induced emf in each conductor, and, as in a dc machine, when summed, the emfs do not cancel. The emfs would cancel if the two axes of coils d and 2 were coincident instead of being in quadrature. The value of a_{32} is the value of the mutual coefficient when the d-winding axis is advanced through $\pi/2$ in the direction of motion. Then the axes of coils d and 2 are coincident but opposed, so the sign of a_{32} will be negative. Windings d and q are similar, so the absolute value of a_{32} is the same as the real mutual inductance coefficient of coils q and 2, that is, M_q. Hence $G_{d2} = -M_{d2}$ if d were advanced through $\pi/2$. That is, $G_{d2} = -M_{q2} = -M_q$. Again the torque becomes

$$T'_e = i_1 i_q M_d - i_2 i_d M_q + i_d i_q L_d - i_d i_q L_q. \tag{10.8.81}$$

The Real Machine Transformed

Let us start with the power in the real machine, that is

$$\underline{i}^t \underline{v} = \underline{i}^t [R] \underline{i} + \underline{i}^t p[L] \underline{i} \tag{10.8.82}$$

and use the transformations

$$\underline{i} = [C] \underline{i}', \quad \underline{v} = [C^{-1}]^t \underline{v}' \quad \text{and} \quad [Z'] = [C]^t [Z][C] \tag{10.8.83}$$

to develop the generalized machine equations. The power equation becomes

$$\begin{aligned} \underline{i}^t \underline{v} &= \underline{i}''^t [C]^t \underline{v} = \underline{i}''^t [C]^t ([R] \underline{i} + p[L] \underline{i}) \\ &= \underline{i}''^t [C]^t [R][C] \underline{i}' + \underline{i}''^t [C]^t p([L][C] \underline{i}'). \end{aligned} \tag{10.8.84}$$

It must be noted that p operates on $[L]$, $[C]$, and $\underline{i}'$ so that, since $\underline{i}^t \underline{v} = \underline{i}''^t \underline{v}'$,

$$\underline{i}''^t \underline{v}' = \underline{i}''^t [R'] \underline{i}' + \underline{i}''^t [C]^t [L][C] p \underline{i}' + \underline{i}''^t [C]^t [pL][C] \underline{i}' + \underline{i}''^t [C]^t [L][pC] \underline{i}' \tag{10.8.85}$$

or

$$\begin{aligned} \underline{i}''^t \underline{v}' &= \underline{i}''^t [R'] \underline{i}' + \underline{i}''^t [L'] p \underline{i}' + \underline{i}''^t [C]^t [dL/d\theta] \dot{\theta} [C] \underline{i}' \\ &\quad + \underline{i}''^t [C]^t [L][dC/d\theta] \dot{\theta} \underline{i}'. \end{aligned} \tag{10.8.86}$$

The first three terms are recognized from the direct circuit analysis of the generalized machine (commutator primitive) as the electric power input, the power dissipated, and the rate of stored energy in the magnetic field, respectively. Accordingly, the gross mechanical power output is

$$p_m = \underline{i}''^t \{[C]^t [dL/d\theta][C] + [C]^t [L][dC/d\theta]\} \dot{\theta} \underline{i}'. \tag{10.8.87}$$

Therefore the torque of electric origin is

$$T_e = \underline{i}''\{[C]'[dL/d\theta][C] + [C]'[L][dC/d\theta]\}\underline{i}'. \tag{10.8.88}$$

Now

$$\left[\frac{dC}{d\theta}\right] = \begin{bmatrix} 0 & 0 & 0 & 0 \\ 0 & 0 & 0 & 0 \\ 0 & 0 & -\sin\theta & -\cos\theta \\ 0 & 0 & \cos\theta & -\sin\theta \end{bmatrix} \tag{10.8.89}$$

so that

$$[C]'[L]\left[\frac{dC}{d\theta}\right] = \begin{bmatrix} 0 & 0 & 0 & -M_d \\ 0 & 0 & M_q & 0 \\ 0 & 0 & 0 & -L_d \\ 0 & 0 & L_q & 0 \end{bmatrix} = -[G_c]' \tag{10.8.90}$$

and

$$[C]'\left[\frac{dL}{d\theta}\right][C] + [C]'[L]\left[\frac{dC}{d\theta}\right] = \begin{bmatrix} 0 & 0 & 0 & 0 \\ 0 & 0 & 0 & 0 \\ 0 & -M_q & 0 & -L_q \\ M_d & 0 & L_d & 0 \end{bmatrix} = [G_c] \tag{10.8.91}$$

or

$$[G_s'] - [G_c]' = [G_c].$$

That is

$$[G_s'] = [G_c] + [G_c]'. \tag{10.8.92}$$

Hence

$$T_e = [i_1\ i_2\ i_d\ i_q] \cdot \begin{bmatrix} 0 & 0 & 0 & 0 \\ 0 & 0 & 0 & 0 \\ 0 & -M_q & 0 & -L_q \\ M_d & 0 & L_d & 0 \end{bmatrix} \cdot \begin{bmatrix} i_1 \\ i_2 \\ i_d \\ i_q \end{bmatrix}. \tag{10.8.93}$$

It is to be noted that this is the same as $T_e' = \underline{i}''[G_c]\underline{i}'$.

E. PERFORMANCE

Most ac machines can be modeled as slip-ring machines. We have developed the most general type which has saliency. If the slip-ring primitive is transformed to a commutator primitive and the speed is constant, the resulting equations are a set of linear differential equations with constant coefficients. A commutator primitive is often referred to as the *generalized machine*.

The general form of solution is to find $\underline{i}'$ from

$$\underline{i}' = [Z']^{-1}\underline{v}' \tag{10.8.94}$$

and substitute in the equation

$$T_e = \underline{i}''[G_c]\underline{i}' \tag{10.8.95}$$

so that the performance equation

$$T_e = T_m + J\ddot{\theta} + D\dot{\theta} \tag{10.8.96}$$

can be investigated. If the values of real currents are required, use is made of $\underline{i} = [C]\underline{i}'$.

Some synchronous machines have saliency, so that the steady-state performance or transient performance can be obtained readily from the above equations. Certain simplifications can be made if the induction machine or the round rotor synchronous machine is being studied. There is no saliency in these cases so that $L_d = L_q$ and $M_d = M_q$.

We will analyze both the induction motor and the synchronous motor in order to compare the generalized theory with the conventional steady-state theory.

10.9. POLYPHASE INDUCTION MOTOR

Whether the rotor be wound or of the cage type, both rotor and stator currents of the induction motor can be transformed so that the winding axes are fixed. That is the induction motor can be modeled as a commutator primitive. The motor is considered to have only two poles until the torque is considered, at which point p pole pairs are taken into account.

The induction motor can be represented diagrammatically in its generalized form as shown typically in Fig. 10.19 and, from an inspection of this, the voltage equation can be written down as $\underline{v}' = [Z']\underline{i}'$. That is

$$\begin{bmatrix} v_D \\ v_Q \\ v_d \\ v_q \end{bmatrix} = \begin{bmatrix} R_D + L_D p & 0 & M_d p & 0 \\ 0 & R_Q + L_Q p & 0 & M_q p \\ M_d p & G_{dQ}\omega_m & R_d + L_d p & G_{dq}\omega_m \\ G_{qD}\omega_m & M_q p & G_{qd}\omega_m & R_q + L_q p \end{bmatrix} \cdot \begin{bmatrix} i_D \\ i_Q \\ i_d \\ i_q \end{bmatrix} \tag{10.9.1}$$

where the uppercase subscripts refer to the stator and the lowercase subscripts refer to the rotor.

The induction motor has particular constraints. The air gap is uniform because the stator is round just like the rotor. This makes the stator self-inductances the same on both axes. The same applies to the rotor coils. There are balanced windings on both members. This means that the mutual inductances are the same on the direct and quadrature axes. Resistance values of the stator phases are identical and so are those of the rotor. Probably the most important constraint of all is the fact that the rotor phases are not excited externally; they are short-circuited. Noting that the rotor speed ω_m is clockwise, the parameter values can be simplified to

$$M_d = M_q \triangleq M, \quad L_D = L_Q \triangleq L_S, \quad L_d = L_q \triangleq L_R$$

$$R_d = R_q \triangleq R_2, \quad R_D = R_Q \triangleq R_1 \tag{10.9.2}$$

$$v_d = v_q = 0, \quad G_{dQ} = M, \quad G_{dq} = L_R, \quad G_{qd} = -L_R, \quad \text{and} \quad G_{qD} = -M.$$

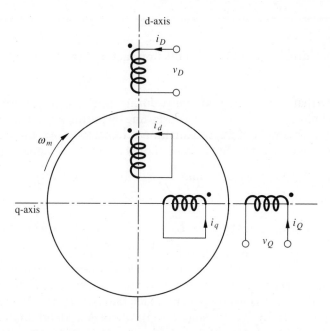

Figure 10.19 Induction motor commutator primitive.

The self-inductance symbols L_R and L_S are used because L_1 and L_2 are used conventionally as leakage-inductance symbols. All the above are phase values no matter what the polyphase system, so that

$$\begin{bmatrix} v_D \\ v_Q \\ 0 \\ 0 \end{bmatrix} = \begin{bmatrix} R_1 + L_S p & 0 & Mp & 0 \\ 0 & R_1 + L_S p & 0 & Mp \\ Mp & M\omega_m & R_2 + L_R p & L_R \omega_m \\ -M\omega_m & Mp & -L_R \omega_m & R_2 + L_R p \end{bmatrix} \cdot \begin{bmatrix} i_D \\ i_Q \\ i_d \\ i_q \end{bmatrix}. \quad \textbf{(10.9.3)}$$

The vector entries are all instantaneous values so that a transient analysis can be studied. The torque equations are

$$T_e = \underline{i}^t[G_c]\underline{i} = M(i_q i_D - i_d i_Q) \quad \text{and} \quad T_e = J\dot{\omega}_m + D\omega_m + T_m. \quad \textbf{(10.9.4)}$$

Electrical and mechanical transients arise in induction motors if the load changes and if the terminal voltage changes. Analysis to determine the transient behavior can provide information for design procedures to satisfy protection criteria and performance specifications, or provide information how the transients affect the system to which the motor is connected. For example, induction motor start-up causes high values of current to be drawn momentarily from the supply. For some applications this may be undesirable because of overheating or system voltage reduction, but in a control application it may be an advantage because of the need for rapid response. In the following two examples we study the transients at motor start-up.

EXAMPLE 10.4

Determine the inrush current to the induction motor at start-up.

Solution

If it is assumed that there is no speed change during the electrical transient, the inrush current to the motor can be calculated at starting by putting ω_m equal to zero in eq. (10.9.3). A further simplifying approximation is to neglect the rotor winding resistance R_2. Because of the symmetry we need only consider the equations

$$v_D = (R_1 + L_S p)i_D + Mpi_d \quad \text{and} \quad v_d = 0 = Mpi_D + L_R pi_d.$$

Elimination of i_d yields

$$v_D = (R_1 + L'p)i_D, \quad \text{where} \quad L' = L_S - M^2/L_R \triangleq \text{transient inductance.}$$

Transform methods may be used to solve for i_D, but in this case a knowledge of ac theory for static networks can be used. This network consists of R_1 and L' to which a sinusoidal voltage is applied. The stator phase current i_D will comprise a steady-state component i_{ss} and a transient component i_t such that $i_D = i_{ss} + i_t$. It is assumed that the voltage v_D is given by $V_D = \hat{V} \sin(\omega t + \phi)$, where ϕ is an arbitrary angle determined by the initial conditions of switching. Thus the component i_{ss} is given by

$$i_{ss} = \frac{\hat{V}}{(R_1^2 + \omega^2 L'^2)^{1/2}} \sin(\omega t + \phi - \phi')$$

where $\phi' = \tan^{-1}\omega L'/R_1$. The component i_t has the exponential form

$$i_t = Ae^{-t/r}$$

where $A =$ constant to be determined and $\tau = L'/R_1 =$ transient time constant. The initial current is zero, so that

$$0 = i_{ss}(0) + i_t(0)$$

and it follows that

$$A = -\frac{\hat{V}}{(R_1^2 + \omega^2 L'^2)^{1/2}} \sin(\phi - \phi').$$

Therefore the complete stator phase transient is given by

$$i_D = \frac{\hat{V}}{(R_1^2 + \omega^2 L'^2)^{1/2}} [\sin(\omega t + \phi - \phi') - \sin(\phi - \phi')e^{-t/r}].$$

This equation has no significance once the rotor begins to move. The time constant of the transient decay is important; practical values are of the order of 0.02 second for 50-kW motors. For the same size motor the mechanical time constant may be about 1 second. So it is reasonable to assume that the electrical transient decays before the speed changes noticeably if the ratio of mechanical to electrical time constants is as high as 50.

EXAMPLE 10.5

A laboratory generalized machine is connected as a 3-phase, 60-Hz, 4-pole, 200-V induction motor with a rating of 2163 VA. Standard tests on the induction motor yielded the following parameter values.

$$L_S = L_R = 0.0883 \text{ H}, \quad M = 0.085 \text{ H}, \quad R_1 = 0.32 \ \Omega, \quad R_2 = 0.27 \ \Omega,$$

$$J = 0.1546 \text{ kg} \cdot \text{m}^2, \quad \text{and} \quad D = 0.00565 \text{ N} \cdot \text{m} \cdot \text{s/rad}.$$

Predict the no-load starting characteristics of the induction motor if rated voltage at rated frequency is applied.

Solution

Whenever electrical and mechanical transients in induction motors occur simultaneously, it is not possible to simplify eqs. (10.9.3) and (10.9.4) in order to solve them analytically. The equations are nonlinear so that it is necessary to use a computer to obtain a solution. It is necessary to reformulate the equations into a set of first-order differential equations for suitable handling.

In general matrix form the voltage equation

$$\underline{v} = [R + G_c \omega_m] \underline{i} + [L] p \underline{i}$$

is arranged to become

$$p \underline{i} = [L]^{-1} \underline{v} - [L]^{-1} [R + G_c \omega_m] \underline{i}.$$

Equations (10.9.4) can be combined to give

$$p \omega_m = -T_m / J - (D/J) \omega_m + (1/J) \underline{i}^t [G_c] \underline{i}.$$

Expansion provides the set of first-order equations

$$p i_D = \frac{M}{L_R L_S - M^2} \left(\frac{L_R}{M} v_D - \frac{L_R}{M} R_1 i_D - M \omega_m i_Q + R_2 i_d - L_R \omega_m i_q \right)$$

$$p i_Q = \frac{M}{L_R L_S - M^2} \left(\frac{L_R}{M} v_Q + M \omega_m i_D - \frac{L_R}{M} R_1 i_Q + L_R \omega_m i_d + R_2 i_q \right)$$

$$p i_d = \frac{L_S}{L_R L_S - M^2} \left(-\frac{M}{L_S} v_D + \frac{M}{L_S} R_1 i_D + M \omega_m i_Q - R_2 i_d + L_R \omega_m i_q \right)$$

$$p i_q = \frac{L_S}{L_R L_S - M^2} \left(-\frac{M}{L_S} v_Q - M \omega_m i_D + \frac{M}{L_S} R_1 i_Q - L_R \omega_m i_d - R_2 i_q \right)$$

$$\text{and} \quad p \omega_m = \frac{T_m}{J} - \frac{D}{J} \omega_m + \frac{M}{J} (i_D i_q - i_Q i_d).$$

Also $v_D = \hat{V} \sin \omega t$, $v_Q = \hat{V} \cos \omega t$, and $\omega = 377$ rad/s.

Let us normalize these equations by employing per unit values.[7] Once the base

---

[7] Refer to Appendix E for a description of per unit values.

values have been chosen, the per unit values are obtained by dividing the actual value by the base value. Suitable base values (with subscripts b) are

base VA = rated VA and base stator-winding voltage $V_{sb} = (m/2)\hat{V}$,

where the factor $m/2$ is for conversion from an m-phase to a 2-phase system and $\hat{V}$ is the amplitude of the rated phase voltage. The remaining base values follow.

Base rotor winding-voltage $V_{rb} = V_{sb}M/kL_S = V_{sb}kL_R/M$,

where M/kL_S = turns ratio and k = coefficient of coupling.

Base stator-winding current I_{sb} = base VA/V_{rb},

base rotor-winding current I_{rb} = base $VA/V_{rb} = kL_S I_{sb}/M$,

base stator-winding impedance $Z_{sb} = V_{sb}/I_{sb}$, and

base torque = base VA/ω, and base speed = ω = angular frequency of the supply.

Accordingly the dimensionless per unit equations are

$$p\bar{i}_D = \frac{1}{1-k^2}\left(\frac{\omega\bar{v}_D}{\bar{x}_S} - \frac{\bar{i}_D}{\tau_D} - k^2\bar{\omega}_m\omega\bar{i}_Q + \frac{k}{\tau_d}\bar{i}_d - k\bar{\omega}_m\omega\bar{i}_q\right)$$

$$p\bar{i}_Q = \frac{1}{1-k^2}\left(\frac{\omega\bar{v}_Q}{\bar{x}_S} + k^2\bar{\omega}_m\omega\bar{i}_D - \frac{\bar{i}_Q}{\tau_D} + k\bar{\omega}_m\omega\bar{i}_q + \frac{k}{\tau_d}\bar{i}_d\right)$$

$$p\bar{i}_d = \frac{1}{1-k^2}\left(-\frac{k\omega}{\bar{x}_S}\bar{v}_D + \frac{k}{\tau_D}\bar{i}_D + k\bar{\omega}_m\omega\bar{i}_Q - \frac{\bar{i}_d}{\tau_d} + \bar{\omega}_m\omega\bar{i}_q\right)$$

$$p\bar{i}_q = \frac{1}{1-k^2}\left(-\frac{k\omega}{\bar{x}_S}\bar{v}_Q - k\bar{\omega}_m\omega\bar{i}_D + \frac{k}{\tau_D}\bar{i}_Q - \bar{\omega}_m\omega\bar{i}_d - \frac{\bar{i}_q}{\tau_d}\right)$$

and $2Hp\bar{\omega}_m = -\bar{P}_f\bar{\omega}_m + k\bar{x}_S(\bar{i}_D\bar{i}_q - \bar{i}_d\bar{i}_Q) - \bar{T}_m$

where $\bar{x}_S = \omega L_S/Z_{sb}$, $\tau_d = L_R/R_2$, $\tau_D = L_S/R_1$, $H = \frac{1}{2}J\omega^2/$base VA = inertia constant, and $\bar{P}_f = D\omega/$base VA = per unit friction loss. The variables with the bar over them represent per unit values. These equations are for the 2-pole commutator primitive. Therefore, the parameters J and D, that are used in the calculations, are equal to the real parameters multiplied by a factor q, where $q = 1/(\text{no. of pole pairs})^2$.

If the principal base values are

base $VA = 2163$ VA   and $v_{sb} = (3/2)\sqrt{2} \times (200/\sqrt{3}) = 245$ V

then $V_{rb} = 245$ V, $I_{rb} = I_{sb} = 2163/245 = 8.84$ A, base speed = 377 rad/s, $H = \frac{1}{2}(0.1546/4) \times (377)^2/2163 = 1.268$, $\bar{P}_f = (0.00565/4) \times (377)^2/2163 = 0.0926$, base torque = $2163/377 = 5.74$ N·m, $Z_{sb} = 245/8.84 = 27.7$ Ω, $1/x_S = 377 \times 0.0883/27.7 = 0.832$, $\tau_D = 0.276$s, and $\tau_d = 0.327$s.

A final set of scaled equations is

$$p\bar{i}_D = 0.942\,\sin\omega t - 0.131\bar{i}_D - 12.6\bar{\omega}_m\bar{i}_Q + 0.106\bar{i}_d - 13.1\bar{\omega}_m\bar{i}_q$$

$$p\bar{i}_Q = 0.942\,\cos\omega t - 0.131\bar{i}_Q + 12.6\bar{\omega}_m\bar{i}_D + 0.106\bar{i}_d - 13.1\bar{\omega}_m\bar{i}_d$$

$$p\bar{i}_d = 0.907\,\sin\omega t + 0.126\bar{i}_D + 13.1\bar{\omega}_m\bar{i}_Q - 0.11\bar{i}_d + 13.6\bar{\omega}_m\bar{i}_q$$

$$p\bar{i}_q = 0.907\,\cos\omega t + 0.126\bar{i}_Q - 13.1\bar{\omega}_m\bar{i}_D - 0.11\bar{i}_q - 13.6\bar{\omega}_m\bar{i}_d$$

and $p\bar{\omega}_m = -0.97 \times 10^{-4}\bar{\omega}_m + 0.175(\bar{i}_D\bar{i}_q - \bar{i}_Q\bar{i}_d)$.

There is, as yet, no general best method to obtain a numerical solution of ordinary differential equations. Once a system has been resolved into a set of simultaneous first-order differential equations with given initial conditions, the method of solution may be either a predictor method incorporating a correction routine, or the single-step method of Runge-Kutta. Each has many variations. Oversimplifying, the

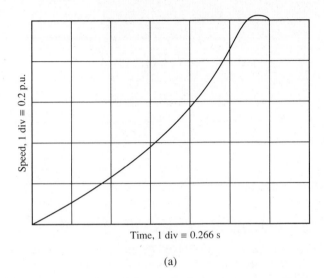

(a)

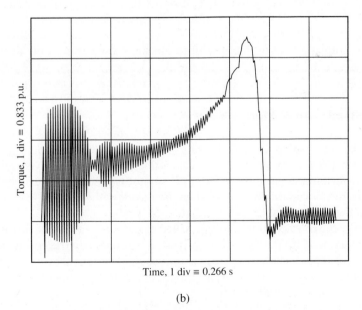

(b)

Figure 10.20 Induction motor starting characteristics. (a) Speed characteristic, (b) torque characteristic, (c) current characteristic (see over).

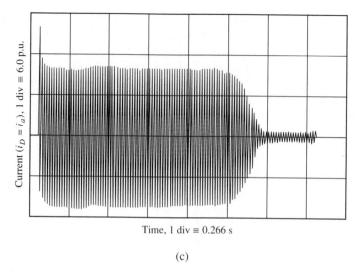

Time, 1 div ≡ 0.266 s

(c)

Figure 10.20 *Continued*

total transient period of interest is divided into discrete time intervals. For each interval the differential equations are solved by numerical integration and the values of the variables are then used in the solution over the next interval. To speed the solution, the time intervals can be lengthened but a reduction of accuracy must be suffered and there is a possibility of instability due to discrete processes.

A fifth-order Runge-Kutta method was utilized to obtain the solution of this induction motor start-up problem. The results are shown graphically in Fig. 10.20.

10.9.1. Steady-state Analysis

It is possible to develop the steady-state performance of an induction motor by starting with the matrix eq. (10.9.3). The values of voltages and currents in eq. (10.9.3) are instantaneous, but the induction motor is excited by sinusoidal voltages and it is desired to work in phasor variables. By means of the rotating axes depicted in Fig. 10.16 the rotor variables can be related.

In reality the rotor conductors, with speed ω_m, move in a sinusoidal distribution of air-gap flux, that rotates synchronously at ω. Accordingly the conductors move with a relative angular speed $(\omega - \omega_m)$ in a sinusoidal field, so the angular frequency of the rotor currents will be $(\omega - \omega_m)$. Thus the actual instantaneous current in phase a of the rotor winding can be written

$$i_a = \hat{I}_2 \cos(\omega - \omega_m)t \qquad \text{or} \qquad i_a = \sqrt{2}I_2 \cos(\omega - \omega_m)t \qquad (10.9.5)$$

where $\hat{I}_2$ and I_2 are the maximum and rms values, respectively, of the rotor winding phase currents. The only difference between currents in windings a and b of Fig.

10.16 is a phase displacement of $\pi/2$ electrical radians, so that the current in winding b is

$$i_b = \sqrt{2}I_2 \sin(\omega - \omega_m)t. \tag{10.9.6}$$

Using the connection matrix for the rotating axes of Fig. 10.16 and the stationary axes of Fig. 10.19 gives

$$i_a = i_d \cos\theta - i_q \sin\theta \tag{10.9.7}$$

where $\theta = -\omega_m t$ (an increase in θ in Fig. 10.16 is opposite to ω_m in Fig. 10.19) for constant speed and is the displacement between winding a and winding d, measured from time $t = 0$. Accordingly

$$i_a = i_d \cos\omega_m t + i_q \sin\omega_m t. \tag{10.9.8}$$

After substituting for i_d and i_q, expansion yields

$$i_a = \sqrt{2}I_2 \cos\omega t \cos\omega_m t + \sqrt{2}I_2 \sin\omega t \sin\omega_m t. \tag{10.9.9}$$

For this to be true, inspection of the two equations of i_a results in

$$i_d = \sqrt{2}I_2 \cos\omega t \quad \text{and} \quad i_q = \sqrt{2}I_2 \sin\omega t. \tag{10.9.10}$$

Therefore

$$i_d + ji_q = \sqrt{2}I_2(\cos\omega t + j \sin\omega t) = \sqrt{2}I_2 e^{j\omega t}. \tag{10.9.11}$$

The angular frequency ω is what would be expected for currents i_q and i_d in a stationary reference frame. But

$$I_2 e^{j\omega t} = \mathbf{I}_2 \tag{10.9.12}$$

which is the phasor current, so

$$\mathbf{I}_2 = \frac{i_d + ji_q}{\sqrt{2}}. \tag{10.9.13}$$

Similarly

$$\mathbf{I}_1 = \frac{i_D + ji_Q}{\sqrt{2}} \quad \text{and} \quad \mathbf{V}_1 = \frac{v_D + jv_Q}{\sqrt{2}}. \tag{10.9.14}$$

The stator alternating current can be defined by

$$i_D = \sqrt{2}I_1 e^{j(\omega t - \psi)} \tag{10.9.15}$$

where ψ is an arbitrary phase angle, so that by differentiating with respect to time

$$pi_D = j\omega\sqrt{2}I_1 e^{j(\omega t - \psi)}. \tag{10.9.16}$$

Therefore $p = j\omega$ for steady-state operation.

Again, for example, from eq. (10.9.10) $i_q = \sqrt{2}I_2 \sin\omega t$. Therefore

$$pi_q = \omega\sqrt{2}I_2 \cos\omega t = j\omega\sqrt{2}I_2 \sin\omega t \tag{10.9.17}$$

and so $p = j\omega$. Thus, in the steady state, the operator p is replaced by $j\omega$ throughout.

A. VOLTAGE AND CURRENT RELATIONS

From eq. (10.9.14) the stator phase voltage is

$$\mathbf{V}_1 = \frac{v_D + jv_Q}{\sqrt{2}}$$

so that, from the matrix eq. (10.9.3) and the fact that $p = j\omega$ for steady conditions

$$\mathbf{V}_1 = (1/\sqrt{2})[(R_1 + j\omega L_S)i_D + j\omega M i_d + j(R_1 + j\omega L_S)i_Q + j \times j\omega M i_q]$$
$$= (1/\sqrt{2})(R_1 + j\omega L_S)(i_D + ji_Q) + (1/\sqrt{2})j\omega M(i_d + ji_q)$$

or

$$\mathbf{V}_1 = (R_1 + j\omega L_S)\mathbf{I}_1 + j\omega M\mathbf{I}_2. \tag{10.9.18}$$

For the rotor circuit

$$0 = (1/\sqrt{2})(v_d + jv_q)$$
$$= j\omega M i_D + M\omega_m i_Q + (R_2 + j\omega L_R)i_d + L_R\omega_m i_q - jM\omega_m i_D$$
$$\quad + j \times j\omega M i_Q - jL_R\omega_m i_d + j(R_2 + j\omega L_R)i_q$$
$$= j\omega M(i_D + ji_Q) - jM\omega_m(i_D + ji_Q) + (R_2 + j\omega L_R)(i_d + ji_q) - jL_R\omega_m(i_d + ji_q)$$
$$= M(j\omega - j\omega_m)\mathbf{I}_1 + \{R_2 + (j\omega - j\omega_m)L_R\}\mathbf{I}_2$$

or

$$0 = Msj\omega\mathbf{I}_1 + (R_2 + sj\omega L_R)\mathbf{I}_2 \tag{10.9.19}$$

where s is the slip defined by $s = (\omega - \omega_m)/\omega$. This provides two simultaneous equations

$$\mathbf{V}_1 = (R_1 + j\omega L_S)\mathbf{I}_1 + j\omega M\mathbf{I}_2 \tag{10.9.20}$$

and

$$0 = j\omega M\mathbf{I}_1 + (R_1/s + j\omega L_R)\mathbf{I}_2 \tag{10.9.21}$$

from which the currents may be calculated.

A circuit diagram can represent these two equations and can be developed further than Fig. 10.21 to arrive at an equivalent circuit, from which the complete steady-state performance can be determined. There are two major points to note. One is that the circuit was developed on a per phase basis, so it is quite general. Second, it will be noted that from eq. (10.9.11)

$$i_d + ji_q = \sqrt{2}I_2 e^{j\omega t}.$$

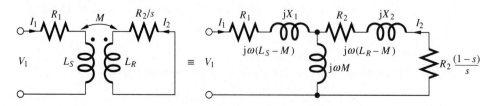

Figure 10.21 Induction motor circuit equivalents.

These quantities have an angular frequency ω, that is, they refer to the stator frequency. Consequently the referral of rotor quantities to the stator side using power invariance, which has to be done in the traditional theory, is accomplished here naturally through the generalized theory concepts.

B. TORQUE AND POWER

The total electromagnetic torque is

$$T_e = \underline{i}''[G_c]\underline{i}'$$

and expanding this from the matrix equations

$$T_e = [i_D\ i_Q\ i_d\ i_q] \cdot \begin{bmatrix} 0 & 0 & 0 & 0 \\ 0 & 0 & 0 & 0 \\ 0 & M & 0 & L_R \\ -M & 0 & -L_R & 0 \end{bmatrix} \cdot \begin{bmatrix} i_D \\ i_Q \\ i_d \\ i_q \end{bmatrix}$$

so

$$T_e = M(i_d i_Q - i_q i_D).$$

Now $\sqrt{2}\mathbf{I}_1 = i_D + ji_Q$ and the rotor winding current conjugate is $\sqrt{2}\mathbf{I}_2^* = i_d - ji_q$. Therefore

$$2\mathbf{I}_1\mathbf{I}_2^* = (i_d i_D + i_q i_Q) + j(i_d i_Q - i_q i_D).$$

Thus

$$T_e = 2\text{Im}(M\mathbf{I}_1\mathbf{I}_2^*)$$

and the torque per phase is

$$T_{eph} = \text{Im}(M\mathbf{I}_1\mathbf{I}_2^*). \tag{10.9.22}$$

From the voltage and current relations of eq. (10.9.21)

$$\mathbf{I}_1 = \frac{-(R_2 + js\omega L_R)\mathbf{I}_2}{js\omega M}.$$

Therefore

$$T_{eph} = \text{Im}\left(M\mathbf{I}_2\mathbf{I}_2^*\frac{(-R_2 - js\omega L_R)}{js\omega M}\right) = \text{Im}\left(\frac{(-R_2 - js\omega L_R)}{js\omega}I_2^2\right).$$

That is

$$T_{eph} = \frac{1}{\omega}I_2^2\frac{R_2}{s}. \tag{10.9.23}$$

In real machines this might be multiplied by the number of pole pairs.

The gross mechanical power output per phase is

$$P_m = T_{eph}\omega_m = I_2^2 R_2\frac{1-s}{s} \tag{10.9.24}$$

and this could have been deduced from the fictitious load resistance in the equivalent circuit of Fig. 10.21. The rotor winding ohmic loss per phase is

$$P_2 = I_2^2 R_2 \tag{10.9.25}$$

and the gap power (the power transferred across the air gap through the media of the electric and magnetic fields) is then

$$P_1 = P_2 + P_m = I_2^2 \frac{R_2}{s} = \frac{P_2}{s}. \tag{10.9.26}$$

This gives the important power relations

$$P_1 : P_2 : P_m = 1 : s : 1 - s. \tag{10.9.27}$$

Accordingly, we have used a generalized machine, applied the constraints of an induction motor, used a transformation, and the results have been the same as those from traditional theory. Further usefulness lies in the fact that transient performance can be determined also, and if power or torque values were required it would not be necessary to use any transformation; power invariance holds.

10.10. POLYPHASE SYNCHRONOUS MOTOR

The polyphase synchronous motor can have its field winding, which is excited by direct current, on either the rotor or stator member. For the convenience of analysis it will be considered to be on the stator. The rotor is then round and its polyphase windings are balanced.

The stator poles can be salient or otherwise. The former, as the more general case, is taken here and the latter will be treated as a special case. By transformations to fixed axes (the commutator primitive) the synchronous machine is reduced to that shown in Fig. 10.22, where it will be noted that there are no damper windings. It is in its simplest form. The transient equations can be written down by inspection to be

$$\begin{bmatrix} v_f \\ v_d \\ v_q \end{bmatrix} = \begin{bmatrix} R_f + L_f p & M_d p & 0 \\ M_d p & R_d + L_d p & -L_q \omega_m \\ M_d \omega_m & L_d \omega_m & R_q + L_q p \end{bmatrix} \cdot \begin{bmatrix} i_f \\ i_d \\ i_q \end{bmatrix}. \tag{10.10.1}$$

The only constraints here are that the rotor winding is balanced, so

$$R_d = R_q \triangleq R. \tag{10.10.2}$$

We want to develop the steady-state voltage and power equations from the generalized theory in order to see that they are the same as those obtained in Chapter 8.

10.10.1. Steady-state Analysis

In the steady state, that is, synchronous speed operation, no currents flow in the damper windings. Thus, if there were damper windings, for this analysis they would be redundant and eq. (10.10.1) can be used.

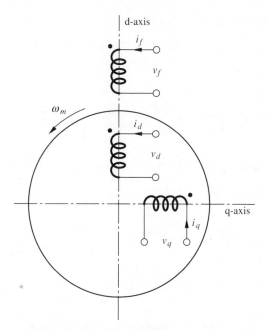

Figure 10.22 Synchronous motor commutator primitive.

The alternating current in the real machine is alternating at supply frequency, so that, in the two-phase, rotating-axis configuration (slip-ring primitive), the phase currents can be written

$$i_a = \hat{I} \cos\omega t, \quad \text{or} \quad i_a = \sqrt{2}I \cos\omega t \quad \text{and} \quad i_b = \sqrt{2}I \sin\omega t \quad \textbf{(10.10.3)}$$

where $\hat{I}$ and I are the maximum and rms values of the ac phase currents, respectively. Using the connection matrix for the rotating axes of Fig. 10.16 and the stationary axes of Fig. 10.22, the commutator-primitive, rotor winding currents are

$$i_d = i_a \cos\theta + i_b \sin\theta \quad \text{and} \quad i_q = -i_a \sin\theta + i_b \cos\theta. \quad \textbf{(10.10.4)}$$

Substituting for i_a and i_b in the above equation

$$i_d = \sqrt{2}I \cos\omega t \cos\theta + \sqrt{2}I \sin\omega t \sin\theta$$

or

$$i_d = \sqrt{2}I \cos(\theta - \omega t) \quad \text{and} \quad i_q = \sqrt{2}I \sin(\theta - \omega t). \quad \textbf{(10.10.5)}$$

For the 2-pole synchronous motor the steady-state speed is given by $\omega_m = \dot{\theta} = \omega$, since the rotor motion is at synchronous speed. Integrating this expression yields

$$\theta = \omega t + \delta \quad \textbf{(10.10.6)}$$

where δ is the phase difference between the stator exciting field and the armature field and is known as the *load angle*. This gives

$$i_d = \sqrt{2}I \cos(\omega t + \delta - \omega t) = \sqrt{2}I \cos\delta. \qquad (10.10.7)$$

Similarly

$$i_q = \sqrt{2}I \sin\delta. \qquad (10.10.8)$$

These are constant values, so that

$$pi_d = pi_q = 0.$$

Therefore the operator p must be put equal to zero in eq. (10.10.1) for steady-state operation.

Using the inverse of the connection matrix and allowing δ to be zero (this occurs at no load),

$$i_a = i_d \cos\theta - i_q \sin\theta = i_d \cos\omega t - i_q \sin\omega t.$$

In exponential form this is

$$i_a = \mathrm{Re}(i_d e^{j\omega t} + ji_q e^{j\omega t}) = \mathrm{Re}((i_d + ji_q)e^{j\omega t})$$

but

$$I_a = \mathrm{Re}(\sqrt{2}Ie^{j\omega t})$$

therefore in phasor form[8]

$$\mathbf{I} = I_d + jI_q \qquad (10.10.9)$$

and these are the rms values used in traditional theory. Here

$$\mathbf{I} = (1/\sqrt{2})(i_d + ji_q), \quad I_d = i_d/\sqrt{2}, \quad \text{and} \quad I_q = i_q/\sqrt{2}. \qquad (10.10.10)$$

Similarly

$$\mathbf{V} = (1/\sqrt{2})(v_d + jv_q). \qquad (10.10.11)$$

Since $\omega_m = \omega$, let $\omega_m L_q = X_q$, and $\omega_m L_d = X_d$, where X_d is the synchronous reactance on the direct axis and X_q is the synchronous reactance on the quadrature axis. The voltage eq. (10.10.1) in the steady state becomes

$$\begin{bmatrix} V_f \\ v_d \\ v_q \end{bmatrix} = \begin{bmatrix} R_f & 0 & 0 \\ 0 & R & -X_q \\ M_d\omega & X_d & R \end{bmatrix} \cdot \begin{bmatrix} I_f \\ i_d \\ i_q \end{bmatrix}. \qquad (10.10.12)$$

A. PHASOR DIAGRAM

The applied ac voltage is given by eq. (10.10.11) and substituting from the matrix eq. (10.10.12)

$$\mathbf{V} = (1/\sqrt{2})(Ri_d - X_q i_q + j\omega M_d I_f + jX_d i_d + jRi_q)$$

---

[8] I_d and I_q are not phasors but are represented on the argand diagram as space vectors.

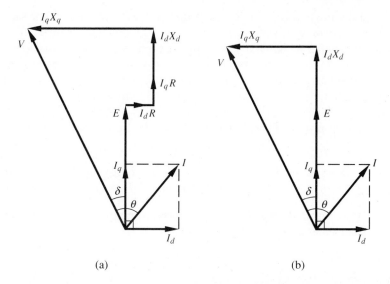

(a)   (b)

Figure 10.23 Synchronous machine phasor diagram.

or

$$V = E + RI_d + jRI_q + jX_dI_d - X_qI_q \qquad (10.10.13)$$

where $E = (1/\sqrt{2})jM_d\omega I_f$. The phasor diagram can be drawn from an inspection of the voltage equation.

This voltage E is the motionally induced emf or back emf per phase, which can only be measured directly when the machine is running as a generator and the ac winding is open-circuited. A phasor diagram is shown in Fig. 10.23a and, as the armature winding resistance drop is much smaller than the reactance drop in practice, the former can be neglected to give the phasor diagram drawn in Fig. 10.23b.

B. TORQUE AND POWER

Power invariance means that it is unnecessary to transform back to real quantities once the power and torque have been obtained from the commutator primitive. The values would be the same for both systems.

The total torque follows from

$$T_e = \underline{i}''[G_c]\underline{i}'$$

or

$$T_e = [I_f \, i_d \, i_q] \cdot \begin{bmatrix} 0 & 0 & 0 \\ 0 & 0 & -L_q \\ M_d & L_d & 0 \end{bmatrix} \cdot \begin{bmatrix} I_f \\ i_d \\ i_q \end{bmatrix}.$$

That is

$$T_e = i_q I_f M_d + i_q i_d (L_d - L_q). \qquad (10.10.14)$$

$i_q i_d(L_d - L_q)$ is known as the reluctance torque and arises because of the saliency. If there were a uniform air gap, that is, no saliency, the reluctances of the direct- and quadrature-axes paths are equal, $L_d = L_q$, and the reluctance torque would be zero. It is also clear to see that, at synchronous speed, there would be a torque output even if there were no field excitation as long as some saliency provides different values for L_d and L_q. With saliency there is a greater torque and hence greater power for a given excitation.

The power output per phase, independent of the number of poles, is

$$P_m = \tfrac{1}{2} T_e \omega = \tfrac{1}{2} \{ i_q M_d \omega I_f + i_q i_d (\omega L_d - \omega L_q) \}$$
$$= \tfrac{1}{2} \{ \sqrt{2} I_q \sqrt{2} E + \sqrt{2} I_q \sqrt{2} I_d (X_d - X_q) \} \qquad (10.10.15)$$
$$= I_q E + I_q I_d (X_d - X_q).$$

It is necessary to eliminate the currents and this is done by obtaining their relations from the phasor diagram of Fig. 10.23b. That is

$$V \sin\delta = X_q I_q \qquad \text{and} \qquad E + X_d I_d = V \cos\delta.$$

Substituting in eq. (10.10.15) for I_d and I_q

$$P_m = \frac{VE \sin\delta}{X_q} + \frac{V \sin\delta}{X_q} \frac{(V \cos\delta - E)}{X_d} (X_d - X_q)$$
$$= \frac{VE \sin\delta}{X_q} + \frac{V^2 \sin 2\delta}{2} \left(\frac{1}{X_q} - \frac{1}{X_d} \right) - VE \sin\delta \left(\frac{1}{X_q} - \frac{1}{X_d} \right) \qquad (10.10.16)$$
$$= \frac{VE \sin\delta}{X_d} + \frac{V^2}{2} \left(\frac{1}{X_q} - \frac{1}{X_d} \right) \sin 2\delta.$$

This can be recognized as the power equation obtained from traditional theory with δ as the load angle. The first term is the power obtained from an equivalent smooth stator machine and the second term represents the power from saliency and is present with or without excitation of the field winding.

C. DAMPER WINDINGS

To attempt to stabilize the speed of a synchronous machine, a winding, like the cage of an induction motor, is often placed on the same member as the field winding. This is the *damper winding*. If the speed were below synchronous speed, an induction torque resulting from currents in the cage winding would tend to accelerate the machine toward synchronism, just as in an induction motor. Similarly, above synchronism, induction generator action would provide a braking torque to reduce the speed. The equilibrium speed is synchronous speed, at which the damping winding is inoperative, because it does not experience a change of flux linkage and accordingly has no currents in the bars. A solid pole shoe would play the same part as the cage.

Just as in the case of the induction motor, the cage can be transformed to have windings on the direct and quadrature axes of the generalized machine. This is

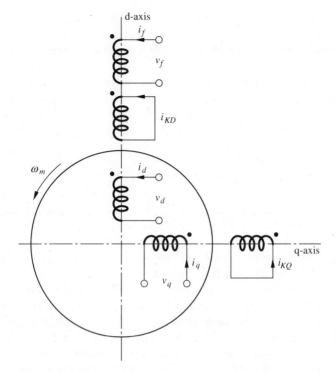

Figure 10.24 Commutator primitive with damper windings.

depicted in Fig. 10.24. From an inspection of this figure we can express the operational impedance as

$$[Z] = \begin{bmatrix} R_f + L_f p & M_d p & 0 & M_d p & 0 \\ M_d p & R_{KD} + L_{KD} p & 0 & M_d p & 0 \\ 0 & 0 & R_{KQ} + L_{KQ} p & 0 & M_q p \\ M_d p & M_d p & -M_q \omega_m & R + L_d p & -L_q \omega_m \\ M_d \omega_m & M_d \omega_m & M_q p & L_d \omega_m & R + L_q p \end{bmatrix}. \quad (10.10.17)$$

For more exact computational purposes the bars of the damper winding would be treated separately and the approximation that the mutual inductances between windings d and KD, and d and f are the same would not be made.

10.11. SUMMARY

The generalized theory has been developed for a particular set of rotating machines by employing the principles of dynamic circuit theory laid down in Chapter 2. The real machines are far too complex, with their conductors placed discreetly in slots or wound around poles of magnetic material, to produce an exact mathematical description. Simplification in the form of primitives has provided a model where calculations can be made. It is perhaps amazing that results from the model agree as well as they do with tests on the real machines.

The aim has been to describe the electrical system by a set of linear differential equations with constant coefficients. With the impedance parameters known by calculation or measurement, the Laplace transform technique can be applied to obtain a solution. This is inverted back into the time domain and, with further transformations if necessary, the solution is obtained in terms of the real machine variables. Although the processes are many they remain simple and work well.

The idealized model employs a number of assumptions. The flux distribution has been assumed to be sinusoidal, the magnetic circuit produces linear relationships, the rotor windings are balanced, the rotor is a smooth cylinder that rotates at constant speed, and transformations, with power invariance, are made to align the winding axes along two fixed axes at right angles. If the speed of the motor is not constant the electrical system equations are nonlinear and operational methods cannot be used to find a solution. It is usual to use the computer in this case.

The basic ac machine slip-ring primitive needs to be transformed to a mathematical model called the commutator primitive. This implies that the rotor winding axis is stationary although the rotor conductors move. It is assumed that the brushes are stationary. If the winding axes are fixed then the inductances are constant. Therefore the resulting differential equations are linear with constant coefficients. It is a relatively simple matter to determine the machine performance from this model.

There are three basic equations governing the behavior of rotating machines. In general terms they are

$$\underline{v} = [R]\underline{i} + [Lp]\underline{i} + [G]\dot{\theta}\underline{i}$$

$$T_e = \underline{i}^t[G]\underline{i}$$

and $$T_e = J\ddot{\theta} + D\dot{\theta} + T_m.$$

The first equation can be formulated systematically.

Most of the common machines can be analyzed this way. Only a few lie too far beyond the imposed limitations to be included within the scope of the generalized theory. Accordingly they must be given individual treatment. The hysteresis motor, as its name implies, relies on the use of hard magnetic materials for the conversion of hysteresis energy to mechanical work. The multivalued functions of the nonlinearity of hysteresis precludes a general treatment. The single-phase alternator has unbalanced windings and the inductor alternator depends on rotor saliency for its operation, so they always produce nonconstant coefficients in the differential equations.

The dc machine has been dealt with in its most elementary form. The polyphase induction and synchronous machines have singular constraints in their common forms and have been used here as examples to consolidate the generalized theory and show the equivalence of results obtained by traditional steady-state theories.

10.12. PROBLEMS

Section 10.3

10.1. From the consideration of the slip-ring primitive and energy principles prove that the force on a conductor is $F_e = Bli$, where $F_e =$ force of electromagnetic origin, $B =$ flux density component perpendicular to the conductor, $l =$ effective length of conductor in the field, and $I =$ current in the conductor.

10.2. Consider the slip-ring primitive shown in Fig. 10.4. Assume the system to be conservative, the self-inductances to be constant, but the mutual inductance to be given by

$$L_{12} = L_{21} = \sum_{n=odd} \frac{M_d}{n^4} \cos n\theta$$

for $n = 1, 3, 5 \ldots,$ where $M_d =$ constant.

The machine constraints are $i_2 = I =$ constant, $\theta = \omega_m t$, $\omega_m =$ constant, and $i_1 = 0$ (i.e., the stator winding is open-circuited). **(a)** What are the winding flux linkages in terms of currents and machine parameters? **(b)** What is the instantaneous value of the stator winding terminal voltage v_1? **(c)** What is the ratio of the amplitudes of the nth harmonic and fundamental components of the stator-winding voltage? **(d)** What is the ratio of the third harmonic and fundamental amplitudes of the stator winding voltage? **(e)** What type of conventional machine does this slip-ring primitive represent? (*Answer:* (c) $1 : n^3$.)

10.3. The multiply-excited, slip-ring primitive in Fig. P10.3 represents a two-phase, round rotor, synchronous machine. The machine constraints are

$$L_a = L_b = \text{constant}, \quad M_{af} = M_d \cos\theta, \quad M_{bf} = -M_d \sin\theta,$$

$$i_f = I_f = \text{constant}, \quad I_a = \hat{I} \sin\omega t, \quad i_b = \hat{I} \cos\omega t, \quad \dot\theta = \omega_m,$$

and $\theta = \omega t - \delta$, where $\delta =$ constant.

Determine **(a)** the values of the stator-winding terminal voltages, **(b)** the instantaneous value of the electromagnetic torque, and **(c)** the average value of the electromagnetic torque. (*Answer:* (c) $-M_d I_f \hat{I} \cos\delta$.)

10.4. Let the slip-ring primitive, shown in Fig. P10.3, represent a salient-pole, two-phase, synchronous machine. The constraints are

$$L_a = L_0 + L_2 \cos 2\theta, \quad L_b = L_0 - L_2 \cos 2\theta, \quad M_{ab} = -L_2 \sin 2\theta,$$

$$M_{af} = M_d \cos\theta, \quad M_{bf} = -M_d \sin\theta, \quad i_f = I_f = \text{constant},$$

$$i_a = \hat{I} \sin\omega t, \quad i_b = \hat{I} \cos\omega t, \quad \dot\theta = \omega_m, \quad \text{and} \quad \theta = \omega t - \delta,$$

where $\delta =$ constant. $L_f, L_0, L_2,$ and M_d are constants.

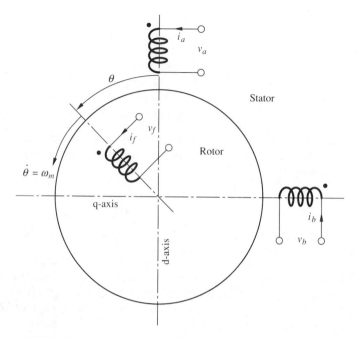

Figure P10.3

(a) Find an expression for the instantaneous value of the electromagnetic torque. (b) What is the average value of the torque? (c) What is the value of the torque if I_f is zero?

(*Answer:* (c) $L_2\hat{I}^2 \sin\delta$.)

10.5. Consider the slip-ring primitive shown in Fig. 10.4. Coil 1 is excited by a current source $I(t)$ and coil 2 is short-circuited. The machine constraints are $L_{12} = L_{21} = M_d \cos\theta$, $R_2 = 0$ and L_{11}, L_{22}, and M_d are constants. (a) Prove that the flux linkages of coil 2 must always be constant and independent of θ, if coil 2 has zero resistance. (b) Find the instantaneous value of the electromagnetic torque as a function of θ and $I(t)$, if the initial conditions are created by (i) $\theta = 0$ and coil 2 open-circuited, (ii) current $I(t)$ in coil 1 raised to some steady value I_1, then (iii) coil 2 short-circuited while $I(t) = I_1$ at $\theta = 0$. Use the result of part (a).

10.6. The slip-ring primitive depicted in Fig. P10.3 represents a constant-energy storage system, if all the coils are superconductive, that is, they exhibit zero resistance. Initially at $\theta = 0$, coils a and b are short-circuited and coil f is supplied from a dc source. When $i_f = I$, coil f is short-circuited and the source is removed. If the constraints are that coils a and b are similar and $L_{af} = M \cos\theta$, $L_{fb} = -M \sin\theta$ and L_a, L_b, L_f, and M are constants, show that the electromagnetic torque is zero for all θ.

10.7. The slip-ring primitive shown in Fig. 10.3 has four coils. The constraints are that the two stator coils are identical but in quadrature. These constraints also apply to the two rotor coils. The medium has constant permeability and the mutual inductance relations are either cosinusoidal or sinusoidal. Determine the flux linkages of each coil in circuit terms and find the electromagnetic torque in terms of currents and rotor position.

10.8. The singly-excited primitive shown in Fig. P10.8 represents a single-phase reluctance machine. Assume the stator coil inductance varies according to the expression $L = L_0 + L_2 \cos2\theta$. **(a)** Determine the electromagnetic torque in terms of rotor position. **(b)** Calculate the average torque for the constraints $\dot\theta = \omega$ and $i = \hat{I} \cos(\omega t + \delta)$. **(c)** For the same conditions set in part (b) find the maximum possible average torque.
(*Answer:* (c) $\hat{I}^2 L_2/4$.)

10.9. The slip-ring primitive shown in Fig. 10.4 has the constraints

$$L_{12} = L_{21} = M_d \cos\theta, \quad \theta = \omega_m t + \delta,$$

$$i_2 = I_2 \sin\omega_2 t, \quad i_1 = I_1 \sin\omega_1 t + I_3 \sin3\omega_1 t.$$

$L_{11}, L_{22}, M_d, I_1, I_2, I_3, \omega_m, \omega_1, \omega_2$, and δ are constants.

Find the values of ω_m at which the machine can produce an average torque. Find the average torque for each value of ω_m.
(*Answer:* $\pm\omega_1 \pm \omega_2$.)

10.10. The slip-ring primitive shown in Fig. P10.3 has two identical stator coils in quadrature. Assume the permeability of the medium to be constant and let the mutual inductance relationships be sinusoidal. The constraints are $i_a =$

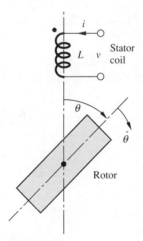

Figure P10.8

$I \cos\omega t$, $i_b = I \sin\omega t$, and $i_f = I_f =$ constant. Find the condition for average power conversion.
(*Answer:* $\omega_m = -\omega$.)

10.11. The slip-ring primitive shown in Fig. 10.3 has balanced windings on both the stator and rotor. The medium has constant permeability and the mutual inductances vary sinusoidally with rotor position. The constraints are

$$i_D = I_s \cos\omega t, \quad i_Q = I_s \sin\omega t, \quad i_a = I_r \cos\omega_r t, \quad i_b = I_r \sin\omega_r t,$$

and $\quad \theta = \omega_m t + \delta$.

Find the condition for average power conversion.
(*Answer:* $\omega_m = -(\omega - \omega_r)$.)

Section 10.5

10.12. From the considerations of the commutator primitive and energy principles prove that the force on a conductor is $F_e = Bli$, where $F_e =$ force of electromagnetic origin, $B =$ flux density component perpendicular to the conductor, $l =$ effective length of conductor in the field, and $i =$ current in the conductor.

10.13. Determine the voltage equations of the commutator primitive shown in Fig. P10.13. From the G matrix find the electromagnetic torque in terms of winding currents and machine parameters.
(*Answer:* $M_d i_a i_f$.)

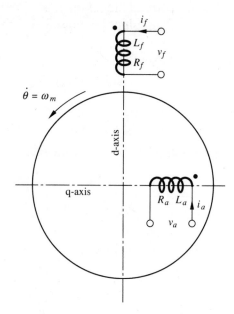

Figure P10.13

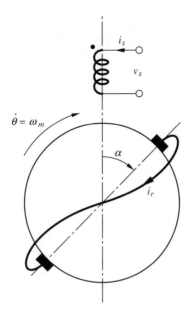

Figure P10.14

10.14. The repulsion motor, illustrated in Fig. P10.14, is a single-phase, ac machine. The rotor winding is provided with a commutator, whose brushes are short-circuited and set at some angle α. Using the assumptions accorded the commutator primitive determine **(a)** the instantaneous value of the electromagnetic torque in terms of currents and circuit parameters. **(b)** For the constraints

$$v_s = V \cos\omega t, \quad i_s = I \cos(\omega t + \theta_s), \quad i_r = I_r \cos(\omega t + \theta_r)$$

and $\quad \dot{\theta} = \omega_m = \text{constant}$

find the condition for average power conversion and the value of the average torque.

(*Answer:* (a) $- M i_s i_r \sin\alpha$, (b) $-\frac{1}{2} M I I_r \sin\alpha \cos(\theta_s - \theta_r)$.)

10.15. A commutator primitive is shown in Fig. P10.15. Neglect saliency and determine **(a)** the voltage equations in matrix form; **(b)** the instantaneous value of the electromagnetic torque as a function of terminal currents, machine parameters, and brush displacement α; **(c)** the brush position for maximum torque; and **(d)** whether the machine acts as a motor or a generator. (*Answer:* (c) 90°, (d) generator.)

Section 10.7

10.16. Figure P10.16 represents a model of a dc machine with the brush axis displaced from the quadrature axis by an angle α. There is no field winding and

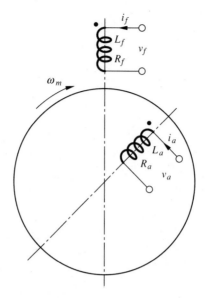

Figure P10.15

the armature winding is fed with a constant direct current I_a. Find an expression for the armature winding self-inductance in terms of α. Determine the torque expression as a function of L_d and L_q, where L_d = winding inductance in the direct axis ($\alpha = \pi/2$) and L_q = winding inductance in the quadrature axis ($\alpha = 0$). Is there an electromagnetic torque to cause motoring action? (*Answer:* Yes.)

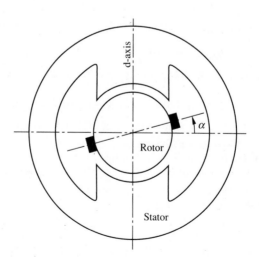

Figure P10.16

10.17. Consider the commutator primitive of a 2-pole cumulatively-compounded dc motor without interpoles. Determine the instantaneous values of the voltages and electromagnetic torque in terms of the currents and machine parameters. Show that the steady-state armature winding current has the form

$$I_a = \frac{V}{(R_a + R_s) + \omega_m M_{as}} \left(1 - \frac{\omega_m M_{af}}{R_f} \right).$$

Section 10.8

10.18. The slip-ring primitive, shown in Fig. P10.18a, is transformed to a commutator primitive, shown in Fig. P10.18b. Derive the relations between the slip-ring primitive and the commutator primitive currents, voltages, and inductances. That is, find the functions

$$\underline{i}_{fab} = f_1(\underline{i}_{fdq}), \quad \underline{v}_{fab} = f_2(\underline{v}_{fdq}) \quad \text{and} \quad [L_{fab}] = f_3[L_{fdq}].$$

Determine by inspection the G matrices of the two primitives.

Section 10.9

10.19. The stator and rotor windings of a 3-phase induction motor are symmetric and the machine has a uniform air gap. Determine the inductance matrix of the commutator primitive in terms of the phase quantities.

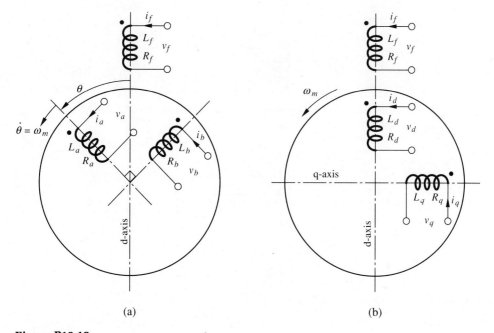

(a)   (b)

Figure P10.18

10.20. Figure 10.3 can represent the slip-ring primitive of a 2-phase induction motor if the rotor coils are short-circuited. The machine constraints are

$$i_D = I\cos\omega t, \quad i_Q = I\sin\omega t, \quad v_a = v_b = 0, \quad \text{and} \quad \theta = \omega_m t + \delta.$$

Derive the differential equations for the rotor coil currents. From these equations and the condition for average power conversion, determine the operating speed ranges.

(*Answer:* $\omega_m \neq \omega$.)

10.21. The commutator primitive depicted in Fig. 10.19 represents a single-phase induction motor if the stator coil on the direct axis is sinusoidally excited while the stator coil on the quadrature axis remains unexcited and open-circuited. For these conditions prove that the steady-state electromagnetic torque has the form

$$T_e = \frac{2I_2^2 R_2}{\omega_m}\frac{1-s}{s} - \frac{2I_2^2 R_2}{\omega_m}\frac{1-s}{s(2-s)}.$$

10.22. The commutator primitive shown in Fig. P10.22 represents a single-phase, series commutator motor. Prove that the steady-state electromagnetic torque has the average value

$$T_e = \frac{V^2 M_d}{(R_f + R_a + M_d\omega_m)^2 + \omega^2(L_f + L_a)^2}.$$

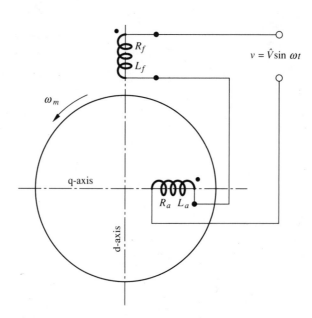

Figure P10.22

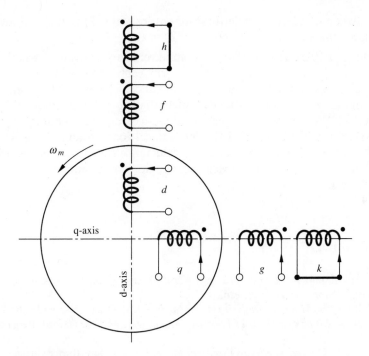

Figure P10.23

Section 10.10

10.23. Figure P10.23 represents the commutator primitive of a synchronous machine. Coils d, q are the armature windings, coils f, g are the excitation windings and coils h, k are the damper windings. **(a)** Find the flux linkages of each winding in terms of the inductances and currents. **(b)** Write down the

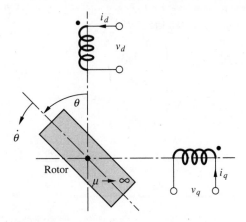

Figure P10.24

voltage equations and determine the entries of the G matrix. **(c)** Show that the electromagnetic torque is given by $T_e = \lambda_d i_q - \lambda_q i_d$.

10.24. Figure P10.24 represents a two-phase reluctance machine whose coils are identical and have 1000 effective turns each. Assume that the coils have negligible resistance and mutual inductance. The machine is connected to a 2-phase, 100-V, 60-Hz supply such that $v_d = -141 \sin\omega t$ and $v_q = 141 \cos\omega t$. The air-gap reluctance for the d-axis coil has the form $\mathcal{R}_d = (6 - 4 \cos 2\theta) \times 10^6$ AT/Wb and the mechanical loading and speed is such that $\theta = \omega t - \pi/4$. Find **(a)** the air-gap reluctance for the q-axis coil, **(b)** the instantaneous value of the electromagnetic torque, and **(c)** the average mechanical power output.

(*Answer:* (c) 105 W.)

10.13. BIBLIOGRAPHY

Adkins, Bernard, and R. G. Harley. *The General Theory of Alternating Current Machines.* London: Chapman and Hall, Ltd., 1975.

Gibbs, W. J. *Electric Machine Analysis Using Matrices.* London: Pitman and Sons, Ltd., 1962.

Hancock, N. N. *Matrix Analysis of Electrical Machinery.* 2nd ed. Oxford: Pergamon Press, 1974.

Jones, C. V. *The Unified Theory of Electrical Machines.* London: Butterworths, 1967.

Krause, Paul C. *Analysis of Electric Machinery.* New York: McGraw-Hill Book Co., 1986.

Kron, G. *Tensors for Circuits.* 2nd ed. New York: Dover, 1959.

Meisel, Jerome. *Principles of Electromechanical Energy Conversion.* New York: McGraw-Hill Book Co., 1966.

Messerle, Hugo K. *Dynamic Circuit Theory.* Oxford: Pergamon Press, 1965.

Morgan, A. T. *General Theory of Electrical Machines.* London: Heyden and Son, Ltd., 1979.

O'Kelly, D., and S. Simmons. *Introduction to Generalized Electrical Machine Theory.* London: McGraw-Hill Publishing Co., Ltd., 1968.

Seely, S. *Electromechanical Energy Conversion.* New York: McGraw-Hill Book Co., 1962.

Woodson, H. H., and J. R. Melcher. *Electromechanical Dynamics, Part I: Discrete Systems.* New York: John Wiley & Sons, Inc., 1968.

General Bibliography

Acarnley, P. P. *Stepping Motors,* 2nd ed. IEE Control Engineering Series 19: Peter Peregrinus, Ltd., 1984.

Adkins, Bernard, and W. J. Gibbs. *Polyphase Commutator Machines.* Cambridge University Press, 1951.

Adkins, Bernard, and R. G. Harley. *The General Theory of Alternating Current Machines.* London: Chapman and Hall, Ltd., 1975.

Alger, P. L. *Induction Machines,* 2nd ed. New York: Gordon and Breach, 1970.

Anderson, Leonard R. *Electric Machines and Transformers.* Reston, Va.: Reston Publishing Co., Inc., 1981.

Bergseth, F. Robert, and S. S. Venkata. *Introduction to Electric Energy Devices.* Englewood Cliffs, N.J.: Prentice-Hall, Inc., 1987.

Bewley, L. V. *Alternating Current Machinery.* New York: Macmillan Co., 1949.

Blume, L. F., et al. *Transformer Engineering.* 2nd ed. New York: John Wiley & Sons, Inc., 1951.

Boldea, I., and S. A. Nasar. *Linear Motion Electromagnetic Systems.* New York: John Wiley & Sons, Inc., 1985.

Bose, B. K. *Power Electronics and AC Drives.* Englewood Cliffs, N.J.: Prentice-Hall, 1986.

Bose, B. K. *Microcomputer Control of Power Electronics and Drives.* New York: IEEE Press, 1987.

Brown, David, and E. P. Hamilton. *Electromechanical Energy Conversion.* New York: Macmillan Publishing Co., 1984.

Bumby, J. R. *Superconducting Rotating Electrical Machines.* Oxford: Clarendon Press, 1983.

Carr, Laurence. *The Testing of Electrical Machines.* London: MacDonald Co., Ltd., 1960.

Chalmers, B. J. *Electromagnetic Problems of AC Machines.* London: Chapman and Hall, Ltd., 1965.

Chapman, S. J. *Electric Machinery Fundamentals.* New York: McGraw-Hill Book Co., 1985.

Chaston, A. N. *Electric Machinery.* Reston, Virginia: Reston Publications Co., Inc., 1986.

Concordia, Charles. *Synchronous Machines Theory and Performance.* Schenectady, N.Y.: General Electric Co., 1951.

Del Toro, Vincent. *Electric Machines and Power Systems.* Englewood Cliffs, N.J.: Prentice-Hall, Inc., 1985.

Dewan, S. B., G. Slemon, and A. Straughen. *Power Semiconductor Drives.* New York: Wiley-Interscience, 1984.

Edwards, J. D. *Electrical Machines,* 2nd ed. London: Macmillan Education, Ltd., 1986.

Elgerd, Olle. *Basic Electric Power Engineering.* Reading, Mass.: Addison-Wesley Publishing Company, 1977.

El-Hawary, Mohamed E. *Principles of Electric Machines with Power Electronic Applications.* Englewood Cliffs, N.J.: Prentice-Hall, 1986.

Ellison, A. J. *Electromechanical Energy Conversion.* London: George G. Harrap & Co., Ltd., 1970.

Fitzgerald, A. E., C. Kingsley, Jr., and S. D. Umans. *Electric Machinery,* 4th ed. New York: McGraw-Hill Book Co., 1983.

Flanagan, William M. *Handbook of Transformer Applications.* New York: McGraw-Hill Book Co., 1986.

Franklin, A. C., and D. P. Franklin. *The J & P Transformer Book,* 11th ed. London: Butterworth's, 1983.

Fransua, A., and R. Magureanu. *Electrical Machines and Drive Systems.* Oxford: Technical Press, 1984.

Garik, M. L., and C. C. Whipple. *Alternating-Current Machines,* 2nd ed. New York: D. Van Nostrand, 1961.

Gehmlich, D. K., and S. B. Hammon. *Electromechanical Systems.* New York: McGraw-Hill Book Co., 1967.

Gibbs, W. J. *Electric Machine Analysis Using Matrices.* London: Pitman and Sons, Ltd., 1962.

Gingrich, H. W. *Electrical Machinery, Transformers and Control.* Englewood Cliffs, N.J.: Prentice-Hall, Inc., 1979.

Halliday, David, and Robert Resnick. *Fundamentals of Physics,* 2nd ed., Extended Version. New York: John Wiley & Sons, Inc., 1986.

Hancock, N. N. *Matrix Analysis of Electrical Machinery,* 2nd ed. Oxford: Pergamon Press, 1974.

Hindmarsh, J. *Electrical Machines and Their Applications,* 4th ed. Oxford: Pergamon Press, 1984.

Hindmarsh, J. *Electrical Machines and Drives: Worked Examples,* 2nd ed. Oxford: Pergamon Press, 1985.

Hochart, Bernhard (Ed). *Power Transformer Handbook.* London: Butterworth's, 1987.

Ireland, J. R. *Ceramic Permanent-Magnet Motors.* McGraw-Hill Book Co., 1968.

Jones, C. V. *The Unified Theory of Electrical Machines.* London: Butterworth's, 1967.

Jordan, H. E. *Energy Efficient Electric Motors and Their Applications.* New York: Van Nostrand Reinhold, 1983.

Karsai, K., D. Kerenyi, and L. Kiss. *Large Power Transformers.* Amsterdam: Elsevier, 1987.

Kenjo, T. *Stepping Motors and Their Microprocessor Controls.* Oxford: Clarendon Press, 1984.

Kenjo, T., and S. Nagamori. *Permanent-Magnet and Brushless DC Motors.* Oxford: Clarendon Press, 1985.

Kimbark, E. W. *Power System Stability: Synchronous Machines.* New York: Dover, 1968.

Kosow, Irving L. *Electric Machinery and Transformers.* Englewood Cliffs, N.J.: Prentice-Hall, Inc., 1972.

Kosow, Irving L. *Control of Electric Machines.* Englewood Cliffs, N.J.: Prentice-Hall, Inc., 1973.

Krause, Paul C. *Analysis of Electric Machinery.* New York: McGraw-Hill Book Co., 1986.

Kron, G. *Tensors for Circuits,* 2nd ed. New York: Dover, 1959.

Ku, Y. H. *Electric Energy Conversion.* New York: Ronald Press, 1959.

Kuo, B. C. *Step Motors.* St. Paul, Minn.: West Publishing Co., 1974.

Kusko, A. *Solid-State DC Motor Drives.* Cambridge, Mass.: MIT Press, 1969.

Laithwaite, E. R. (Ed). *Transport Without Wheels.* Boulder, Colo.: Westview Press, Inc., 1977.

Laithwaite, E. R. *A History of Linear Electric Motors.* London: Macmillan Educational, Ltd., 1987.

Langsdorf, A. S. *Principles of Direct-Current Machines.* New York: McGraw-Hill Book Co., 1959.

Lawrence, R. R., and H. E. Richard. *Principles of Alternating-Current Machinery.* 4th ed. New York: McGraw-Hill Book Co., 1953.

Leonhard, W. *Control of Electric Drives.* Berlin: Springer-Verlag, 1985.

Lindsay, J. F., and M. H. Rashid. *Electromechanics and Electric Machinery.* Englewood Cliffs, N.J.: Prentice-Hall, Inc., 1986.

Mablekos, Van E. *Electric Machine Theory for Power Engineers.* New York: Harper & Row Publishers, Inc., 1980.

Majmudar, Harit. *Electromechanical Energy Converters.* Boston: Allyn and Bacon, Inc., 1965.

Matsch, Leander W. *Capacitors, Magnetic Circuits and Transformers.* Englewood Cliffs, N.J.: Prentice-Hall, Inc., 1964.

Matsch, L. W., and J. D. Morgan. *Electromagnetic and Electromechanical Machines,* 3rd ed. New York: Harper & Row Publishers, Inc., 1986.

Meisel, Jerome. *Principles of Electromechanical Energy Conversion.* New York: McGraw-Hill Book Co., 1966.

Messerle, Hugo K. *Dynamic Circuit Theory.* Oxford: Pergamon Press, 1965.

Morgan, A. T. *General Theory of Electrical Machines.* London: Heyden and Son, Ltd., 1979.

Murphy, J. M. D. *Thyristor Control of AC Motors.* Oxford: Pergamon Press, 1973.

Nasar, S. A. *Electric Machines and Electromechanics.* Schaum's Outline Series in Engineering. New York: McGraw-Hill Book Co., 1981.

Nasar, S. A. *Electric Energy Conversion and Transmission.* New York: Macmillan Publishing Co., 1985.

Nasar, S. A. (Ed). *Handbook of Electric Machines.* New York: McGraw-Hill Book Co., 1987.

Nasar, S. A., and Boldea, I. *Linear Motion Electric Machines.* New York: Wiley-Interscience, 1976.

Nasar, S. A., and L. E. Unnewehr. *Electromechanics and Electric Machines,* 2nd ed. New York: John Wiley & Sons, Inc., 1983.

O'Kelly, D., and S. Simmons. *Introduction to Generalized Electrical Machine Theory.* London: McGraw-Hill Publishing Co., Ltd., 1968.

Poloujadoff, M. *The Theory of Linear Induction Machinery.* Oxford: Clarendon Press, 1980.

Sarma, M. S. *Synchronous Machines.* New York: Gordon and Breach, 1979.

Sarma, M. S. *Electric Machines.* Dubuque, Iowa: W.C.B. Publishers, 1985.

Say, M. G. *Introduction to the Unified Theory of Electromagnetic Machines.* London: Pitman Publishing, 1971.

Say, M. G. *Alternating Current Machines.* 5th ed. London: Pitman Publishing, 1983.

Say, M. G., and E. O. Taylor. *Direct Current Machines.* London: Pitman Publishing, 1980.

Seely, S. *Electromechanical Energy Conversion.* New York: McGraw-Hill Book Co., 1962.

Sen, P. C. *Thyristor DC Drives.* New York: Wiley-Interscience, 1981.

Siskind, C. S. *Direct-Current Machinery.* New York: McGraw-Hill Book Co., 1952.

Skilling, H. H. *Electromechanics.* New York: John Wiley & Sons, Inc., 1962.

Slemon, Gordon, and A. Straughen. *Electric Machines.* Reading, Mass.: Addison-Wesley Publishing Co., Inc., 1980.

Smeaton, Robert W. *Motor Applications and Maintenance Handbook,* 2nd ed. New York: McGraw-Hill Book Co., 1987.

Smith, Richard T. *Analysis of Electrical Machines.* New York: Pergamon Press, 1982.

Smith, Steve. *Magnetic Components: Design and Applications.* New York: Van Nostrand Reinhold Co., Inc., 1985.

Still, Alfred, and Charles S. Siskind. *Elements of Electrical Machine Design.* Tokyo: McGraw-Hill Book Co., 1954.

Teago, F. J. *The Commutator Motor.* London: Methuen & Co., Ltd., 1930.

Upson, A. R., and J. H. Batchelor. *Synchro Engineering Handbook.* London: Hutchison, 1965.

Van Valkenburgh, Nooger, and Neville, Inc. *Basic Synchros and Servomechanisms.* New York: John F. Rider Publisher, Inc., 1955.

Veinott, G. G. *Theory and Design of Small Induction Motors.* New York: McGraw-Hill Book Co., 1959.

Veinott, G. G., and J. E. Martin. *Fractional- and Subfractional-Horsepower Electric Motors.* 4th ed. New York: McGraw-Hill Book Co., 1986.

Walker, J. H. *Large Synchronous Machines.* Oxford: Oxford University Press, 1981.

White, David C., and H. H. Woodson. *Electromechanical Energy Conversion.* New York: John Wiley & Sons, Inc., 1959.

Woodson, H. H., and J. R. Melcher. *Electromechanical Dynamics, Part I: Discrete Systems.* New York: John Wiley & Sons, Inc., 1968.

Yamamura, S. *Theory of Linear Induction Motors,* 2nd ed. Tokyo: University of Tokyo Press, 1978.

Yang, S. J. *Low-Noise Electrical Motors.* Oxford: Clarendon Press, 1981.

——. *Carbon Brushes and Electrical Machines.* Morganite Carbonite, Ltd., 1961.

——. *Electrical Variable Speed Drives.* IEE Conference No. 93, 1972; Ibid, IEE Conference No. 179, 1979.

——. *Drives: Motors: Controls.* IEE Conference Proceedings. London, 1983.

——. *Electrical Machines: Design and Applications.* IEE Second International Conference. London, 1985.

——. *Test Code for Direct-Current Machines.* IEEE Standard No. 113. New York, 1962.

——. *General Requirements for Synchronous Machines.* ANSI Standard C50.10, 1977.

——. *Test Procedures for Synchronous Machines.* IEEE Standard 115, 1983.

——. *Test Procedures for Polyphase Induction Motors and Generators.* IEEE Standard 112, 1984.

APPENDIX A

Magnetic Circuits

A1. INTRODUCTION

Together with the subjects *electric circuits* and *mechanics, magnetic circuits* form the fundamental base upon which the principles of electromechanical energy conversion are built. This appendix reviews some concepts of magnetism and links the elements that are important for the engineering calculations found in this text.

We will describe the quantities used in magnetic-circuit calculations. The concepts behind these quantities are not always easy to understand, but frequent practice with calculations does provide familiarity. From a description of the relationships between these quantities and from a knowledge of magnetic material characteristics we can set about performing magnetic-circuit computations. These computations are done to find the flux linkage λ in terms of a current i. This is needed to apply the energy conversion principles, that are discussed in the text, to find the operation and performance of electromechanical devices.

A2. DESCRIPTION OF TERMS

Table A1 illustrates symbols and units of terms that are used to link current to the magnetic flux linkage. A description of the symbols in Table A1 follows. The arrows in the table indicate that, if one parameter is known, we can proceed to determine the next parameter to the right by analysis or calculation.

I: Wherever there is a flow of charge there is a magnetic field. The current I is considered to be the source of the magnetic field.

F: One turn of a coil with a current produces a magnetic field. Each additional turn with a current will increase the magnetic field. Therefore the *cur-*

Table A1   ELECTRIC/MAGNETIC CIRCUIT LINKS

Terms	Current	Magneto-motive force	Magnetic field intensity	Magnetic flux density	Magnetic flux	Flux linkage
Symbols	$I \rightarrow\!\rightarrow$	$\rightarrow\!\rightarrow F \rightarrow\!\rightarrow$	$\rightarrow\!\rightarrow H \rightarrow\!\rightarrow$	$\rightarrow\!\rightarrow B \rightarrow\!\rightarrow$	$\rightarrow\!\rightarrow \Phi \rightarrow\!\rightarrow$	$\rightarrow\!\rightarrow \lambda$
Units	amperes (A)	ampere-turns (A)	ampere-turns/meter (A/m)	tesla (T)	webers (Wb)	webers (Wb)

rent × *turns* is the source of the magnetic field. This source is called the *magnetomotive force* (or *mmf*) F. It is analogous to *emf*. Thus

$$F = NI. \tag{A2.1}$$

H: The magnetic field intensity H describes the field produced by the *mmf*, and has units amperes per meter, although ampere-turns per meter is the term that is frequently used.

V_{mag}: V_{mag} is a quantity that was not included in Table A1, but it is useful for relating current and magnetic field. V_{mag} is the symbol for magnetic potential difference and has units amperes, although the term ampere-turns is often used. It is defined by

$$V_{mag} = -\int_a^b \mathbf{H} \cdot d\mathbf{l}. \tag{A2.2}$$

This is analogous to the relation between the electric potential difference V volts and the electric field intensity E volts per meter

$$V = -\int_a^b \mathbf{E} \cdot d\mathbf{l}. \tag{A2.3}$$

B: The magnetic flux density B is mathematically related to H by

$$\mathbf{B} = \mu\mathbf{H}. \tag{A2.4}$$

The value of B depends not only on H (and hence the current) but also on the medium in which H is found.

μ: The influence of the medium is presented by the permeability μ (henrys per meter), which is given by

$$\mu = \mu_0\mu_r. \tag{A2.5}$$

μ_0:   The permeability of free space μ_0 has a value $4\pi \times 10^{-7}$H/m.

μ_r: Any material's magnetic characteristic is described by how much larger the flux density is compared with the value in free space for the same value of H. So relative permeability μ_r (dimensionless) is used. For air μ_r is approximately equal to unity, while some ferromagnetic materials have μ_r with a value of the order of 10^4. The value of μ_r of ferromagnetic materials depends on the magnitude of H.

Φ: The magnetic flux Φ (webers) is related to the flux density B (or flux per unit area). If B is not uniform

$$\phi = \int \mathbf{B} \cdot d\mathbf{A}. \qquad \text{(A2.6)}$$

If B is uniform and normal to the area A,

$$\phi = BA. \qquad \text{(A2.7)}$$

The flux entering a closed surface or volume also leaves it. That is, the net flux through a closed surface is zero, or

$$\phi = \oint \mathbf{B} \cdot d\mathbf{A} = 0. \qquad \text{(A2.8)}$$

λ: If a flux Φ links a coil of N turns, it is convenient to define flux linkage λ as

$$\lambda \triangleq N\Phi. \qquad \text{(A2.9)}$$

Currents in coils magnetize materials. That is, a magnetic field of intensity H, of flux Φ, and density B is produced by an mmf F. Because of the material permeability μ, for the same mmf different values of B will be produced in different materials. This is characterized by experimental magnetization curves of B against H (or saturation curves of λ against I) to relate the magnetic effects for a given coil excitation (H or I) for different materials. Figures A1 and A2 exemplify the B-H characteristics for different steels.

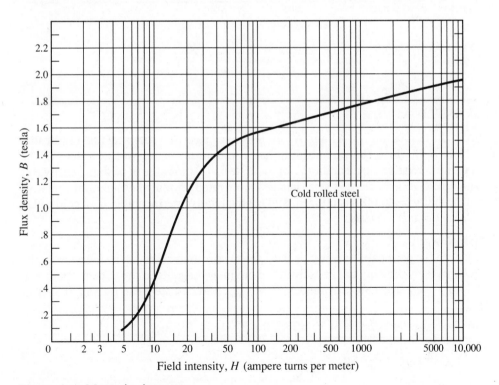

Figure A.1 Magnetization curve.

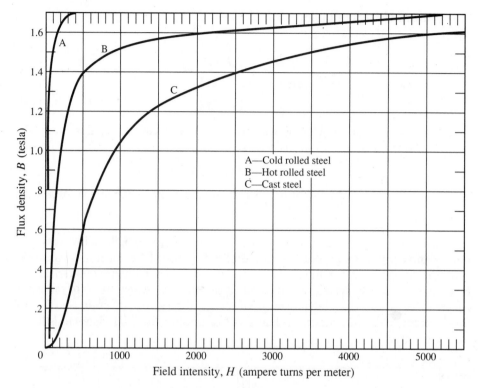

Figure A.2 Comparison of magnetization curves.

EXAMPLE A1

From the curves in Figs. A1 and A2 for hot rolled steel and cast steel find **(a)** the flux density B at which the permeability has a maximum value, **(b)** the values of the maximum relative permeability μ_r for these steels, **(c)** the value of the flux density B at which cold-rolled oriented steel has to be operated so that the maximum relative permeability ($\sim 46{,}000$) is utilized, and **(d)** the relative permeability μ_r of cold rolled steel at 1.5 and 2.0 tesla.

Solution

(a) Since the permeability μ equals B/H, the maximum values can be found from the BH curves by drawing the lines from the origin as tangents to the curves. From Fig. A2 hot-rolled steel can provide 0.54 T at μ_{max}, and cast steel can provide 0.72 T at μ_{max}.

(b) For hot-rolled steel $H \approx 70$ AT/m for $B = 0.54$ T. Since

$$\mu_{max} = \mu_0 \mu_r = \frac{B}{H}, \quad \mu_r = \frac{B}{\mu_0 H}.$$

On substitution, the maximum value of μ_r equals 6138. For cast steel $H \approx 560$ AT/m for $B = 0.72$ T. Therefore μ_r has a maximum value of 1023.

(c) $\dfrac{B}{H} = \mu_r \mu_0 = 46000 \times 4\pi \times 10^{-7} = 0.058$ H/m.

By trial and error, or from a plot of B against B/H, the value of B, for the ratio B/H equal to 0.58, is 0.9 T.

(d) For $B = 1.5$T, Fig. A1 gives H to be 62 AT/m, so that

$$\mu_r = \frac{B}{\mu_0 H} = 19{,}250.$$

For $B = 2.0$ T, H is 2×10^4 AT/m, and $\mu_r = 80$.

A3. AMPERE'S MAGNETIC CIRCUITAL LAW

Figure A3 represents a coil of N turns fed by a current I. The dotted line is an arbitrary closed path. At any point on the path the magnetic field intensity has some value H over an elementary length dl. Keeping this figure in mind Ampere's law is as follows.

• The magnetic potential drop around a closed path is balanced by the mmf giving rise to the field (i.e., the mmf encircled by the closed path).
In mathematical terms

$$\oint \mathbf{H} \cdot d\mathbf{l} = F = NI. \tag{A3.1}$$

The equation describing Ampere's law contains an integral that is easiest to evaluate if the closed path of integration is chosen along the direction of H, and H has a

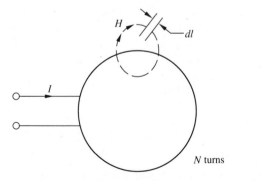

Figure A.3 An N turn coil.

constant value along l. For this case

$$NI = Hl. \tag{A3.2}$$

The next simplest case occurs if H has different but constant values over different sections of the closed magnetic path. This allows eq. (A3.1) to be simplified to

$$NI = \oint \mathbf{H} \cdot d\mathbf{l} = H_1 l_1 + H_2 l_2 + H_3 l_3 + \cdots = \sum_{n=1}^{n} H_n l_n. \tag{A3.3}$$

The cases found in this book are covered by eqs. (A3.2) and (A3.3) and represent the first step in the magnetic circuit calculations to determine how much current it takes to produce a particular value of field (B, Φ, or λ).

Ampere's magnetic circuital law is similar to Kirchhoff's voltage law for an electric circuit.

A4. RELUCTANCE

The concept of reluctance of a magnetic circuit is similar to that of resistance of an electric circuit. It leads to an Ohm's law equivalent for magnetic circuits. Figure A4 represents a ferromagnetic circuit of mean length l, and constant cross section of area A. The aim is to find the flux Φ as a function of the current I or the mmf NI.

The mmf NI produces a flux Φ, which is confined to the core, if it is assumed that the permeability μ is high. Since A is constant for the average path length, $B = \Phi/A = $ constant, and H is the same at every point because $B = \mu H$. $\mathbf{H}$ and $d\mathbf{l}$ are collinear vectors so that Ampere's law can be written as

$$\oint H \, dl = NI. \tag{A4.1}$$

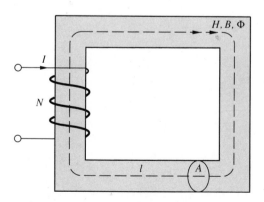

Figure A.4   A simple magnetic circuit.

Because H is constant around the path

$$H \oint dl = Hl = NI \tag{A4.2}$$

For routine calculations of reasonable accuracy it is acceptable to consider the value of H at the center line of the flux path. The corresponding B and μ are assumed to exist over the entire cross section A. The center-line length is l in the above equations. Accordingly, $H = B/\mu$ and $B = \Phi/A$. Equation (A4.2) becomes

$$\phi \frac{l}{\mu A} = NI. \tag{A4.3}$$

Putting $NI = F$, the mmf, and rearranging we have

$$F = \left(\frac{l}{\mu A}\right)\Phi. \tag{A4.4}$$

If we allow $\mathscr{R}$ to be equal to $(l/\mu A)$ then we have a magnetic Ohm's law

$$F = \mathscr{R}\Phi \tag{A4.5}$$

where $\mathscr{R}$ is known as the *reluctance* of the magnetic circuit. Compare this with $V = RI$ for an electric circuit. The unit of reluctance is H^{-1}.

A4.1. Reluctances in Series and Parallel

In an electromechanical device the magnetic circuit can comprise two parts in series, an iron part and an air gap. The iron is used as an efficient means to direct the magnetic flux to the air gap where the action takes place. Figure A5 depicts this kind of magnetic circuit.

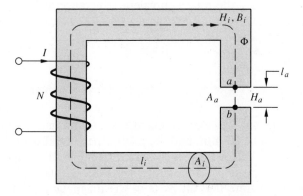

Figure A.5 A magnetic circuit with an air gap.

The mmf $(F = NI)$ produces the flux Φ, which is assumed to be confined to the iron core and the air gap. From Ampere's law it follows that

$$
\begin{aligned}
NI = \oint H\, dl &= \int_b^a H_i\, dl_i + \int_a^b H_a\, dl_a \\
&= H_i \int_b^a dl_i + H_a \int_a^b dl_a = H_i l_i + H_a l_a \\
&= \frac{B_i l_i}{\mu_i} + \frac{B_a l_a}{\mu_0} = \frac{\Phi l_i}{\mu_i A_i} + \frac{\Phi l_a}{\mu_0 A_a}.
\end{aligned}
\tag{A4.6}
$$

Reluctance has the form $l/\mu A$ so eq. (A4.6) can be written as

$$
F = \Phi(\mathcal{R}_i + \mathcal{R}_a)
\tag{A4.7}
$$

where $\mathcal{R}_i$ and $\mathcal{R}_a$ are the reluctances of the iron and air-gap paths, respectively. For an equivalent single-reluctance circuit

$$
F = \Phi\mathcal{R}.
\tag{A4.8}
$$

Hence for this series magnetic circuit the equivalent reluctance of the flux path is

$$
\mathcal{R} = \mathcal{R}_i + \mathcal{R}_a.
\tag{A4.9}
$$

In general, a series magnetic path of n different reluctance parts has a total equivalent reluctance given by

$$
\mathcal{R} = \sum_{i=1}^{i=n} \mathcal{R}_i.
\tag{A4.10}
$$

It can be shown that for a magnetic circuit of n parallel parts the total equivalent reluctance can be calculated from

$$
\frac{1}{\mathcal{R}} = \sum_{i=1}^{i=n} \frac{1}{\mathcal{R}_i}.
\tag{A4.11}
$$

The way we calculate the total equivalent reluctance of a complex magnetic circuit of series and parallel paths is the same way we tackle complex electric circuits comprising series and parallel resistances.

EXAMPLE A2

Consider the magnetic circuit shown in Fig. A5. Let the air-gap length be 1 mm, the mean iron-path length be 50 mm, and the relative permeability of the iron core be 5000. These are typical values for a small device. Compare the ampere turns necessary to provide the magnetic potential drops across the air gap and the iron parts of the magnetic path. Assume that all parts of the magnetic circuit have the same cross-sectional area A.

Solution

From the result of the application of Ampere's law, given in eq. (A4.6),

$$NI = \frac{\Phi}{\mu_0 A} \frac{l_i}{\mu_r} + \frac{\Phi}{\mu_0 A} l_a.$$

The equation has split the total number of ampere-turns in two parts. The first term on the right-hand side gives the number of ampere-turns to produce the flux Φ in the iron, while the second term gives the number of ampere-turns to produce the flux Φ in the air gap. A comparison of the two terms is provided by taking their ratio. The value of l_i/μ_r is 10^{-5} m and the value of l_a is 10^{-3} m. Therefore the number of ampere turns to produce Φ in the air gap is 100 times the number of ampere turns to produce Φ in the iron alone.

Because this result is general in practice, it is common to neglect the iron effects when doing approximate magnetic circuit calculations.

If it is assumed that $\mu_r \rightarrow \infty$ then $H_i \rightarrow 0$ and $(NI)_{iron} \rightarrow 0$. So $F = NI \approx H_a l_a$. If only the air-gap conditions are considered, the application of the much simplified Ampere's law gives a reasonable approximation of the number of ampere-turns necessary to satisfy the field requirements.

A5. MAGNETIC CIRCUIT COMPUTATIONS

In order to design an electromechanical device, or to predict its performance, one of the first steps is to calculate the excitation (I, F, or H) that is necessary to produce the required magnetic field (λ, Φ, or B). Ferromagnetic materials are used in devices, because (1) they allow B thousands of times as large as that without these materials for the same excitation, and (2) they restrict the magnetic field to the most desirable path. However, the calculations are complicated by the complex characteristics of ferromagnetic materials.

A single cycle[1] of excitation for a circuit like that shown in Fig. A4 produces a BH characteristic that may look like one of the loops depicted in Fig. A6.[2] The function $B = f(H)$ is nonlinear. It is double valued. It depends on the value of H_{max} attained and on the history of excitation. If we are to make calculations some approximations have to be made.

---

[1] More details are given in Appendix C on ac excitation.

[2] B lags H in the excitation cycling. The loops are called *hysteresis* loops and are the result of effort needed to orient the magnetic domains under the influence of H. Work is done to reorient the domains during one cycle of H variation. Part of this work is converted into heat and is referred to as *hysteresis* loss.

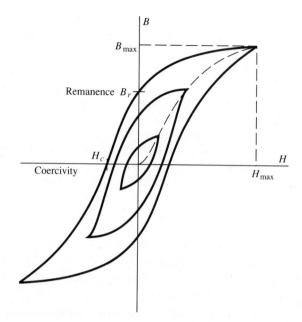

Figure A.6   Typical *BH* characteristics for a
ferromagnetic material.

Approximation 1

The first approximation to make magnetic circuit calculations easier is to use what is
called the normal magnetization curve of a material. This is formed from the *BH*
characteristics by drawing a curve through the tips of increasingly larger hysteresis
loops. The broken curve in Fig. A6 illustrates the creation of the normal magnetization curves. Practical examples are shown in Fig. A1 and A2. This approximation
reduces the *BH* curve to a single-valued, nonlinear function, in which the permeability μ is not a constant.

Approximation 2

It is desirable to work the magnetic material at the maximum value of *B* for the
minimum excitation *H*. The usual operating value is at the knee of the normal
magnetization curve.

If the magnetic operating region is close to the knee, where the permeability is
high, we can use the approximation that the flux Φ is confined to the path provided by
the ferromagnetic materials. The flux density distribution is not uniform and its
determination is complex.

Approximation 3

In practical cases, such as those found in this book, **B** and **H** have the same direction.
H follows the center-line of the magnetic flux path and **B** is perpendicular to the cross
section through which the flux Φ passes. This allows the scalar form of eqs. (A2.4),

(A2.8), and (A3.1) to be used. That is

$$B = \mu H, \quad \oint H \, dl = NI \quad \text{and} \quad \oint B \, dA = 0.$$

Approximation 4

Great simplification is made if we use the fact that a knowledge of center-line flux densities in the different parts of the magnetic structure is normally all that is required. A final approximation is to assume that the permeability of the ferromagnetic material has a constant value. This value corresponds to the B and H at the center line of the magnetic circuit. This simplifies the analytical solution.

We study two worked examples to compare the complexity of the magnetic calculations for different sets of approximations.

EXAMPLE A3

A toroid, made of hot-rolled steel strip, has a square cross section of 36 cm². The inner radius R_i of the toroid is 0.14 m and the outer radius R_o is 0.2 m. A coil of 1000 turns, carrying a current of 0.107 A, is wound around the toroid. Calculate the amount of magnetic flux Φ in the toroid.

Solution 1

The nonlinear BH curve is drawn in Figure A2. Mean values of B and H over the whole core cross section are used, and the magnetic path is taken to have a mean radius R_m, where $R_m = (R_o + R_i)/2$. From Ampere's law

$$\oint H \, dl = H \oint dl = H \times 2\pi R_m = NI$$

$$H = \frac{NI}{2\pi R_m} = 100 \text{ AT/m}.$$

From the BH curve, for $H = 100$ AT/m, $B = 0.7$ T. Hence the flux is

$$\phi = BA = 0.7 \times 36 \times 10^{-4} = 2.52 \times 10^{-3} \text{ Wb}.$$

This solution takes into account the nonlinearity of the magnetic-circuit characteristic. However we have taken the value of B to be constant over the cross-sectional area of the core. The geometry of the toroid allows us to relax this constraint. We can divide the core into any number of sections, make calculations as above for each section to find the fluxes, combine the results, and compare with the value already obtained.

Solution 2

We can divide the cross section of the toroidal core into two parts of equal area. This creates an inner toroidal ring and an outer toroidal ring. The flux is calculated for each ring and the total flux for the toroid is the sum of the two components Φ_o and Φ_i.

For the inner ring the mean radius $R_{mi} = \dfrac{17 + 14}{2} \times 10^{-2}$ m,

so $H_i = \dfrac{NI}{2\pi R_{mi}} = 110$ AT/m.

From the BH curve B_i is 0.77 T, so

$\phi_i = B_i A_i = 1.386 \times 10^{-3}$ Wb.

For the outer ring the mean radius $R_{mo} = \dfrac{20 + 17}{2} \times 10^{-2}$ m,

so $H_o = \dfrac{NI}{2\pi R_{mo}} = 92$ AT/m.

From the BH curve B_o is 0.64T, so

$\phi_o = B_o A_o = 1.152 \times 10^{-3}$ Wb.

The total flux $\Phi = \Phi_i + \Phi_o = 2.538 \times 10^{-3}$ Wb.

Solution 2 is more accurate than solution 1 but the difference is not much more than 1%. This method of solution is used if $(R_o - R_i)/R_m$ has a large value (usually > 6),
and if better accuracy is required.

Solution 3
At any radius r within the core of the toroid Ampere's circuital law gives $H_r \times 2\pi r = NI$. Therefore the flux density B_r at radius r is a function of the radius, given by

$$B_r = \frac{\mu NI}{2\pi r}.$$

Consider an elementary flux path of width dr at r and of height h equal to the core height ($h = 6$ cm). The flux $d\phi$ in the path is

$$d\phi = B_r \, dA = B_r h \, dr = \frac{\mu NIh}{2\pi r} \, dr.$$

Assuming μ to be constant, the total flux in the core is

$$\phi = \int d\phi = \int_{R_i}^{R_o} \frac{\mu NIh}{2\pi} \frac{dr}{r} = \frac{\mu NIh}{2\pi} \ln \frac{R_o}{R_i}.$$

The constant μ is taken from Fig. A2 for the mean values $B = 0.7$ T and $H = 100$ AT/m, that is, $\mu = 7 \times 10^{-3}$ H/m. On substitution of the parameter values in the flux equation, the total flux ϕ is

$\phi = 7.15 \times 10^{-3} \ln 1.43 = 2.56 \times 10^{-3}$ Wb.

Any error here is produced by the approximation that the permeability is taken to be a constant. The result is not much different from Solution 1. A problem is that we had to integrate a function. This can be overcome if we resort to the use of the magnetic circuit reluctance.

Solution 4

Use can be made of Ohm's law equivalent, $F = \Phi\mathcal{R}$, where $\mathcal{R}$ is the mean reluctance of the magnetic path. Thus

$$\phi = \frac{NI}{\mathcal{R}} = \frac{\mu NIh}{\pi} \frac{R_o - R_i}{R_o + R_i} = 2.52 \times 10^{-3}\ \text{Wb}.$$

This answer is, and must be, the same as that of Solution 1.

From the four methods of solution the values of flux were found to be 2.52×10^{-3} Wb, 2.538×10^{-3} Wb, 2.56×10^{-3} Wb, and 2.52×10^{-3} Wb. These values are close. Therefore it can be concluded that the method entailing the least work is normally satisfactory. In this particular example, Solution 4 appears to be the best.

EXAMPLE A4

Consider a parallel magnetic circuit with an air gap as shown in Fig. A7. The area of cross section A of the core is constant and equal to 25 cm². If the core is cast steel, find the mmf to produce an air-gap flux of 5×10^{-4} Wb.

Solution

Approximations 1 (normal magnetization curve, Fig. A2), 2 (flux confined within core boundaries), and 3 (average values of B, l, and H are sufficient) can be used to get

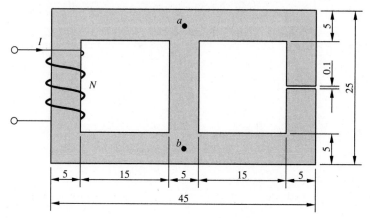

Note: Dimensions shown in cm

Figure A.7 Parallel magnetic circuit with an air gap.

Table A2

Path	Φ $Wb \times 10^3$	A $m^2 \times 10^4$	L m	B T	H AT/m	$V_{mag}(Hl)$ AT
Left iron $a-b$	3.6 (12)	25 (1)	0.6 (2)	1.44 (13)	2875 (14)	1725 (15)
Center $a-b$	3.1 (11)	25 (1)	0.2 (2)	1.23 (10)	1515 (9)	303 (8)
Right iron $a-b$	0.5 (1)	25 (1)	0.6 (2)	0.2 (3)	240 (6)	144 (7)
Air gap	0.5 (1)	25 (1)	10^{-3} (1)	0.2 (3)	1.6×10^5 (4)	159 (5)

good results. Table A2 lists the results of the calculations to find the magnetic potential drops across the various lengths constituting the magnetic paths. The numbers in parentheses indicate the order in which the calculations are made.

Step 1.
We are given the air-gap flux $\Phi = 5 \times 10^{-4}$ Wb, the core cross section $A = 25$ cm^2, and the air-gap length $l_a = 10^{-3}$ m.
The average length of the yoke $= (15 + 2 \times 2.5) \times 10^{-2} = 0.2$ m.
The average length of any leg $= (25 - 2 \times 2.5) \times 10^{-2} = 0.2$ m.

Step 2.
$$l_{i(left)} = 1 \text{ leg} + 2 \text{ yokes} = 0.6 \text{ m.}$$
$$l_{i(right)} = l_{i(left)} - l_a \approx l_{i(left)} \approx 0.6 \text{ m.}$$
$$l_{i(center)} = 1 \text{ leg} = 0.2 \text{ m.}$$

Step 3.

$$B_a = \frac{\Phi_a}{A} = 0.2 \text{ T.}$$

Therefore $B_{i(right)} = B_a = 0.2$ T.

Step 4.

$$H_a = \frac{B_a}{\mu_0} = 1.59 \times 10^5 \text{ AT/m.}$$

Step 5.

$$V_{mag\,a} = H_a l_a = 159 \text{ AT.}$$

Step 6.
From the *BH* curve for cast steel in Fig. A2, for $B_{i(right)} = 0.2$ T, the value of $H_{i(right)}$ is 240 AT/m.

Step 7.

$$V_{mag\,i(right)} = H_{i(right)}l_{i(right)} = 240 \times 0.6 = 144 \text{ AT.}$$

The total magnetic potential drop for the path *right a–b* is $159 + 144 = 303$ AT.

Step 8.
Since the center leg $a–b$ is parallel to the path to the right of $a–b$, it is subject to the same magnetic potential drop 303 AT.

Step 9.

$$H_{(center)} = \frac{V_{mag\,a-b}}{l_{i(center)}} = \frac{303}{0.2} = 1515 \text{ AT/m.}$$

Step 10.
From the *BH* curve for $H_{(center)} = 1515$ AT/m, the value of $B_{(center)}$ is 1.23 T.

Step 11.

$$\Phi_{(center)} = B_{(center)} \times A = 3.1 \times 10^{-3} \text{ Wb.}$$

Step 12.
The total flux $\Phi_{(left)} = \Phi_{(center)} + \Phi_a = 3.6 \times 10^{-3}$ Wb.

Step 13.

$$B_{(left)} = \frac{\Phi_{(left)}}{A} = 1.44 \text{ T.}$$

Step 14.
From the *BH* curve, for $B_{(left)} = 1.44$ T, the value of $H_{(left)}$ is 2875 AT/m.

Step 15.

$$V_{mag(left)} = H_{(left)} \times l_{i(left)} = 2875 \times 0.6 = 1725 \text{ AT.}$$

Applying Ampere's law around a closed magnetic circuit is equivalent to applying Kirchhoff's voltage law around a closed electric circuit. In the case of Ampere's law the sum of the mmf rises around a closed loop must equal the sum of the magnetic potential drops. Applying this around the loop (*left*) $a–b$ and (*center*) $a–b$, the total coil ampere turns required is equal to $(1725 + 303)$ AT, that is, 2028 AT.

Most magnetic circuit problems can be solved in this way.

A6. MAGNETIC FLUX LEAKAGE AND FRINGING

Figure A8 shows a simple series electric circuit and a simple series magnetic circuit. Table A3 indicates the duality between the two circuits.

The reader might like to find the magnetic dual equations that correspond to Ohm's law, Kirchhoff's law, and the relation between resistance and conductivity σ.

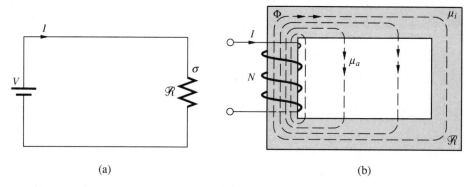

(a)   (b)

Figure A.8   Electric and magnetic circuit duals.

There are some differences between the two systems. The ratio of the conductivities of the copper conductor and its insulation is of the order of 10^{20}. This value is high enough to confine current entirely to the conductor. The ratio of the permeabilities of iron and air (its magnetic insulation) is of the order of 10^4. That is, the air-path reluctance of the magnetic insulation is not very great compared with that of the iron. Consequently there is some flux, called leakage flux, that goes beyond the boundaries of the iron. If there are no air gaps in the iron path, ignoring leakage and applying Approximation 2 to the magnetic-circuit calculations may not produce significant errors.

The leakage flux Φ_l can be considerable if the magnetic circuit has an air gap as shown in Fig. A9. This is because the magnetic potential drop across the air gap has the same order of magnitude as the magnetic potential difference between the yoke faces. In Fig. A9a the air gap and the path between the yoke faces are paths in parallel so

$$\frac{\Phi_l}{\Phi_a} \approx \frac{\mathscr{R}_a}{\mathscr{R}_l}.$$

This ratio can be as much as 10% to 15% in rotating machines.

One way to minimize leakage flux for a given magnetic configuration is to have the excitation coil close to the air gap, as shown in Fig. A9b. This reduces greatly the magnetic potential difference between the yoke faces. Further, the air-gap flux is greater in this case than that in Fig. A9a for the same mmf and it is the air gap where the flux is needed.

Flux fringing at air gaps as shown in Fig. A10 is difficult to avoid. As with leakage flux paths the reluctance $\mathscr{R}_{fringe}$ of the fringe paths is of the same order of magnitude as

Table A3   DUALS

Electrical quantity	V	I	R	σ
Magnetic dual	F	Φ	$\mathscr{R}$	μ

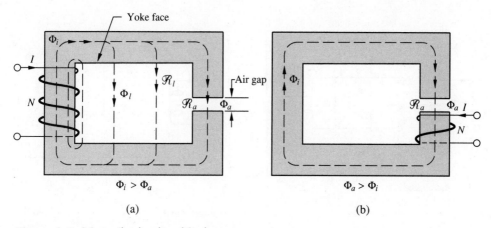

(a)   (b)

Figure A.9 Magnetic circuits with air gaps.

the reluctance $\mathcal{R}_a$ of the air gap where the flux is intended to be concentrated. Thus the flux crossing the air gap tends to spread, the effective area of the air gap increases, and the effective air-gap reluctance decreases. This has a significant effect on the value of the mmf required for a certain flux in the air gap.

Fringing is minimized by keeping the air-gap length as short as possible in comparison with the area of cross section of the pole faces. A method to account for fringing in calculations is to use an effective air-gap area made up by increasing each of the cross-section dimensions by an amount equal to the air-gap length.

EXAMPLE A5

A magnetic circuit is depicted in Fig. A11a. The core is cold-rolled steel (refer to Fig. A1). If the air-gap flux is 4×10^{-3} Wb, estimate the leakage flux and the source mmf. Neglect fringing.

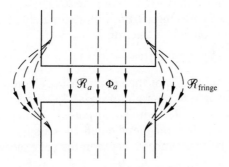

Figure A.10 Fringing.

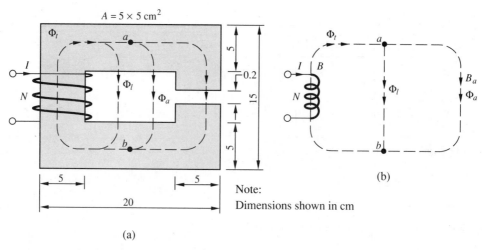

Figure A.11   Flux leakage.

Solution

This problem can be treated in the same way as the problem in Example A4. The difference is that the center leg in Example A4 is replaced by an equivalent leakage path between points *ab* as shown in Fig. A11b. The reader may like to repeat the steps set out in the Example A4 solution.

Since the air-gap flux Φ_a is 4×10^{-3} Wb, the flux density B_a in the air gap is

$$B_a = \frac{\Phi_a}{A} = \frac{4 \times 10^{-3}}{25 \times 10^{-4}} = 1.6 \text{ T}.$$

The magnetic potential drop V_{maga} across the air gap is

$$V_{maga} = H_a l_a = \frac{B_a l_a}{\mu_0} = 2550 \text{ AT}.$$

Between a, the air gap, and b, the flux density in the steel core is assumed to be the same as that in the air gap, that is 1.6 T. From the BH curve $H_i = 150$ AT/m for this part of the core. Since the core length l_i ab is approximately 25 cm for this part, then the magnetic potential drop across the iron is

$$V_{magiab} = H_i l_{iab} = 150 \times 25 \times 10^{-2} = 38 \text{ AT}.$$

This potential drop is only 1.5% of the air-gap drop. The total magnetic potential drop V_{magab} around the path to the right of ab is $2550 + 38 = 2588$ AT.

The leakage flux Φ_l is due to the magnetic potential difference between the yoke surfaces facing each other. This magnetic potential is assumed to be that across the parallel path V_{magab}. The reluctance $\mathcal{R}_l$ between the yoke faces is

$$\mathcal{R}_l = \frac{l_i}{\mu_0 A_l} = \frac{5 \times 10^{-2}}{\mu_0 \times 5 \times 10 \times 10^{-4}} = 7.96 \times 10^6 \text{ H}^{-1}.$$

Thus the leakage flux is given by

$$\phi_l = \frac{V_{mag\,ab}}{\mathcal{R}_l} = \frac{2588}{7.96 \times 10^6} = 0.3325 \times 10^{-3}\ \text{Wb}$$

and amounts to approximately 8.3% of the air-gap flux.

The total flux Φ_t linking the coil and the left leg of the magnetic circuit is

$$\Phi_t = \Phi_a + \Phi_l = 4 \times 10^{-3} + 0.325 \times 10^{-3} = 4.325 \times 10^{-3}\ \text{Wb}.$$

The flux density B in this part of the circuit is

$$B = \frac{\Phi_t}{A} = \frac{4.325 \times 10^{-3}}{25 \times 10^{-4}} = 1.73\ \text{T}.$$

For this value of B the BH curve gives a corresponding $H = 500$ AT/m. Therefore the magnetic potential drop around the core to the left of ab is approximately

$$V_{mag(left)} = Hl_{(left)} = 500 \times 25 \times 10^{-2} = 125\ \text{AT}.$$

The total magnetic potential drop around either of the closed paths incorporating the coil is

$$\text{Total } V_{mag\,loop} = 125 + 2588 = 2713\ \text{AT}.$$

Around a closed loop the sum of the mmf rises equals the sum of the magnetic potential drops (Ampere's law), so the source mmf is 2713 AT.

A good exercise is to repeat this problem with the coil moved to the position shown in Fig. A9b.

A7. SUMMARY

For most purposes in this text Ampere's law in the form $\Sigma Hl = F$ is used to relate the sum of magnetic potential drops around a closed flux path to the mmf of the coils wound on that path. Some magnetic relations are $B = \mu H$, $\Phi = BA$, and $F = NI$.

The magnetic equivalent of Ohm's law is $F = \Phi\mathcal{R}$.

The reluctance $\mathcal{R}$ of a section of a magnetic circuit is given by $\mathcal{R} = l/\mu A$. Reluctances in series and parallel are treated in the same way as resistances in series and parallel in order to find a single reluctance equivalent.

The above equations, together with a combination of approximations, enable calculations to be made to find the number of ampere-turns necessary to produce a given value of magnetic field in a defined configuration. The approximations are

1. the normal magnetization curve can be used
2. the flux is confined to the path provided by the magnetic materials
3. the value of flux density B at the center line of the magnetic circuit is adequate
4. the permeability μ has a constant value corresponding to the value of B at the center line of the magnetic circuit.

A8. PROBLEMS

Note: The magnetization curves needed to solve the following problems are shown in Figs. A1 and A2.

A.1. A ring of cast steel, shown in Fig. PA.1, contains an air gap of 1 mm length. The average length and cross section of the ring are 0.2 m and 0.001 m^2, respectively. A coil of 100 turns is wound on the ring. **(a)** Determine the value of the current required to establish a flux of 1.06×10^{-3} webers in the air gap. **(b)** Determine the value of the current required to establish the same air-gap flux, if the ring is made of cold rolled steel strip. **(c)** What are the reluctances of the airgap and the steel rings for each type of ferromagnetic material used? (*Answer:* (a) 10.4 A, (b) 8.5 A, (c) 7.96×10^5 A/Wb, 1.89×10^5 A/Wb, 3.58×10^3 A/Wb.) Notice the difference in values of the ampere-turns required for the air gap, the cast steel ring and the ring made of cold rolled steel.

A.2. The front and side views of a padded yoke magnetic circuit for a single-phase transformer are shown in Fig. PA.2. The mean length of the magnetic circuit is indicated by the broken line. All dimensions are in centimeters. The flux in this magnetic circuit is 4.8×10^{-3} webers. The material of the magnetic circuit is cold rolled sheet steel. **(a)** Determine the mmf required to produce this flux, using the magnetization curve. **(b)** Determine the mmf using the magnetic-circuit concept of reluctances in series.
(*Answer:* (a) 106 AT, (b) 106 AT.)

A.3. The width of both yokes of the core, shown in Fig. PA.2, is reduced from 8 to 6 cm so that the magnetic circuit has a uniform cross section. **(a)** What is the mmf required to produce a flux of 4.8×10^{-3} webers? **(b)** What is the value of the flux if the mmf is increased to 2400 ampere-turns? **(c)** For a core with the

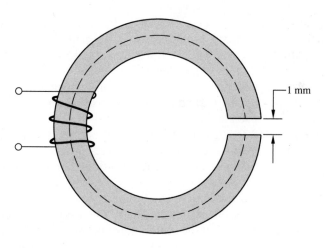

Figure PA.1

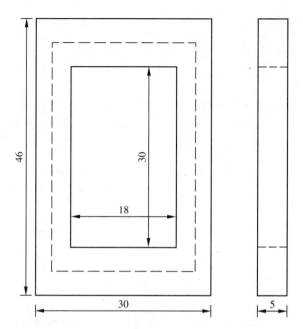

Figure PA.2

same window width (18 cm) and the same uniform cross section (30 cm²), what window height is required, if an mmf of 10,200 AT is to produce a flux of 5.88×10^{-3} webers?

(*Answer:* (a) 150 AT, (b) 5.52×10^{-3} webers, (c) 21 cm.)

A.4. A three-legged ferromagnetic circuit is shown in Fig. PA.4. The coil, wound on the center limb, has 250 turns and has a resistance of 3 ohms. The uniform cross section of the magnetic circuit is 10 cm². The assumption is made that the relative permeability of the iron part of the circuit is infinitely high. **(a)** Determine the value of the dc voltage that must be applied to the coil terminals to produce a flux of 2.515×10^{-2} webers in the air gap that has a length of 0.5

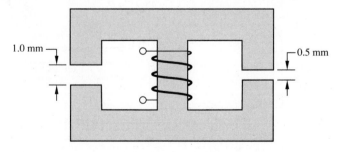

Figure PA.4

mm. **(b)** What is the value of the flux in the air gap of length 1 mm, if the voltage, found in part (a), is applied to the coil terminals?
(*Answer:* (a) 120 volts, (b) 1.26×10^{-2} webers.)

A.5. Consider the ferromagnetic circuit shown in Fig. PA.4. Let the coil be removed from the center limb and placed on the limb that contains the 1.0 mm air gap. A dc voltage of 120 V is applied to the coil. **(a)** Determine the value of the flux in the 0.5-mm air gap and **(b)** determine the value of the flux in the 1.0-mm air gap.
(*Answer:* (a) 0 webers, (b) 1.26×10^{-2} webers.)

A.6. The following information is given for the magnetic circuit shown in Fig. PA.6.

Path	Material	Area meter^2	Length meters	Flux webers
c f a	Cast steel	0.05	0.8	
a b c	Cast steel	0.03	0.4	4.2×10^{-2}
a d	Cast steel	0.0266	0.15	
d e	Air	0.0266		2.8×10^{-2}
e c	Cast steel	0.0266	0.15	

Determine **(a)** the total number of ampere-turns required to produce the given fluxes and **(b)** the length of the airgap.
(*Answer:* (a) 3000 AT, (b) 8.54×10^{-4} m.)

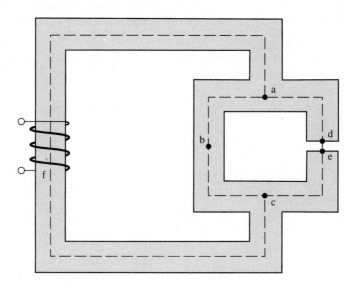

Figure PA.6

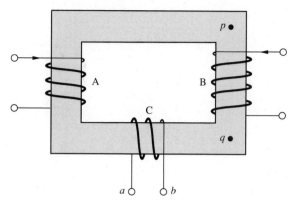

Figure PA.7

A.7. The magnetic circuit shown in Fig. PA.7 has a uniform cross-sectional area of 5×10^{-4} m² and a mean length of 0.4 m. Three coils (A, B, C) are wound on the cast steel core. Coil A has 200 turns and carries a current of 0.5 A. Coil B has 400 turns and carries a current of 0.75 A. The current directions in these coils are shown in the figure. Coil C has 100 turns. What is the magnitude and direction of the current in coil C in order to produce a flux of 0.45×10^{-3} webers in a counterclockwise direction?
(*Answer:* 7 A into *a*.)

A.8. The magnetic circuit, shown in Fig. PA.8 is made of cold rolled steel and has a uniform cross-sectional area of 50×10^{-4} m². The center leg of the magnetic circuit contains an air gap of 1.0 mm length. An air-gap flux of 5×10^{-3} webers has to flow in the center leg from B to A. Determine the magnitude and

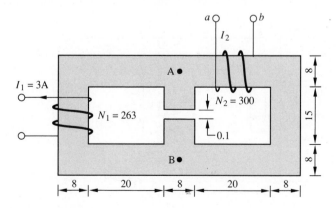

Figure PA.8

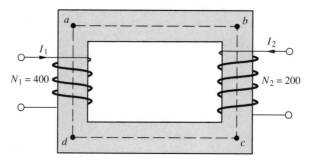

Figure PA.9

direction of the current I_2 required in the 300-turn coil, if a current $I_1 = 3$ A flows through the 263-turn coil in the direction indicated in the figure. (*Answer:* 6.1 A entering terminal *a*.)

A.9. The magnetic circuit shown in Fig. PA.9 is made of hot rolled steel laminations. The mean length of each of the two legs is 8 cm and their cross-sectional area is 30 cm². The mean length of each of the yokes is 10 cm and their cross-sectional area is 40 cm². Coil N_1 has 400 turns, coil N_2 has 200 turns. The positive current directions are as indicated in the figure. The flux in the circuit is 4.8×10^{-3} webers. **(a)** What is the value of I_2 and what is the magnetic potential difference $V_b - V_c$, if the current $I_1 = 0$? **(b)** What is the value of I_1, if $I_2 = 0$ and what is the magnetic potential difference $V_b - V_c$ in this case? **(c)** What are the values of I_1 and I_2 to make the magnetic potential difference $V_a - V_d = 0$? For this case what is the magnetic potential difference $V_b - V_c$? **(d)** If $I_2 = 0.5$ A, what should I_1 be? (*Answer:* (a) 2.06 A, −232 AT, (b) 1.03 A, +180 AT, (c) $I_1 = 0.45$ A, $I_2 = 1.16$ A, −52 AT, (d) 0.78 A.)

A.10. The magnetic circuit shown in Fig. PA.10 has a uniform cross section. All legs and yokes have the same average length. Coil 1 on the left leg has $N_1 = 100$

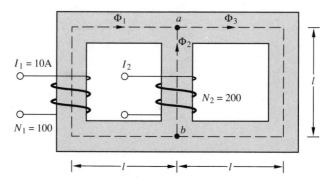

Figure PA.10

turns and carries a current of 10 A. The coil on the center leg has $N_2 = 200$ turns and carries a current I_2. The indicated direction for I_2 is taken as the direction of the positive current. **(a)** Calculate the value of I_2 required for zero flux in the left leg. **(b)** What is the magnetic potential difference $V_b - V_a$, if the flux in the left leg is zero? **(c)** Calculate the value of I_2 required for zero flux in the right leg.

(*Answer:* (a) 6.67 A, (b) -1000 AT, (c) -1.67 A.)

APPENDIX B

Mechanical System Relations

B1. INTRODUCTION

Both electrical circuits and mechanical systems can be modeled by lumped parameters to ease the solution of a problem. An electric circuit has inductance L and capacitance C for stored energy, resistance R for dissipation, and current i in response to voltage v. Analogous to these a translational mechanical system has mass M and elastance[3] $1/K$ for stored energy, D for dissipation by friction and velocity $\dot{x}$ in response to a force f. Table B1 indicates this particular analog and includes the rotational system.

B2. INERTIAL EFFECTS

Inertia is a property of matter by which it continues in its existing state of rest or uniform motion in a straight line, unless that state is changed by an external force.

Figure B1 shows bodies of mass M and moment of inertia J for translational and rotational movement respectively. Newton's law states that the rate of change of momentum associated with a body is equal to and in the same direction as the applied

---

[3] K is called the *stiffness*.

Table B1   ANALOGS

Electrical	v	i	q	R	L	C
Mechanical (translational)	f	$\dot{x}$	x	D	M	$1/K$
Mechanical (rotational)	T	$\dot{\theta}$	θ	D	J	$1/K$

force or torque. That is

$$f_M = \frac{d}{dt}\left(M\frac{dx}{dt}\right) = M\ddot{x} \quad \text{and} \quad T_J = \frac{d}{dt}\left(J\frac{d\theta}{dt}\right) = J\ddot{\theta}. \quad \text{(B2.1)}$$

A body in motion stores kinetic energy. For the translational and rotational systems this energy is

$$W_M = \int f_M\, dx = \int f_M \dot{x}\, dt \quad \text{and} \quad W_J = \int T_J\, d\theta = \int T_J \dot{\theta}\, dt. \quad \text{(B2.2)}$$

That is

$$W_M = \int M\frac{d\dot{x}}{dt}\, \dot{x}\, dt = \int M\dot{x}\, d\dot{x} \quad \text{and}$$

$$W_J = \int J\frac{d\dot{\theta}}{dt}\, \dot{\theta}\, dt = \int J\dot{\theta}\, d\dot{\theta}. \quad \text{(B2.3)}$$

Hence

$$W_M = \tfrac{1}{2} M\dot{x}^2 \quad \text{and} \quad W_J = \tfrac{1}{2} J\dot{\theta}^2. \quad \text{(B2.4)}$$

It is assumed that there is neither initial energy nor initial velocity.

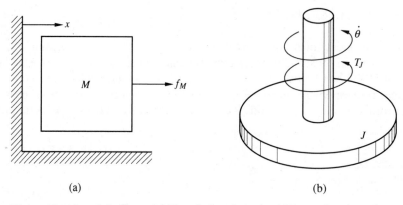

(a) (b)

Figure B.1   Inertial effects. (a) Translational motion, (b) rotational motion.

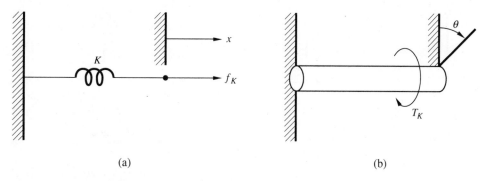

(a)   (b)

Figure B.2 Elastic effects. (a) Translational motion, (b) rotational motion.

B3. ELASTIC EFFECTS

Elasticity is a general name to that property of a body by virtue of which it resists and recovers from a change of shape. A spring and a shaft are examples illustrated in Fig. B2.

Hooke's law states that the restoring force or torque of an elastic material is proportional to the displacement. For a particular displacement the applied force and torque balance the restoring force and torque. So, if the spring and shaft are initially unstrained,

$$f_K = Kx \qquad \text{and} \qquad T_K = K\theta \tag{B3.1}$$

where K is the stiffness of the material and is usually a constant within the elastic limit.

Deformation of an elastic material stores potential energy which can be expressed by

$$W_K = \int f_K \, dx = \int Kx \, dx = \tfrac{1}{2}Kx^2 \qquad \text{and}$$

$$W_K = \int T_K \, d\theta = \int K\theta \, d\theta = \tfrac{1}{2}K\theta^2. \tag{B3.2}$$

B4. DAMPING EFFECTS

Damping is a term given to the resistance offered to the motion of a body. It is associated with energy dissipation in the form of heat, because damping arises from eddy currents or mechanical friction such as in bearings or between the body and the surrounding medium. It is usual to design mechanical systems to have minimum resistance to motion, but there are many cases where damping is built into a system for the purposes of stability and controlled motion. Shock absorbers are examples in

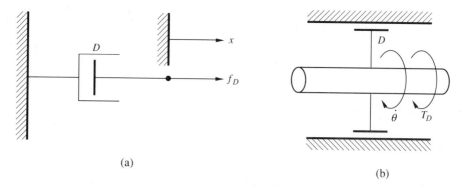

Figure B.3   Damping effects. (a) Translational motion, (b) rotational motion.

which the energy of a sudden disturbance is absorbed quickly in a spring and dissipated slowly by the friction of oil through a small aperture, thus preventing sustained oscillations of the body.

Figure B3 depicts the mechanical circuit symbols used for translational and rotational systems. The relation between the damping force or torque and speed is complicated and nonlinear. However, test results for electromechanical systems prove that it is a reasonable approximation to assume that the relations are given by

$$f_D = D\dot{x} \quad \text{and} \quad T_D = D\dot{\theta} \tag{B4.1}$$

where D is the friction coefficient.

The energy that is irreversibly dissipated is

$$W_D = \int f_D \, dx = \int f_D \dot{x} \, dt \quad \text{and} \quad W_D = \int T_D \, d\theta = \int T_D \dot{\theta} \, dt. \tag{B4.2}$$

Hence

$$W_D = \int D\dot{x}^2 \, dt \quad \text{and} \quad W_D = \int D\dot{\theta}^2 \, dt. \tag{B4.3}$$

B5. MECHANICAL SYSTEMS

Mechanical systems may be represented by circuits comprising lumped elements in series and parallel. In most cases electromechanical systems have a parallel arrangement of elements. We want to find the dynamic equation of motion relating the applied forcing function to the combined effects of inertia, elasticity, damping, and the load.

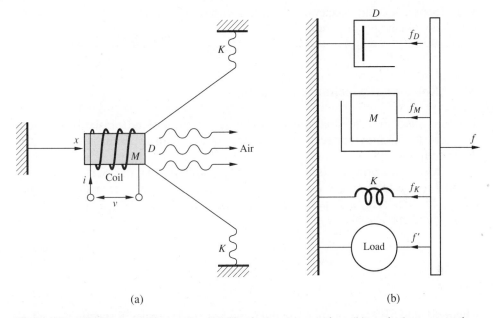

Figure B.4   Electromagnetic speaker. (a) Physical representation, (b) equivalent magnetic circuit.

B5.1.   Translational Mechanical System

Figure B4 shows a sketch of an electromagnetic loudspeaker and its lumped element equivalent circuit. The applied force f_e, which is electromagnetic in origin, is considered to act at the centroid. Opposing motion are the forces of inertia, elasticity, damping and the air, which is the load to which the energy is transmitted in the electrical/mechanical/acoustic energy conversion process. In this case the load could have been lumped with the damping if only mechanical responses were required, but it is kept separate for the sake of generality.

Analogous to Kirchhoff's current law at a node is d'Alembert's principle for a mechanical system in equilibrium. D'Alembert's principle states that at the centroid of a body the sum of the input forces is equal to the sum of the output forces. That is, for the case of motoring action,

$$f_e = f_M + f_D + f_K + f'. \qquad \textbf{(B5.1)}$$

Hence

$$f_e = M\frac{d^2x}{dt^2} + D\frac{dx}{dt} + Kx + f'. \qquad \textbf{(B5.2)}$$

If the applied force f_e and the load force f' are known functions of time, then the solution of this differential equation provides the information how the displacement

x or the speed $\dot{x}$ or the acceleration $\ddot{x}$ varies with time. This is a general equation for translational energy converters.

B5.2. Rotational Mechanical System

The concepts of the rotational system are the same as those of the translational system discussed in the previous section. A general system with rotational movement is depicted in Fig. B5.

D'Alembert's principle is an alternative way of stating that the total energy input is equal to the total energy output, which has contributions in terms of potential energy, kinetic energy and damping energy. For an incremental displacement $d\theta$ the incremental energy balance is

$$dW = dW_J + dW_D + dW_K + dW' \tag{B5.3}$$

or

$$T_e d\theta = T_J d\theta + T_D d\theta + T_K d\theta + T' d\theta \tag{B5.4}$$

where dW and dW' are the incremental input and load energies, respectively, and T_e

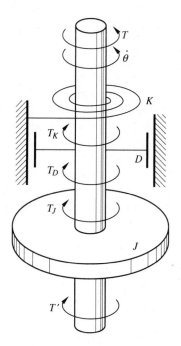

Figure B.5 Rotational mechanical system.

and T' are the applied and load torques, respectively. Hence

$$T_e = T_J + T_D + T_K + T'. \tag{B5.5}$$

In terms of the lumped elements

$$T_e = J\ddot{\theta} + D\dot{\theta} + K\theta + T'. \tag{B5.6}$$

A solution of this differential equation gives the motional response as a function of time.

In this dynamic equation of motion the elastic effect K models a torsional spring such as that found in the moving coil movement of an ammeter. For motors with continuous motion the elastic effect of the shaft is usually neglected unless torsional oscillations are being studied. Load characteristics are often nonlinear and make the solution of eq. (B5.6) complicated. However, the equation is simplified if the load torque is constant or can be combined with either the inertial torque or the damping torque.

AC Characteristics of Ferromagnetic Circuits

C1. INTRODUCTION

Transformers as well as motors, generators, magnetic relays, phonograph pickups, tape recorder heads, and moving coil and moving-iron ammeters and voltmeters have magnetic circuits that are well defined. The circuits have to be designed to have high efficiency. *High efficiency* means that a required flux density in a particular region of the circuit is produced by a low value of magnetomotive force. This is achieved by using low-reluctance paths provided by ferromagnetic materials.

The topic, ac characteristics of ferromagnetic circuits, involves the study of the effects associated with ac excitation of these circuits. This is treated in some detail, because it provides a basis for understanding the loss mechanisms and the approximations that are made in the development of models for transformers and ac machines.

C2. MAGNETIC MATERIAL CHARACTERISTICS

Assume that we have a coil wound around a sample of virgin ferromagnetic material, perhaps like that shown in Figure A4. We can vary the value of the coil current and measure the value of magnetic flux density B that corresponds to a particular value of magnetic field intensity H.

If the magnetic field intensity is increased from zero to a value $+H_{max}$ we will measure values of B which are located on the nonlinear curve oab' as in Figure C1.

547

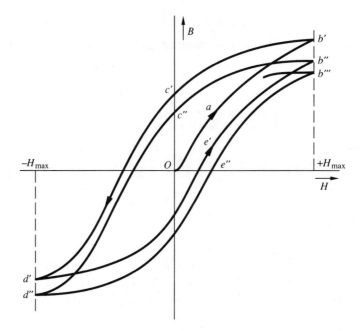

Figure C.1   Magnetization cycling.

We can decrease the value of the field intensity from $+H_{max}$ through zero to $-H_{max}$ and find that the measured values of B do not retrace the previously measured curve but form such a curve as $b'c'd'$. Increasing the field intensity from $-H_{max}$ through zero back to $+H_{max}$ results in the curve $d'e'b''$, where point b'' is lower than point b', which was previously measured at the same value $+H_{max}$. Decreasing the field intensity again from $+H_{max}$ through zero to $-H_{max}$ will result in the curve $b''c''d''$. Taking the field intensity up again to $+H_{max}$ we obtain curve $d''e''b'''$. Point b''' is closer to point b'' than point b'' is to point b'. After several cycles of field intensity variations between the limits $+H_{max}$ and $-H_{max}$, the measured value of flux density will retrace a fixed closed loop. This closed loop is symmetrical about the origin if $+H_{max}$ and $-H_{max}$ have the same absolute value. The steel is then said to be in its cyclic condition for a particular value of H_{max}.

Smaller closed loops will be measured in the cyclic condition, if the ferromagnetic material is subjected to cyclic variations of the magnetic field intensity between extreme values $\pm H_{max}$ that are smaller than the foregoing ones. This is illustrated in Fig. A6. The loops are called hysteresis loops.[4]

---

[4] The Greek word *hysteresis* means *coming late* and for magnetic materials *hysteresis* refers to the magnetic effect B lagging its cause H.

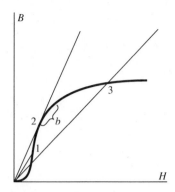

Figure C.2 Normal magnetization curve.

Magnetic-field calculations are based on a single-valued relation for the $B - H$ relation, which is called the *normal magnetization curve.*[5] For soft magnetic materials, that is, non-permanent magnetic materials, the errors introduced are negligible because the hysteresis loops are very narrow. Typical normal magnetization curves are shown in Figs. A1 and A2.

If B and H are related to each other by a normal magnetization curve, the permeability obtained from this curve is called the *static permeability.* From Fig. C2 we see that the value of the static permeability for different points on the normal magnetization curve varies. The permeability increases from point 1 to point 2 on the curve and decreases again from point 2 to point 3 and beyond. Thus the permeability of a ferromagnetic material is not a constant, but changes with the excitation (I, F, or H). It is desirable for economic reasons to work the magnetic materials at high flux densities, preferably with as small an mmf as possible. Therefore the normal operating flux densities in practical devices are in the region b of Fig. C2. This region is called the *knee of the magnetization curve.*

C3. VOLTAGE AND TIME-VARYING MAGNETIC FLUX

We need to know the general equation for the rms (or effective) value of the voltage that is induced in a coil by a time-varying flux. Let us assume an induced emf with a symmetric and periodic waveshape as shown in Fig. C3. From the waveshape of the emf the general form of the magnetic flux linking the coil of N turns can be deduced. Use is made of Faraday's law.

---

[5] See section A5, Approximation 1 of Appendix A, and Fig. A6.

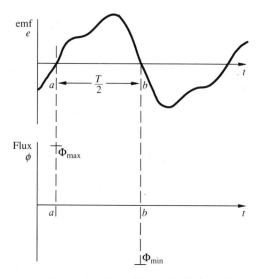

Figure C.3   Nonsinusoidal periodic waveform.

At the instants a and b, shown in Fig. C3, the induced emf is zero. From Faraday's law $N\dfrac{d\phi}{dt} = -e = 0$, it follows that the slope $d\phi/dt$ is zero. The flux on the ϕ-time curve has a maximum or minimum value. We differentiate Faraday's equation to obtain

$$\frac{d^2\phi}{dt^2} = -\frac{1}{N}\frac{de}{dt}. \tag{C3.1}$$

At instant a the slope de/dt is positive. Therefore $d^2\phi/dt^2$ is negative. This indicates a maximum value of ϕ at time a. A similar argument shows that the flux ϕ is a minimum at time b.

The waveform of the flux is symmetrical because the emf waveform is symmetrical (from Faraday's law). Accordingly

$$|\Phi_{\min}| = |\Phi_{\max}| \triangleq \hat{\Phi}. \tag{C3.2}$$

Integration of Faraday's equation between the two time limits a and b yields

$$\int_a^b e\,dt = -N\int_a^b \frac{d\phi}{dt}\,dt = -N\int_{\hat{\Phi}}^{-\hat{\Phi}} d\phi = 2N\hat{\Phi}. \tag{C3.3}$$

The average value of the emf is

$$e_{av} = \frac{1}{t_b - t_a}\int_a^b e\,dt = \frac{2}{T}\int_a^b e\,dt \tag{C3.4}$$

where T is the period (in seconds) of the waveform. It follows that

$$e_{av} = 4\hat{\Phi}\,\frac{1}{T}\,N = 4\hat{\Phi}fN \tag{C3.5}$$

where f is the fundamental frequency of the voltage waveform. Since the form factor of a periodic wave is defined as

$$\text{Form factor} \triangleq \frac{\text{rms value}}{\text{average value}} \tag{C3.6}$$

the rms value E of the induced emf is

$$E = e_{av} \times \text{form factor.} \tag{C3.7}$$

On substitution, the general equation for the rms value of the emf induced in a coil of N turns by a periodically varying flux becomes

$$E = 4(\text{form factor})\hat{\Phi}fN. \tag{C3.8}$$

The average value of a periodic function is defined as

$$\text{average value} \triangleq \frac{2}{T}\int_0^{T/2} f(t)\,dt$$

and the rms (or effective) value is defined by

$$\text{rms value} = \left[\frac{1}{T}\int_0^T t^2(t)\,dt\right]^{1/2}$$

Hence the form factor can be determined for any waveshape. For a sinusoidal voltage of amplitude V_{max} the rms value is $0.707 V_{max}$, the average value is $\dfrac{2}{\pi}V_{max}$, so that the form factor is 1.11.

In the performance analysis of transformers and machines we consider only sinusoidal voltage waveshapes. For these waveshapes eq. (C3.8) becomes

$$E = 4.44\hat{\Phi}fN. \tag{C3.9}$$

It is of interest to note that the geometry and magnetic characteristics of the magnetic circuit have no effect on the rms value of the emf induced by a time-varying flux. This important equation is encountered again in the study of transformers, induction motors, and synchronous machines.

EXAMPLE C1

A source of voltage with a square waveform is connected to the terminals of a coil wound on a ferromagnetic core of constant cross section. For a supply voltage of 115 V rms at a frequency of 60 Hz, the peak flux density B_{max} is 1.0 T, for which the magnetic field strength H is 300 AT/m. The cross-sectional area of the core is $4.8 \times 10^{-4} m^2$. What are the number of turns N on the winding and the maximum value of the magnetizing current $\hat{I}_m$? The core length is 2.4 m.

Solution

If the resistance of the winding is negligibly small, from eq. (C3.9) we have

$$V \approx E = 4 \times \text{form factor} \times \hat{\Phi} fN.$$

The form factor for a square wave is 1.0, so on substitution

$$115 = 4 \times 1.0 \times (1.0 \times 4.8 \times 10^{-4}) \times 60 \ N.$$

Therefore $N = 1000$ turns.
From Ampere's circuital law

$$\text{``} H = \frac{NI}{l} . \text{''}$$

On substitution $300 = \dfrac{1000 \hat{I}_m}{2.4}$, so $\hat{I}_m = 0.72$ A.

C4. IRON LOSSES

Time-varying fluxes produce losses in their ferromagnetic circuits. These losses, which are called *iron losses,* are energy that is irreversibly converted into heat. The occurrence of these losses has an important effect on the design of devices because

- they lower the operating efficiency
- they increase the operating temperature. In order to prevent the temperature rising beyond a certain limit, cooling is required, and this increases the capital cost.

The operating temperature has to be limited to a certain value to prevent the electrical insulating materials from losing their insulating and mechanical properties through accelerated aging. Thus the rated volt-ampere capacity of a device is limited by the permissible temperature rise. Insulating materials are grouped into classes according to the maximum temperature they can withstand. If the rated voltages or currents of a device are exceeded on a continuous basis, the lifetime of the insulation is drastically reduced.

The losses produced in ferromagnetic materials by time-varying fluxes consist of hysteresis loss and eddy current loss.

C4.1. Hysteresis Loss

The interpretation of the behavior of ferromagnetic materials is based on the assumption that there are small, local regions (or domains), in which adjacent atomic magnetic dipoles are aligned in parallel. In the absence of an externally applied magnetic field these domains can be randomly oriented. While subjected to an mmf the domains tend to align themselves with the magnetic field intensity **H**. Consequently, when the excitation is cyclic, the domains experience a periodic reorientation. Work has to be done to orient the domains. This work manifests itself as heat, and represents a loss that is called the *hysteresis* loss. The energy loss is supplied from the excitation source.

In order to derive an expression for the hysteresis loss we can consider a toroidal magnetic circuit excited by a coil which has N turns, as shown in Fig. C4. A time-varying voltage $v_1(t)$ is applied to the coil. From an inspection of the figure, Kirchhoff's voltage law can be written as

$$v_1(t) - i(t)R = v(t) \tag{C4.1}$$

where $v(t)$ is the inductive voltage drop across the coil with the coil resistance R placed external to it. Using Faraday's law, the voltage $v(t)$ can be expressed in the form

$$v(t) = N\frac{d\phi(t)}{dt} = NA\frac{dB(t)}{dt}. \tag{C4.2}$$

The flux is produced by the excitation current $i(t)$. By Ampere's circuital law

$$i(t) = \frac{H(t)l}{N}. \tag{C4.3}$$

The net electric energy supplied by the source to the ferromagnetic circuit during a time interval dt is

$$p\,dt = v(t)i(t)dt = AlH(t)dB(t) \tag{C4.4}$$

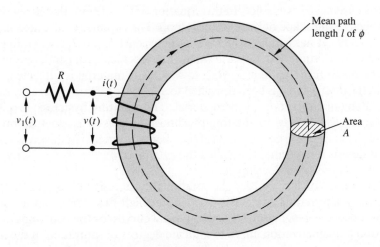

Figure C.4   Toroidal magnetic circuit.

where p is the instantaneous power. During one complete cycle (T seconds) of the applied voltage, the energy input per unit volume of the magnetic circuit (or iron core) is

$$\frac{1}{Al}\int_{t}^{t+T} p\, dt = \int_{B_1}^{B_2} H(t)dB(t). \tag{C4.5}$$

The limits of integration, B_1 at $t = t$ and B_2 at $t = t + T$, have the same numerical value, because the integration takes place around a complete hysteresis loop. The value of the integral is equal to the area enclosed by the hysteresis loop. Therefore the energy input per unit volume of iron for one cycle is equal to the area of the hysteresis loop.

The power input per unit volume is equal to the energy input per unit volume per second. Hence for the entire magnetic circuit the power input is

$$P_h = V_{vol} f \oint H\, dB \tag{C4.6}$$

where P_h = power loss due to hysteresis (W)

V_{vol} = volume of the iron core (m^3)

f = supply frequency (Hz)

and $\oint HdB$ = area of the hysteresis loop (J/m^3).

This integration can be done by graphical methods, since the relation between B and H of a hysteresis loop cannot be expressed analytically with any degree of accuracy.

A search for a more useful method to calculate hysteresis loss led C. P. Steinmetz, in 1892, to the empirical relation

$$P_h = \eta f B_{max}^{x} V_{vol} \tag{C4.7}$$

and this has been used ever since. In this equation η is a constant that depends on the factors affecting the shape of the hysteresis loop. The exponent x is called the *Steinmetz coefficient*. It has a value which varies between 1.5 and 2.5 depending on the quality and type of material. Also, the value of the coefficient depends on the magnitude of the maximum flux density B_{max} to which the material is subjected.

So far the discussion has been restricted to alternating fluxes, where the magnetization is along one direction in the material. The resulting hysteresis loss is called *alternating hysteresis loss.* In rotating machines the axis of magnetization rotates relative to the material in certain parts of the magnetic circuit. In this case a different type of hysteresis loss occurs. It is called the *rotational hysteresis loss,* for which no simple formula has been developed.

At low values of flux density the rotational hysteresis loss is greater than the corresponding alternating hysteresis loss, and at high flux densities the rotational hysteresis loss decreases. Figure C5 shows typical curves for the two kinds of hysteresis loss and a magnetization curve to indicate degrees of saturation of the material corresponding to the different loss levels.

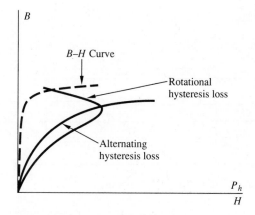

Figure C.5 Alternating and rotational hysteresis losses.

C4.2. Eddy Current Loss

An electric field is induced in any medium in which a magnetic field varies as a function of time. If the medium is an insulator, there are no further consequences, but if the medium is a conductor (and this is what all ferromagnetic materials are), currents will flow as a result of the induced voltages. These currents produce ohmic losses that are called eddy current losses.

Consider a bar of ferromagnetic material as shown in Fig. C6. Assume that a time-varying magnetic flux $\phi(t)$ is directed perpendicular to the cross section of the bar. Induced eddy currents will flow in planes perpendicular to the axis of the magnetic field. These currents set up their own magnetic fields which oppose the

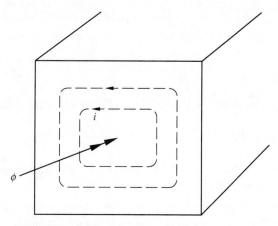

Figure C.6 Eddy currents.

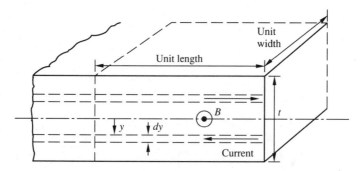

Figure C.7 Lamination eddy current.

change of the original field and thus cause the resultant magnetic flux density to decrease progressively from the surface of the bar inward. The effect of the eddy currents is to crowd the flux toward the surface of the magnetic core.

An approximate expression for the eddy current loss can be derived if it is assumed that a uniform flux density exists in the core material. Consider a sheet of steel of thickness t, which is small compared to the other dimensions. See Fig. C7. At a distance y from the center-line of the ferromagnetic sheet consider an elementary path of width dy. Let the uniformly distributed flux vary as a function of time according to $B(t) = B_{max}\sin\omega t$. The flux enclosed by the elementary path of unit length is

$$\phi(t) = 2yB_{max} \sin\omega t. \tag{C4.8}$$

The voltage drop induced in this part of the elementary path is

$$v(t) = \frac{d\phi(t)}{dt} = 2\omega y B_{max} \cos\omega t. \tag{C4.9}$$

Neglecting the end portions of the path, the resistance of the elementary path is $R = 2\rho/dy$, where ρ is the resistivity of the material and the width of the lamination is taken to be one unit of length. The eddy current in the elementary path is

$$i(t) = \frac{v(t)}{R} = \frac{\omega y B_{max} \cos\omega t}{\rho} dy. \tag{C4.10}$$

The energy per unit time converted to heat in the elementary path is

$$P_1 = \frac{1}{T} \int_0^T v(t)i(t) \, dt = \frac{\omega^2 y^2 B_{max}^2}{\rho} dy. \tag{C4.11}$$

Integrating between the limits $y = \frac{1}{2}t$ and $y = 0$, the total eddy-current loss per unit length and width of the sheet of thickness t is

$$P_2 = \int_{y=0}^{y=\frac{1}{2}t} \frac{\omega^2 B_{max}^2 y^2}{\rho} \, dy = \frac{\omega^2 B_{max}^2 t^3}{24 \rho}. \tag{C4.12}$$

Substituting $\omega = 2\pi f$, the eddy current loss per unit of volume V_{vol} is

$$P_3 = \frac{\pi^2}{6\rho} t^2 f^2 B_{\max}^2.$$ (C4.13)

Hence the total eddy current loss is

$$P_e = \frac{\pi^2}{6\rho} t^2 f^2 B_{\max}^2 V_{vol}.$$ (C4.14)

Note that the eddy current loss varies as the square of the sheet thickness t. To keep the eddy current loss low, magnetic circuits are built with thin laminations. The eddy current loss also varies as the square of the frequency f. This calls for the use of even thinner laminations for higher-frequency applications. At very high frequencies powdered cores are used.

In practice, account must be taken of the nonuniform flux density distribution and material inhomogeneity, so that we arrive at the more practical equation

$$P_e = \sigma t^2 f^2 B_{\max}^2 V_{vol}$$ (C4.15)

where σ is a constant depending on the type of material and has units siemens per meter.

C4.3. Separation of Iron Losses

The total power loss in a magnetic circuit is the sum of the component losses, the hysteresis loss, and the eddy current loss, and is known as the *iron loss*. This loss is present in any apparatus used for electric energy conversion or transformation if it is connected to an ac supply. Therefore this loss[6] has an important bearing on the operating cost of the device.

The iron loss can be resolved into its two components by the following method. The total loss is the sum of the hysteresis and eddy current losses, that is

$$P_h + P_e = (\eta B_{\max}^x V_{vol})f + (\sigma t^2 B_{\max}^2 V_{vol})f^2 = K_1 f + K_2 f^2$$ (C4.16)

where, for a given maximum flux density, K_1 and K_2 are constants. This can be rewritten as

$$\frac{P_h + P_e}{f} = K_1 + K_2 f$$ (C4.17)

and is the equation of a straight line relating $(P_h + P_e)/f$ and the frequency f. The slope of the straight line is K_2, which, if multiplied by f^2, gives the eddy current loss. The intercept of the line with the ordinate axis is K_1, which, if multiplied by f, gives the hysteresis loss.

---

[6] Iron loss is present whether the apparatus has no load or is delivering power. This loss is often called the *no-load* loss, because other losses are minimal for this condition.

In practice the iron loss is obtained from experiments at different frequencies but at constant maximum flux density.

EXAMPLE C2

A 60-Hz voltage of 120V that is applied to a reactor winding produces a loss of 100W in the ferromagnetic circuit. If both the frequency and voltage are doubled, the core loss is 300W. Calculate the eddy current loss and the hysteresis loss at 120V, 60Hz.

Solution
For the conditions $V_1 = 120V$ and $f_1 = 60Hz$,

$$V_1 = 4.44\hat{\Phi}_1 f_1 N.$$

For the conditions $V_2 = 240V$ and $f_2 = 120Hz$, then $V_2 = 2V_1$, $f_2 = 2f_1$ and $V_2 = 4.44\hat{\Phi}_2 f_2 N$, or $2V_1 = 4.44\hat{\Phi}_2 \times 2f_1 N$.
Therefore, $\hat{\Phi}_1 = \hat{\Phi}_2$ and B_{max} is the same for both conditions.
The core losses are given by

$$P_1 = P_{h1} + P_{e1} = k_h f_1 B_{max}^x + k_e f_1^2 B_{max}^2 = 100W$$

and

$$P_2 = P_{h2} + P_{e2} = k_h f_2 B_{max}^x + k_e f_2^2 B_{max}^2 = 300W.$$

Thus $\dfrac{P_1}{f_1} = \dfrac{100}{60} = k_h B_{max}^x + k_e f_1 B_{max}^2$

and $\dfrac{P_2}{f_2} = \dfrac{300}{120} = k_h B_{max}^x + 2k_e f_1 B_{max}^2$.

Accordingly, $k_e f_1 B_{max}^2 = \dfrac{P_2}{f_2} - \dfrac{P_1}{f_1} = \dfrac{P_{e1}}{f_1} = \dfrac{300}{120} - \dfrac{100}{60}$.

Hence, $P_{e1} = 50W$ and $P_{h1} = P_1 - P_{e1} = 50W$.

C4.4. Dynamic Hysteresis Loop

The hysteresis loop introduced earlier is called the static hysteresis loop.

If the magnetic field intensity H alternates at the power frequency, an alternating flux results. This causes eddy currents to flow in the iron and these have an effect on the resultant flux which, in turn, affects the shape of the hysteresis loop. To determine this effect we start with the static loop of flux versus current, shown by the solid line curve in Fig. C8. While the flux is increasing in the ascending branch of the static loop, the eddy currents try to oppose this increase in flux. In order to maintain a particular value of flux, for instance ϕ_1, an additional amount of coil current Δi_1 will be required to offset the opposing effect of the eddy currents. Thus we find that for an alternating flux, point a is a point of the ascending branch of a different loop. Similar arguments apply to other points on the ascending branch, but the amount of addi-

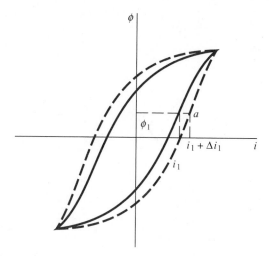

Figure C.8 Eddy current effect on the hysteresis loop.

tional coil current Δi required depends on the rate of change of the flux. If a sinusoidally varying flux is assumed, the largest rate of flux change takes place at zero flux, and when the flux has its maximum value the rate of change is zero.

The presence of eddy currents is to increase the width of the hysteresis loop as is indicated by the broken line curve in Fig. C8. The higher the frequency of the alternating flux, the larger the eddy currents, the wider the hysteresis loop.

A hysteresis loop that takes the effect of eddy currents into consideration is called a *dynamic hysteresis loop*. We found that the area of the static hysteresis loop is equal to the hysteresis loss in the material per cycle and per unit volume. The area of a dynamic hysteresis loop is equal to the total iron loss in the material per cycle and per unit volume for a particular frequency.

C5. AC EXCITATION CHARACTERISTICS OF FERROMAGNETIC CIRCUITS

The purpose of this section is to determine the current response to a sinusoidally varying voltage that is applied to a coil wound on a ferromagnetic circuit.

If an alternating voltage is applied to a coil, a current will flow, which, with the N turns of the coil, forms the magnetomotive force. This mmf produces an alternating flux in the magnetic circuit. The alternating flux induces a voltage drop v in the coil. If the applied voltage varies sinusoidally with time, $d\phi/dt$ has to vary sinusoidally.[7] Therefore the flux ϕ varies cosinusoidally with time. A linear flux-current relation

---

[7] This accords with Faraday's law.

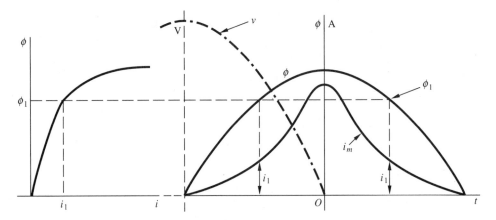

Figure C.9   Magnetization without hysteresis.

would mean that the current would have a cosinusoidal variation also. However, the relation between flux ϕ and current i in the case of a ferromagnetic circuit is not linear because of the magnetic saturation of the material, the effect of magnetic hysteresis, and the presence of eddy currents. It follows that the current will have a distorted waveshape.

Consider the single-valued, normalized saturation curve shown in Fig. C9. The effects of hysteresis and eddy current losses are absent. Also shown in the figure are a half cycle of the applied voltage v and the corresponding flux ϕ as a function of time. A current response to the applied voltage can be constructed graphically from these curves.

For a certain value of flux, say ϕ_1, we can find from the saturation curve the value of the current i_1 that is required to produce this flux. The value of flux ϕ_1 occurs twice during half a cycle and is caused by the same value of current i_1 each time. In this same manner we can find as many points as are required to draw the current waveform, which is indicated in Fig. C9 by the curve labeled i_m.[8] Notice that the current waveform is peaked. This is the result of the maximum value of flux being in the saturated region. If the maximum value of the flux had not exceeded the linear part of the saturation curve, the current waveform would have been sinusoidal.

We can conclude that the saturation phenomenon causes the exciting current to depart from a sinusoidal waveshape. The higher the level of saturation the sharper is the peak of the current waveform. Zero crossings of flux and current waveforms coincide. This means that the flux and the fundamental component of the current are in phase. Since the flux lags the applied voltage by 90°, the fundamental current component lags the voltage by 90° also and is thus a reactive (or wattless) current. The higher harmonic components are also wattless currents since the applied voltage

---

[8] The subscript m is used to denote a magnetizing current since no real power is drawn from the supply in this case.

is sinusoidal of fundamental frequency. Thus, if the normalized saturation curve is used, the derived current is totally reactive (or wattless). This agrees with the initial assumption that hysteresis and eddy current losses were absent.

The actual current waveform is affected by the presence of hysteresis and eddy current losses. We must base the construction of the current waveform on the dynamic flux-current relation. Assume that this relation is given by the loop, of which the top half is shown in Fig. C10. The known sinusoidal voltage and flux waveforms are shown also. For a certain value of flux, say ϕ_1, which occurs at two instants during one half cycle, the corresponding current values can be found from the dynamic saturation curve. Two different current values correspond to the same value of flux, because the flux-current relation is double valued. When the flux is increasing the current i_1 is found from the ascending branch of the loop. When the flux is decreasing the current i_2 is found from the descending branch of the loop. These two current values can be plotted at the appropriate instants when the flux has the value ϕ_1. In a similar manner a sufficient number of points can be found to draw the current waveshape, shown in Fig. C10 and labeled i_{NL}.

Note that the current waveform is peaked but not symmetrical, also the zero values of the current and the flux are displaced in time. Subtracting the instantaneous values of the current i_m of Fig. C9 (also shown in Fig. C10 by the broken-line curve) from the corresponding instantaneous current values of the current i_{NL} results in a sinusoidal current, which is labeled i_{h+e}. The current component i_{h+e} leads the flux wave by 90° and thus is in phase with the applied voltage wave.

Thus the actual exciting current i_{NL}, which results from a sinusoidal applied voltage, can be resolved into two component currents. These are

1. a sinusoidal current component i_{h+e} in phase with the applied voltage. It supplies the power losses in the magnetic circuit and is called the *iron loss current component*.

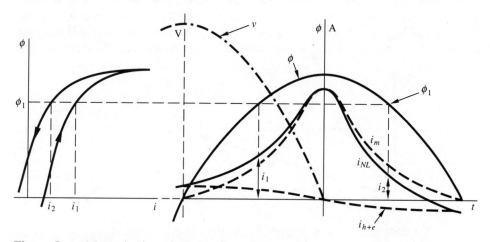

Figure C.10 Magnetization with saturation and hysteresis.

2. a peaked current component i_m, which is reactive and produces the mmf necessary to create the magnetic flux. This is called the *magnetizing current component.*

The complete current waveform is symmetrical, so that a Fourier representation would contain only odd harmonics. Therefore we can write the following expression for the exciting current.

$$i_{NL}(t) = \sqrt{2}I'_1 \sin\omega t + \sqrt{2}I'_3 \sin3\omega t + \cdots$$
$$+ \sqrt{2}I''_1 \cos\omega t + \sqrt{2}I''_3 \cos3\omega t + \cdots \tag{C5.1}$$

This exciting current is the result of an applied voltage $v(t) = \sqrt{2}V \sin\omega t$. The total average power is equal to the sum of the average powers due to voltage and current components of the same frequency. In this case $P = VI'_1$, where I'_1 is the rms value of the current component in phase with and of the same frequency as $v(t)$.

Since we assumed the resistance of the coil to be zero, the power supplied to the coil has to be absorbed by the core. That is

$$P_{NL} = VI'_1 = VI_{h+e}. \tag{C5.2}$$

Consequently $I'_1 = I_{h+e}$. The rms value of the exciting current can be written as

$$I_{NL} = (I^2_{h+e} + I^2_3 + \cdots + I''^2_1 + I''^2_3 + \cdots)^{1/2}. \tag{C5.3}$$

We can write the rms value of the peaked current (see Fig. C9) as

$$I_m = (I^2_3 + \cdots + I''^2_1 + I''^2_3 + \cdots)^{1/2}. \tag{C5.4}$$

Thus

$$I_{NL} = (I^2_{h+e} + I^2_m)^{1/2}. \tag{C5.5}$$

Since I_{h+e} is a current component in phase with the voltage and the current component I_m together with the flux ϕ are both lagging the applied voltage by 90°, it appears that we can draw a phasor diagram for these quantities as shown in Fig. C11. This is not really correct, because in a phasor diagram we may draw only those quantities that have the same frequency in order to maintain the proper phase angles. Therefore V, I_{h+e}, and Φ can be plotted on one phasor diagram, but this diagram should not include I_m and I_{NL} which contain higher harmonics. However, for the

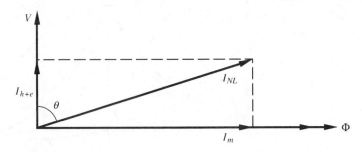

Figure C.11 Phasor diagram of excitation current components.

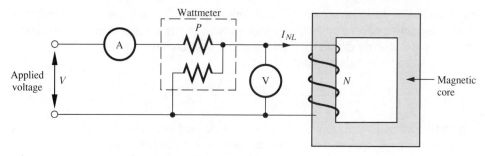

Figure C.12 Measurement of excitation components.

development of phasor diagrams of ac machines it is necessary to represent all currents on one diagram. To do this we introduce the concept of the *equivalent sinusoidal current*. Such a current has the same rms value as the current it represents and is assumed to have the fundamental frequency.

The rms values of the phasors shown in Fig. C11 can be determined from the measurements indicated in Fig. C12. If the corrected wattmeter reading is P, the corrected voltmeter reading is V, and the corrected ammeter reading is I_{NL}, we find

$$I_{h+e} = \frac{P}{V}, \quad I_m = (I_{NL}^2 - I_{h+e}^2)^{1/2} \quad \text{and} \quad \cos\theta = \frac{P}{VI_{NL}}. \tag{C5.6}$$

$\cos\theta$ is defined as the power factor and θ is the phase angle between the phasors V and I_{NL}. The value of the flux is determined from

$$\Phi = \frac{\hat{\Phi}}{\sqrt{2}} = \frac{V}{4.44 \, fN\sqrt{2}}. \tag{C5.7}$$

The development of phasor diagrams for transformers and machines involves adding a true sinusoidal current I to I_{NL}. Again, we may not perform a standard phasor addition because I_{NL} is not a sinusoidal current.

In the case of transformers, which have a closed ferromagnetic circuit, I_{NL} is normally only a few percent of other currents and a straightforward phasor addition will result in a negligible error. In the case of rotating machines the magnetic circuit contains air gaps, which have relatively large and linear reluctances. This has the effect of increasing the fundamental component of I_{NL} considerably. The waveshape of I_{NL} is then very close to being sinusoidal and a simple phasor addition with other sinusoidal currents will again result in negligible error.

C6. ELECTRIC CIRCUIT REPRESENTATION

Analysis of an electromagnetic device, or analysis of a system of which the device is a component, can be accomplished by representing the device by an equivalent electric circuit. Electromagnetic devices have a magnetic circuit that is excited by one or more coils. For the moment we will deal with a single coil. Such a coil has a predominantly inductive effect and as a first approximation the arrangement can be represented by the electric circuit symbol for an ideal inductance.

Across an ideal inductor the voltage drop is

$$v = \frac{d\lambda}{dt} = \frac{d(N\phi)}{dt} = \frac{d(N\phi)}{di}\frac{di}{dt} = L\frac{di}{dt} \tag{C6.1}$$

where the inductance L, in a linear system, is defined as the total flux linkage of all turns per ampere, or $L = N\phi/i$. This can be rewritten as follows.

$$L = \frac{N\phi}{i} = \frac{NBA}{i} = \frac{N\mu HA}{i} = \frac{N\mu(Ni/l)A}{i} = \mu\frac{N^2A}{l}. \tag{C6.2}$$

Equation (C6.2) gives us the information that the magnitude of the inductance depends on the value of the permeability and the magnetic circuit geometry.

In a nonlinear system we must take into account the fact that the permeability μ is a function of the magnetic field intensity H (or the current i). So, for the voltage drop, eq. (C6.1) should be written as $v = d(Li)/dt$.

If a coil is wound on a ferromagnetic circuit, the coil current has a power component I_{h+e}, that accounts for the hysteresis and eddy current losses in the ferromagnetic material. Therefore an equivalent circuit should contain a component to represent the losses. These losses are expressed by

$$P_{h+e} = VI_{h+e} = V_{vol}(\eta f B_{max}^2 + \sigma f^2 t^2 B_{max}^2). \tag{C6.3}$$

Steinmetz's coefficient has been assigned the value $x = 2$. This is a realistic assumption. The maximum flux density is

$$B_{max} = \frac{\hat{\Phi}}{A} = \frac{V}{4.44\,fNA}. \tag{C6.4}$$

Substituting this expression into eq. (C6.3) yields

$$I_{h+e} = \frac{V_{vol}\eta f V^2}{V(4.44\,NfA)^2} + \frac{V_{vol}\sigma f^2 t^2 V^2}{V(4.44\,NFA)^2}$$

$$= C_1\frac{V}{f} + C_2 V \tag{C6.5}$$

where C_1 and C_2 are constants for a fixed value of B_{max}. For a particular value of frequency

$$I_{h+e} = (C_1' + C_2)V \tag{C6.6}$$

where C_1' is a new constant. The constant $(C_1' + C_2)$ has the dimensions of conductance. Thus we can write

$$I_{h+e} = G_m V \tag{C6.7}$$

and, for a particular frequency and maximum flux density, the hysteresis and eddy

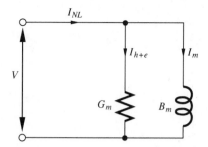

Figure C.13 Equivalent circuit of coil on ferromagnetic core.

current losses can be modeled by a conductance. Since

$$I_{NL} = I_m + I_{h+e} \tag{C6.8}$$

the logical equivalent circuit is one in which the conductance G_m is in parallel with a circuit element that represents the inductive behavior of the device, that is, a susceptance B_m.

This equivalent electric circuit of a coil wound on a ferromagnetic circuit is shown in Fig. C13. Any equivalent circuit is only of value if the magnitudes of the circuit elements can be determined experimentally. The values of G_m and B_m can be calculated from measurements taken from a test circuit as shown in Fig. C12. At particular values of voltage and frequency the corrected meter readings are taken to be P watts, I_{NL} amperes, and V volts. The conductance value follows directly from

$$G_m = \frac{P}{V^2}. \tag{C6.9}$$

The circuit admittance Y_m and susceptance B_m follow from

$$Y_m = \frac{I_{NL}}{V} \quad \text{and} \quad B_m = (Y_m^2 - G_m^2)^{1/2}. \tag{C6.10}$$

In the above discussion it was tacitly assumed that the resistance of the coil was negligible. This assumption is justified, since in all actual devices the coil resistance is made as small as economically acceptable. The exciting current I_{NL} is made as small as possible also. Consequently, any effect of voltage drop in the coil $I_{NL}R$ and of ohmic coil loss $I_{NL}^2 R$ is negligible.

Instead of using the parallel elements G_m and B_m it is possible to use the equivalent series elements as shown in Fig. C14. From the same test values of power, current, and voltage, the resistance R_c to account for core losses, the circuit impedance Z_m, and the reactance X_c to account for magnetization are

$$R_c = \frac{P}{I_{NL}^2}, \quad Z_m = \frac{V}{I_{NL}} \quad \text{and} \quad X_c = (Z_m^2 - R_c^2)^{1/2}. \tag{C6.11}$$

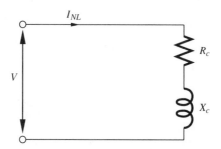

Figure C.14 Series equivalent circuit.

The relations between the two circuit models, depicted in Figs. C13 and C14, can be obtained by writing the admittance for both cases. That is

$$\mathbf{Y}_m = \frac{1}{\mathbf{Z}_m} = \frac{1}{R_c + jX_c} = \frac{R_c}{Z_m^2} - j\frac{X_c}{Z_m^2} = G_m - jB_m. \qquad \textbf{(6.12)}$$

EXAMPLE C3

An exciting current, which can be expressed by

$$i_{NL}(t) = 10\sin(377t - 0.52) + 12\sin(1131t - 1.05)\text{A}$$

flows through a coil of 10 turns wound on a ferromagnetic circuit. The resulting flux is

$$\phi(t) = 0.053\sin(377t - 1.57)\text{Wb}.$$

Neglecting the coil resistance, determine the iron losses, the rms value of the exciting current, the rms value of the core loss current, and the rms value of the magnetizing current.

Solution
The instantaneous value of the applied voltage is

$$v(t) = N\frac{d\phi(t)}{dt} \approx 200\sin 377t \text{ volts}.$$

Only the fundamental component of $i_{NL}(t)$ will contribute to the power, since it is the term corresponding to the frequency of $v(t)$.

Hence the iron losses are $\dfrac{200}{\sqrt{2}} \times \dfrac{10}{\sqrt{2}}\cos 0.52 = 867$ W.

The rms value of the exciting current is

$$I_{NL} = [(10/\sqrt{2})^2 + (12/\sqrt{2})^2]^{1/2} = 11 \text{ A}.$$

We need to find the fundamental frequency component of current that is in phase with the voltage. The fundamental frequency current component $i_1(t)$ is[9]

$$i_1(t) = 10 \sin(377t - 0.52) = 10(0.87 \sin 377t - 0.5 \cos 377t).$$

The effective value of the component in phase with the voltage is $I_{h+e} = 8.7/\sqrt{2} = 6.15$ A.

Finally, the rms value of the magnetizing current is

$$I_m = (I_{NL}^2 - I_{h+e}^2)^{1/2} = 9.2 \text{ A.}$$

C7. INRUSH CURRENT

The *inrush current* is the transient current that flows when a coil, wound on a ferromagnetic circuit, is connected to a supply. This current can have a magnitude many times the steady-state value and its decay time can be a few seconds. A common example is the transformer. The inrush phenomenon poses a problem in that the high transient current can trip overcurrent relays.

The steady-state variation of applied voltage and the associated flux are shown in Fig. C15. For each instantaneous value of voltage there corresponds only one value of flux. Let us look at the transient case.

When a transformer is disconnected from the voltage supply the core flux does not necessarily attain a zero value but may maintain a value ϕ_r, which is the remanent flux corresponding to the zero excitation point of the hysteresis loop. Depending on the core material and construction, ϕ_r may be as high as 60% of the maximum value $\hat{\Phi}$. The presence of a remanent core flux has an effect on the transient flux and current.

The most severe current transient will take place when, at the moment of switching, the impressed voltage is passing through its zero value. The flux would normally vary from $-\hat{\Phi}$ to $+\hat{\Phi}$ during the first half-cycle. To induce the proper voltage in the winding after switching, the flux must vary over the range $2\hat{\Phi}$, if it is assumed that the winding voltage drop is negligible. Under the worst condition the initial flux has a value $+\phi_r$. Then the flux must increase to a peak value $2\hat{\Phi} + \phi_r$. See Fig. C16. A practical transformer is designed to have a steady-state value $\hat{\Phi}$ that is slightly beyond the knee of the saturation curve. Therefore the core will be fully saturated at the flux value $2\hat{\Phi} + \phi_r$, and the permeability will be reduced to the permeability of air. If the core goes into saturation first at a value ϕ_{sat}, an additional flux $\phi_a = 2\hat{\Phi} + \phi_r - \phi_{sat}$ has to be produced in a medium that has the permeability of air.

It can be concluded that, when the applied voltage passes through a zero value at the moment of switching, the exciting current can reach a peak value that is several

---

[9] $\sin(A - B) = \sin A \cos B - \cos A \sin B.$

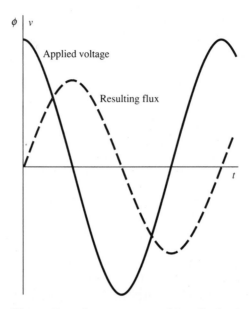

Figure C.15 Instantaneous values of voltage and flux.

hundred times as large as the steady-state exciting current. For such a large inrush current the leakage impedance voltage in the winding cannot be neglected. The voltage induced by the flux must at every instant equal the difference between the instantaneous values of the applied voltage and the winding voltage drop. As a result the induced voltage will be smaller than the applied voltage and the flux will not be

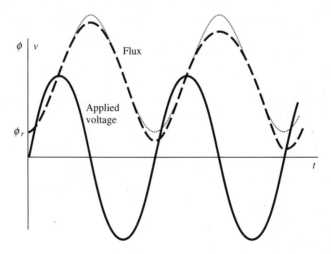

Figure C.16 Flux variation. - - - - actual flux variation, ... theoretical flux variation.

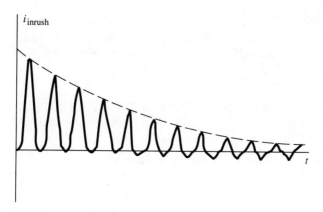

Figure C.17 Typical inrush current waveform.

required to reach a peak value as large as $2\hat{\Phi} + \phi_r$ after the first half cycle of voltage. See Fig. C16. Therefore, during each subsequent half-cycle the flux variation will not be $2\hat{\Phi}$, as would be the case if the winding voltage drop were neglected, but will be less to an extent depending on the value of the exciting current. Thus the positive flux peaks, shown in Fig. C16, reduce gradually in value. The peak value of the exciting current also reduces gradually and eventually the flux and the exciting current will reach steady-state conditions. During the first few cycles after switching, it can be seen from Fig. C16 that the flux is virtually unidirectional. The same is true for the inrush current, for which a typical waveform is shown in Fig. C17.

C8. SUMMARY

For sinusoidal excitation the rms induced emf in a coil is $E = 4.44\hat{\Phi}\ fN$. This equation is important for transformers and ac machines.

Iron losses have hysteresis and eddy current component losses. The hysteresis loss is $P_h = \eta f B^x_{max}\ V_{vol}$ and the eddy current loss is $P_e = \sigma t^2 f^2 B^2_{max}\ V_{vol}$.

A sinusoidally excited coil wound on a ferromagnetic core can be modeled by a reactance to represent the magnetization and a resistance to account for the iron loss. Transformer and induction motor analysis is aided by the use of this model.

C9. PROBLEMS

C.1. Calculate the effective length and cross-sectional area of the air gap in a reactor that has 300 turns and is required to have an induced emf of 100 V rms with 10 A rms at 50 Hz. The maximum permissible core density is 1.0 tesla and the iron requires 10% of the total ampere turns. Leakage and fringing may be neglected.
(*Answer:* 4.8 mm, 15 cm^2.)

C.2. When a 60-Hz voltage of 100 V is applied to a coil wound on a ferromagnetic circuit, half the losses (50 W) are found to be hysteresis losses. What are the eddy current loss and the hysteresis loss at half the applied voltage, if the Steinmetz coefficient is assumed to be 2?
(*Answer:* 12.5 W, 12.5 W.)

C.3. An iron core 15 cm long and 10 cm^2 in cross section is magnetized uniformly by a winding of 1000 turns carrying an alternating current. The following measurements apply.

Frequency	Voltage	Power loss in core
40 Hz	88.8 V	8 W
60 Hz	239.8 V	40 W

Determine the area of the hysteresis loop for the test at 40 Hz assuming a Steinmetz coefficient of 1.7.
(*Answer:* 950 joules/m^3.)

C.4. Two different voltages are applied successively to the exciting coil wound on a ferromagnetic circuit with a mean length of 1.02 m and a cross-sectional area of 32.3 cm^2. The resistance of the exciting winding is negligible. The Steinmetz coefficient has the value of 1.58. The total core loss values measured for the two applied voltages are **(a)** 100 W for the voltage $v_1(t) = 140 \sin 377t$ and **(b)** 360 W for the voltage $v_2(t) = 280 \sin 377t$. Determine the hysteresis loss and eddy current loss components for each of the applied voltages.
(*Answer:* (a) 40 W, 60 W, (b) 120 W, 240 W.)

C.5. The eddy current loss is 12 W and the hysteresis loss is 0.3 W per cycle, when the rated, 60-Hz voltage is applied to a winding on a ferromagnetic core. What are the eddy current and hysteresis losses if the same voltage but at 50 Hz is applied to the coil? Assume a Steinmetz coefficient of 2.
(*Answer:* 12 W, 21.6 W.)

C.6. An inductor has a coil designed for 120 V, 60 Hz and is wound on a toroid of ferromagnetic material for which $\mu_r = 10,000$. An air gap that is 1% of the length of the ferromagnetic circuit is subsequently introduced in the toroid by a cross-sectional saw cut. **(a)** Determine the ratio of the inductance values without and with the air gap. **(b)** What is the change in core loss?
(*Answer:* (a) $L_1/L_2 = 100$, (b) 1% reduction.)

C.7. When measuring the core loss of a power transformer, the test standards call for a true sinusoidal voltage to be applied to the terminals of one of the windings. Assume that we cannot comply with this requirement. Instead, we apply a voltage that, apart from the fundamental, also contains a third harmonic. **(a)** Sketch both the applied voltage wave $v(t) = \hat{V}_1 \sin \omega t + \hat{V}_3 \sin 3\omega t$ and the corresponding core flux variation with time. **(b)** Sketch both the applied voltage wave $v(t) = \hat{V}_1 \sin \omega t - V_3 \sin 3\omega t$ and the corresponding flux wave. **(c)** What is the voltage measured by an rms voltmeter in the above two cases? **(d)** What would the qualitative difference in eddy current loss in the core be for the two applied voltage waveforms of parts (a) and (b)?

C.8. If both the frequency and the voltage applied to the coil of a magnetic core reactor are increased by 10%, what will be the percentage change of **(a)** flux density, **(b)** eddy current loss, and **(c)** hysteresis loss?
(*Answer:* (a) 0, (b) 10% increase, (c) 21% increase.)

C.9. A reactor, with a winding of negligible resistance, has an exciting current that can be expressed by

$$i_{NL}(t) = \sqrt{2}[10.4 \cos(377t - 1.38) + 4.5 \cos(1131t + 1.57)$$
$$+ 1.10 \cos(1885t - 1.57)]$$

if the applied 60-Hz voltage has a value of 1200 V (cosinusoidal). Determine **(a)** the core loss in watts, **(b)** the effective value of the exciting current, **(c)** the effective value of the core-loss current, and **(d)** the effective value of the magnetizing current.
(*Answer:* (a) 2380 W, (b) 11.4 A, (c) 1.97 A, (d) 11.21 A.)

C.10. A current

$$i_L(t) = 806 \cos(377t - 0.52) \text{ has to be added to the exciting current}$$

$$i_{NL}(t) = \sqrt{2}[10.4 \cos(377t - 1.38) + 4.5 \cos(1131t + 1.57)$$
$$+ 1.10 \cos(1885t - 1.57)]$$

that resulted from applying the voltage $v(t) = 1697 \cos 377t$ to a coil wound on a ferromagnetic circuit. (*Note:* $v(t)$ and $i_{NL}(t)$ are the same as those in the previous problem.) **(a)** Determine what percentage I_{NL} is of I_L (rms values). **(b)** Determine accurately the effective value of the resultant current. **(c)** Determine the effective value of the resultant current using the equivalent sinusoidal current of $i_{NL}(t)$. **(d)** What is the percentage error made when the method of part (c) is used?
(*Answer:* (a) 2%, (b) 578 A, (c) 578.3 A, (d) 0.05%.)

C.11. Construct the induced voltage waveshape resulting from a sinusoidal magnetization current that takes the ferromagnetic circuit well into saturation. (To simplify matters, consider a saturation curve rather than a saturation loop.)

APPENDIX D

Three-Phase Circuits

D1. INTRODUCTION

A polyphase system has a number of equal-voltage, ac sources operating at fixed but different phase angles; the sources supply power to loads connected across the supply lines. The reason for using a polyphase system to transmit large amounts of power is one of economics.

In a balanced n-phase system there are n sources connected together. Each phase voltage (or source) varies sinusoidally, has the same magnitude, and has a phase difference of $2\pi/n$ degrees from its adjacent voltage phasors except in the case of 2-phase systems.[10] There is a load for each phase. For balanced systems the loads are equal, but this is only a desired requirement. In practice it cannot be guaranteed.

There can be many generators feeding power into a polyphase system, but they must all have the same number of phase sources operating at the same frequency and voltage. Only similar phase voltages are connected in parallel, otherwise circulating currents ensue without useful work being done.

Two-phase systems are uncommon. Three-phase systems are used in power systems extensively. Six, twelve or more phases are used in rectifier systems in order to reduce the voltage ripple on the dc side.

The advantages of ac systems in general and of three-phase operation in particular are as follows.

1. An ac supply is used in preference to a dc supply because its voltage can be changed to any desired value by means of transformers. This permits genera-

---

[10] The voltages have a phase difference of 90° in a 2-phase system.

tion, transmission, distribution and utilization at the most economical or desirable voltage level. A 60-Hz frequency in North America (50 Hz is most common in the rest of the world) is a compromise between the economy of machines and transformers at higher frequencies and the decrease of reactance of transmission at lower frequencies.

2. Alternating currents can be interrupted more easily than direct currents.

One disadvantage of ac power is that it cannot be stored in batteries as dc power can. Three-phase operation is preferable to single-phase operation for the following reasons.

1. Three-phase windings make more efficient use of the iron and copper in generators and motors. For a given frame size a three-phase machine has a greater output than a single-phase machine.
2. A three-phase transmission line requires less copper than a single-phase line to transmit a given amount of power at a given voltage over a given distance.
3. Three-phase motors have a constant torque, whereas single-phase motors (except commutator motors) have a pulsating torque.
4. Single-phase motors (except commutator motors) are not self-starting, whereas three-phase motors are.
5. Instantaneous single-phase power is pulsating, whereas the instantaneous value of three-phase power is constant.
6. Three-phase power can be transformed by means of appropriate transformers to power of any number of phases, whereas single-phase power cannot be transformed to steady power of any number of phases by simple transformation.

Three-phase systems are used almost exclusively for the distribution of electric power. The complications of additional phases are not compensated by further increases in operating efficiency. Single-phase power is obtained from one phase of a three-phase system and the balance of a three-phase system is maintained approximately by distributing the single-phase loads equally among the three phases.

D2. GENERATION OF THREE-PHASE VOLTAGES

A simple sketch of a three-phase generator is shown in Fig. D1a. Assume that the flux density in the air gap has a sinusoidal distribution. Consider a concentrated coil $A'A$ on the rotor.[11] If the rotor speed ω is constant, a sinusoidal emf will be induced in coil $A'A$. That is, $e_{A'A} = e_{max} \sin\omega t$. Such a sinusoidal emf can be considered to be supplied

---

[11] The flux linking the coil $A'A$, as a function of rotor position $\theta = \omega t$, is $\phi_{A'A}(\theta) = \Phi_{max} \cos\theta$.

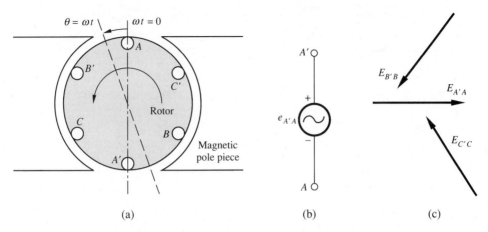

Figure D.1 (a) Three-phase generator, (b) generator equivalent, (c) voltage phasors.

by a single-phase generator as shown in Fig. D1b. The corresponding voltage phasor $\mathbf{E}_{A'A}$ is indicated in Fig. D1c. $E_{A'A}$ is the rms value of the emf.

The double subscript notation in $e_{A'A}$, as specified by the polarity markings in the figure, has the following meanings.

- When $e_{A'A}$ has a positive value during the positive half-cycle (from $\omega t = 0$ to $\omega t = \pi$), the terminal A' has the positive polarity.

- If the coil voltage from A' to A is $e_{A'A}$, the coil voltage from A to A' is $e_{AA'}$ and $e_{AA'} = -e_{A'A}$. That is, a change in the positions of the subscripts corresponds to a change in sign of the original quantity.

Figure D1a shows two other concentrated coils $B'B$ and $C'C$, which are wound in the same direction and have the same number of turns as coil $A'A$. Coil $B'B$ is displaced from coil $A'A$ by 120 electrical degrees along the rotor periphery and coil $C'C$ is displaced 240 electrical degrees from coil $A'A$ (or 120 electrical degrees from coil $B'B$). Since the rotor speed is in a counterclockwise direction, the voltage induced in coil $B'B$ lags the voltage induced in coil $A'A$ by 120° and the voltage induced in coil $C'C$ lags the voltage induced in coil $A'A$ by 240°. The amplitudes of the coil voltages will be the same. Thus expressions for the coil voltages may be written as follows.

$$e_{A'A} = e_{\max} \sin \omega t \tag{D2.1}$$

$$e_{B'B} = e_{\max} \sin\left(\omega t - \frac{2\pi}{3}\right) \tag{D2.2}$$

$$e_{C'C} = e_{\max} \sin\left(\omega t - \frac{4\pi}{3}\right). \tag{D2.3}$$

Their phasor representation is shown in Fig. D1c.

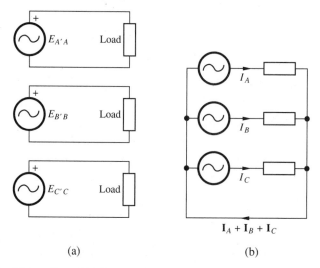

(a) (b)

Figure D.2 (a) Three 1-phase systems, (b) one 3-phase system.

D2.1. Three-phase Generator and Load Connections

Let us represent the three coils $A'A$, $B'B$ and $C'C$ of Fig. D1a by three single-phase ac generators. We can connect a three-phase load to the three coils as indicated in Fig. D2a. This requires six wires.

Instead of having a return wire from each load, the three return wires can be combined into one for the sake of economy. See Fig. D2b. If the three coil voltages are equal in magnitude and 120° apart in phase, and if the three load impedances are equal (equal magnitude and phase), the three currents will be equal in magnitude and 120° apart in phase. The three coil voltages form a *balanced voltage supply* and the three loads are called a *balanced three-phase load*. The three load currents, that result from applying balanced voltages to balanced loads, are also balanced. The current in the common return wire is the phasor sum of these three currents. Under the ideal conditions of balanced voltages and balanced loads, the sum of the balanced currents is zero and the return wire carries no current at all.

Generators are designed to produce balanced voltages. Because of practical loading conditions in a three-phase power system, the phasor sum of the three line currents is rarely zero. However, the phasor sum is usually quite small and the common return wire of a *four-wire, three-phase system* shown in Fig. D3a can have a smaller cross section than the other three lines. The connection shown in Fig. D3a is important in three-phase power systems. The generator coils and the loads are said to be connected in *star, wye,* or *Y.* The common return wire is called the *neutral.*

Can the common return wire be left out entirely as in the three-wire, three-phase system shown in Fig. D3b? This can be and is done, if there is assurance that the load

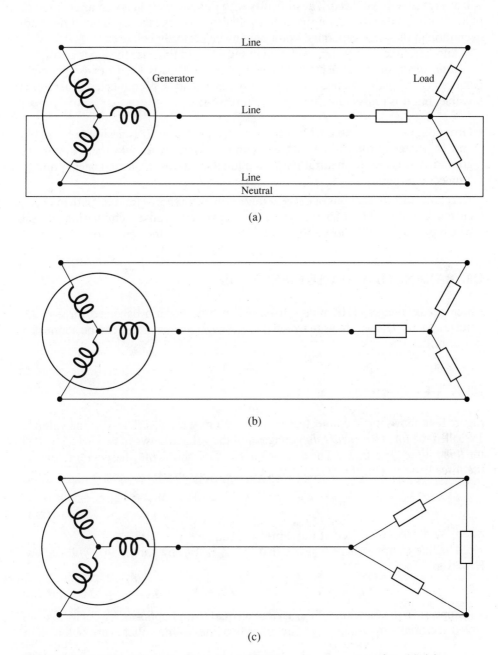

Figure D.3   (a) Four-wire, star connection, (b) three-wire, star connection, (c) delta connection.

will always be well balanced. The difficulty with this connection is that an imbalance of load impedance produces unbalanced voltages across the three phases of the load, even though the three generated voltages may be perfectly balanced.

Another three-wire system is shown in Fig. D3c. In this case the loads are said to be *mesh, delta,* or Δ connected. These Δ-connected loads need not be perfectly balanced to maintain balanced voltages across each load. A Δ-connected generator is possible, but it is undesirable for two main reasons.

- Grounding is not possible with a Δ-connected generator. It is desirable for safety and for system protection to have a connection from the electrical system to ground. The generator neutral point (the junction of the Y) is the logical point to connect to ground.

- A Δ connection of the coils of the generator provides a short-circuited path in which current can flow. Third harmonics in the coil voltages cause a circulating current which produces power loss and lowers the efficiency of the generator.

D3. BALANCED VOLTAGE SYSTEMS

Three equal voltages, 120° phase displaced in time, form a balanced three-phase voltage system. The windings in which these voltages are generated can be connected in star or delta.

D3.1. Y Connection

Figure D4a shows the Y connection of a generator's windings. The winding voltages are called the *line-to-neutral (ph)* voltages and the voltage between two lines is called the *line-to-line (l)* voltage.[12] The diagram in Fig. D4b shows the phasors representing the three line-to-neutral voltages that have the expressions

$$\mathbf{V}_{NA} = \mathbf{V}_{ph} \angle 0°, \quad \mathbf{V}_{NB} = V_{ph} \angle -120° \quad \text{and} \quad \mathbf{V}_{NC} = V_{ph} \angle -240° \tag{D3.1}$$

where N, A', B', and C' are at the same potential.

The line-to-line voltage $\mathbf{V}_{BA}$ is equal to $\mathbf{V}_{BN} + \mathbf{V}_{NA}$ by Kirchhoff's voltage law. However

$$\mathbf{V}_{BN} = -\mathbf{V}_{NB} \quad \text{so} \quad \mathbf{V}_{BA} = \mathbf{V}_{NA} - \mathbf{V}_{NB}. \tag{D3.2}$$

$\mathbf{V}_{NA}$ appears in the diagram; $-\mathbf{V}_{NB}$ is drawn equal to and opposite $\mathbf{V}_{NB}$. The two are added vectorially to obtain $\mathbf{V}_{BA}$. The other two line-to-line voltages are obtained in

[12] The line-to-neutral voltage is also called the *phase* voltage and is often given the symbol V_{ph}. Often the line-to-line voltage is shortened to *line* voltage and given the symbol V_l.

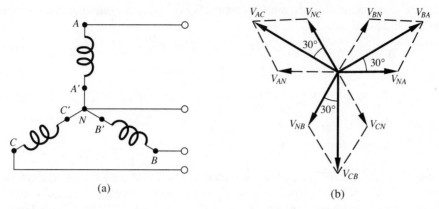

Figure D.4 (a) Star-connected generator, (b) voltage-phasor diagram.

the same way to give

$$\mathbf{V}_{CB} = \mathbf{V}_{CN} + \mathbf{V}_{NB} = \mathbf{V}_{NB} - \mathbf{V}_{NC} \qquad \textbf{(D3.3)}$$

and

$$\mathbf{V}_{AC} = \mathbf{V}_{AN} + \mathbf{V}_{NC} = \mathbf{V}_{NC} - \mathbf{V}_{NA}. \qquad \textbf{(D3.4)}$$

The line-to-neutral and the line-to-line voltages with their proper phase relations are shown in Fig. D5. From the geometry of the phasor diagram it can be seen that

$$V_{CB} = \sqrt{3} V_{NB} \qquad \textbf{(D3.5)}$$

that is, the magnitude of the line-to-line voltages is equal to $\sqrt{3}$ times the magnitude of the line-to-neutral voltages. Further, $\mathbf{V}_{BA}$ leads $\mathbf{V}_{NA}$ by 30°, $\mathbf{V}_{AC}$ leads $\mathbf{V}_{NC}$ by 30°, and $\mathbf{V}_{CB}$ leads $\mathbf{V}_{NB}$ by 30°. The three line-to-line voltages have equal magnitudes

$$V_{BA} = V_{AC} = V_{CB} = V_l \qquad \textbf{(D3.6)}$$

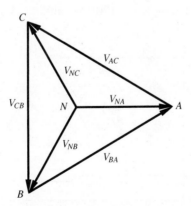

Figure D.5 Phase and line voltages.

and these voltages, as phasors, are $120°$ apart. With $\mathbf{V}_{NA}$ as the reference phasor, their phasor expressions are

$$\mathbf{V}_{BA} = \sqrt{3}V_{ph} \angle 30° = V_l \angle 30° \qquad \text{(D3.7)}$$

$$\mathbf{V}_{CB} = \sqrt{3}V_{ph} \angle -90° = V_l \angle -90° \qquad \text{(D3.8)}$$

$$\mathbf{V}_{AC} = \sqrt{3}V_{ph} \angle 150° = V_l \angle 150°. \qquad \text{(D3.9)}$$

D3.2. Δ Connection

If the coil terminal A of Fig. D1a is connected to C', C to B' and B to A', the coils are connected in the delta (Δ) connection and the line-to-line voltage is equal in magnitude to the coil (or phase) voltage.

D4. THREE-PHASE LOADS

A three-phase load can be connected to the supply in *delta,* or in *wye* with or without the neutral brought out. These three possible connections will be examined for balanced loads, assuming a balanced supply system.

For a load impedance Z_{AB} there will be a voltage drop V_{AB} across that impedance and a current I_{AB} through that impedance. The convention followed in this book is as follows. The voltage $\mathbf{V}_{AB}$ indicates that point A is at the higher potential. The current $\mathbf{I}_{AB}$ would then be a current directed from A to B through the impedance $\mathbf{Z}_{AB}$.

D4.1. Delta-connected Load

Figure D6 shows the load impedances connected in a *delta* configuration. The line voltages may be represented by

$$\mathbf{V}_{CB} = V \angle 0°, \quad \mathbf{V}_{AC} = V \angle -120° \qquad \text{and} \qquad \mathbf{V}_{BA} = V \angle -240°. \quad \text{(D4.1)}$$

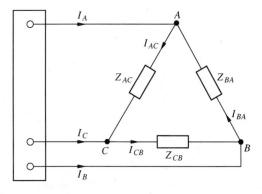

Figure D.6 Delta-connected load.

The individual load currents are

$$I_{CB} = \frac{V_{CB}}{Z_{CB}}, \quad I_{AC} = \frac{V_{AC}}{Z_{AC}} \quad \text{and} \quad I_{BA} = \frac{V_{BA}}{Z_{BA}}. \tag{D4.2}$$

The line currents are

$$I_A = I_{AC} - I_{BA}, \quad I_B = I_{BA} - I_{CB} \quad \text{and} \quad I_C = I_{CB} - I_{AC}. \tag{D4.3}$$

Note that by adding the last three current equations we get

$$I_A + I_B + I_C = 0. \tag{D4.4}$$

In general, the phasor sum of the three line currents in a three-phase, three-wire system is zero.

The total power taken by the load is the sum of powers taken by each individual phase load. That is, $P = P_{CB} + P_{AC} + P_{BA}$. The powers taken by the three phases are

$$P_{CB} = V_{CB}I_{CB} \cos\angle (V_{CB}, I_{CB}) \tag{D4.5}$$

$$P_{AC} = V_{AC}I_{AC} \cos\angle (V_{AC}, I_{AC}) \tag{D4.6}$$

$$P_{BA} = V_{BA}I_{BA} \cos\angle (V_{BA}, I_{BA}). \tag{D4.7}$$

A balanced load is defined by three equal and separate phase impedances. The properties associated with a balanced load are as follows.

1. $Z_{BA} = Z_{CB} = Z_{AC} = Z = Z \angle \theta$.
2. The phase currents will be equal and 120° apart.
3. The line currents will be equal and 120° apart.
4. The line current is $\sqrt{3}$ times the phase current ($I_l = \sqrt{3}I_{ph}$).
5. The powers are the same for each phase. That is

$$P_{CB} = P_{AC} = P_{BA} = V_{ph} I_{ph} \cos\theta. \tag{D4.8}$$

Therefore the total power P is three times the power of one phase.

$$P = P_{CB} + P_{AC} + P_{BA} = 3P_{CB} = 3V_{CB} I_{CB} \cos\angle (V_{CB}, I_{CB}) \tag{D4.9}$$

but for a Δ connection

$$V_{CB} = V_{ph} = V_l \quad \text{and} \quad I_{CB} = I_{ph} = \frac{I_l}{\sqrt{3}}. \tag{D4.10}$$

Therefore

$$P = 3 V_l \frac{I_l}{\sqrt{3}} \cos\angle (V_{CB}, I_{CB}) = \sqrt{3}V_lI_l \cos\theta \tag{D4.11}$$

where $\cos\theta$ is the power factor[13] of the load in any phase branch.

---

[13] *Power factor* is defined as the ratio of average real power ($P = VI \cos\theta$) to apparent power *(S = VI)* for any phase. This definition is applied to the three-phase system also.

EXAMPLE D1

Consider a Δ-connected load whose impedances are given, with reference to Fig. D6, to be

$$\mathbf{Z}_{CB} = 10 \angle 0° \ \Omega, \quad \mathbf{Z}_{BA} = 5 \angle 30° \ \Omega \quad \text{and} \quad \mathbf{Z}_{AC} = 20 \angle -30° \ \Omega.$$

This is not a balanced load. The line-to-line voltages with $\mathbf{V}_{CB}$ as reference are

$$\mathbf{V}_{CB} = 100 \angle 0° \ \text{V}, \quad \mathbf{V}_{AC} = 100 \angle -120° \ \text{V} \quad \text{and} \quad \mathbf{V}_{BA} = 100 \angle -240° \ \text{V}.$$

Determine the phase and line currents.

Solution
The phase currents are

$$\mathbf{I}_{CB} = \frac{\mathbf{V}_{CB}}{\mathbf{Z}_{CB}} = \frac{100 \angle 0°}{10 \angle 0°} = 10 \angle 0° \ \text{A}$$

$$\mathbf{I}_{AC} = \frac{\mathbf{V}_{AC}}{\mathbf{Z}_{AC}} = \frac{100 \angle -120°}{20 \angle -30°} = 5 \angle -90° \ \text{A}$$

$$\mathbf{I}_{BA} = \frac{\mathbf{V}_{BA}}{\mathbf{Z}_{BA}} = \frac{100 \angle -240°}{5 \angle 30°} = 20 \angle -270° = 20 \angle +90° \ \text{A}.$$

The line currents are

$$\mathbf{I}_A = \mathbf{I}_{AC} - \mathbf{I}_{BA} = 5 \angle -90° - 20 \angle 90° = 25 \angle -90° \ \text{A}$$

$$\mathbf{I}_B = \mathbf{I}_{BA} - \mathbf{I}_{CB} = 20 \angle 90° - 10 \angle 0° = 22.4 \angle 116.6° \ \text{A}$$

$$\mathbf{I}_C = \mathbf{I}_{CB} - \mathbf{I}_{AC} = 10 \angle 0° - 5 \angle -90° = 11.2 \angle 26.6° \ \text{A}.$$

Note that the phasor sum of the three line currents is zero.

D4.2. Four-wire Y-connected Load

Figure D7 shows wye-connected load impedances and indicates the notation. With $\mathbf{V}_{CB}$ as the reference the line and phase voltages are

$$\mathbf{V}_{CB} = V_l \angle 0°, \quad \mathbf{V}_{AC} = V_l \angle -120° \quad \text{and} \quad \mathbf{V}_{BA} = V_l \angle +120°. \quad \textbf{(D4.12)}$$

$$\mathbf{V}_{AN} = V_l/\sqrt{3} \ \angle -90°, \quad \mathbf{V}_{BN} = V_l/\sqrt{3} \ \angle 150° \quad \text{and} \quad \mathbf{V}_{CN} = V_l/\sqrt{3} \ \angle 30°. \quad \textbf{(D4.13)}$$

The phase and line currents have the same value and are given by

$$\mathbf{I}_{AN} = \mathbf{I}_A = \frac{\mathbf{V}_{AN}}{\mathbf{Z}_A}, \quad \mathbf{I}_{BN} = \mathbf{I}_B = \frac{\mathbf{V}_{BN}}{\mathbf{Z}_B} \quad \text{and} \quad \mathbf{I}_{CN} = \mathbf{I}_C = \frac{\mathbf{V}_{CN}}{\mathbf{Z}_C}. \quad \textbf{(D4.14)}$$

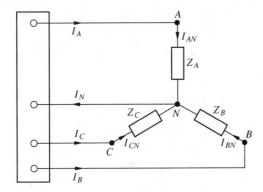

Figure D.7 Star-connected load.

The neutral current is

$$\mathbf{I}_N = \mathbf{I}_A + \mathbf{I}_B + \mathbf{I}_C. \tag{D4.15}$$

The total power P is the sum of the phase powers, that is

$$P = P_{AN} + P_{BN} + P_{CN} \tag{D4.16}$$

where
$$P_{AN} = V_{AN}I_{AN} \cos\angle(\mathbf{V}_{AN},\mathbf{I}_{AN})$$
$$P_{BN} = V_{BN}I_{BN} \cos\angle(\mathbf{V}_{BN},\mathbf{I}_{BN})$$
and
$$P_{CN} = V_{CN}I_{CN} \cos\angle(\mathbf{V}_{CN},\mathbf{I}_{CN}).$$

The properties following from equal load impedances are as follows.

1. $\mathbf{Z}_A = \mathbf{Z}_B = \mathbf{Z}_C = \mathbf{Z} = Z \angle \theta$.
2. The phase currents, which are also the line currents, are equal in magnitude and 120° apart.
3. The neutral current is zero.
4. The same power is absorbed in each phase of the load. The total power is three times the phase power. That is

$$P = P_{AN} + P_{BN} + P_{CN} = 3 P_{AN} = 3 V_{AN}I_{AN} \cos\angle(\mathbf{V}_{AN},\mathbf{I}_{AN}). \tag{D4.17}$$

Since

$$V_{AN} = V_{ph} = \frac{V_l}{\sqrt{3}} \quad \text{and} \quad I_{AN} = I_A = I_l \tag{D4.18}$$

$$P = 3 V_{ph}I_l \cos\angle(\mathbf{V}_{AN},\mathbf{I}_{AN}) = \sqrt{3}V_lI_l \cos\theta \tag{D4.19}$$

where $\cos\theta$ is the power factor of the load in any phase branch.

EXAMPLE D2

Three identical impedances, $\mathbf{Z}_A = \mathbf{Z}_B = \mathbf{Z}_C = \mathbf{Z} = 10 \angle 53.1° = (6 + j8)\Omega$, are connected in wye to a three-phase, four-wire, 260-V supply. Taking $\mathbf{V}_{CB}$ as the reference phasor, find the line currents and the neutral current. Determine the total power taken by the load.

Solution

The line-to-line voltages are

$$\mathbf{V}_{CB} = 260 \angle 0° \text{ V}, \quad \mathbf{V}_{AC} = 260 \angle -120° \text{ V}, \quad \text{and} \quad \mathbf{V}_{BA} = 260 \angle -240° \text{ V}.$$

The line-to-neutral (or phase) voltages are

$$\mathbf{V}_{AN} = 150 \angle -90° \text{ V}, \quad \mathbf{V}_{BN} = 150 \angle 150° \text{ V}, \quad \text{and} \quad \mathbf{V}_{CN} = 150 \angle 30° \text{ V}.$$

The phase and line currents are

$$\mathbf{I}_{AN} = \mathbf{I}_A = \frac{\mathbf{V}_{AN}}{\mathbf{Z}_A} = \frac{150 \angle -90°}{10 \angle 53.1°} = 15 \angle -143.1° \text{ A}$$

$$\mathbf{I}_{BN} = \mathbf{I}_B = \frac{\mathbf{V}_{BN}}{\mathbf{Z}_B} = \frac{150 \angle 150°}{10 \angle 53.1°} = 15 \angle 96.9° \text{ A}$$

$$\mathbf{I}_{CN} = \mathbf{I}_C = \frac{\mathbf{V}_{CN}}{\mathbf{Z}_C} = \frac{150 \angle 30°}{10 \angle 53.1°} = 15 \angle -23.1° \text{ A}.$$

The neutral current is $\mathbf{I}_N = \mathbf{I}_A + \mathbf{I}_B + \mathbf{I}_C = 0$, because the phase currents are equal in magnitude and 120° apart. The neutral wire may be removed without disturbing the network relations, if the load is balanced.

The total power is

$$P = \sqrt{3} \, V_l \, I_l \cos\theta = \sqrt{3} \times 260 \times 15 \times \cos 53.1° = 4050 \text{ W}.$$

D4.3. Three-wire Y-connected Load

The phasor diagram for an unbalanced load is shown in Fig. D8b, from which we see that the common point O is at a different voltage from N, the geometric neutral point. For instance, if Z_C approaches zero, the point O approaches C, and as Z_C becomes larger, the point O moves away from C. If all three impedances are equal (that is, balanced), point O will coincide with N (the ideal neutral point). Due to the presence of this variable *neutral voltage ON,* unbalanced Y loads are avoided in three-phase, three-wire systems.

The procedure for determining the line currents (which are the same as the phase currents) and the phase powers is similar to that for the three-phase, four-wire, balanced Y load, once the phase voltages $\mathbf{V}_{AO}$, $\mathbf{V}_{BO}$, and $\mathbf{V}_{CO}$ are found.

If the load is balanced there is no difference between the three-wire and the four-wire Y-connected systems, since the neutral current is zero in the latter case.

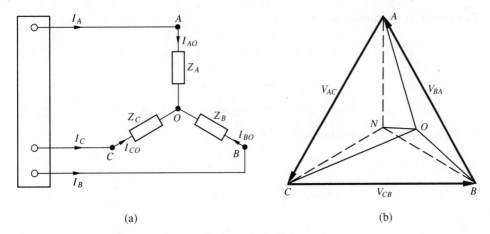

Figure D.8 (a) Star-connected, (b) unbalanced load-phasor diagram.

D5. SINGLE-PHASE REPRESENTATION

If a three-phase system is balanced, the method of analysis normally used is to determine the voltage, current and power absorbed in one phase only. The other phases will give corresponding quantities that are exactly the same except for phase displacements of 120°. This method simplifies the calculations.

One phase of a balanced system, in which the load is Y-connected, can be represented as in Fig. D9a. The line-to-neutral voltage is represented by a single-phase generator, $\mathbf{Z}$ is the impedance of the load in one phase, and $\mathbf{Z}_l$ is the impedance of the transmission line. To this extent only one third of the three-phase system is drawn. The actual situation in a balanced, three-phase system is that no current passes from n to n' through any neutral wire or in the ground. If there is no neutral current, there is no voltage drop from n to n'. The circuit of Fig. D9a is therefore closed by a hypothetical conductor of zero impedance as shown in Fig. D9b.

In balanced Y-connected systems the current and voltage relations between phase (ph) and line (l) quantities are

$$V_l = \sqrt{3}V_{ph} \qquad \text{and} \qquad I_l = I_{ph}. \qquad \textbf{(D5.1)}$$

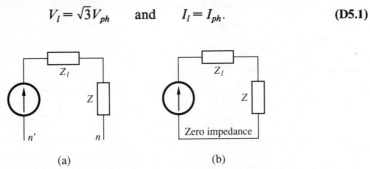

Figure D.9 (a) One phase, (b) balanced condition.

The power in one phase is

$$P_{ph} = V_{ph} I_{ph} \cos\theta = \frac{V_l}{\sqrt{3}} I_l \cos\theta. \tag{D5.2}$$

Therefore the total power P in the 3-phase circuit is

$$P = 3P_{ph} = 3V_{ph} I_{ph} \cos\theta = \sqrt{3} V_l I_l \cos\theta. \tag{D5.3}$$

In balanced Δ-connected systems the relations are

$$V_l = V_{ph} \quad \text{and} \quad I_l = \sqrt{3} I_{ph} \tag{D5.4}$$

$$P_{ph} = V_{ph} I_{ph} \cos\theta = V_l \frac{I_l}{\sqrt{3}} \cos\theta \tag{D5.5}$$

and

$$P = 3P_{ph} = 3V_{ph} I_{ph} \cos\theta = \sqrt{3} V_l I_l \cos\theta. \tag{D5.6}$$

D6. EQUIVALENT Y AND Δ LOADS

If a balanced load is Δ-connected and we wish to solve a problem on a single-phase basis, it is convenient to substitute an equivalent Y load for the actual Δ load. See Fig. D10. For the two circuits to be equivalent, the impedances that can be measured between terminals A and B, B and C, and C and A have to be the same for both connections. This requirement leads to the following equations

$$\frac{Z_A (Z_B + Z_C)}{Z_A + Z_B + Z_C} = Z_1 + Z_2 \tag{D6.1}$$

$$\frac{Z_B (Z_C + Z_A)}{Z_A + Z_B + Z_C} = Z_2 + Z_3 \tag{D6.2}$$

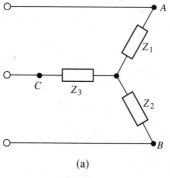
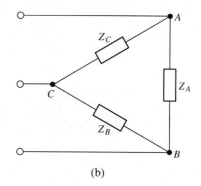

(a)   (b)

Figure D.10   (a) Star to (b) delta connection.

and

$$\frac{Z_C\,(Z_A + Z_B)}{Z_A + Z_B + Z_C} = Z_3 + Z_1. \tag{D6.3}$$

From these equations

$$Z_1 = \frac{Z_A Z_C}{Z_A + Z_B + Z_C} \tag{D6.4}$$

$$Z_2 = \frac{Z_A Z_B}{Z_A + Z_B + Z_C} \tag{D6.5}$$

and

$$Z_3 = \frac{Z_B Z_C}{Z_A + Z_B + Z_C}. \tag{D6.6}$$

With a balanced load

$$Z_A = Z_B = Z_C = Z_\Delta, \quad Z_1 = Z_2 = Z_3 = Z_Y, \quad \text{and} \quad Z_Y = \frac{Z_\Delta}{3}. \tag{D6.7}$$

The change from an actual Δ-connected load to an equivalent Y-connected load is commonly known as the Δ-Y transformation.

EXAMPLE D3

Figure D11 shows part of a three-phase distribution system with two balanced loads. One load consists of three impedances, $Z_s = (12 + j16)$ ohms per phase connected in

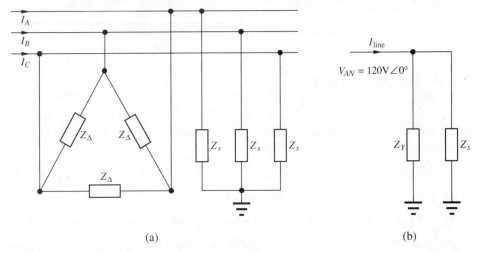

(a)   (b)

Figure D.11   (a) Two balanced 3-phase loads, (b) single-phase equivalent.

wye. The other load consists of three impedances, $Z_\Delta = (20 + j80)$ ohms per phase, connected in delta. The line-to-line supply voltage is 208 volts. Determine the line current and the power dissipated by the total load.

Solution

To deal with the two loads we convert the delta-connected load to an equivalent wye-connected load. This enables calculations to be made on a per phase basis for all of the system shown. The line-to-neutral voltage

$$V_{AN} = \frac{V_l}{\sqrt{3}} = \frac{208}{\sqrt{3}} = 120 \text{ volts}$$

is chosen as reference.

The transformed delta-to-wye load has a line-to-neutral impedance

$$Z_Y = \frac{Z_\Delta}{3} = \frac{20 + j80}{3} = 27.5 \angle 76° \text{ ohms.}$$

Since each three-phase load is balanced, both neutral points are at the same potential. Consequently the two wye loads can be considered to be in parallel. The total impedance per phase is

$$Z = \frac{Z_s Z_Y}{Z_s + Z_Y} = \frac{(12 + j16)(6.67 + j26.67)}{(12 + j16) + (6.67 + j26.67)} = 11.8 \angle 74.5° \text{ ohms.}$$

The line current is found from

$$\mathbf{I}_l = \mathbf{I}_{AN} = \frac{\mathbf{V}_{AN}}{\mathbf{Z}} = \frac{120 \angle 0°}{11.8 \angle 74.5°} = 10.2 \angle -74.5° \text{ amperes.}$$

The power absorbed by the total load is

$$P = 3V_{AN}I_{AN} \cos\angle(\mathbf{V}_{AN}, \mathbf{I}_{AN}) = 3 \times 120 \times 10.2 \cos 74.5° = 981 \text{ watts.}$$

D7. PHASE SEQUENCE

Phase sequence is defined as the order in which the voltages are considered in time phase with respect to each other. Thus, with $\mathbf{V}_{CB}$ as the reference phasor and phase sequence *ABC*, the line-to-neutral voltages are

$$\mathbf{V}_{AN} = V \angle -90°, \quad \mathbf{V}_{BN} = V \angle 150°, \quad \text{and} \quad \mathbf{V}_{CN} = V \angle 30°. \quad \textbf{(D7.1)}$$

In the phase sequence *ACB* these voltages are

$$\mathbf{V}_{AN} = V \angle -90°, \quad \mathbf{V}_{CN} = V \angle 150°, \quad \text{and} \quad \mathbf{V}_{BN} = V \angle 30°. \quad \textbf{(D7.2)}$$

Notice that the phase sequence is reversed by interchanging the positions of any two of the three voltage phasors in the phasor diagram.

D8. MEASUREMENT OF THREE-PHASE POWER

In three-phase systems a knowledge of the total power delivered to the load is important. If the system is balanced and the power delivered to one phase can be measured by means of a wattmeter, the total power is three times the single wattmeter reading. For the case of an unbalanced load, it is not possible to do this. A solution is to use two wattmeters that are connected in a special way. The sum of the two wattmeter readings gives the value of the total power, whether the system is balanced or not. This method is developed as follows.

D8.1. The Wattmeter

The important elements in a wattmeter are the current coil and the potential coil. The current coil, which is normally fixed, carries the load current i(t). The potential coil is made to carry a small current, that is proportional to the voltage v(t) across the load. This potential coil is pivoted so that it can rotate with respect to the current coil, against the restraining torque of a spring.

The magnetic fields, produced by the currents in the two coils, interact and give rise to a time-varying torque that is proportional to the product $v(t) \cdot i(t)$. The angle of deflection of the pivoted coil is proportional to the average value of the torque. This means that the deflection is proportional to the average value of $v(t) \cdot i(t)$, which is the electric power taken by the load. The meter scale is calibrated to read the power in watts.

D8.2. Three-wattmeter Method

An obvious method to measure total power in a three-phase system is to use three wattmeters so that each meter measures the power of one phase. If the load were Δ-connected, each wattmeter would have its current coil connected in one side of the Δ and its potential coil connected line to line. If the load were Y-connected, the connections of the wattmeters 1, 2, and 3 would be as shown in Fig. D12.

The total power P consumed by the load, whether it be Δ- or Y-connected, balanced or unbalanced, is the sum of the three wattmeter readings. That is

$$P = P_1 + P_2 + P_3. \tag{D8.1}$$

A three-wire, Y-connected load does not always have an accessible neutral point; nor is it always practical to break into a Δ-connected load to connect a wattmeter in each of the phases. In this case, three wattmeters can be connected as shown in Fig. D13.[14] The common point O is a floating potential point.

---

[14] Since a Δ-connected load may be transformed into an equivalent Y-connected load, only Y-connected loads will be considered.

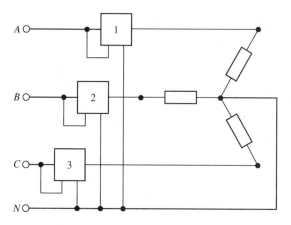

Figure D.12   Power measurement with three wattmeters.

If the three wattmeter potential coils have identical impedances and the load is balanced, point O has the same potential as point N and $V_{ON} = 0$. The three watt-meter readings are identical. However, if the impedances of the potential coils are different, or the load is unbalanced, V_{ON} will not be equal to zero. In this case, the wattmeters will have different readings. Figure D14 is for the case that the potential coils and the load are unbalanced. The total power to the load is

$$P = \frac{1}{T}\int_0^T (v_{NA}i_A + v_{NB}i_B + v_{NC}i_C)dt \tag{D8.2}$$

where $v_{NA}i_A$ is the instantaneous power absorbed by the load of phase NA. Each wattmeter indicates the average of the product of the line current and the voltage to

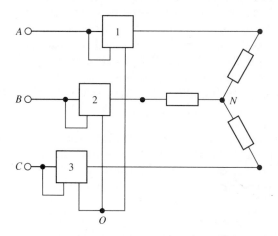

Figure D.13   Alternative wattmeter connections.

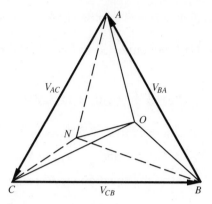

Figure D.14 Phasor diagram for an
unbalanced system.

point O. The sum of the three wattmeter readings is

$$P' = P_1 + P_2 + P_3 = \frac{1}{T} \int_0^T (v_{OA}i_A + v_{OB}i_B + v_{OC}i_C)dt. \qquad \textbf{(D8.3)}$$

From Fig. D14 it follows that

$$\mathbf{V}_{OA} = \mathbf{V}_{ON} + \mathbf{V}_{NA}, \quad \mathbf{V}_{OB} = \mathbf{V}_{ON} + \mathbf{V}_{NB}, \quad \text{and} \quad \mathbf{V}_{OC} = \mathbf{V}_{ON} + \mathbf{V}_{NC}. \quad \textbf{(D8.4)}$$

Substituting these relations into eq. (D8.3) for P' results in

$$P' = \frac{1}{T} \int_0^T [(v_{NA}i_A + v_{NB}i_B + v_{NC}i_C) + v_{ON}(i_A + i_B + i_C)]dt. \qquad \textbf{(D8.5)}$$

In a three-wire system Kirchhoff's current law indicates that

$$i_A + i_B + i_C = 0. \qquad \textbf{(D8.6)}$$

So the last term of eq. (D8.5) for P' is zero. This leaves the equation for P' identical to
the equation for P. This proves that the sum of the three wattmeter readings is always
equal to the power supplied to a three-phase load.

D8.3. Two-wattmeter Method

Since point O of Fig. D13 can have any potential, it can be given the potential of one
of the lines. If point O is connected to line C, the reading of wattmeter P_3 is zero
because the voltage across the potential coil of this meter is zero. Wattmeter P_3 can
then be removed. The algebraic sum of the readings of the two remaining wattmeters
will give the total average power taken by a three-phase load. The term *algebraic* is
used because it is possible under certain conditions for one of the wattmeter readings
to be negative. The two-wattmeter method is the standard method for measuring

power in a three-wire system supplied to a balanced or unbalanced, Y- or Δ- connected load.

Let point O in Fig. D13 be connected to line C. The sum of the readings of wattmeters P_1 and P_2 is

$$P'' = \frac{1}{T} \int_0^T (v_{CA}i_A + v_{CB}i_B)dt. \qquad \text{(D8.7)}$$

From Fig. D14, with point O moved to C, it follows that

$$\mathbf{V}_{CA} = \mathbf{V}_{CN} + \mathbf{V}_{NA} \qquad \text{and} \qquad \mathbf{V}_{CB} = \mathbf{V}_{CN} + \mathbf{V}_{NB}. \qquad \text{(D8.8)}$$

Substitution of these relations in the equation for P'' results in

$$P'' = \frac{1}{T} \int_0^T [v_{NA}i_A + v_{NB}i_B + v_{CN}(i_A + i_B)]dt. \qquad \text{(D8.9)}$$

In a three-wire system eq. (D8.6) gives

$$i_A + i_B = -i_C \qquad \text{(D8.10)}$$

so that the expression for the sum of the two wattmeter readings becomes

$$P'' = \frac{1}{T} \int_0^T (v_{NA}i_A + v_{NB}i_B + v_{NC}i_C)dt. \qquad \text{(D8.11)}$$

This is the equation for the power supplied to a three-phase load.

A. TWO-WATTMETER READINGS

Consider two wattmeters connected as in Fig. D15 to measure the power supplied to a balanced Y-connected load.

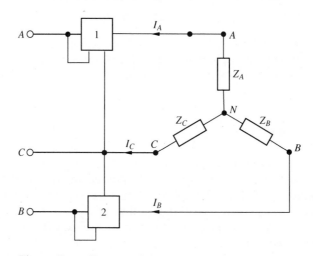

Figure D.15   Two-wattmeter method.

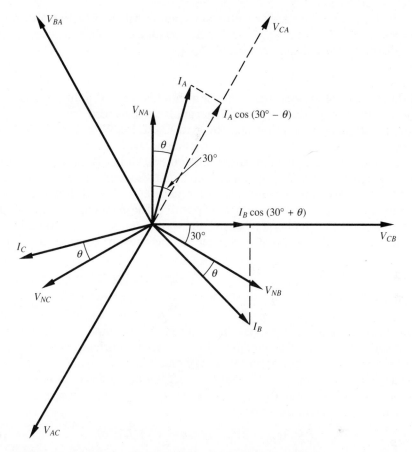

Figure D.16 Phasor diagram for the two-wattmeter method.

If $v_{NA} = \hat{V} \sin\omega t$, then $i_{NA} = i_A = \hat{I} \sin(\omega t - \theta)$. The phasor diagram for θ positive, lagging power factor, with $\mathbf{V}_{CB}$ as the reference is developed from Fig. D15 and shown in Fig. D16. It can be seen that the wattmeter readings are

$$P_1 = V_{CA}I_A \cos(\angle \mathbf{V}_{CA} - \angle \mathbf{I}_A) = V_{CA}I_A \cos(30° - \theta) \tag{D8.12}$$

and

$$P_2 = V_{CB}I_B \cos(\angle \mathbf{V}_{CB} - \angle \mathbf{I}_B) = V_{CB}I_B \cos(30° + \theta). \tag{D8.13}$$

For leading power factors the angle θ assumes negative values.

The equations for P_1 and P_2 show the following.

1. If $\theta = 0°$, the two wattmeter readings will be the same.
2. If $0° < \theta < \pm 60°$, both wattmeter readings will be positive.
3. If $\theta = +60°$ (lagging), P_2 will be zero and P_1 will represent the total power.
4. If $\theta = -60°$ (leading), P_1 will be zero and P_2 will represent the total power.

5. If $\theta > \pm 60°$, the smaller wattmeter reading will be negative and must be subtracted from the larger wattmeter reading to obtain the correct power dissipated by the balanced load.

From the foregoing it follows that the smaller of the two wattmeter readings needs to be checked to determine whether it should be added to or subtracted from the larger wattmeter reading. A method that applies to balanced or unbalanced, Δ- or Y- connected loads is the following. Refer to Fig. D15. With only lines B and C connected to the supply, connect P_2 in such a manner that it reads upscale, indicating the power taken by the impedance between points B and C. Next, with only lines A and C connected to the supply, connect P_1 so that it also reads upscale, indicating the power taken by the impedance between points A and C. With lines A, B, and C connected to the supply, both readings are positive if both meters read upscale. If the connections of the current coil of one of the wattmeters must be reversed for that meter to read upscale, then that wattmeter reading must be considered negative.

D9. SUMMARY

A three-phase generator output has three voltages of the same rms magnitude, but phase displaced by $120°$. The windings of generators are normally connected in wye but the load phases can be connected in wye or delta.

In a delta connection $V_l = V_{ph}$ and $I_l = \sqrt{3} I_{ph}$ for balanced conditions. The relations for a wye connection are $V_l = \sqrt{3} V_{ph}$ and $I_l = I_{ph}$. For both connections in a balanced 3-phase system total power

$$P = 3P_{ph} = 3V_{ph}I_{ph} \cos\theta = \sqrt{3}\ V_l I_l \cos\theta. \qquad \textbf{(D9.1)}$$

The analysis of a balanced 3-phase system is normally reduced to the analysis of a single-phase circuit. A delta-connected system is first transformed to a wye-connected system before making calculations on a per phase basis.

The value of the power delivered to a 3-phase load, balanced or unbalanced, can be obtained from the readings of two wattmeters.

D10. PROBLEMS

Note: The answers to the following problems are based on the phase sequence ABC and the assumption that the positive direction of the line current is toward the nodes of the three-phase Y- or Δ-connected load.

D.1. Three impedances $Z_{CB} = 10 \angle 0°$ ohms, $Z_{BA} = 10 \angle 30°$ ohms, and $Z_{AC} = 10 \angle -30°$ ohms are connected in Δ to a balanced three-phase, 100-V supply. The line-to-line voltages are $V_{CB} = 100 \angle 0°$ volts, $V_{BA} = 100 \angle 120°$ volts, and

$V_{AC} = 100 \angle -120°$ volts. Determine **(a)** the phase currents, **(b)** the line currents, and **(c)** draw the phasor diagram.
(*Answer:* **(a)** $I_{CB} = 10 \angle 0°$ A, $I_{AC} = 10 \angle -90°$ A, $I_{BA} = 10 \angle 90°$ A, **(b)** $I_A = 20 \angle -90°$ A, $I_B = 14.1 \angle 135°$ A, $I_C = 14.1 \angle 45°$ A.)

D.2. Three impedances $Z = (4 + j3)$ ohms are connected in Y. For balanced line-to-line voltages of 208 V and with V_{CB} as reference, find **(a)** the line currents, **(b)** the power factor, **(c)** the real power, **(d)** the reactive power, and **(e)** the apparent power.
(*Answer:* **(a)** $I_A = 24 \angle -126.9°$ A, $I_B = 24 \angle 113.1°$ A, $I_C = 24 \angle -6.9°$ A, **(b)** 0.8 lagging, **(c)** 6912 W, **(d)** 5184 VAr, **(e)** 8640 VA.)

D.3. Three impedances $Z = (4 - j3)$ ohms are connected in Δ. For balanced line-to-line voltages of 208 V and with V_{CB} as reference, find **(a)** the line currents, **(b)** the power factor, **(c)** the real power, **(d)** the reactive power, and **(e)** the apparent power.
(*Answer:* **(a)** $I_A = 72 \angle -53°$ A, $I_B = 72 \angle -173°$ A, $I_C = 72 \angle 67°$ A, **(b)** 0.8 leading, **(c)** 20,767 W, **(d)** 15,575 VAr, **(e)** 25,958 VA.)

D.4. Consider a 3-phase balanced source of 400 V (line voltage). A 3-phase, Δ-connected, unbalanced load is connected to the source. The phase impedances are $Z_{BA} = (5 + j8.66)$ ohms, $Z_{AC} = (8.66 + j5)$ ohms, and $Z_{CB} = (8.66 - j5)$ ohms. With V_{BA} as reference, determine **(a)** the phase currents, **(b)** the total power dissipated, and **(c)** the line currents.
(*Answer:* **(a)** $I_{BA} = 40 \angle -60°$ A, $I_{AC} = 40 \angle 90°$ A, $I_{CB} = 40 \angle -90°$ A, **(b)** 35,712 W, **(c)** $I_A = 77.3 \angle 105°$ A, $I_B = 20.7 \angle 15°$ A, $I_C = 80 \angle -90°$ A.)

D.5. Consider a 3-phase balanced source of 100 V (line voltage). A 3-phase, Δ-connected, unbalanced load is connected to the source. The phase impedances are $Z_{BA} = 10$ ohms (resistive), $Z_{AC} = \omega L = 10$ ohms, and $Z_{CB} = 6$ ohms (resistive) in series with a reactance X ohms. With V_{CB} as reference, determine the voltage
V_{AD}, **(a)** if $X = \omega L = 8$ ohms and **(b)** if $X = \dfrac{1}{\omega C} = 8$ ohms.
(*Answer:* **(a)** $V_{AD} = 135.3 \angle -84.1$ V, **(b)** $V_{AD} = 41 \angle -110°$ V.)

D.6. A balanced, Δ-connected, 3-phase load is formed by three identical impedances, each of which has two circuit elements in series. With $V_{CB} = 100 \angle 0°$ volts, the current in line A toward the load is $I_A = 17.32 \angle -150°$ A. **(a)** Determine the values and types of element of one impedance, **(b)** calculate the total real power and **(c)** calculate the total reactive power.
(*Answer:* **(a)** $R = 5$ ohms, $X = 8.66$ ohms inductive, **(b)** 1500 W, **(c)** 2598 VAr.)

D.7. A 1.732 kV 3-phase generator supplies a balanced Δ-connected load that is formed by three impedances, each of magnitude $Z = (22.5 + j15)$ ohms. Each of the lines, that connects the generator to the load, has an impedance $Z = (0.5 + j1.0)$ ohms. With the generator voltage V_{CB} as reference, determine **(a)** the line currents to the load, **(b)** the power supplied to the 3-phase, Δ-connected load, and **(c)** the reactive power supplied by the generator.

(*Answer:* (a) $I_A = 100 \angle -127°$ A, $I_B = 100 \angle 113°$ A, $I_C = 100 \angle -7°$ A, (b) 225kW, (c) 180 kVAr.)

D.8. Three, single-phase loads are connected to a 3-phase, four-wire outlet. The voltage between lines C and B is $V_{CB} = 173.2 \angle 0°$ volts. Two of the single-phase loads are $Z_B = 10 \angle 90°$ ohms and $Z_C = 10 \angle -90°$ ohms. Determine the value of Z_A such that the voltages from the common starpoint N to the lines remain the same if the neutral wire is disconnected from N.
(*Answer:* $Z_A = 5.77 \angle 0°$ ohms.)

APPENDIX E

Per Unit Values

E1. INTRODUCTION

No significance is obtained from the information that transformer A has a no-load current of 5 A while transformer B has a no-load current of 20 A. However, if we were told that the no-load current of transformer A is 1% of its full-load current, while transformer B has a no-load current that is 3% of its full-load current, a useful comparison can be made. The conclusion can be made that transformer A is of better construction, or that the magnetic material is of better quality, or that the design value of the flux density is much lower than that of transformer B.

Instead of working with percentage values, per unit values are often used. The advantage of using per unit values is twofold. A meaningful comparison between design and performance characteristics of different transformers can be made and a larger system with transformers can be analyzed without the need to refer primary quantities to the secondary side, or vice versa.

E2. DEFINITIONS

A *per unit value* (pu) is defined by the expression

$$\text{Per unit value of quantity} \triangleq \frac{\text{actual value of quantity}}{\text{base value of quantity}}. \tag{E2.1}$$

Application of the per unit system requires the selection of a proper set of base or reference values. There is no restriction to the choice, but, customarily, rated quantities are selected because of the significance that can be attached to the resulting per unit values as demonstrated above. If rated voltage and full-load current are chosen for the base values, the per unit current and per unit voltage values of a specific current and voltage follow from

$$\text{pu}I = \frac{I \text{ in amperes}}{\text{full-load } I \text{ in amperes}} = \frac{I \text{ in amperes}}{\text{base amperes}} \qquad (E2.2)$$

and

$$\text{pu}V = \frac{V \text{ in volts}}{\text{rated } V \text{ in volts}} = \frac{V \text{ in volts}}{\text{base volts}}. \qquad (E2.3)$$

Having chosen independent base values for voltage and current the base power is defined by the product of the two and cannot be chosen arbitrarily. With the choice of full-load values for base voltage and base current it follows that the base power is the full-load VA rating. Thus

$$\text{pu power} = \frac{\text{actual power in watts}}{\text{base } VA}. \qquad (E2.4)$$

Similarly the base value for impedance follows from the base value for voltage and the base value for current.

$$\text{Base impedance in ohms} = \frac{\text{base voltage in volts}}{\text{base current in amperes}}. \qquad (E2.5)$$

An impedance, resistance, or reactance is expressed as a per unit value as follows.

$$\text{pu impedance} = \frac{\text{actual impedance in ohms}}{\text{base impedance in ohms}}. \qquad (E2.6)$$

$$\text{pu resistance} = \frac{\text{actual resistance in ohms}}{\text{base impedance in ohms}}. \qquad (E2.7)$$

$$\text{pu reactance} = \frac{\text{actual reactance in ohms}}{\text{base impedance in ohms}}. \qquad (E2.8)$$

E3. APPLICATION TO TRANSFORMERS

Once the base values for voltage and current on the primary side of a transformer have been selected, those on the secondary side follow from a knowledge of the turns ratio a of the transformer. Giving the base value symbols a subscript b, so that, for example, base $V_1 \triangleq V_{b1}$, then

$$V_{b_1} = aV_{b2}, \quad I_{b1} = \frac{1}{a} I_{b2} \qquad (E3.1)$$

and therefore

$$Z_{b1} = \frac{V_{b1}}{I_{b1}} = \frac{aV_{b2}}{I_{b2}/a} = \frac{a^2 V_{b2}}{I_{b2}} = a^2 Z_{b2}. \qquad \text{(E3.2)}$$

It is to be noted that the base quantities have units like the actual quantities whereas the per unit values are dimensionless (or normalized) values. If the rated values are chosen for the base values, the per unit resistance of the primary winding is

$$\text{pu}R_1 = \frac{R_1}{V_{b1}/I_{b1}}. \qquad \text{(E3.3)}$$

An important feature of the per unit system can be demonstrated if the expression for per unit secondary winding resistance is considered.

$$\text{pu}R_2 = \frac{R_2}{V_{b2}/I_{b2}} = \frac{R_2}{Z_{b2}}. \qquad \text{(E3.4)}$$

Since

$$V_{b2} = \frac{1}{a} V_{b1}, \quad I_{b2} = aI_{b1} \quad \text{and} \quad Z_{b2} = \frac{1}{a^2} Z_{b1} \qquad \text{(E3.5)}$$

we can write

$$\text{pu}R_2 = \frac{R_2}{\frac{1}{a} V_{b1}/aI_{b1}} = \frac{R_2}{\frac{1}{a^2} Z_{b1}} = \frac{a^2 R_2}{Z_{b1}} \quad \text{or} \quad \text{pu}R_2 = \frac{R_2'}{Z_{b1}} = \text{pu}R_2'. \quad \text{(E3.6)}$$

Therefore the per unit resistance referred to the primary side is the same as the per unit resistance on the secondary side. In general, per unit impedance referred to one side of a transformer is the same as the per unit impedance on the original side.

Another important feature of the per unit system can be illustrated by modifying the expression for per unit primary winding resistance.

$$\begin{aligned}
\text{pu}R_1 &= \frac{R_1}{V_{b1}/I_{b1}} = \frac{I_{b1}R_1}{V_{b1}} = \frac{I_{1FL}R_1}{V_{b1}} = \text{pu resistive voltage drop} \\
&= \frac{I_{1FL}^2 R_1}{V_{b1}I_{1FL}} = \frac{I_{1FL}^2 R_1}{\text{base}VA} = \text{pu ohmic loss.}
\end{aligned} \qquad \text{(E3.7)}$$

Thus, if rated quantities are chosen for the base values, the per unit primary winding resistance, the per unit resistive voltage drop and the per unit resistive loss in the primary winding, when full-load current flows, are all represented by the same numerical per unit value. Similarly the per unit reactance or impedance is equal to the per unit reactance or impedance voltage drop at full-load current.

The foregoing applies equally well to the secondary winding and to quantities referred from one winding side to the other.

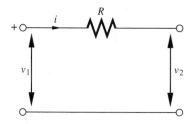

Figure E.1   Simple circuit.

E4. TRANSIENT CALCULATIONS

Per unit quantities were used with reference to the steady-state rms values in the previous sections, but they can be used equally well in transient calculations with instantaneous values. Consider the simple circuit illustrated in Fig. E1. The voltage equation for this circuit is

$$v_1 = Ri + v_2. \tag{E4.1}$$

Let arbitrary values V_b and I_b be chosen as reference or base values of voltage and current, respectively. Divide through by V_b so that

$$\frac{v_1}{V_b} = \frac{Ri}{V_b} + \frac{v_2}{V_b} \tag{E4.2}$$

and rearranging the Ri term

$$\frac{v_1}{V_b} = \frac{RI_b}{V_b} \frac{i}{I_b} + \frac{v_2}{V_b} \tag{E4.3}$$

we have an equation in which the terms are dimensionless quantities and per unit quantities. v_1/V_b and v_2/V_b are pu voltages, i/I_b is a pu current, and $RI_b/V_b = R/Z_b$ is a pu impedance. The equation for this circuit is now

$$\mathrm{pu}v_1 = (\mathrm{pu}R \times \mathrm{pu}i) + \mathrm{pu}v_2. \tag{E4.4}$$

E5. SUMMARY

In a per unit system each electrical quantity is expressed as a fraction of a base unit. Two base quantities are needed to define a given per unit system.

Meaningful comparisons can be made if the defining quantities are rated values of a machine or system.

In a system, the use of rated values of one side of a transformer to define per unit quantities eliminates the need for turns-ratio transformations at every transformer in the system.

APPENDIX F

Rotating Magnetic Field

F1. INTRODUCTION

A characteristic feature of a polyphase winding is that the exciting currents give rise to a magnetic field pattern, which travels along the winding. This traveling wave is usually called a rotating magnetic field and is a useful concept from which to develop the steady-state theory of polyphase ac machines.

A polyphase winding[15] comprises sets of coils that are placed in slots of the stator or rotor of a machine. Each set is called a phase winding and is geometrically displaced from other phases. Every phase winding is excited by an alternating current that is time displaced from every other phase-winding current. If the polyphase winding and the polyphase currents are symmetrical and balanced, we obtain the desired rotating magnetic field of constant amplitude.

In this appendix we develop the mathematical equations[16] that describe the behavior of the magnetic field. We start with a simple concentrated coil and apply Ampere's circuital law[17] to find the relationship between current and the spatial distribution of mmf in the air gap. The magnetic field pattern follows. In progression we examine a distributed winding and a three-phase winding in the same way. The final equation can be used to investigate the action of induction and synchronous machines.

Some basic assumptions of the theory of a rotating field are that the air gap is uniform and the permeability of the iron parts approaches an infinite value. This

---

[15] Polyphase systems are described in a general manner in Appendix D.

[16] A physical picture of how a rotating field is created is described in Chapter 4, subsection B under section 4.2.1.

[17] See Appendix A, Section A3 for information how to use Ampere's circuital law.

601

implies that both the slot effects and saturation effects can be neglected. Further, the diameter of the conductors are assumed to be small compared with the coil pitch, so that the variation of field within a conductor need not be considered. Hysteresis effects are small and are neglected except when determining efficiency. Only two-pole machines are illustrated, because it is simpler to identify the principles.

F2. MMF PATTERN OF A CONCENTRATED COIL

Figure F1a depicts a single, concentrated, full-pitch coil on the perimeter of a rotor or stator member. Figure F1b shows the coil in the air gap in a developed form.

Let us apply Ampere's circuital law to a closed path linking the coil. If the instantaneous current in a conductor is i, and if the coil has N turns, then

$$Ni = \oint H \, dl. \tag{F2.1}$$

The magnetic field intensity H for the iron is zero since the permeability is considered to have an infinite value. H for the air gap, in the radial direction, is constant since the flux does not change radially. Therefore for the complete path, which includes two air-gap lengths

$$Ni = H \times 2g. \tag{F2.2}$$

Hence, across one air gap the mmf F needed to produce a magnetic field intensity H is

$$F = Hg = \frac{Ni}{2}. \tag{F2.3}$$

This value is constant, since Ampere's law for any path linking the coil will give the same result. Thus the mmf wave for a single coil is rectangular with an amplitude $Ni/2$, as illustrated in Fig. F1b. The flux density pattern follows the same form as the mmf in the uniform air gap, and its magnitude is obtained from $B = \mu_0 H$.

F3. MMF PATTERN FOR A DISTRIBUTED WINDING

The more conductors in a winding the more energy can be converted, so it is reasonable to fill all available space around the perimeter of an air gap with conductors. Figure F2a shows, in symbolic form, a distributed winding with current directions as shown. A developed form of the winding is portrayed in Fig. F2b. Each coil has a full pole pitch and all coils are connected in series. Each separate coil has a rectangular mmf pattern, as shown in Fig. F1b. At any position along the air gap the net mmf is found by adding algebraically the contribution of every coil.

The result is an mmf pattern that is stepped, with each step being Ni. The resultant wave amplitude is $\sigma Ni/2$, where σ is the number of separate coils per pole pair. The stepped mmf wave is shown in Fig. F2c. Eventually, as the coil sides get closer and closer together, they would touch as depicted in Fig. F2d. Ideally this can be represented by uniform current sheets, which are shown in Fig. F2e. The corresponding mmf wave, as in Fig. F2f, has no steps and is triangular.

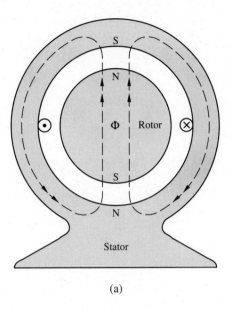

(a)

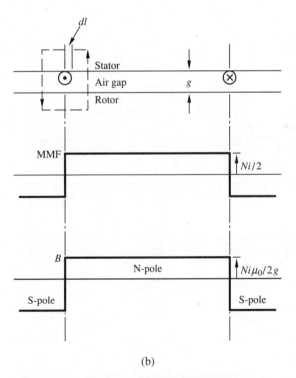

(b)

Figure F.1 Field pattern of a single coil. (a)
Physical configuration, (b) developed winding and
air-gap distribution of mmf and flux density.

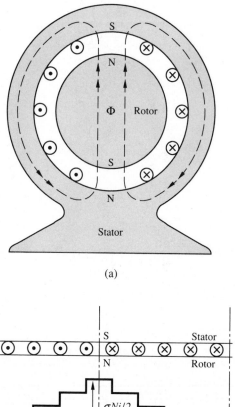

(a)

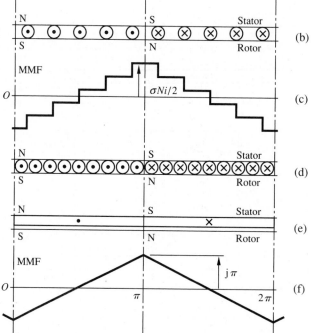

Figure F.2 Distributed winding. (a) Physical configuration, (b) developed winding, (c) stepped mmf pattern, (d) uniformly distributed winding, (e) current sheets, (f) ideal mmf pattern.

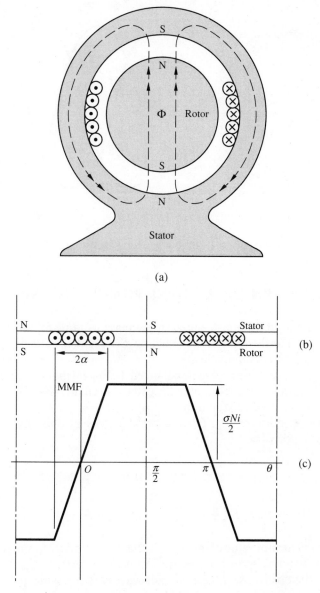

Figure F.3 Limited spread winding. (a) Physical configuration, (b) developed winding, (c) ideal mmf waveform.

If a polyphase winding is placed on the stator or rotor of an ac machine, one phase of the winding can be allotted only a fraction of the total space available. This limits the spread and alters the shape of the ideal mmf wave. If a single phase of a 3-phase winding is considered alone, the triangular wave of Fig. F2f becomes a trapezium. This is shown in Fig. F3.

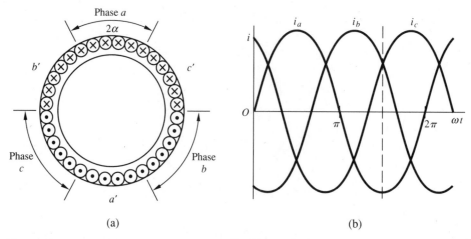

Figure F.4 Three-phase winding. (a) Winding with 60 spread, (b) phase currents.

F4. MMF PATTERN OF A THREE-PHASE WINDING

Let us consider a uniformly distributed three-phase winding of full-pitch coils, that are in slots parallel to the shaft of the machine. See Fig. F4a. The three phase bands (each with a spread of 2α radians) are a–a', b–b', and c–c'. The 3-phase currents exciting the winding are assumed to be balanced. In Fig. F4 the current distribution in the winding coincides with an instant in time that current in phase a is maximum and negative as shown by the broken line.

From an inspection of Fig. F3, the mmf pattern of one phase (phase a, for instance) can be described by

$$F(\theta) = \hat{F}\frac{\theta}{\alpha} \text{ for } 0 \le \theta \le \alpha, \quad F(\theta) = \hat{F} \text{ for } \alpha \le \theta \le \pi/2 \qquad \textbf{(F4.1)}$$

$$F\left(\frac{\pi}{2} - \theta\right) = F\left(\frac{\pi}{2} + \theta\right) \quad \text{and} \quad F(\theta - \pi) = -F(\theta) \qquad \textbf{(F4.2)}$$

where $\hat{F}$ is the maximum value of the mmf F, that is, $F = \sigma Ni/2$ (ampere-turns) and α equals half the winding spread in electrical radians. A further inspection of Fig. F3 indicates that the positive and negative half-waves are the same and that the wave function is odd. Thus, there are only odd harmonics, only sine terms, and the average value is zero in the Fourier series. Therefore, the general Fourier expressions for the single-phase mmf waves are

$$F_a(\theta) = \sum_{n=1}^{\infty} b_n \sin n\theta \quad F_b(\theta) = \sum_{n=1}^{\infty} b_n \sin n\left(\theta - \frac{2\pi}{3}\right)$$

$$F_c(\theta) = \sum_{n=1}^{\infty} b_n \sin n\left(\theta + \frac{2\pi}{3}\right) \qquad \textbf{(F4.3)}$$

where $b_n = 2/\pi \int_0^\pi f(\theta) \sin n\theta \, d\theta = 4/\pi \int_0^{\pi/2} f(\theta) \sin n\theta \, d\theta$.

In this case

$$b_n = 4 \frac{\hat{F}}{\alpha\pi} \int_0^\alpha \theta \sin n\theta \, d\theta + 4 \frac{\hat{F}}{\pi} \int_\alpha^{\pi/2} \sin n\theta \, d\theta$$

$$= 4 \frac{\hat{F}}{\pi n} \frac{\sin n\alpha}{n\alpha} = 4 \frac{\hat{F}}{\pi} \frac{k_n}{n} \tag{F4.4}$$

where $k_n = \sin n\alpha / n\alpha$ is known as the winding distribution factor for the nth harmonic. Had we considered windings with fractional pole pitches and skewed conductors,[18] then b_n and hence the mmf would be reduced slightly, and the effects could be taken into account by factors like k_n without altering the general form.

The currents in the three phases can be given by

$$i_a = I_m \cos\omega t, \quad i_b = I_m \cos\left(\omega t - \frac{2\pi}{3}\right) \quad \text{and} \quad i_c = I_m \cos\left(\omega t + \frac{2\pi}{3}\right). \tag{F4.5}$$

Since, in general, $\hat{F} = \sigma N i/2$, then the amplitude of a phase mmf in space varies with time and has a maximum value $F_m = \sigma N I_m/2$. Therefore for each phase the instantaneous value of mmf is given by

$$F_a(\theta,t) = \frac{4F_m}{\pi} \cos\omega t \sum_{n=1}^\infty \frac{k_n}{n} \sin n\theta \tag{F4.6}$$

$$F_b(\theta,t) = \frac{4F_m}{\pi} \cos\left(\omega t - \frac{2\pi}{3}\right) \sum_{n=1}^\infty \frac{k_n}{n} \sin n\left(\theta - \frac{2\pi}{3}\right) \tag{F4.7}$$

and

$$F_c(\theta,t) = \frac{4F_m}{\pi} \cos\left(\omega t + \frac{2\pi}{3}\right) \sum_{n=1}^\infty \frac{k_n}{n} \sin n\left(\theta + \frac{2\pi}{3}\right). \tag{F4.8}$$

The resultant mmf F in space and time is obtained by superposition of the three-phase components; that is

$$F = F(\theta,t) = F_a + F_b + F_c. \tag{F4.9}$$

For the nth harmonic term of the resultant mmf

$$F_n = F_n(\theta,t) = \frac{4F_m}{\pi} \frac{k_n}{n} \left[\cos\omega t \sin n\theta + \cos\left(\omega t - \frac{2\pi}{3}\right) \sin n\left(\theta - \frac{2\pi}{3}\right) \right.$$
$$\left. + \cos\left(\omega t + \frac{2\pi}{3}\right) \sin n\left(\theta + \frac{2\pi}{3}\right) \right]. \tag{F4.10}$$

---

[18] A fractional pole pitch indicates that the conductors of a turn span somewhat less than 180°. Skewed conductors indicate that the slot is *skewed*. That is, the slot is not quite parallel with the shaft. The purpose of fractional pole pitches and skewing is to reduce harmonics in the mmf waveform.

Putting $n = 1, 3, 5, \ldots$, etc. in turn, we get the series

$$F = F_1 + F_3 + F_5 + F_7 + \cdots \qquad \text{(F4.11)}$$

That is

$$F = \frac{6F_m}{\pi} [k_1 \sin(\theta - \omega t) + \frac{k_5}{5} \sin(5\theta + \omega t) + \frac{k_7}{7} \sin(7\theta - \omega t) + \cdots]. \qquad \text{(F4.12)}$$

F has an amplitude 1.5 times the amplitude of any phase mmf.

For the case where there is no conductor skewing, a full pole-pitch winding and a $60°$ winding spread (2α), the mmf pattern is

$$F = \frac{18F_m}{\pi^2} [\sin(\theta - \omega t) + \frac{1}{25} \sin(5\theta + \omega t) - \frac{1}{49} \sin(7\theta - \omega t) + \cdots] \qquad \text{(F4.13)}$$

where the fundamental mmf value is

$$F_1 = \frac{18F_m}{\pi^2} \sin(\theta - \omega t). \qquad \text{(F4.14)}$$

The sinewave distribution of eq. (F4.14) has a constant amplitude $18F_m/\pi^2$. However, the location of the amplitude of this mmf wave moves, because the value of the phase angle ωt changes with time. A particular value or point on this fundamental sinusoid is given by $\theta - \omega t = $ constant. By differentiation $\dot\theta = \omega$, therefore this particular point is traveling with an angular speed ω. In fact all particular points on this sinusoid are traveling with the same speed of ω electrical radians per second. Consequently, it can be considered that eq. (F4.14) is an equation of a traveling wave of constant amplitude $18F_m/\pi^2$, that moves with synchronous speed $\dot\theta = \omega$ elec. rad/s (or ω/p mech. rad/s, where p is the number of pole pairs).

Considering the complete mmf pattern in eq. (F4.13), the third harmonic is zero and the next mmf harmonic is the fifth, which is also a traveling wave. It has a constant amplitude and moves in the opposite direction with a constant speed one fifth of the fundamental speed. There is rapid convergence of amplitude in the series, so that there is no need to go further than the seventh mmf harmonic, which travels in the same direction as the fundamental component with constant amplitude and at a speed one seventh of the synchronous speed.

In a uniform air gap the magnetic flux pattern is the same as the mmf pattern, if there is no saturation of the iron parts. So a flux wave rotates in the air gap. The designer wants a fundamental sinusoidal distribution of flux in the air gap, and by judicious winding arrangement (distribution, pitch, skew, and connection) the harmonic fluxes (5th and 7th) can be reduced to negligible proportions. Therefore we can proceed on the basis of a synchronously rotating field that has a sinusoidal distribution in space.

EXAMPLE F1

A two-phase induction motor has the axes of the two stator windings displaced in space by 90 electrical degrees. The currents in the two windings are phase displaced in

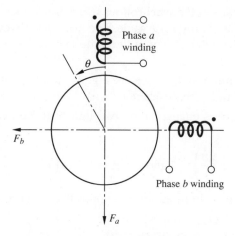

Figure F.5 Two-phase induction motor.

time by 90 electrical degrees and vary sinusoidally with time. Show that there is a sinusoidal mmf pattern in the air gap with a fixed amplitude and rotating at ω rad/s for a 2-pole machine, where ω is the angular frequency of the currents.

Solution

The 2-phase, stator winding is represented by coils a and b displaced by 90 electrical degrees and shown in Fig. F.5. The mmf component axes are shown and the distribution of each is sinusoidal.

At some angle θ, measured along the air gap from the phase a axis, the instantaneous value of the total mmf is given by

$$F(\theta) = -F_a\cos\theta + F_b\sin\theta.$$

However, the phase mmfs are time varying. That is

$$F_a = \hat{F}\cos\omega t \quad \text{and} \quad F_b = \hat{F}\sin\omega t.$$

Therefore

$$F(\theta) = -\hat{F}(\cos\omega t\,\cos\theta - \sin\omega t\,\sin\theta) = -\hat{F}\cos(\omega t + \theta).$$

This is the equation of a traveling wave. Thus the mmf pattern has a fixed amplitude $\hat{F}$ and rotates at an angular speed of ω radians per second in a clockwise direction.

F5. SUMMARY

The rotating magnetic field set up by a set of three-phase currents in a machine winding is summarized by the following.

• What holds for mmf, holds for flux.

- The flux pattern rotates at the synchronous speed ω electrical rad/s with an amplitude 1.5 times the maximum phase value. The speed is with respect to the coils, and the flux rotates around the air gap in the same manner as if there were revolving field poles.
- The direction of rotation (in this case, see Fig. F4) is such that the field travels past the winding phases in the sequence *abc*.
- If the phase sequence is changed (i.e., two line connections are changed to the supply) such that

$$i_b = I_m \cos\left(\omega t + \frac{2\pi}{3}\right) \quad \text{and} \quad i_c = I_m \cos\left(\omega t - \frac{2\pi}{3}\right)$$

for example, then the field rotates in the opposite direction in the sequence *acb*.
- If the coils rotate at $\pm \omega_m$ with respect to the sequence *abc*, then the field rotates in space at a speed $\pm \omega_m + \omega$ electrical rad/s.
- If the phase angle of the winding impedance is changed, then the position of the flux amplitude changes by the same amount.

APPENDIX G

Windings

G1. INTRODUCTION

Windings are groups of coils that are usually insulated copper wires in special configurations. There are two main kinds of windings. One kind is used to form an electromagnet; the current in the winding produces a magnetic field that is required for energy conversion to take place. The other kind of winding is the one associated with the actual electric energy that is converted to or from mechanical energy, sometimes known as an armature winding.

The greater are the ampere-conductors per meter of air-gap periphery (specific electric loading), the greater is the machine output power for a given geometry and flux density. So, instead of having conductors connected in series to form the coils of a concentrated winding, the conductors are distributed in slots adjacent to the air gap. In this way the greatest amount of copper, and hence greatest electric loading, is made available for the given peripheral space.

In this text we have simplified the view of machine windings in order to derive the performance equations as easily as possible. The general view has been to consider a winding as a concentrated coil. This is a reasonable approximation for transformers and for the excitation windings on the pole pieces of dc and synchronous machines. Windings of induction motors and of armatures of dc machines and synchronous machines have a complex distribution in the iron core around the air gap. Consequently we have tended to use factors in the equations to compensate for the differences between the effects of distributed and concentrated windings.

In this appendix we will give a short description of windings and winding factors. The purpose is really one of awareness of geometric configurations and how winding factors are derived. This is an extension of the discussion in Chapter 4 and is useful but not an absolute necessity for understanding Chapters 5, 6, and 7.

611

G2. WINDING TERMS

The following is a brief glossary of terms characterizing a winding.

Circuits

Between the terminal connections of a winding the path for the current can be made single or multiple by joining the coils in series or parallel. If the voltage is low and the current high for a given power, the coils would have a number of parallel connections (e.g., lap winding).

Coils

A coil consists of at least a single turn of two conductors, which are displaced about one pole pitch. Usually a coil has multiturns.

Coil Span

A *coil span* is the displacement (in radians) of the two coil sides around the periphery of the air gap. If the coil sides span one pole-pitch, the span has a full pitch. Otherwise the coil span is said to be chorded or have a fractional pitch. Short chording indicates a coil span less than a pole pitch, and this is usual, both to eliminate harmonic effects and reduce the length of the end connection to the coil.

Commutator Pitch

The commutator pitch y_c is the number of commutator segments between coil end connections of the commutator. For a simple lap winding $y_c = \pm 1$. The positive value relates to a progressive winding (the coils overlap in a forward direction). See Fig. G5. The negative value relates to a retrogressive winding (the coils overlap in a backward direction). For a simple wave winding $y_c = (C \pm 1)/p$, where C is the number of coils or commutator segments and p is the number of pole pairs. See Fig. G7.

Duplex Winding

A duplex winding (lap or wave) is a winding which has twice as many parallel paths as the simple (simplex) winding. The duplex winding may consist of a single closed winding or two separately closed windings.

Lap Winding

A simple lap winding has the *finish* of each coil connected to the *start* of the next coil, so that the winding or commutator pitch is unity. See Figs. G5 and G6. All the pole-groups of coils that generate emf in the same direction at any instant are connected in parallel by the brushes.

Layers

A slot may contain one coil side, or two coil sides of different coils. Such configurations are called single- and double-layer windings, respectively. The latter is usual for all motors except small induction motors, because it allows short chording and identical coils for the whole winding.

Multiplex Winding

A multiplex winding has m times as many parallel paths as a simplex winding. For example, $m = 2$ for a duplex winding.

Overhang

Overhang relates the shape of connections at the ends of the slots, where the conductors are joined to form coils. The shapes for fitting the coils close together are diamond, involute, and multiplane and these depend on the type of coil end connections.

Phases

A phase is one of a set of identical coil groupings, whose emfs are symmetrically time-phase displaced. The number of winding phases is usually three, where each phase axis is $2\pi/3$ elec. rads displaced in space from the others.

Phase Connection

A three-phase winding has its phase groups connected in star (wye) or delta (mesh) as shown in Fig. G1. For the delta connection the phase voltage V_{ph} is the same as the line voltage V_l, but for the star connection $V_l = \sqrt{3} V_{ph}$. Of the six ends of the three phases it is usual that only three connections are brought out to the terminal box. The other connections are made internally. An advantage of the star connection is that the line-to-line emfs have no triplen harmonics in the currents.

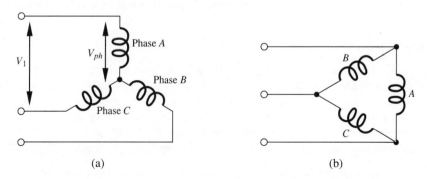

(a)   (b)

Figure G.1 Phase connections. (a) Star connection, (b) delta connection.

Phase Spread

In practice an ac winding has coils, which are not concentrated. They are distributed over a pole pitch. Three phase groups, distributed over a pole pitch, indicate a 60 elec. degree spread for the distribution of each set of coil sides. However, the winding spread can be 120 elec. degrees, although it is infrequent.

Pole Pitch

The distance, measured in radians, between the axes of adjacent poles. A pole pitch is π electrical radians or π/p mechanical radians, if there are p pole pairs on the machine.

Slotting

Slots are punched out of the laminations, which make up the iron core, and accommodate the winding. The ideal place for the winding would be in the air gap. However, such a winding could not be made strong enough to withstand the stresses or transmit the torque to the shaft, except in a few special cases. With the winding placed in slots, the electromagnetic torque acts on the core teeth.

Slotting itself relates whether the number of slots per pole is an integer or is fractional. An integral number would seem to be the most logical to create a balanced winding, but the fractional number of slots per pole has two advantages. First, it is economic to stamp slots in different diameter laminations with the same machine, so it is not always possible to punch an integral number of slots per pole. Secondly, it helps to reduce the high frequency slot harmonics caused by flux *tufting,* which is brought about by the tooth reluctance being less than the slot reluctance. Skewing slots, so that they are not quite axial, is also used to allow gradual entry of conductors into the field in order that slot harmonic effects are diminished.

Slots

Slots in the laminations, adjacent to the air gap, accommodate the winding. They can be open, closed, or semiclosed. Figure G2 shows semiclosed slots. The conductors are pushed into the opening and then a wedge is inserted to hold the coil sides in place.

Type of Coil

See Fig. G3. The type of coil is classified by the end connections, which may be lap, wave, or concentric. The lap winding is for low voltage and high current, while the wave winding suits high voltage and low current. The former is used most and is associated with diamond coils, but the latter, if used at all on ac machines, can be found on wound rotors of induction motors. Concentric windings can be called chain or basket windings, which can also have a spiral form. They are found in high-voltage generators.

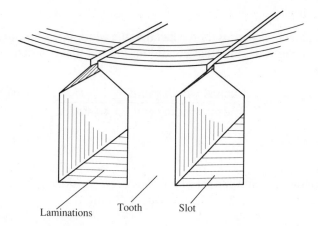

Figure G.2 Semiclosed slots.

Wave Winding

A simple wave winding has all coils carrying current in the same direction connected in series. Since there are only two directions for the flow of current in the coils there are only two parallel paths in the winding, independent of the number of poles. See Figs. G7 and G8.

Winding

In general, the ac winding is required to produce a polyphase system of emfs, whose magnitudes, frequencies, and waveshapes should be identical but phase displaced in time. In particular, the most common polyphase winding is three phase, whose emfs are displaced by 120 electrical degrees.

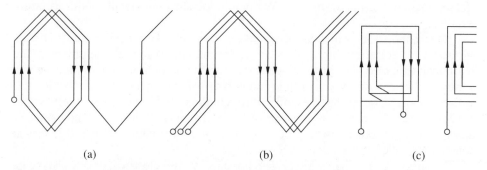

(a)   (b)   (c)

Figure G.3 End connections. (a) One phase of lap winding, (b) one phase of wave winding, (c) one phase of concentric winding.

Winding Factor

The winding factor k_w accounts for the fact that the winding is not concentrated with a full pole-pitch coil span, and that conditions are not ideal. A definition of the ac winding factor is

$$k_\omega \triangleq \frac{\text{actual winding emf}}{\text{ideal winding emf}} = \frac{E}{4.44\Phi f N.'} \tag{G2.1}$$

so the actual emf is

$$E = 4.44 k_\omega \Phi f N.' \tag{G2.2}$$

Only the fundamental component of emf has been considered here, but there is a winding factor for every harmonic component of emf.

G3. BASIC TYPES OF MODERN DC ARMATURE WINDINGS

DC machines have two basic windings. A dc field winding, concentrated on salient poles on the stator, produces the main magnetic field. A uniformly distributed winding is formed on the drum-type rotor or armature and this allows the process of electromechanical energy conversion to take place.

There is a commutator on the armature. This comprises many copper segments, each connected to a coil of the armature winding. There are stationary brushes in contact with the commutator and the purpose of the brushes is to allow current to flow from a stationary dc source to the winding in a special fashion. The copper segments of the commutator in conjunction with the brushes are switches. As the armature and commutator rotate the brushes allow current to be fed to the winding through the segments that are moving under those brushes. The winding of the armature is continuous and the special fashion that results from the continual switching is that, no matter what the armature speed is, the current distribution remains virtually unaltered in space. Thus the armature field is virtually fixed in space. The interaction between the main field and the armature field provides the electromagnetic torque for motion. We should look at some details of the armature winding.

If two conductors, which together make up one turn of a coil, are mounted on the surface of the armature, both will be active conductors since both then move in the main magnetic field. The instantaneous magnitude and polarity of emf induced in a conductor depend on the magnitude and polarity of the magnetic field in which it is situated. Thus, in order that the emfs induced in the straight axial surface conductors are additive at all times, they must be 180 electrical degrees or one full pole pitch τ_p apart. Turns, or coils consisting of more than one turn, which span 180 electrical degrees, are called full-pitch coils.

Sometimes a coil span which is less but close to 180 electrical degrees has to be used. These reduced-width coils are called fractional-pitch coils and the resulting armature winding is known as a chorded winding. The only moments in time during which the emfs induced in the two sides of one such coil oppose each other is when the two coils sides are located close to the axis between poles (called the magnetic neutral). The coil-side emfs are then so small that the effect of fractional-pitch coils on the total generated winding voltage is negligible. All modern machines use windings of which the coils have both their sides on the rotor surface subjected to the field of the magnetic poles. This type of winding is generally called a drum winding (named after the shape of the rotor).

The individual coils of a winding may consist of one or more turns. The turns of a coil must be electrically insulated from each other. To reduce the air gap between the rotor and the stator pole shoes and thus reduce the reluctance of the magnetic circuit, the coil sides are placed in axial slots, which are provided on the rotor surface. Round, insulated, copper conductor, which is used in machines of low current rating, can be wound directly in the slots of the rotor. For larger current ratings, copper conductors of rectangular cross section are used. The rectangular cross section gives a better utilization of the slot space and results in a mechanically stronger assembly. The coil is wound on a winding jig and then expanded and formed into the shape shown in Fig. G4 on a special forming machine. The turns, which are all taped together, receive a further insulating treatment so that the diamond-shaped coil becomes a rigid entity before it is slipped into a rotor slot lined with insulating material. All coils are identical in size, shape, and weight. When mounted in the rotor slots they will nest together in a neat, regular and mechanically interlocking fashion due to the special shape of the ends between the straight coil sides. One coil side will be in the upper part of a slot, the other coil side will be in the bottom half of a slot. This coil design, invented by R. Eichemeyer, is used for all form-wound, dc machine windings because of the simple and economical manufacturing process involved. It may be necessary to place several coils side by side in one rotor slot. These coil sides are first taped together before insertion in the slot. Since each slot contains two layers of coil sides this type of winding is also known as a two-layer winding. The coil sides are held in the slots by insulating wedges which may be further reinforced by a circumferential wrapping of

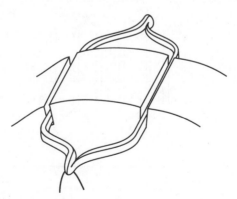

Figure G.4 Diamond-shaped coil of a drum winding.

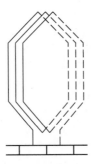

Figure G.5
Coil connected
to adjacent
commutator
segments.

nonmagnetic banding wire at regular intervals in the axial direction of the rotor surface. The slot wedges and banding are to prevent the coil sides from being forced out of the slots by centrifugal forces.

The coils, which are uniformly distributed around the rotor surface, are connected in series by connecting the coil ends or coil terminal leads to commutator segments. The ends of individual coils can be connected to commutator segments in two basically different ways, resulting in two different basic types of windings.

In the first manner, the coil ends are made to emerge close to midway between the coil sides so that they can readily be connected to two adjacent commutator segments as is shown in Fig. G5 for a three-turn coil. When drawing a winding diagram, which shows the interconnection of all coils of the winding, it is customary not to show the individual turns of each coil. For the sake of clarity only coil sides and the connec-

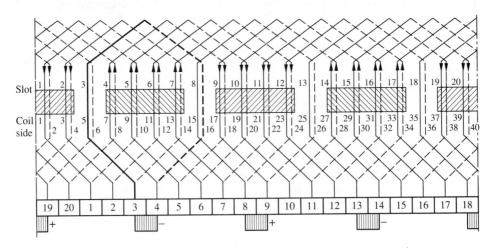

Figure G.6 Simplex lap winding.

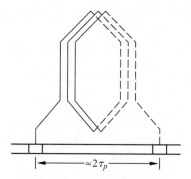

Figure G.7 Coil connected to segments about 2 pole pitches apart.

tions to the commutator are shown. It is general practice to draw the left-hand coil side, which is assumed to be in the upper half of a rotor slot as a solid line. The right-hand coil side, which is then in the bottom half of a rotor slot, is shown by a broken line. When the coils of the type shown in Fig. G5 are inserted in the rotor slots and series connected through commutator segments, an armature winding results. This is shown in a developed manner by the winding diagram of Fig. G6. Because of the manner in which successive coils overlap each other when mounted on the rotor, this type of winding is called a *lap winding*.

In the second manner of connecting coil ends to the commutator, the ends are brought out near the coil sides so that they can be connected to commutator segments, which are approximately 360 electrical degrees apart, as shown in Fig. G7 for a three-turn coil. When coils of this type are mounted on the rotor and series connected through commutator segments a winding results as shown in Fig. G8. Because of the

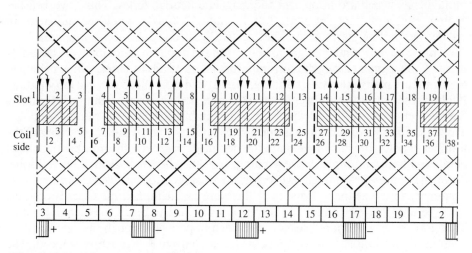

Figure G.8 Simplex wave winding.

wave-like fashion in which successive coils appear to progress around the rotor, this type of winding is called a *wave winding.*

If the stationary brushes are placed on commutator segments connected to coils, which are moving through magnetic neutral zones, the generated terminal voltage will be maximum. In a p pole-pair machine there are $2p$ magnetic neutral zones and thus $2p$ brushes should be required.

To determine the number of parallel paths of a *simplex* lap winding, we turn to the winding diagram shown in Fig. G6. Starting at a negative brush, for instance, the brush resting on commutator segments 3 and 4, we can trace the winding by following successive coils in a clockwise or in a counterclockwise direction. Tracing in a clockwise direction we end at the positive brush resting on commutator segments 8 and 9. Tracing the coils in a counterclockwise direction we end at the positive brush resting on commutator segments 18 and 19. Since equal polarity brushes are interconnected externally to the winding, we in fact traced two parallel paths going through half of the total number of coils or coil sides on the rotor. We can trace the winding in a similar fashion by starting at the negative brush, which rests on commutator segments 13 and 14, and going first in a clockwise direction and then in a counterclockwise direction through the coils, until we end at adjacent brushes of positive polarity. In doing so we trace two other parallel paths through the remaining half of the coils or coil sides. Having traced through all the coils on the rotor, we can conclude that a lap winding requires a number of brushes equal to the number of poles and that the number of pairs of parallel paths a in a simplex lap winding is equal to the number of pole pairs p. In a duplex lap winding the number of pairs of parallel paths is twice as large as the number of pole pairs. In general, we can state for a lap winding that $a = mp$, where m is the multiplicity of the winding.

Consider the simplex wave winding of Fig. G8. Since there are two pole pairs, there are four magnetic neutral zones. However, because coils which are under subsequent pole pairs are connected in series, we only require two brushes to commutate coils which are in or near the magnetic neutral zones. Thus, two brushes resting on commutator segments 7 and 8, and 12 and 13 would be sufficient. In tracing the winding, starting from the negative brush on commutator segments 7 and 8, we move once around the rotor, passing through p coils in series for every advancement of one segment along the commutator. These p coils in series, which belong to one armature path of the wave winding, would be in p parallel paths in a lap winding. It follows then that the number of pairs of parallel paths in a wave winding is $1/p$ times the number of pairs of parallel paths in a lap winding for the same number of poles. Thus, a simplex wave winding has one pair of parallel winding paths, regardless of the number of poles for which the winding is designed. In general the number of pairs of parallel paths in a multiplex wave winding is $a = m$, where m indicates the multiplicity of the winding. Although two brushes will suffice for any wave winding, it is customary to use as many brushes as there are magnetic neutral zones if physical space permits this to be done. The current density which a brush can carry is limited. Therefore, by having more brushes connected in parallel through the external connection between brushes of equal polarity, the current per brush is reduced. The brushes and thus the commutator, which is an expensive component of a dc machine, can be made shorter.

For given voltage and current ratings, the total amount of winding copper is the same for a lap and a wave winding if the number of poles and the rotor speed are the same. Where a lap winding requires p times as many conductors as a wave winding to generate the same voltage, the cross section of the conductors of the lap winding is p times smaller than that required for a wave winding. The smaller number of conductors required for a wave winding to generate the same voltage results in a saving in labor and production cost. Also less insulation is required, which means a better slot-space utilization or a more economical design. This winding type is the preferred one for current ratings up to 400 or 500 amperes. Beyond this rating lap windings for lower voltages or duplex wave windings for higher voltages are to be used. Lap windings give better commutation than wave windings but the voltage rating of lap windings is restricted by the maximum permissible voltage between commutator segments.

G4. AC WINDINGS

In general, ac windings are similar to dc windings. There are notable exceptions. The ac windings of the induction motor and the synchronous machine have no commutator. They have slip rings if they are wound on the rotor. There are phase bands and the distribution is created to minimize harmonics in the space distribution of the magnetic field.

The flux distribution in the air gap is not a simple sinusoid, but is arranged to be as close to it as is practicably possible. The harmonic components of the flux create torque oscillations, which increase losses, cause increased wear in the bearings and may produce a braking effect. It is usual for the winding to be arranged to minimize the harmonic effects. Arrangements such as distributing the coils in a number of slots to form a phase spread, making the coils span a fraction of a pole pitch, skewing the slots and special connections between phases are four means to reduce or eliminate the harmonic-induced emfs. These added complications have to be paid for by a slight reduction of the magnitude of the fundamental emf, but the total effect is beneficial.

G4.1. Reduction of Harmonic Effects

Windings are placed such that the conductors fill all the space available around the air-gap periphery. The purpose is to make the energy density associated with conversion as large as possible. Also within that purpose, the aim is to make the converted energy do useful work, and this is accomplished by special arrangement of the windings.

The winding configuration is designed to reduce as much as possible the emf harmonic components that are induced by the harmonic fluxes. These harmonic emfs can be the cause of energy dissipation in the form of heat. The rms value of the nth harmonic emf of a concentrated winding with a pole-pitch span is expressed as

$$E'_n = 4.44\Phi_n(nf)N' \tag{G4.1}$$

For a practical, distributed winding, the emf is less by a factor $k_{\omega n}$. That is

$$E_n = 4.44k_{\omega n}\Phi_n(nf)N' \tag{G4.2}$$

where $k_{\omega n}$ is the winding factor E_n/E'_n of the nth harmonic. We can separate the effects of the winding arrangement into coil distribution (factor k_{dn}), coil span (factor k_{sn}), skewing (factor k_{skn}), and phase connection (factor k_{cn}). Since all these factors in the winding arrangement are cumulative, then it follows that the overall winding factor is

$$k_{\omega n} = k_{dn} \times k_{sn} \times k_{skn} \times k_{cn}. \tag{G4.3}$$

In particular, the rms fundamental phase component of emf of an induction motor stator, which has a concentrated winding of N'_1 turns, fully pitched and without skewing, the expression is

$$E'_1 = 4.44\Phi_1 f N'_1. \tag{G4.4}$$

However, in a practical winding the actual induced emf is reduced by a factor $k_{\omega 1}$, which accounts for the winding topology. Consequently the actual emf is

$$E_1 = k_{\omega 1}E'_1 = 4.44k_{\omega 1}\Phi_1 f N'_1 \tag{G4.5}$$

where $k_{\omega 1} = k_{d1} \times k_{s1} \times k_{sk1} \times k_{c1}$.
This emf is often written as

$$E_1 = 4.44\Phi_1 f N_1 \tag{G4.6}$$

where N_1 is the effective number of turns of an equivalent, ideal, concentrated winding and given by $N_1 = k_{\omega 1}N'_1$. This equation also applies to each phase winding of the synchronous machine.

If the resultant air-gap field pattern due to all winding currents is nonsinusoidal but is periodic and symmetrical, a Fourier series can be used to express the flux density in the form

$$B = B_1 \sin(\omega t + \theta_1) + B_3 \sin(3\omega t + \theta_3) + B_5 \sin(5\omega t + \theta_5) + \cdots \tag{G4.7}$$

From this expression the fundamental component B_1 gives rise to p pole pairs having a sinewave distribution and generating an emf of frequency $f = \omega/2\pi$ in the stator windings. This fundamental rms phase component of emf is

$$E_1 = 4.44k_{\omega 1}\Phi_1 f N'_1. \tag{G4.8}$$

Superimposed on the fundamental component of flux density there will be a third harmonic of $3p$ pole pairs of sinewave distribution and amplitude B_3, that will generate an emf of frequency $f_3 = 3\omega/2\pi$ in the stator windings. This third-harmonic rms phase component of emf is

$$E_3 = 4.44k_{\omega 3}\Phi_3(3f)N_1'. \tag{G4.9}$$

In general, for the *nth* harmonic

$$E_n = 4.44k_{\omega n}\Phi_n(nf)N_1' \tag{G4.10}$$

and the resultant rms phase emf is

$$E = (E_1^2 + E_3^2 + E_5^2 + \cdots + E_n^2)^{1/2}. \tag{G4.11}$$

In practice the effective value of emf E will be very little different from the fundamental rms value of emf E_1, because of the attempts through winding distribution, span, skew, and phase connection to minimize the harmonic values. Also the total flux is very nearly equal to Φ_1, the total fundamental component, so that calculations based on

$$E_1 = 4.44k_{\omega 1}\Phi_1 f N_1' = 4.44\Phi_1 f N_1 \tag{G4.12}$$

can usually be used without producing much error. In this text, we assume that this is so.

APPENDIX H

Constants and Conversion Factors

H1. CONSTANTS

Gravitational acceleration	$g = 9.806$ m/s^2.
Speed of light	$c = 3.00 \times 10^8$ m/s.
Electron charge	$e = 1.61 \times 10^{-19}$ C.
Electron rest mass	$m = 9.11 \times 10^{-31}$ kg.
Proton rest mass	$m = 1.67 \times 10^{-27}$ kg.
Permeability of free space	$\mu_0 = 4\pi \times 10^{-7}$ H/m.
Permittivity of free space	$\epsilon_0 = 8.85 \times 10^{-12}$ F/m.

H2. PREFIXES

giga	$G = 10^9$
mega	$M = 10^6$
kilo	$k = 10^3$
hecto	$h = 10^2$
deca	$da = 10$
deci	$d = 10^{-1}$
centi	$c = 10^{-2}$
milli	$m = 10^{-3}$
micro	$\mu = 10^{-6}$
nano	$n = 10^{-9}$
pico	$p = 10^{-12}$
atto	$a = 10^{-18}$.

H3. CONVERSION FACTORS

Length	1 micron = 1 μm
	1 inch = 25.4 mm
	1 yard = 0.9144 m
	1 mile = 1.609 km
Mass	1 ounce = 28.35 g
	1 pound = 0.45359 kg
	1 slug = 14.59 kg
Moment of inertia	1 lb·ft^2 = 0.0422 kg·m^2
Force	1 kilogram-force = 9.806 N
	1 ounce-force = 0.278 N
	1 pound-force = 4.448 N
	1 poundal = 0.138 N
	1 dyne = 10 μN
Torque	1 ounce-force inch = 7.062 mN·m
	1 pound-force foot = 1.356 N·m
Energy	1 calorie = 4.1868 J
	1 British thermal unit = 1.055 kJ
	1 foot poundal = 42.14 mJ
	1 foot pound-force = 1.3558 J
	1 erg = 0.1 μJ
	1 electronvolt = 0.1602 aJ
	1 watt second = 1 J
	1 kilowatt hour = 3.6 MJ
Power	1 watt = 1 J/s
	1 horsepower (electric) = 746 W
Pressure	1 atmosphere = 101.3 kPa
	1 bar = 100 kPa
	1 pound-force per square inch = 6.895 kPa
	1 pound-force per square foot = 47.88 Pa
	1 N/m^2 = 1 Pa
Charge	1 ampere hour = 3.6 kC
Conductance	1 mho = 1 S
Mmf (ampere turns)	1 gilbert = 0.7958 A·turns
Magnetic field strength	1 oersted = 1000/4π A·turns/m
Magnetic flux	1 maxwell = 1 line = 0.01 μWb
Magnetic flux density	1 gauss = 0.1 mWb/m^2 = 0.1 mT

Note that 1 kilogram-force is the force that accelerates 1 kilogram through the pull of gravity (9.816 m/s^2). Equally well 1 pound-force is the force that accelerates 1 pound (mass) through 32.2 ft/s^2. In the SI units the force of 1 newton accelerates 1 kilogram (mass) through 1 m/s^2.

Index

AC commutator motor, 166, 417, 418, 504
AC control motor, 313
AC excitation, 155, 159, 559
AC ferromagnetic circuits, 547, 559
AC motor speed control, 234
AC series commutator motor, 417
AC series motor analysis, 419
AC tachogenerator, 315
AC windings, 621
Actuator, 27, 38
Acyclic machine, 164
Air gap, 31
Air-gap leakage flux, 353
Alnico magnet, 233
Alternating hysteresis loss, 554
Ammeter, 21, 88
Amortisseur winding, 402
Ampere, 1
Ampere's law, 517
Approximate equivalent circuit, 276
Approximate equivalent transformer circuit, 105
Arago, 256
Armature mmf, 187, 188
Armature reaction, 213, 265
Armature reaction reactance, 358
Armature reaction salient poles, 382
Armature reaction synchronous machine, 353, 354
Armature windings, 177
ASA method regulation, 374
Asynchronous motor, 255
Automobile motors, 233
Autotransformer, 138
Autotransformer starter, 297, 298
Average power conversion, 438

Back emf, 183, 458
Balanced system, 573, 576
Barium-ferrite magnet, 233
Base value, 597

Behrend, 330
Bibliography, general, 507
Braking, 200
Braking of dc motors, 230
Braking induction motors, 312
Braking torque, 154, 259
Brush chatter, 194
Brush shifting, 191, 199
Brushless dc motor, 234

Cage rotor, 299
Cage winding, 257
Capacitor motor, 337
Cathode-ray tube, 8
Ceramic magnet, 233
Characteristics of dc motors, 218
Chatter, 194
Circuit analysis, 1
Circuits, 612
Coenergy, 48, 54, 77, 349
Coercivity, 233
Cogging, 289
Coil span, 612
Coil types, 614
Coils, 612
Commutating emf, 198
Commutation, 175, 194, 213
Commutation period, 194, 198
Commutator, 30, 175
Commutator motor, 166, 417, 502
Commutator motor analysis, 419
Commutator pitch, 612
Commutator primitive, 443, 475, 500
Commutator primitive equations, 445
Commutator primitive G matrix, 446
Commutator primitive power, 447
Commutator primitive torque, 447
Commutator primitive transformation, 465, 475
Commutator segments, 178

627

Compound excitation, 175
Computer motors, 233
Concentrated coil, 30, 602
Concentrated winding, 440
Concentric windings, 99
Consequent poles, 305
Conservation of energy, 27, 41
Conservative system, 34, 42
Constant flux theory, 270
Constants, 625
Contactor, 52, 64
Control dc motor speed, 225
Conversion factors, 626
Conversion to mechanical energy, 49
Converter, 200
Copper losses, 240
Core-type core, 136
Coreless dc motor, 237
Counter emf, 183
Coupled coils, 429
Coupling field, 42
Coupling field energy, 45
Crawling, 289
Cross-field theory, 327
Current direction, 29
Current transformer, 141
Cyclotron, 6

Dahlander, 305
D'Alembert's principle, 543
Damper windings, 402, 494
Damping, 541
DC excitation, 153, 158, 164
DC generator, 171, 200
DC generator characteristics, 210
DC generator equations, 207
DC machine, 165, 166, 171, 453, 501
DC machine back emf, 458
DC machine efficiency, 240
DC machine losses, 240
DC machine power, 457
DC machine rating, 243
DC motor, 171, 211
DC motor applications, 239
DC motor braking, 230
DC motor characteristics, 218
DC motor equations, 214
DC motor separately excited, 455
DC motor speed, 212, 218
DC motor speed characteristics, 218
DC motor speed control, 225
DC motor starting, 227
DC motor torque characteristics, 222
DC shunt motor, 459
DC windings, 616
Delta connection, 132, 578, 580

Delta-star transformation, 586
Demagnetization, 190
Design, 9
Developed torque, 181, 184
Differential leakage flux, 353
Differential synchro, 317
Direct axis, 56
Direct axis synchronous reactance, 383
Direct current machines, 171
Direct-on-line starting, 297
Disc motor, 237
Distortion, 97
Distributed winding, 30, 440, 602
Distribution factor, 607
Domains, 552
Dot convention, 32, 156, 163
Double revolving field, 330
Double-cage rotor, 299
Doubly-excited rotating system, 36
Doubly-excited system, 32, 36, 61, 63, 429
Drag-cup motor, 315
Drum winding, 178, 180, 617
Duplex winding, 179, 183, 612
Dynamic braking 231, 312, 313
Dynamic circuit(s), 26, 76
Dynamic circuit analysis, 28, 72, 439
Dynamic equation of motion, 74, 77
Dynamic equations, 72
Dynamic hysteresis loop, 558

Eddy current brake, 154
Eddy current loss, 552, 555
Eddy currents, 35, 42
Efficiency, 123, 240, 290
Efficiency synchronous machine, 408
Eichemeyer, 617
Elasticity, 541
Electric circuit relations, 33, 74, 76
Electric energy, 42, 53
Electric energy processes, 2
Electric radians, 434
Electric shaver, 166
Electric system energy, 42
Electromagnet, 65, 78, 83
Electromagnetic force, 74
Electromagnetic induction, 91
Electromagnetic relay, 84
Electromagnetic torque, 32, 53, 76, 181, 184
Electromagnetics, 1
Electromechanical energy conversion, 4, 15, 42
Electromotive force, 549
Electron beam, 8
Electron gun, 7
Electrostatic devices, 66
Electrostatic force, 67
Electrostatic machine, 69

Electrostatic voltmeter, 87
Electrostatics, 1
Elementary synchronous machines, 344
Emf, 10, 259, 549
Emf method regulation, 370
Emf polarity, 12
End-turn leakage flux, 353
Energy, 2, 3
Energy balance, 41, 42
Energy balance equation, 53, 76
Energy conversion, 1, 26, 41
Energy density, 9
Energy losses, 41
Energy relations, 41
Energy storage system, 498
Equivalent circuit induction motor, 273, 331
Equivalent circuit synchronous machine, 360
Equivalent circuits, 105
Equivalent flux, 100
Exact equivalent circuit, 273
Excitation, 28, 152
Excitation methods, 173
Excitation torque, 161

Faraday, 1, 256
Faraday's law, 1, 10, 33, 151
Ferrari, 256
Ferraris, 330
Ferrite magnet, 233
Ferromagnetic circuits, 547
Field energy, 45
Field resistance line, 205
Fleming's rule, 12
Floppy disk drives, 237, 408
Flux, 514
Flux density, 514
Flux linkage, 10, 33, 515
Force, 5, 67, 72, 74
Force density, 9
Form factor, 551
Fractional pitch, 607
Frequency, 17
Frequency change, 301
Frequency converter, 255, 263
Frequency response, 108
Friction coefficient, 72
Friction losses, 42

G matrix, 432, 446
G matrix rules, 433
General bibliography, 507
General method regulation, 372
Generalized dc machine, 453
Generalized machine, 475, 477, 483

Generalized theory, 28
  of machines, 428
Generated voltage, 181
Generating action, 43, 55
Generator, 1, 9
Generator action, 9, 151
Generator connections, 576
Generator operation, 14
Generator voltage build-up, 203
Geometric neutral plane, 176
Gramme-ring winding, 165, 178

Hard-disk drives, 237
Harmonic effects, 621
Harmonic torques, 289
Harmonics synchronous machine, 352, 383
Henry, 1
Homopolar generator, 21
Homopolar machine, 164
Hooke's law, 541
Hunting, 402
Hysteresis, 35, 42, 521, 548
Hysteresis loop, 558
Hysteresis loss, 552
Hysteresis motor, 402, 403

Ideal transformer, 93
Impedance matching, 92
Inching, 239
Incremental energy balance, 43
Incremental inductance, 35
Induced efm, 10, 34, 93, 259, 549
Induced emf synchronous machine, 353
Inductance, 35, 41, 74, 471
Induction, 1, 91
Induction braking, 284
Induction generator, 255, 259, 285, 313
Induction motor, 255, 436, 480, 503
Induction motor analysis, 267, 486
Induction motor braking, 312
Induction motor efficiency, 290
Induction motor equivalent circuit, 273, 331
Induction motor inrush current, 482
Induction motor loading factors, 295
Induction motor losses, 290
Induction motor parameter measurement, 332
Induction motor parameters, 277
Induction motor performance, 280, 333
Induction motor power, 271, 489
Induction motor rating, 295
Induction motor rotor power, 280
Induction motor speed control, 301
Induction motor starting, 296, 336, 483
Induction motor starting torque, 286
Induction motor torque, 282, 493

Induction motor transients, 482, 483
Induction torque, 155
Inertia, 72, 539
Inrush current, 482, 567
Instrument transformer, 92, 140
Interpoles, 199, 213
Iron loss separation, 557
Iron losses, 43, 552
Isolating transformer, 92

Jogging, 239

Kinetic energy, 3
Kron, 428, 438, 451

Lap winding, 183, 612, 619
Layers, 613
Leakage flux, 98
Leakage flux synchronous machine, 353
Leakage reactance synchronous machine, 353, 368
Lenz's law, 11, 15, 94
Line voltage, 578
Linear commutation, 194
Linear induction motor, 315
Linear synchronous motor, 316
Linear transformations, 461
Line-to-neutral voltage, 578
Liquid metal pump, 22, 316
Load angle, 378
Load angle synchronous machine, 389
Load angle synchronous motor, 399
Load connections, 576
Loading factors, 244, 295
Locked rotor test, 279
Lorentz, 1
Lorentz force, 5, 12
Lorentz's law, 9, 151
Loss separation synchronous machine, 361
Losses, 35, 42, 240
Loudspeaker, 28

Machine inductances, 471
Machine performance, 479
Magnet field, 28
Magnet flux, 98
Magnetic characteristics, 547
Magnetic circuit representation, 563
Magnetic circuits, 513
Magnetic domains, 552
Magnetic field, 601
Magnetic field intensity, 514
Magnetic flux, 514

Magnetic flux density, 514
Magnetic flux linkage, 515
Magnetic force, 65
Magnetic poles, 29
Magnetic potential, 514
Magnetic relay, 28
Magnetic stored energy, 53
Magnetization curves, 515
Magnetizing current, 30, 94
Magnetomotive force, 514
Matrix equation, 37
Maximum power transfer, 92, 145
Maxwell-Lorentz voltage equation, 429
Mechanical energy conversion, 49
Mechanical power, 56
Mechanical relations, 71
Mechanical systems, 72, 539, 542
Mesh connection, 578
Microphone, 28
Mmf, 188, 514
Mmf method regulation, 371
Mmf pattern, 602, 606
Moment of inertia, 72
Motional emf, 430, 445, 458
Motional inductance matrix, 432
Motional voltage, 35, 37, 56
Motor, 9
Motor action, 10, 152
Motor operation, 15
Motoring action, 41, 43, 56
Moving coil, 21
Multicoil primitive, 436
Multiplex winding, 613
Multiply-excited rotating system, 36, 59
Multiply-excited system, 37
Mutual flux, 98
Mutual inductance, 440
Mutual inductance measurement, 461

Neutral wire, 576
Normal magnetization curve, 549
Numerically-controlled tools, 408

Oersted, 1
Ohmic loss, 44, 240, 241
Open-circuit characteristic, 362
Operational impedance, 430
Output equation, 295
Overcommutation, 198
Overexcited synchronous machine, 399
Overhang, 613

Parallel paths, 179
Parameter determination, 116

Parameter measurement induction motor, 332
Parameters of induction motor, 277
Parameters synchronous machine, 361
Park, 438
Particle accelerator, 6
Per unit efficiency, 125
Per unit regulation, 130
Per unit values, 597
Performance, 256, 479
Performance induction motor, 333
Permanent magnet, 172, 233
Permanent magnet de motor, 233
Permeability, 514
Permeance synchronous machine, 382
Phase connection, 613
Phase converter, 251
Phase spread, 614
Phase voltage, 578
Phases, 613
Phasor diagram synchronous generator, 388
Phasor diagrams synchronous machine, 350, 360, 508
Phasor diagrams of transformer, 114
Plasma motor, 13
Plugging, 231, 284, 312
Pole amplitude modulation, 305
Pole changing, 304
Pole pitch, 614
Pole shoe, 173
Poles, 29
Polyphase excitation, 155, 160
Polyphase induction motor, 255, 480
Polyphase synchronous motor, 168
Position controller, 73, 75, 88
Potential energy, 3
Potential transformer, 141
Potier reactance, 369
Potier triangle, 368
Power, 185, 493
Power chart synchronous generator, 408
Power commutator primitive, 447
Power in dc machine, 457
Power invariance, 95, 481
Power measurement, 589
Power primitive, 435
Power synchronous machine, 378, 389, 399, 493
Power system, 3
Primary winding, 92
Primitive, 428, 431, 443
Primitive machine, 34
Primitive power, 435
Primitive torque, 435
Primitive voltage equation, 431
Pull-out torque, 300, 400

Quadrature axis, 56
Quadrature axis synchronous reactance, 383
Quadruplex winding, 179

Radians electric, 434
Rare earth magnet, 233
Rating of dc machines, 244
Rawcliffe, 305
Reactance voltage, 197
Reactive power synchronous machine, 392
Reactor, 79
Real machine, 452
Real transformer, 96
Regeneration, 200
Regenerative braking, 231, 284, 312
Regulation, 126, 208
Regulation ASA method, 374
Regulation emf method, 370
Regulation general method, 372
Regulation mmf method, 371
Regulation of speed, 212
Regulation of voltage, 126, 208, 370, 394
Relay, 28, 84
Reluctance, 56, 518
Reluctance machine, 56, 160, 499, 506
Reluctance motor, 56, 402
Reluctance torque, 29, 159, 160
Reluctance-start motor, 340
Remanent magnetism, 233
Repulsion, 66
Repulsion motor, 417, 423
Response characteristics, 110
Revolving-field theory, 330
Rheostatic braking, 231
Robots, 408
Rotating field, 156, 259, 261
Rotating flux, 156
Rotating machines, 152
Rotating magnetic field, 601
Rotational hysteresis loss, 554
Rotational losses, 240, 242
Rotor, 28, 152
Rotor configurations, 32
Rotor excitation, 152, 161
Rotor injected voltage, 309
Rotor power, 280
Rotor resistance adjustment, 301
Rotor resistance starting, 300
Round rotor synchronous generator, 353
Rules G matrix, 433
Runge-Kutta, 485

Saliency, 157
Salient poles, 31, 152, 172
Salient-pole synchronous generator, 381
Samarium-cobalt magnet, 233
Saturated synchronous reactance, 365
Saturation, 45, 190
Saturation curve, 46, 202, 515
Saturation factor, 365
Schrage motor, 318
Secondary winding, 93

Self-excitation, 174
Self-excited generator, 203
Semiconductor converter, 200
Separate excitation, 173
Separately-excited dc motor, 455
Series commutator motor, 417, 504
Series commutator motor analysis, 419
Series excitation, 174
Servo motor, 232, 313
Shaded-pole motor, 339
Shaft torque, 282
Shell-type core, 135
Short-circuit characteristic, 362
Shunt excitation, 174
Shunt generator, 203
Shunt motor, 186, 459
Simplex winding, 179, 620
Simulation, 73
Single-cage rotor, 299
Single-phase commutator motor, 417
Single-phase generator, 344
Single-phase induction motor, 335, 504
Single-phase transformer, 92, 96
Singly-excited rotating system, 33
Singly-excited system, 30, 43
Singly-excited translation system, 66
Skewing, 607, 622
Slip, 258
Slip frequency, 344
Slip rings, 30
Slip-ring primitive, 431, 470, 497
Slip-ring primitive transformations, 461
Slot leakage flux, 353
Slots, 181, 614
Slotting, 614
Solenoid, 38, 39, 51, 73, 78, 85
Sources of energy, 2
Space motors, 237
Special dc motors, 233
Specific electric loading, 244
Specific loading factors, 244, 295
Speed control de motors, 212, 225
Speed control induction motors, 301
Speed regulation, 212
Speed voltage, 35
Split-phase induction motor, 336
Spread, 605
Spring constant, 72
Squirrel cage, 256
Star connection, 132, 576, 578, 582, 584
Star-delta transformation, 586
Star-delta starter, 299
Starting dc motors, 227
Starting induction motors, 296, 336, 483
Starting synchronous motor, 401
Starting torque, 286
Static hysteresis loop, 558
Static permeability, 549
Stator, 29, 152

Stator, configurations, 32
Stator excitation, 150, 161
Stator-resistance starting, 298
Steady-state emf, 264
Steinmetz coefficient, 554
Stepper motor, 236, 402, 406
Stiffness, 541
Stored energy, 45, 46, 53, 55, 77
Stray flux, 98
Stray losses, 241
Structure of synchronous machines, 342
Superconducting coil, 34
Superconductors, 498
Supersynchronous speed, 281
Supply frequency adjustment, 301
Symbols, 28
Synchro transformer, 316
Synchronizing power, 381, 400
Synchronizing power coefficient, 391, 400
Synchronous compensator, 399
Synchronous generator, 259, 342
Synchronous generator power chart, 408
Synchronous generator round rotor, 353
Synchronous generator salient poles, 381
Synchronous impedance, 359
    unsaturated, 365
Synchronous machine, 167, 497, 505
Synchronous machine armature reaction, 354, 382
Synchronous machine efficiency, 408
Synchronous machine equivalent circuit, 360
Synchronous machine harmonics, 383
Synchronous machine induced emf, 353
Synchronous machine leakage reactance, 353, 368
Synchronous machine load angle, 378, 389
Synchronous machine loss separation, 361
Synchronous machine OCC, 362
Synchronous machine parameter determination, 393
Synchronous machine parameters, 361
Synchronous machine phasor diagram, 350, 360, 388
Synchronous machine power, 378, 389
Synchronous machine reactive power, 392
Synchronous machine SCC, 362
Synchronous machine structure, 342
Synchronous machine torque, 348
Synchronous machine voltage regulation, 370, 394
Synchronous machine ZPFC, 366
Synchronous motor, 168, 236, 342, 396, 490
Synchronous motor analysis, 397
Synchronous motor load angle, 399
Synchronous motor phasor diagram, 492
Synchronous motor power, 399, 493
Synchronous motor starting, 401
Synchronous motor torque, 399, 493
Synchronous reactance, 359, 365
Synchronous reactance determination, 392
Synchronous reactance direct axis, 385
Synchronous reactance quadrature axis, 385

Synchronous reactance saturated, 365
Synchronous reluctance motor, 161
Synchronous speed, 156, 258
Synchronous watts, 282
Synchros, 316

Tachogenerator, 315
Tape drives, 408
Teeth, 173
Tesla, 256
Three-phase circuits, 573
Three-phase connections, 576
Three-phase generator, 346
Three-phase loads, 580
Three-phase power, 589
Three-phase transformer, 131
Three-phase winding, 606
Three-wattmeter method, 589
Torque, 53, 54, 56, 57, 77, 181, 184, 282, 434, 493
Torque commutator primitive, 447
Torque primitive, 434
Torque synchronous machine, 348
Torque synchronous motor, 399, 493
Traction, 418
Transducer, 28
Transform methods, 455
Transformation, 461
Transformation ratio, 93
Transformer, 85, 91
Transformer approximate equivalent circuit, 111
Transformer connections, 131
Transformer efficiency, 123
Transformer equivalent circuits, 105
Transformer frequency response, 108
Transformer open-circuit test, 117
Transformer parameters, 116
Transformer performance, 123
Transformer phasor diagrams, 113
Transformer regulation, 126
Transformer short-circuit test, 120
Transformer synchro, 317
Transformer voltage, 35, 37
Transformer winding resistances, 119
Transients, 482
Translational magnetic devices, 64
Translational system, 38, 543

Traveling wave, 601
Triplex winding, 179
Two-phase induction motor, 436
Two-phase motor, 313
Two-reactance theory, 383
Two-wattmeter method, 591

Undercommutation, 198
Underexcited synchronous machine, 399
Unified theory of machines, 428
Universal motor, 166, 418
Unsaturated synchronous impedance, 365

Vibrator, 86
Voltage adjustment, 301, 309
Voltage build-up, 208
Voltage equation, 35
Voltage equation primitive, 431
Voltage regulation, 126, 208, 370, 394
Voltage transformation, 464
Voltmeter, 87

WATAND, 75
Wattmeter, 589
Wave winding, 615, 619
Waveshape, 18
Wind generation, 285
Windage and friction losses, 42
Winding distribution factor, 607
Winding factor, 616
Winding resistance, 97
Winding spread, 605
Winding terms, 612
Windings, 172, 177, 611
Wye connection, 132, 576

Yoke, 17

Zero power factor characteristic, 366